APPLYING ENGINEERING THERMODYNAMICS

A Case Study Approach

Related Titles in Thermodynamics

Molecular Thermodynamics of Electrolyte Solutions
Second Edition
by Lloyd L Lee
ISBN: 978-981-123-299-2
ISBN: 978-981-123-417-0 (pbk)

Thermodynamics and Statistical Mechanics
by Richard Fitzpatrick
ISBN: 978-981-122-335-8
ISBN: 978-981-122-423-2 (pbk)

Thermodynamics: Principles and Applications
Second Edition
by İsmail Tosun
ISBN: 978-981-121-706-7
ISBN: 978-0-00-098953-6 (pbk)

Non-equilibrium Thermodynamics of Heterogeneous Systems
Second Edition
by Signe Kjelstrup and Dick Bedeaux
ISBN: 978-981-121-676-3

Adiabatic Thermodynamics of Fluids: From Hydrodynamics to General Relativity
by Christian Fronsdal
ISBN: 978-981-120-067-0

Chemical Thermodynamics: Reversible and Irreversible Thermodynamics
Second Edition
by Byung Chan Eu and Mazen Al-Ghoul

APPLYING ENGINEERING THERMODYNAMICS

A Case Study Approach

Francis A Di Bella

Boston University, USA

World Scientific

NEW JERSEY · LONDON · SINGAPORE · BEIJING · SHANGHAI · HONG KONG · TAIPEI · CHENNAI · TOKYO

Published by

World Scientific Publishing Co. Pte. Ltd.

5 Toh Tuck Link, Singapore 596224

USA office: 27 Warren Street, Suite 401-402, Hackensack, NJ 07601

UK office: 57 Shelton Street, Covent Garden, London WC2H 9HE

British Library Cataloguing-in-Publication Data
A catalogue record for this book is available from the British Library.

APPLYING ENGINEERING THERMODYNAMICS
A Case Study Approach

ISBN 978-981-120-523-1 (hardcover)
ISBN 978-981-120-524-8 (ebook for institutions)
ISBN 978-981-120-525-5 (ebook for individuals)

For any available supplementary material, please visit
https://www.worldscientific.com/worldscibooks/10.1142/11409#t=suppl

Desk Editor: Amanda Yun

Typeset by Stallion Press
Email: enquiries@stallionpress.com

Ieri (Yesterday)

To: Mom (Stella Lauria Di Bella), Dad (Antonino Di Bella)
for more than everything

Brothers: Vincent and Joseph

and in all ways: Julia

Oggi (Today)

Rose Anne, Christina Marie,

Patricia, Debra

and in all ways: Julia

Domani (Tomorrow)

Anthony, Jeana, David B, Megan, David O, Eddy

Gabriella, Aidan, Emily, Gianna, Nolan Francis,
AJ, Juliana Rose, Olivia Grace

and always: Julia

About the Author

Francis A Di Bella was a Program Manager for Large Product Development at Concepts NREC (Woburn, MA office) from May 2008 until he semic — Retired in Jan., 2020 where he is now the principal of his engineering consulting firm: ARGO — Millennium Engineering Ventures continues to work on the development of energy related systems; ranging from water wave energy systems to supercritical CO_2 power generation and CO_2 sequestration to a variety of waste heat recovery system. Mr Di Bella is a Registered Professional Engineer in Massachusetts and the past Director of the School of Engineering Technology at Northeastern University where he also taught as an Academic Specialist in Mechanical Engineering Technology until May 2008. Mr Di Bella has a B.S. Mechanical Engineering from Northeastern University, a Masters in Mechanical Engineering from Rensselaer Polytechnique Institute and he has taken Graduate classes at Northeastern University in Electrical Power Systems. He remains an Adjunct instructor at Northeastern University and Boston University for the evening engineering program where he teaches a Senior Design Project courses when necessary. More recently, his professional engineering interests involve the practical, engineering applications of innovative, alternative energy systems that utilize renewable or recoverable energy resources in novel ways. Mr Di Bella continues to teach a variety of engineering courses at Boston University as a Senior Lecturer.

Introduction

The inevitable question has been asked of almost any instructor who has ventured into the teaching of a tough subject such as math, physics, chemistry, or thermodynamics. That question can be from a first-grader, or a senior in college — although it is mostly likely asked by the youngest of us all. The question? "Where will I use this material? Why do I need to study this subject if I'm not going to be _____?" Here the reader must fill in the blank after an honest introspection.

A similar question can be asked of why I'm writing this book — or more accurately: How is this book different from the many good and very professional texts on the subject of engineering thermodynamics? Is it necessary? The answer, at least from the author's perspective, is very basic. The book is intended to help new students of thermodynamics as well as entrepreneurs with perhaps a general science education but without a background in engineering, venturing into the world of energy and power who need a little substance and solid foundation to write convincing supporting arguments or sales pitches for their "best invention since sliced bread".

This textbook in thermodynamics is intended to cut through the unnecessary — or to phrase it more politely and certainly more positively — to focus on the 20% of thermodynamics that is used 80% of the time in solving engineering thermodynamics problems, with apologies duly noted to Mr. Pareto who formulated the now famous "80/20" rule. This textbook will present the basic theories of thermodynamics by focusing on the application of the subject matter to the most common applications of thermodynamics, hence the title: *Applying Engineering Thermodynamics*.

With due respect to the traditionalist among us who may be reading this book for fun, pleasure or work, or to honestly quantify a degree of skepticism, this author suggests patience and an open mind, qualities that a true engineer most often exemplifies. This author would say that for the renaissance person in all of us, a basic understanding of thermodynamics is as important for the "liberal arts person" to master as it is for that same person to know and appreciate the work of Shakespeare or Hemingway.

As an example of what is meant by the author as "··· cutting through the unnecessary" while achieving what this book is truly intended to do (i.e., "to provide instruction through worked example") consider the following: the concept of compressibility factor, Z, is always among the first concepts that the student learns in the first introductory chapters in thermodynamics or chemistry. The **compressibility**, Z, comes in the chapter that teaches the student about the common properties of a material that is in a gaseous or vapor state but that acts almost perfectly with respect to **pressure**, volume, mass and **temperature**. The compressibility is like a fudge-factor when trying to define the vapor properties of density, pressure or temperature for a particular known substance. It is part of the equation shown here that expresses the relationship between the pressure temperature and density (M/V) of a particular fluid.

$$P \times V = Z \times \left[\frac{M \times R_u \times T}{Mole.\,Weight} \right]$$

The compressibility, Z, was used in another era when it was necessary to try to best approximate the fluid's relationship with the properties of P, T and **V/M**. It could be "looked up" for any fluid (vapor) once the **critical pressure and critical temperatures** were identified. In the 21st century, because of the ubiquitous availability of laptops and tablets, look-up tables for any property and certainly the compressibility factor, Z, has almost become obsolete for any serious thermodynamics study. Technology has made it such that there is no longer a need to have much classroom or reading time devoted to mastering the art and patience of using thermodynamics tables like "look-up" property tables. These days, computer software can help engineers model or calculate the necessary information by imbedding information like fluid properties from databases (like the Thermophysical Properties of Fluid Systems database by the National Institute of Standards and Technology (NIST), for example) into programs like Excel, Visual Basic, or FORTRAN. This author feels that textbook space should be used to present the significance of the compressibility factor, Z, for perhaps historical purposes, but certainly not to expend an entire class period or an entire chapter to its application when more important concepts can be occupying the student's attention span.

Here's another example that could lead to considerable differences of opinion: when discussing **incompressible** flow, the notion that a vapor will accelerate to very high velocities in a tube that has a changing cross-sectional area, proceeding with a converging section and then a diverging section, is often a very challenging concept. There are entire textbooks on the thermodynamics and fluid dynamics of compressible flow, and careers spent on properly deriving the equations for compressible fluid dynamics — and these are certainly worth reading and should not be avoided by the serious student. And most introductory textbooks in thermodynamics typically have at least one chapter, usually about two-thirds of the way through the textbook, where some time is spent reviewing how a converging and diverging nozzle works and its application to propulsion. Now, this is important in that a jet aircraft depends on this thermodynamics science. However, the main point of how a converging — diverging nozzle works can be understood by the student once the concept of **isentropic** (or **adiabatic**) processes is discussed, and with the helpful aid of the instructor who can use the utility of a spreadsheet to do what Sir Isaac Newton could only dream of doing — construct a thermodynamic and mathematical model of a converging — diverging nozzle using numerical integration. The student can immediately demonstrate to themselves why compressible fluid dynamics is a subject onto itself, BUT a subject that can be understood quickly once the foundation has been laid with a through explanation of isentropic processes for an open system.

How about one more: try explaining the concept of **energy** (or **availability**) of an open or a **closed system** so that the student has a working knowledge of the concept, and how it can quickly and accurately be applied to a real-world problem at hand. Once again, the use of several worked examples and the instructor's suggestion that a little imagination of an Ideal World beyond ours that employs perfect, infallible engineers who work exclusively with ideal heat engines and refrigerators/heat pumps (i.e., reverse heat engines) brings the concept of exergy or availability to life and with only the use of the very elegant and concise **Carnot equation**.

The worked examples offered in this textbook are also what make this book different from others. Yes, it is true that most good textbooks have challenging problems that are basically also challenging projects through which each student can quickly experience the need for thermodynamics as well as the logic and the beauty of the subject. But these great thermodynamics texts often leave the solution to the student as a challenging project to be done with a group of students or with an instructor's help. This textbook takes real-world problems that the author has experienced in over 40 years as a professional engineer and provides an in-depth solution to each problem, using all of what the student has learned from reading the chapters preceding the problem. While this method did not originate from the author (this methodology has been applied since Socrates expounded the technique that now shares his name: the Socratic Method of Instruction), this method has found utility in some of the newest engineering colleges or colleges that have only begun to offer an engineering curriculum. It seems that the newest engineering curriculums offer such Socratic instruction in order to keep the student engaged and not merely blindly reciting equations, or even worse, blindly using equations that may not be appropriate for the immediate application wherein the most typical assumptions used in the derivation of the equation do not apply. This textbook thus offers this Socratic Approach with problems that have been selected from the author's experience and are thus truly "real-world" and likely to be experienced in whole or in part by the reader.

This textbook will give the senior high school student, the first-year college student and the entrepreneurial businessman/woman a foundation borne from experience, of the basics of thermodynamics. It is the intention of the author to enable the reader to be able to immediately apply the contents of this textbook to a real-world problem that the reader can solve. This includes the likelihood that the student can use the contents of this book with at least a 2-degrees of freedom or a 2-step approach. That is, if the reader has a problem to solve that is not similar enough to one of the case study problems presented in this textbook, the author suggests that combining the basic principles presented in any of the chapters with the worked examples in each chapter and/or with the case studies, the reader will reach a methodology for solving the problem at hand. Thus, if a pathway to the solution of a thermo problem can be considered as a maze with the reader at the start of the maze and the thermo solution at the end with several circuitous pathways between, then the worked examples in each chapter and the case studies will quickly provide the reader with the correct path to take. The case studies and the worked examples accomplish this by actually placing the reader at the end of the maze where the solution is readily available and the path to the solution from the beginning of the problem statement is then more easily found — just as it is always easier to find the correct path through any maze by actually starting at the end and working backwards, towards the start of the maze.

The textbook does one more thing to help the reader that may already have been found to be an issue. The reader is now asked to look back at the last two pages and observe the many words in bold that have been used by the author. Each of these words has a very specific meaning when applied to the subject of thermodynamics and yet the typical text of thermodynamics does not remind the student often enough that the English dictionary need not apply to the terms used in thermodynamics. There is in fact another dictionary, perhaps from the same Ideal World heretofore mentioned, that applies and only this dictionary must be used faithfully to solve a thermo problem. It is the experience of this author that the sooner that the student starts using the Thermo dictionary and almost completely discarding the English (or Italian or French) dictionary, the more quickly the student will start to solve a problem even as he or she is reading or listening to the problem statement. For example, the word "slow" in thermodynamics means "constant temperature", while the word "fast" can mean "no heat transfer" and "rigid" could mean "a constant volume system".

Having said this, the author is careful to also point out that the student need not feel or be concerned that he or she has already missed something. They haven't. All of the thermo dictionary terms will be explained in the first chapter along with the author's panoramic tour of the subject matter. And this is the last difference that is noted between this text and others. The author provides a very quick, "grand tour" of the thermodynamics subjects that will be covered in more detail in the following chapters.

A word of caution is offered here: because the tour is so quick, the reader must not be dismayed in the first chapter and give up. The purpose of the quick tour is as simple as it is important; by providing a quick explanation without too many equations, the reader will have suffered through the first of three introductions of the subject matter. The author is a believer in the rule of three. That is, by the time the reader has seen, heard, or experienced a new fact three times, it is clear (or should be) that the topic is important. By getting the first of the three exposures out of the way very quickly, whether it is understood by the reader or not, the reader's brain is being prepared to get used to and eventually use the concepts. Neuron pathways are being formed — much like what happens when any new language is being learned — and these pathways get stronger each time the same concept is repeated.

The author emphasizes that the utility of engineering thermodynamics instruction is to be able to solve engineering problems that utilize one form of energy (usually heat energy) and transform that energy as efficiently as possible into useful work. The author is quick to point out that there are basically five process paths and seven mechanical engineering devices that are utilized to form every type of useful engine that can be constructed in this real world of applied engineering. These five processes are: constant pressure, constant temperature, constant volume, no heat transfer and almost no heat transfer. The seven mechanical elements that constitute the inventory of viable mechanical devices are: the turbine, compressor, pump, heat exchanger, valve, nozzle, diffuser and a mixing tank. The student is made aware that there are an infinite number of process paths and an infinite number of ways that these process paths can be arranged: head-to-tail and tail-to-head. However, the seven mechanical devices are the only devices that need to be clearly and thoroughly understood in order to construct any of the possible process configurations. This is analogous to the world of biology wherein it has been discovered that four base chemicals: adenine, A, pyrimidine thymine, T, pyrimidine cytosine, C and purine guanine, G, are sufficient to construct the DNA molecule that makes everyone so different but also so similar. The arrangement of these bases is the key to constructing lifeforms in biology, just as the seven mechanical devices are the key to constructing any functional and worthwhile system. This textbook focuses the student's attention toward the mastering of these seven devices and how the First and Second Laws of Thermodynamics define not only the performance but also the limitations of each of these devices.

The instruction in thermodynamics is thought to be aided by the reader's use of their imagination to consider a world outside our real-world — an Ideal World in which the machines and the resident engineers work flawlessly. The machines achieve maximum efficiency and the workers are specialists in applying these perfect machines to find the solutions to thermodynamics problems, if only to determine the best possible solution that can be achieved if indeed these perfect machines actually existed. With this maximum performance now known, the "real-world" engineering solution at least has an upper bound that the real world engineer recognizes as being as good as it can get, and with judicious application of their real world experience, the real world solution of the problem can at least be corralled and identified as 80%, or 60% or 40% of the best possible. Of course, the reader who has already gained insight into thermodynamics knows this percentage as the **second law efficiency** and thus there perhaps is nothing new for that experienced reader. However, for a new reader only now just being immersed into the subject, the ability to determine the upper bounds of a thermo problem using these "ideal world" engineers operating ideal machines is perhaps a means of formulating a solution much quicker. What are these perfect machines? They are simply the perfect Carnot Heat Engine or the perfect Carnot Refrigerator. Ah, be careful again here. Don't be confused with the English dictionary's definition of the word "refrigerator", because that word was borrowed (perhaps best to say "permanently on loan") from the thermo dictionary where it was meant to refer to a heat engine that works in reverse and thus produce a cooling (heat absorption) effect rather than generate the maximum amount of power using a heat input. The perfect world and infallible engineers who inhabit this perfect world work effortlessly without friction and thus without increasing the entropy of the Universe. Their only prerequisite for the job is to be able to see in a thermodynamics problem a solution that can be achieved with the judicious placement of either a heat engine or refrigerator at the appropriate place. Then by tallying the network performed or absorbed to achieve the desired end and comparing this network (positive or negative) with what the real-world engineer can/did accomplish, one quickly has determined the exergy destruction or unavailability of the process. More importantly, the real-world reader and future practical engineer has more clearly seen how the exergy of the system has been lost and is thus in a better place to improve upon the availability of the system process and, thereafter, increase the first law efficiency, trying to make it approach the second law efficiency but certainly never exceeding it.

Acknowledgements

The engineering experience and the practical use of thermodynamics cannot be practiced in a vacuum. This author's experience in Industry — at Tecogen and Thermedics (once subsidiaries of the Thermo Electron Corporation) and Concepts NREC LLC — has literally provided an opportunity to apply thermodynamics to innovative energy systems throughout the world. As a full-time Assistant Professor and Adjunct Professor at Northeastern University and now a Senior Lecturer at Boston University, the author is indebted to the many students who have asked what may seemingly appear to be innocent questions to complex concepts in thermodynamics. Such questions often compelled the author to re-think old notions about thermodynamics concepts, to the benefit of the author as well as the more silent students.

In the course of producing of this textbook, the author's family has grown by three beautiful grandchildren who have inspired the author with their natural inquisitiveness of all things large and small. In the most relevant ways, this textbook is for them.

This book would not be possible without the enormous help and guidance of the editing team (including copy editors Gregory Lee and Konstanze Tan) led by Senior Editor Amanda Yun at World Scientific. That help is greatly appreciated and a truly "eye opening" and learning experience for me on what effort is needed to edit a textbook so that it is readable, informative and enjoyable.

The practical application of thermodynamics is confidently pursued because there are universal truths espoused as the First and Second Law of Thermodynamics; truths that can be depended upon as a constant and solid foundation upon which the path to a solution may be found. In much the same way the author's confidence has been bolstered by the universal truth and solid foundation of a happy family at home; a constancy that has its source of strength from Julia and for which no amount of gratitude can compensate for this gift.

Lastly, any new effort often requires the perseverance and fortitude to overcome difficulties that impede the path forward. There has been no better example of perseverance and fortitude that has guided this author in all things as exhibited by my brother Vincent who defines these terms for me and his family.

Contents

About the Author   vi

Introduction   vii

Chapter 1. A Panoramic Tour of Thermodynamics
and this Textbook's Treatment   1

Chapter 2. Thermodynamics Units, Energy,
the Property of Fluids and the
Thermodynamics Process   44

Chapter 3. Thermodynamics First Law Applications
to Closed Mass and Open Flow Analyses   89

Chapter 4. Entropy and Thermodynamics Analysis
using the Second Law of Thermodynamics   164

Chapter 5. Heat Engines and Thermodynamics Cycles   213

Chapter 6. Thermodynamics of Reacting and
Non-Reacting Mixtures   284

Case Studies   319

Case Study 1: The Thermodynamics of an Espresso
Coffee Pot   321

Case Study 2:   Organic Rankine Cycle Heat Recovery
and Power Generation System   330

Case Study 3:   Total Recoverable Energy
Cycle (TREC): An Advanced
Thermodynamics Cycle to
Simultaneously Recover Low- and
High-Grade Waste Energy   345

Case Study 4:   Enhanced Gas Turbine with Water
Injection   360

Case Study 5:   Supercritical CO_2(sCO_2)Systems   379

Case Study 6:   Heat Recovery from Reciprocating
Engine for Ramjet Power Augmentation   387

Case Study 7:   Solar Powered Air Conditioning
System Using Water as the Working Fluid   406

Case Study 8:   Utilization of R134a and R1234ze
for the Development of New Chiller
Systems and as "Drop-In fluids"
in Existing Systems   413

Case Study 9:   Power Generation from Geothermal
Energy Using a Closed Thermosyphon
Heat Pipe   424

Case Study 10: Applications of Mechanical Vapor
Recompression   434

Case Study 11: Desiccant Air Conditioning Using
Solar Heat Energy   444

Case Study 12: Carbon Dioxide Capture and
Sequestration by Integrating Pressure
Swing Adsorption with an Open
Supercritical CO_2(sCO_2) Brayton Cycle   452

Case Study 13: Thermodynamics Principles Applied
to the "Drinking Bird" Toy   462

Case Study 14: Refrigeration Cycle with an Ejector   469

Case Study 15: Analysis of an Advanced Compressed
Air Energy System with Continuous
Onsite Power Augmentation via the
Air-Brayton Cycle   475

Case Study 16: Numerical Solution to Compressed
Air Energy Storage (CAES): An Open
Cycle Example Demonstrating $\Delta U = 0$ 494

Case Study 17: Concept Analysis for Increasing
Storage of Hydrogen Onboard Vehicles
Using a Cryogenic Matrix   503

Case Study 18: Compressible Gas Dynamics: Design
of a Converging-Diverging Nozzle   510

Case Study 19: A New Concept for a Thermal Air
Power Tube Used with Waste Heat
Energy Sources and Large Man-Made
or Natural Landforms   524

Case Study 20: Ground Source Heat Energy Storage
for Power Generation   539

Case Study 21: Analysis of Heat Transfer and Pressure
Drop for a Heat Exchanger with the
Mach Numbers of the Heated,
Compressible Fluid Approaching One   545

Index   556

Chapter 1

A Panoramic Tour of Thermodynamics and this Textbook's Treatment

Neuroscientists remind us that the neuron synapses that constitute the gray matter of our brain are continually forming circuits and pathways. These pathways become the memories that allow us to recall events in our history. The sooner these pathways are formed and the more often they are called upon to help with daily tasks, the stronger they get as time moves on. Thus, it will be easier to learn a new language if you begin studying it when you are young and receptive. Of course, it also helps to be young when learning something new because we are less inhibitive about making a mistake while applying what we learn — thus, the quicker we are to recover and learn from those mistakes and become more experienced.

As indicated in the Preface, thermodynamics is literally a new language the reader must become familiar with as soon as possible, in order to move forward in gaining the experience that is the goal of any new endeavor. The sooner we begin hearing, seeing, saying and using the new language, the sooner we become proficient in it. For that reason, this first chapter is intended to be the reader's first contact with the many new words and concepts that thermodynamics has to offer. The only caveat offered by the author is for the reader to read this chapter very quickly, just as one would watch a good movie or read a good mystery novel. But do not read the last paragraph before you have read through the entire text preceding it! In the study of thermodynamics, it is necessary to read the entire book and actually "play out" the scenes (i.e., do more and more worked examples). This procedure is thought to be best at answering the mystery that is sometimes thermodynamics study. The purpose of the chapter is to eliminate this mystery by forming the first neuron synapse, which will be strengthened by later parts of the text as the concepts and words become more familiar to the reader in their application, and not just in their definition.

With this in mind, let us begin our tour of thermodynamics application in the real world and allow this author to very quickly point out the many incredibly useful and elegant features of thermodynamics, but with the understanding that this tour will later return to the beginning and repeat at a much slower pace (and with the intent of actually stopping to ascertain landmarks; to touch, smell, feel, see and taste their characteristics in order to fortify the neurons in the brain) for the purpose of making future applications of what the reader sees in the book more readily available for solving thermodynamics problems.

1.1   Energy

Thermodynamics consists of the words "thermo" and "dynamics" combined. It presents the idea that thermo (or temperature) and dynamics (movement or activity) are the subject matter that is the focus of thermodynamics. Thus, thermal energy reigns almost supreme, at least for most engineering applications of the thermodynamics subject matter. Converting one form of energy into another within the constraints and guidelines of the laws of thermodynamics is the purpose of studying the new language.

Energy is simply and elegantly defined as the ability to do work. Work can simply and elegantly be defined as that "thing" that allows you to raise a weight in a gravity field. The concept of energy is the ability to express the expenditure of this thing we call energy to move a force through distance. Think about this for a moment: energy is that universal entity whose universal quantity is elegantly described as a simple product of a force moved through a distance (in the precise direction of that force). That simple product is used in a myriad of science and engineering disciplines, and in all these disciplines the energy derived from any number of sources can be summed up and confidently stated as being conserved from the beginning to the end of an event. Thus, energy, in whatever discipline it is found (physics, chemistry, neural science, biology, astrophysics, nuclear, etc.), will have units of a force (newtons or pounds) move through a distance (feet or meters). Whether it is called watt-hours (Wh), electron volts, Btu, Joules, etc., all of these are "man-made" measures for what can always be related back to a force moved through a distance.

In thermodynamics there is another very important form of energy called heat or heat transfer (Q). Heat transfer can simply and elegantly be defined as energy transfer due to a temperature difference. If there is no temperature difference, then there cannot be the energy exchange that we commonly refer to as heat energy. Heat energy or heat transfer is very important in thermodynamics because it is usually the latter that is available through the combustion of fuels (a substance that is made up of carbon hydrogen bonds) or bio-fuels, from sun (i.e. solar) energy or geothermal energy, or by rubbing two sticks (or your hands) together to get friction converted into heat. It is this energy exchange called heat that is transformed into something that is more universally useful for the general human population: the rotation of a shaft — and that rotating shaft can be attached to a generator, a wheel, a gear, a fan or anything that is useful for that moment in time as may be required for any human endeavor. Converting the heat energy into useful work (raising a weight in a gravity field) is a basic task in applied thermodynamics. Of course, you can also go the other way and convert work into thermal energy, and this is done every moment of a typical day in the form of friction. Usually friction is not desirable when found in a machine. However, friction is desirable when the task is to walk across the floor or have your vehicle climb a hill, or for that matter, even to simply move. Without friction, a vehicle's wheels would just spin as if on ice. Interestingly, friction between the tire and the road is good, while friction between two gears in a gearbox is not; friction between a pulley and the groove that holds it is good while the friction between the rope and the capstan that enables a man to hold a luxury liner from moving away from the pier is very good.

There is a third category that is much broader — internal energy (E). But this is again properly and sufficiently defined as the ability to do work regardless of the many forms of energy transformations within a thermodynamics process that can be realized. But now, with Q and W well pictured in the reader's mind, the remaining forms of energy (E) can be witnessed to make up all the rest of the different forms of transforming energy. These forms of energy are listed in Table 1.1. Examples of the forms of energy listed are: chemical, potential, kinetic, spring, nuclear, electrical and strain — all are forms of energy in that they all have some measure of being able to do work or raise a weight in a gravity field, but these are distinguished from heat transfer and a rotating shaft. The equations shown in Table 1.1 will be reviewed in more detail in Chapter 2, where a review of the First Law will be provided. Each of these forms of energy can be shown to be a force moved through a distance. In a very real sense, all these energy forms should be considered when proceeding to solve an engineering problem by using a very thorough first law analysis. However, it is clear that with experience, this list of possible energy transformations can be shortened to only those where the energy transformations are numerically significant to the solution of the problem. That is, while all of these energy transformations may be changing, most only change a very small amount and therefore the changes

Table 1.1.   Different forms of internal energy.

Energy type	Equation
Linear kinetic energy	$KE = Mv2/2gc$
Rotational kinetic energy	$KE = I\omega2/2gc$
Potential energy	$PE = \dfrac{Mgg}{gc}\Delta h$
Thermal internal energy	$U_t = Cv \times \Delta T$
Mechanical spring energy	$Es = kx2/2$
Chemical energy	$Ec = L_{hhv}\Delta M$
Nuclear energy	$En = Mc^2$
Electrical energy	$Ee = V \times I \times \Delta time$
Stress-strain energy	$Es = \sigma \times \epsilon \times Vol.$
Magnetic energy	
Atomic energy	$Ea = q \times Volt$
Surface tension energy	$E_{surface} = \sigma \times Area$
Friction energy	$E_f = F_{friction} \times \Delta x$
Drag friction	$E_{drag} = F_{drag\ friction} \times \Delta x;$ where: $F_{drag} = Cd \times V^2 \times \rho \times A/2g_c$
Thermal energy	$E_t = C_v \times \Delta T$

can be ignored for the effort that it can save in quickly arriving at a true solution.

Among the energy transformations given in Table 1.1, the most often used is thermal internal energy, E_t. It is worth repeating here that E_t is determined by multiplying the specific heat capacity with the system's temperature. Thus, the higher the temperature of a system, the more internal thermal energy that system has. It is also important to note that the system can be made up of many different elements and that each of these elements will have a different specific heat capacity,[1] Cp, with units of energy/mass/temperature (or Btu/lbm/R or kJ/kg/K). Thus E becomes a large filing cabinet where each drawer of this cabinet is one form of the previously mentioned types of energy that can be encountered at any time and that can be used to reduce heat, or work, or to just be converted from one form to another. The refrain often heard when two (or more) different things are compared in a rational debate, that someone is comparing "apples to oranges", does not apply in thermodynamics. Any form of the energies shown in Table 1.1 can be added together (or subtracted) in one grand book-keeping scheme to try to account for where the energy has

---

[1]Specific heat will be defined in greater detail in Chapter 2. Suffice to say here that it is a measure of how much energy can be input into the material (in a liquid, solid, vapor or plasma phase) per unit of mass and per unit of temperature change.

gone to, or how the energy of any type has been transformed into one or other forms.

1.2 Units of energy

Energy as defined above is the product of a force multiplied by the distance it moves through and only the product of these two parameters if they are in the same direction. The classic example of this is to observe the dragging of a box across the floor by a string that is at an angle to the floor. Only in the component of the force in the direction of the motion is there any work being done in the sense that this product is expected to be conserved as stated in the First Law of Thermodynamics. This can be identified as the dot product of Force with distance or in a more familiar form:

$$F \cdot \Delta S = F \times (S \times \cos(\theta)). \tag{1.1}$$

The units of energy then are clearly and simply identified as a force distance of Lbf-ft or N-m.

The units of energy are as varied as there are branches of physics. Whether energy is expressed as electron-volts or kwH or Btu, each of the units can be resolved into the basic units of a force multiplied by a distance. Table 1.2 provides a summary of the different units that are used in thermodynamics in the SI and Imperial system.

1.3 Thermodynamics of transforming energy

The preeminent law of the Universe is that the sum of ANY of these energies that can be defined and quantified at the beginning of an event must be kept the same in terms of the magnitude of the total summation of the constituents, in some form, at the end of the event. This is the Law of Conservation of Energy, and because it is so powerful, elegant and absolute,

Table 1.2. Summary of the different units used in thermodynamics in the SI and Imperial system.

Variable	Metric (SI)	Imperial
Mass	kg	lbm
Length	Meter (m)	Feet
Force	Newton (N)	lbf
Time	Second	Second
Temperature	Kelvin ($\Delta 1\,K = 1°C$)	Rankine ($\Delta R = 1°F$)
Angle	radian	radian
g_c	$1\,kg\text{-}m/N\text{-}s^2$	$32.2\,lbm\text{-}ft/lbf\text{-}s^2$

it is given the place of being the First Law of Thermodynamics. According to this law, you can change all (or some) potential energy into kinetic energy, some into heat or some other fraction into work or nuclear energy, or into any other form of energy — the Law of the Conservation of Energy only requires that the magnitude of the sum of all energy forms at the beginning of an event must be equal to the sum of all energy forms at the end of the process: i.e. that Energy is conserved. The classic example of the application of the Conservation of Energy that was likely first seen in a Physics class is shown in Figure 1.1. The problem statement was simple: Given the potential energy of the object on the incline plane, what is the velocity (from the kinetic energy) at the bottom of the incline plane? Of course, what you end up with may not be useful; However, this is not the result of poor thermodynamics but of engineers who did not make the conversion device efficient or make it do what it was supposed to do. You can make a round cylinder roll down an incline and convert all its potential energy into kinetic energy, or you can prevent it from rolling and only make it slide. The result is that some of its potential energy will be transformed into heat and less into kinetic energy. If the ball is forced to impact a spring at the bottom of the ramp, the energy can be stored into the spring. In fact, this is a design principle for the crash barriers that you can see installed at the entrances of the exit ramps on high-speed highways. Converting the potential energy into nuclear energy may sound like a neat trick but the First Law says it can be done (though the Second Law, which has yet to be described, will prevent this). But nuclear conversion is allowed, and potential nuclear energy is usually and much more easily converted into heat, and thus nuclear power generating stations are the typical path.

The notion of work being transformed from the available heat transfer is often pictured in the mind's eye as a rotating shaft and will be represented as such in this text. Thus, when the notion of achieving work (W) is made in any future reference in this text the reader is asked to picture a rotating

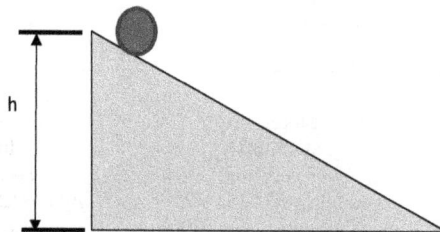

Figure 1.1.   Cylinder on an incline plane.

shaft. The two forms of energy — heat transfer and work — described in the previous section are rather distinct. Heat transfer needs a temperature difference to be created, while work can be represented as a rotating shaft to which a pulley with a string holding a weight has been attached. As the shaft turns (perhaps from a transformed heat energy process), the weight is raised, thus completing the definition of work.

There is a very elegant relationship between these three energy terms, and it is the essence of the First Law of Thermodynamics. It is given here as:

$$\sum Q = \sum W + \Delta E_{internal}. \tag{1.2}$$

But a very important caveat once again must be identified. This equation is strictly true for a CLOSED System.

Ah, another new definition is needed. In thermodynamics a closed system is a collection of things of interest that DOES NOT CHANGE MASS. Once this collection within a closed system has been defined, then it is assumed that the mass within the system does not change. Of course, the type of mass within the system can change, but the total mass (kg or lbm) cannot. Thus, a pound (or kilogram) of feathers somehow can be transformed into a pound (or kilogram) of ashes plus some gases, perhaps as a result of combustion but the collection has the same mass before and after the event has occurred.

Now, there are two things that must be recognized immediately by the reader that has not been defined: how do you define the "collection of things" and what is the significance of the event that has already been used twice in this chapter?

This collection of things is completely under the control of the engineer or analyst. This authority should not be used without some aforethought, which is honed by experience. But basically if you are interested in any collection of things, to which there is some interest to apply the principles of thermodynamics because there is going to be some transformation of one form of energy into another, then go ahead, put the collection of things on your paper and begin the thermodynamics analysis of this collection. The reader should always draw a boundary around the collection of things. This boundary is called a Control Volume (CV) and it is literally the first thing that any experienced engineer will ALWAYS do when confronted by any problem. Draw the CV around the elements that are important to the problem at hand. These elements are usually selected because they will undergo some sort of process that will transform one form of energy into another for some useful purpose.

The collection of elements that are thus constrained is closed to any other admittance or release of mass during the process of interest. What is a process in the thermodynamics dictionary? In thermodynamics, it is safe to say that anything of interest (like life) has a beginning and an end. The difference can be either large or small. Regardless, there is a start and end to every event, and in thermodynamics, what happens between these two moments is called the process. The events begin and end at any prescribed period that the engineer desires. It is noted here that though the engineer has incredible power, he too has responsibilities. The engineer is the person who selects the collection of entities that are to be the system using the CV. He then prescribes the time interval between the start and the end of an event. He must always use this power professionally and ethically. Of course, if the selection of a CV and the start and end of an event was perhaps not judicious and does not lead to a solution, then the fault probably lies with the engineer, who may have lacked the maturity in applying the thermodynamics.

It is the purpose of this textbook to have the reader quickly become a proficient engineer. Of course, based on experience with the principles that are being outlined here, every engineer should first learn to crawl, then walk and, finally, run.

Thus, once selected, we have a closed collection of things. We have the start and end points of the event, the process that this collection will undergo and the First Law of Thermodynamics that elegantly states: if one does an expert accounting of the energy in any form, of all of this collection of things BEFORE the event, then that energy for this collection of things must be accounted for after the event has occurred. With this principle, the engineer can be assured that work, heat transfer or potential energy conversions have been done properly within the framework of thermodynamics, AND maybe there is something of benefit that has been found from this analysis.

The positive and negative signs of the change in internal energy or heat transfer or work that occurs during the event or process is critical to the use of the basic first law equation. Heat transfer out of a system is considered negative (hence heat transfer into a system is considered positive). Work output is considered positive and work into a system is considered negative. In engineering, a change in a system or parameter is ALWAYS judged by looking at the quantity of the parameter AFTER the event and comparing it to the quantity of the parameter BEFORE the event. Thus if the internal energy of a system is 100 ft-lbf at the start of an event but is found to be 150 ft-lbf after the event has concluded, the net change on the internal energy of the system is (150–100) equal to +50 ft-lbf. That is, the system's internal energy has increased.

Let us apply this to a well-known physics problem to see how this formulation is applied.

1.3.1 *Example 1.1: The Sun-Earth system*

Consider Figure 1.2, which displays the Earth and Sun system. The illustration shows a control valve encircling the earth. Applying the First Law of Thermodynamics to this system, assuming that it is a Closed Mass System. The equation is as follows:

$$\sum Q = \sum W + \Delta E_{internal}. \tag{1.3}$$

The equation has the work term equal to zero. The heat transfer into the CV is due to only radiation. The First Law of Thermodynamics requires that the internal energy of the Earth system increases with the input of heat transfer. The internal energy change can take any form of many internal energies, including potential, kinetic, thermal and chemical. The chemical energy change can be of the form of any number of different fossil fuel types. It is interesting to consider that in all the years of Earth's existence, all of its energy has been derived from radiation heat transfer. Earth's fossil fuels have been created from 5 billion years' worth of this energy transfer — the transformation from heat to chemical to fossil fuel; the latter bulging with carbon-hydrogen molecules that have stored a fraction of this radiation energy in the form of chemical energy, ready to be oxidized for at least another 200 to 400 years over the last five billion years.

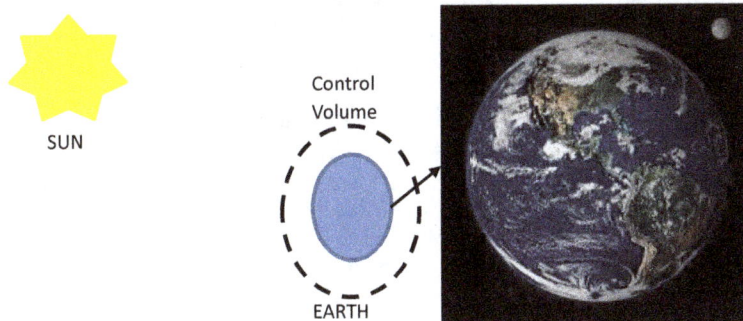

Figure 1.2. Sun-Earth system showing the Control Volume (CV) placed around the earth as the system of interest. A classic and perhaps the first true Closed Mass System (allowing for an asteroid or two).

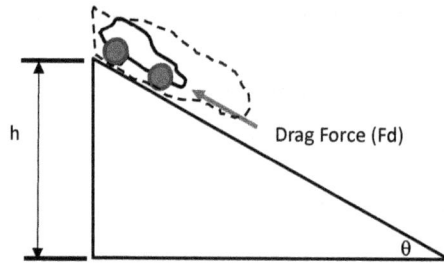

Figure 1.3. Vehicle on an inclined plane that is used to demonstrate the application of the First Law of Thermodynamics.

1.3.2   *Example 1.2: The inclined plane with a car*

The inclined plane shown in Figure 1.3 should be well known to all students who have studied physics. Galileo (1564–1642) was the first to use the inclined plane to study the acceleration of gravity. In a brilliant experimental observation, Galileo reasoned that the angle of the inclined slope enabled gravity to be reduced and thus allowed him to explore the effects of gravity on light and heavy objects. It is also a perfect tool to study the application of the First Law of Thermodynamics. In this example, the vehicle at the top of the inclined plane is released to travel to the bottom of the plane. The problem is to determine the speed of the vehicle at the bottom. However, in this example you must also consider the presence of drag on the vehicle. The drag may be assumed to be constant even though this assumption is not very accurate in the real world, where the aerodynamic drag will increase with the square of the velocity of the vehicle. The student should first place a CV around the vehicle (as shown in the diagram) to start the problem. The placement of a CV is always the first step in the solution of a thermodynamics problem, as noted in this chapter. Next, apply the First Law as shown in Equation (1.2), which is repeated here:

$$\sum Q = \sum W + \Delta E_{internal}.$$

The heat transfer into the CV is clearly equal to zero ($\Sigma Q = 0$)...or is it? If the student is cautious and attentive in the use of the CV, the drag force in fact causes the air to heat up, which is the cause of the drag on the vehicle. In this case the amount of energy consumed as heat is given as the product of the drag force and the distance traveled along the incline plane. That distance is the height (h) divided by the sine of theta, as given in the equation.

$$X = \frac{h}{\sin(\theta)}.$$

Continuing with the First Law (as applied to this problem), it may be observed that there is no "rotating shaft" breaching the CV; therefore it may be stated that the work done on (or by the system) is equal to zero $(\Sigma W = 0)$. That leaves the changes to the system's internal energy to be equal to zero $(\Delta E = 0)$. However, this internal energy may include all of the different types of internal energies given in Table 1.1. A reasonable statement about the potential and kinetic changes in the system from the beginning of the problem (when the vehicle is at the top of the inclined plane) to the end (when the vehicle is at the bottom of the inclined plane) is that of the significant internal energy only. Note the inclusion of rotational kinetic energy in this equation. The First Law of Thermodynamics can now be written as shown in Equation (1.4). The student should carefully note the positive and negative signs and how they are used to solve for the velocity.

$$0 = \Delta P.E. + \Delta K.E. + \Delta Friction\ work \tag{1.4}$$

$$\Delta P.E. = M\frac{gg}{gc}\Delta h; \ \Delta K.E. = M\frac{\Delta V^2}{2gc} + I\frac{\Delta \omega^2}{2gc};$$

$$\Delta Friction\ work = F_{drag} \times \Delta x$$

$$\Delta P.E. = M\frac{gg}{gc}\Delta h - 0; \ \Delta K.E. = M\frac{V_{after}^2}{2gc} - M\frac{V_{before}^2}{2gc};$$

$$\Delta Friction\ work = F_{drag} \times \Delta x;$$

$$where\ \Delta x = x - 0; \ \Delta E_{spring} = k \times (0 - x^2)/2.$$

For an angle (θ) equal to 45 degrees $(\pi/2$ radians), a height of 35 feet, a constant drag force of 100 lbf and a vehicle mass, M, the equation for the final velocity of the car can now be determined. It is found to be 466 ft/s. But wait a moment! The answer of 455 ft/s is suspect. After all, the height of the incline plane is only 35 feet. From the experience of operating an automobile downhill, it is clear that a vehicle travelling at 466 ft/s is more like an aircraft than an automobile. What happened here then? Clearly the answer is wrong and may have been due to some simple calculation mistake; perhaps an incorrect value typed into the calculator or spreadsheet cell. A review of the equation indicates that the equation is, in fact, correct. A re-calculation gives a new answer of 46.6 ft/s. Now this makes more sense. That is the correct answer. What has just now been done is a reality check (or "sanity check") of the answer. In other words: does the answer make sense from a practical point of view or from the intuition or experience of the analyst who might have worked with similar problems? Every "final" answer is never final; it should never be "turned-in" by the engineer until it has been checked two to three times.

It is also worth noting how the equations using "delta", Δ, are written correctly with respect to the differences between the terms. For example, in the equation shown below, the correct way to write it is to always have the first term in the difference equation as the parameter (the velocity, V, in this case) that is at the "end" of the process. It is essential that this "rule" be used regardless of the mathematical sign of the parameter, as it is seemingly human nature to write the difference with the larger parameter first and thus to obtain a positive result when the difference is taken. This tendency should be ignored, and the equation should always be written with the parameter "before" the process being subtracted from the parameter "after" the process.

$$\Delta K.E. = M\frac{V_{after}^2}{2gc} - M\frac{V_{before}^2}{2gc}. \tag{1.5}$$

1.3.3   *Example 1.3: Inclined plane with a spring inside vehicle*

Let us now look at the same problem, shown in Figure 1.4, but with a spring designed into the propulsion system of the vehicle. The spring is compressed at the beginning of the vehicle's trip down the inclined plane. The compressed spring means that energy has been stored in the spring. The first law equation for this system is shown in Equation (1.6). For this problem, the rotational kinetic energy is neglected (note: in Example 1.2, it was found that it was only a small amount of energy compared to the linear kinetic energy).

$$0 = \Delta P.E. + \Delta K.E. + \Delta Friction\ work \tag{1.6}$$

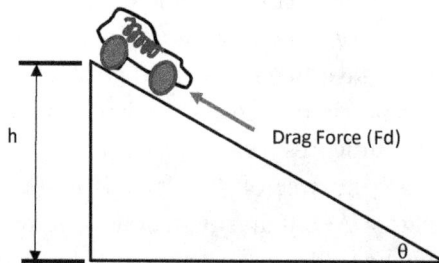

Figure 1.4.   Vehicle with a spring for stored energy on an inclined plane that is used to demonstrate the application of the First Law of Thermodynamics.

$$\Delta P.E. = M\frac{g_g}{gc}\,\Delta h;\ \Delta K.E. = M\frac{\Delta V^2}{gc};$$

$$\Delta Friction\ work = F_{drag} \times \Delta x;\ \Delta E_{spring} = k \times \Delta x^2/2$$

$$0 = \Delta P.E. + \Delta K.E. + \Delta Friction\ work$$

$$\Delta P.E. = M\frac{g_g}{gc}\,\Delta h - 0;\ \Delta K.E. = M\frac{V_{after}^2}{2gc} - M\frac{V_{before}^2}{2gc};$$

$$\Delta Friction\ work = F_{drag} \times \Delta x;\ \Delta E_{spring} = k \times (0 - x^2)/2.$$

Putting in the values of the spring constant, $k = 1{,}250\,\text{lbf/in}$ and the compressed spring length of 12 inches, the value of the velocity, $47.9\,\text{ft/s}$, is now observed to increase. For this example, there is a physical sense as to what should be a reasonable change to the speed of the vehicle. The spring's energy has been released and has apparently contributed to powering the vehicle down the hill, thus helping gravity to achieve a higher velocity.

1.3.4 *Example 1.4: Inclined plane with vehicle with nuclear matter and gas tank*

In the fourth example, shown in Figure 1.5, a nuclear energy pile or liquid fuel is used in the vehicle, which affects the vehicle's final velocity due to the energy content. The same form of the equation shown in Equation (1.5) can substitute for the liquid fuel (or the spring energy), the energy content in the form of liquid fuel. The same intuitive result could have been achieved if the fuel in the fuel tank was measured before and after the vehicle's trip. If the fuel tank has less fuel, the engine could have burned fuel during the downhill trip and thus end with a higher velocity. The energy that would have been released is measured by knowing the change in the mass of the fuel and multiplying this by the amount of chemical energy contained in each lbm of the fuel. This heat content of the fuel is expressed as Btu/lbm. For

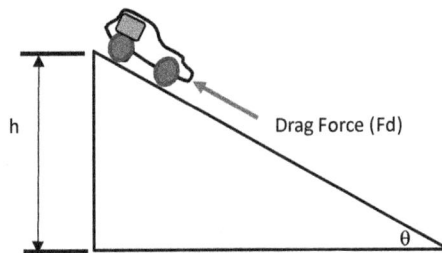

Figure 1.5.   Inclined plane with a nuclear energy source on board.

example, natural gas has an energy content of 21,000 Btu/lbm. Hydrogen has a heat content of 61,000 Btu/lbm. Every carbon-hydrogen based fuel will have some level of chemical energy stored in the chemical. This includes a ham sandwich or a hamburger as well. Of course, in that case, the energy content is quantified by the amount of calories. Careful, however! A thermodynamics unit of energy is a calorie with a lower case "c". The dietary Calorie (with a capital "C") is actually 1,000 times larger. For that reason, you may be tricked into thinking that a one-hour exercise that burns off 200 calories from you can offset eating one dish of ice cream consisting of 200 Calories. In fact, the dish of ice cream actually contains 200 kilocalories (i.e. 200,000 calories), and you would need to do 1,000 constant hours of the 200-calorie-burning exercise to counter the energy value from the ice cream.

If the reader substitutes the liquid fuel with a nuclear fuel energy source, the equations used for the liquid (chemical) stored energy can very easily identify how the vehicle's speed can change. Basically, the energy content of the nuclear fuel is obtained by converting the mass of uranium into energy by using $E = MC^2$, where mass, M, is multiplied by the velocity of light, C, in a vacuum squared.

The liquid fuel and the nuclear mass contain energy that is used to produce motive force that could propel the vehicle with additional speed. However, the amount of work that can be used to propel the vehicle must also consider the efficiency of the energy conversion from "pure" energy to an energy form that can actually mechanically propel the vehicle's wheels. This aspect of the conversion and use of energy from its stored mass/liquid/mechanical form was absent from the third example (Example 1.3), which considered the spring energy.

It was perhaps assumed by the reader that the spring unravels or pushes or pulls a mechanical linkage that ultimately involves turning the wheels (or the crankshaft) and driving the train of the vehicle, which also ends by turning the drive wheels. But how does the energy content of a liquid or a solid actually turn the wheels? The energy content and the subsequent release of energy is not enough. The conversion efficiency from its "pure" form to a method that can actually do the work that has to be done must be known or determined from the application of physics, thermodynamics, or good mechanical or electrical engineering. For example, the liquid form of energy content must first be oxidized with oxygen and the heat energy that is released at a very high temperature must then be used as an input to a thermodynamics heat engine that conforms to the First and Second Laws of Thermodynamics. Such an engine will be designed to be able to mechanically turn a shaft, which in turn can be used to turn the wheels of the vehicle that causes the vehicle to increase speed. The very same methodology is used to convert the energy released from the nuclear material: from mass to heat energy to rotating shaft to wheel power.

1.4   Piston-cylinder thought experiment

Before proceeding with this part of the panoramic tour of thermodynamics, it is necessary to remind the reader that this is only a very quick tour of the very important features and landmarks that will be seen on the slower revisit of the same material in the next chapters. The reader should be reading these sections very quickly and have confidence that the same material will be reviewed in much more detail in later chapters, and that the next reading of the detailed material explanations will make the subject matter's concept much clearer.

The piston-cylinder system is the perfect tool to teach many of the principles of thermodynamics because the system can be treated as a closed system (i.e., no mass transfer into or out of the system) or as an open-system (i.e., mass transfer into or out of the system is allowed) and because there is internal energy, work and heat transfer exchanges for a variety of processes.

Figure 1.6 is a detailed cross section of a very common thermodynamic heat engine. It is one of many piston-cylinders that compose a reciprocating

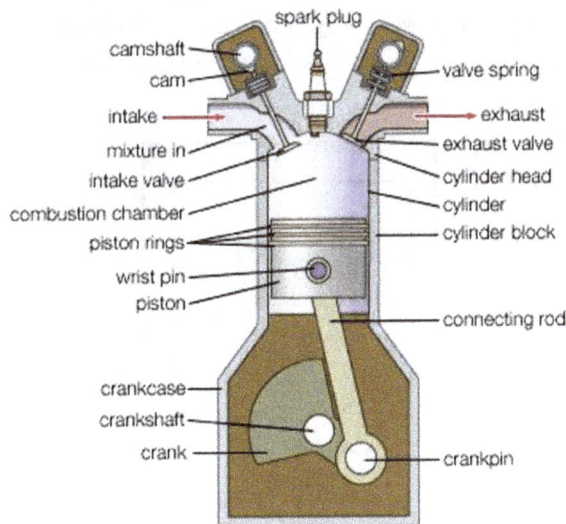

Figure 1.6.   Piston engine.

engine, typically used to power an automobile but also to generate electric power if the shaft were connected to an electric generator.

Completing a thermodynamic analysis often starts with drawing a representation of the problem with as clear a graphic as is possible. Fortunately, it is not always necessary to draw as detailed a piston-cylinder system diagram as shown in Figure 1.6. The much simpler diagram of the piston-cylinder system shown in Figure 1.7 suffices to achieve the start of the thermodynamics analysis; enabling the labeling of the most important geometric features while also identifying the heat input as well as providing a clear visual of the work output from the piston-cylinder system.

The closed mass form of the First Law of Thermodynamics is most appropriately modeled by considering a piston cylinder system. The piston cylinder has been the mechanical workhorse of vehicular engines since the 1880s, once it was realized that liquid fuel oil could be used to power a vehicle using a 2- or 4-cylinder engine. The piston cylinder engine appears in many real-world thermodynamics problems and thus should be considered here, immediately after the First Law of Thermodynamics for a closed system presented to the reader. The piston-cylinder application quickly begins the student's exposure to a real application of the First Law. The system is actually the air or the air and fuel mixture that is contained in the closed piston cylinder system after the intake and exhaust valves are closed. The actual mechanical design of the piston cylinder engine need not be of concern here. Rather the reader should simply accept the fact that the piston-cylinder is a perfect close mass system, where only heat transfer, work and internal energy change can take place. For simplicity purposes, only air (and no fuel) should be considered as the system. The reason for this is because the Perfect Gas Law that the

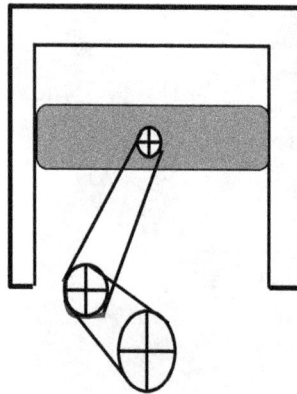

Figure 1.7.   A simple illustration of a piston-cylinder.

reader may have studied in chemistry or physics is very applicable and defines the relationship between the system pressure, temperature and volume. The Perfect Gas Law equation is shown below:

$$\textit{The Perfect Gas Law Equation: } PV = M \times R_u \times \frac{T}{M_{mole\ weight}}.$$

It is also correct to identify the Perfect Gas Law equation as a simple Equation of State.

The first operation to be considered for this equation is to divide both sides of the equation by the mass. The ratio of V/M is a very important parameter in thermodynamics: it is one of many intrinsic properties. An intrinsic property in thermodynamics is a quantity that is specific to the fluid regardless of the size of the mass that it has or the size it might have. Intrinsic properties are critical and necessary to complete a thermodynamics analysis of any size system that may be encountered. It will be shown in the next chapter that only intrinsic properties can be tabulated for every substance. For example, color may be considered a property of a system, as is the ratio of V to M. Pressure and temperature are the most common intrinsic properties of a system. A simple check quickly shows that when a chamber is filled with a substance that has a temperature of T and a pressure of P, all parts of that chamber will have the same pressure and temperature. But if half of the substance in the chamber is removed and the volume of the chamber is reduced by half, the ratio of V/M does not change. Thus, the V/M as a parameter is an intrinsic property of the system. However, when both the volume and the mass have been reduced by half then clearly these parameters taken individually are only considered extrinsic properties of the system; That is, the volume (V) and mass (M) are extrinsic properties that depend on quantity. The ratio of a parameter with mass is called a specific version of that parameter. Thus $v = V/M$ is a specific volume, and specific properties are usually depicted with a lowercase letter. The three properties, P, T and $v = V/M$ are the most common and useful properties of a simple system, if only because the instruments needed to measure these properties are inexpensive, commercially readily-available and very accurate. When the piston-cylinder is opened to the atmosphere for as long as it takes to fill the volume of the cylinder with atmospheric pressure and temperature, the volume/mass $= v$ is actually the reciprocal of the density of the air.

The piston-cylinder closed to any additional amount of mass entering or leaving can perform whatever thermodynamics processes (paths) desired by the engineer/inventor/entrepreneur. Thus, starting with the atmospheric pressure, temperature and density, the closed piston-cylinder can be squeezed very slowly to occupy a smaller space. If this is done very slowly, with sufficient time to allow the piston-cylinder air system to have heat transfer out of the system so as to not change the temperature, then the relationship between the two states — the beginning (1) and the end state (2) — is defined with the equation below:

$$P_1v_1 = P_2v_2; \text{ from perfect Gas law: } Pv = R_u \times \frac{T}{M},$$

$$\text{where } R_u \text{ is the universal gas constant} = \frac{8314Nm}{Kgmole - K} = 1545\frac{ft - lbf}{LbmoleR}.$$

T is in units of Kevin, K, or Rankine, R, and pressure is in units of N/m^2 or lbf/ft^2 for metric and imperial measurements, respectively.

If the same system is compressed quickly, the rapidity of the compression may not have allowed enough time for the heat to enter or leave the system. In that case it can be shown that the relationship between the inlet and outlet pressure and specific volume is given in the equation below. The constant k is simply the ratio of two other air properties called specific heat — one with respect to pressure, C$_p$, and the other, with respect to volume, C$_v$. More of these parameters will be presented in Chapters 2 and 3, but for now the reader only needs to recognize those having the ability of a substance to absorb energy with respect to mass and temperature change. Thus, its units are Btu/lbm/R or kJ/kg/K. The relationship can be derived from the application of the Perfect Gas Law when used with the First Law of Thermodynamics.

$$P_1v_1^k = P_2v_2^k.$$

The same piston-cylinder could follow a process path that does not change pressure or start volume at its beginning and ending. In fact, there are an infinite number of ways that the piston-cylinder system of air can be squeezed, expanded/heated or cooled to follow an infinite number of pathways from a starting to a finishing point. Each of these paths is correctly called a process, as it connects the beginning to the end of any step or task. Any number of processes can begin at point 1 and finish at point n. Point 1 and point n need not be the same position, but if they are, for example, the start and the stop of the process is at the same point, then the process part is considered to be a cycle. That is, you begin at an initial state point and then follow the number of processes or pathways that connect different state points. If you end where you began, then the process is a cycle. Obviously, in an imperfect world such as ours, there must be at least three distinct pathways. In a perfect world, it may be possible to use only two state points.

For this simple system process, the forward and backward processes may trace over themselves without losing entropy. Suffice to say, entropy change is always greater than zero in the real world. A very clear example of this is witnessing what happens when you uncork a bottle of carbonated liquid. Some of the CO_2 in the solution comes out without any assistance from the outside world, other than the act of loosening the bottle cap. Once the CO_2 has escaped the solution in the liquid, it can never be returned to re-carbonate the drink, unless a very specific external work is expended on the CO_2 that escaped — assuming you have captured the escaped CO_2. The expansion and compression processes do not trace over themselves due to the increase in the entropy of the system.

Although there are an infinite number of process paths that could be considered, there are only really five processes that can be produced with real-world engineering. These are summarized in Table 1.3 as: constant pressure, constant volume, constant temperature, no heat transfer, and almost no heat transfer. The relationships between the pressure, temperature and volume for these pathways are also shown in Table 1.3. Chapters 2 and 3 show the work done and/or the heat transfer occurring for each process path in Table 1.2, which is also shown in Table 1.3. In all cases for a perfect gas, the internal thermal energy change is ONLY a function of the temperature change, as shown by using the following formula:

$$\Delta u = C_v \times \Delta T$$

Before leaving this section, it is necessary to remind the reader of the old adage that "a picture is worth a thousand words". The 1,000 words that you have just read on the subject matter of piston-cylinders and Closed Mass Analysis can be visualized very easily by using a graph depicting pressure versus specific volume. The reader will learn later that any two properties can be plotted against itself and help visualize the problem. The visualization technique is by no means unique to the subject of thermodynamics; rather, it is almost an essential requirement for helping to solve any engineering problem. The step-by-step procedure to solve a problem in engineering is to first translate the verbiage into a diagram picture of the problem. The artistic quality of the engineer gets better with time and experience and makes visualizing even more powerful in solving a problem.

If you have not had the need to use drawings to solve problems in the past, you can start here. The P-v diagram shown in Figure 1.8 (left) shows a single process path; one that is very arbitrary and is one of an infinite number that could have been drawn to define a process or path from state point "A" (or "1" in the figure) to state point "B" (or "2"). Figure 1.8 (right) illustrates

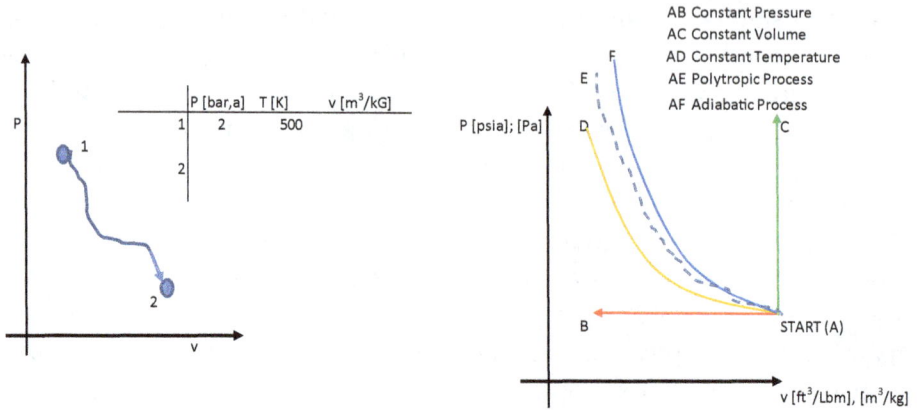

Figure 1.8.   P-v diagrams. (left) Arbitrary process path from state point 1 to state point 2; (right) Four common processes used in engineering.

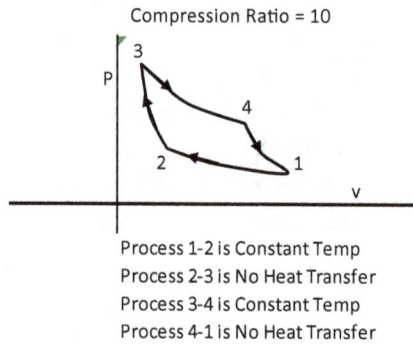

Figure 1.9.   Thermodynamic cycle showing four processes The Carnot Cycle. The most efficient heat engine cycle operating between two operating temperatures.

the five more common process paths that can actually be made by real-world machines. A cycle is shown in Figure 1.9, evidenced by a process starting and stopping at the same point after proceeding to stop at one or more state points along the way. The process paths shown in Table 1.1 and Table 1.3 are shown together in Figure 1.9. The five processes are illustrated to start at the same point in Figure 1.8 only to show the relative paths taken, as mentioned above. Do also note that the area under any of these P-v diagram paths can be expressed as the work done on or by the piston. Take the path's process to any mathematician who may or may not care "what the P and the v" mean, and that they can determine the area under the process path very easily by using Calculus. Suffice to say here, the area found is the work done, because the product of pressure multiplied by volume is easily shown to be ft-lbf per lbm of material that constitutes the system.

Now, for the first time, it is easily seen that a plot of pressure versus volume can easily determine one of the three energies that make up the First Law of Thermodynamics. Then when one considers that thermal internal energy is always expressed as the difference in temperature before and after a process event multiplied by the specific heat with respect to volume, then the third energy (heat transfer) term in the First Law can be found by adding the two other energies. The five basic processes are shown in Table 1.3 along with the derived equations for the work, heat transfer and internal thermal energy change for each process. Chapter 3 will present a detailed analysis of each of these processes and the derivations of the equations shown.

The processes shown in Table 1.3 can begin to solve the most common of engine cycles that are used in the world today. For example, Figures 1.10 and 1.11 display the two most common reciprocating engine cycles: The Diesel and the Otto Cycles. Basically, every passenger and commercial vehicle operates on either one or the other cycle. In Chapter 2, the reader will be able

Table 1.3.   Five basic thermodynamics processes.

Process	Heat transfer	Internal energy	Work
Constant temperature	$\sum U + W$	$Cv \times \Delta T$	$Pv \, ln \left(\dfrac{v2}{v1} \right)$
Constant pressure	$\sum U + W$	$Cv \times \Delta T$	$P \times \Delta V$
Constant volume	$\sum U + W$	$Cv \times \Delta T$	0
No heat transfer (Adiabatic)	$\sum U + W$	$Cv \times \Delta T$	$(p2V2 - P1V1)/(1 - k)$
Small heat transfer (polytropic process)	$\sum U + W$	$Cv \times \Delta T$	$(p2V2 - P1V1)/(1 - n)$

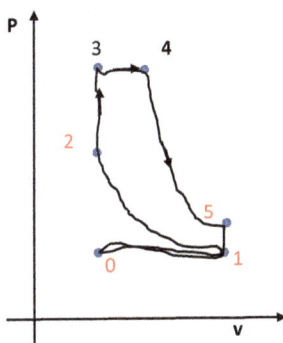

Figure 1.10.   Dual pressure, modified diesel cycle.

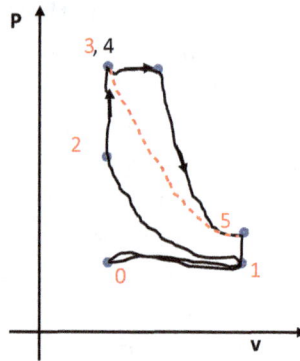

Figure 1.11.   The OTTO Cycle by Setting Vc = 1 and thus v3 = v4.

to follow instructions after a more detailed description of the energy, work and heat transfer that occur in each process, and then, by summing up these work and heat transfer terms they will be able to model both these engines with an acceptable degree of accuracy; comparing the results with published data from engine manufacturers. The reader will notice that these cycles look very similar in terms of their processes, except for Process 3–4. For the Otto Cycle, Process 3–4 is not used. It will be explained in detail in Chapter 3 that this process in the diesel cycle is due to the continuous feeding of fuel to the engine via fuel injectors, which causes the piston to continue moving downward, resulting in a change in the specific volume of the system until the fuel injection is stopped. The Otto Cycle does not include Process 3–4 due to the very distinguishing difference between the Diesel and Otto Cycles — namely the use of fuel injection in the diesel engine versus the spark ignition of the air-fuel mixture in the Otto engine.

1.5   Entropy and Reference to Stephen Hawking's Theorem

Let us continue our panoramic tour of the thermodynamics world. The First Law of Thermodynamics is very rational and elegant. Some would say that the first law for a Closed Mass System is simple to a fault. This is because most newcomers to the study of thermodynamics feel very comfortable with the concept that we call energy, where its unit of force multiplied by distance is conserved between two defined events. It may be less satisfying but nevertheless true that different forms of energy can be added to form a whole quantity of energy that must be conserved in whole or in part, as one or more of the different types of energy during an event or process in thermodynamics.

What is by far much less rationally satisfying is the notion that cold things can get colder by having energy removed from them, and that when that removed energy is input into a hotter system, it makes that hot system even hotter without some external effort being performed on either systems. Let us be clear that the removal of energy certainly makes a system cooler while the addition of energy makes a system hotter, and that neither of these processes or events violate the First Law. It is when you try to put them together that rationality breaks down. The magnitude of energy removed from the cold cannot simply be added to a hot element to make that element hotter; this has never been observed in the real world before and will likely never occur anywhere in the Universe. (We must/can only use the term "likely" here because we haven't been everywhere in the Universe to test the hypothesis; but based on solid thermodynamics in the micro and macro world on this planet you would be silly to bet against that probability happening anywhere in the Universe.) Early practitioners of thermodynamics noted this dilemma and it took a young researcher named Sadi Carnot to develop a concept of entropy that establishes a means of quantifying the Second Law of Thermodynamics. Entropy is "simply" defined as the ratio of the amount of heat transfer occurring during a process (or event) divided by the constant temperature at which this heat transfer is occurring. This last condition is critical to understanding and applying entropy to solve thermodynamics problems. The Second Law simply states that this quantity of entropy change — for both where the heat comes from and the entropy change of where the heat is going to, added together — must be greater than zero. This is elegant and simple as it is a powerful description of how the world/universe works. Entropy's two terms are better described as the entropy change of the "surroundings" and the system, and when taken together the sum is the entropy change of the UNIVERSE... or simply, the inside and outside of a system. BOTH these terms must be used in order to apply the Second Law of Thermodynamics. Now, if the heat extracted from the cold system is added to the hot system, the entropy change of the Universe will be found to be greater than zero, as demonstrated in Example 1.5.

The Second Law actually has a few different "faces" or statements that can either directly or indirectly be derived from the basic concept described above.

The most immediate consequence of the Second Law when applied to heat engines is the derivation of the Carnot Efficiency Equation. This equation determines the maximum efficiency that a thermodynamics cycle can achieve if it is receiving heat from a high temperature heat source and then rejecting a lesser amount to a colder thermal energy reservoir. The Carnot equation for the efficiency of a perfect heat engine is shown here:

$$\eta = 1 - \frac{T_c}{T_h}.$$

It is derived by witnessing that the heat recovered from a high temperature reservoir (i.e., the hot reservoir is losing heat energy), Th, causes an entropy change of:

$$\Delta S_{hot\ reservoir} = \frac{-Q}{T_h}.$$

Similarly, the entropy change of the lower temperature reservoir (i.e., the low temperature reservoir gains heat energy) has an entropy change of:

$$\Delta S_{cold\ reservoir} = \frac{+Q}{T_c}.$$

But notice the negative and positive signs, respectively, based on the direction the heat energy is heading: "out from" is depicted with the negative symbol and "into" with the positive. Thus the equation $(Q/T)_{hot} = (Q/T)_{cold}$, because $(Q/T)_{hot} + (Q/T)_{cold} >= 0$, and in the BEST case, the equal sign is used. From the First Law for a closed system it is true that: $\Sigma Q = \Delta E + \Sigma W$ and for a system of processes that closes to form a cycle, it is also true that the internal energy of the system, $\Delta E = 0$. The consequence is that the net exchange of heat transfer into and out of the system will be equal to the work that is performed by the thermodynamic cycle. That is, a thermodynamics cycle is a process that can be comprised of two or more pathways or processes that are sequential (that is: the end of one process is the start of the next second) until the process paths all lead back to the beginning of the system; in other words, the system (eventually) returns to its initial condition, and is therefore ready to start again. Thus, a heat engine has at least two processes wherein the material in the system starts at point 1 and finishes at the same point 1. Yes, the material literally goes in circles, returning to where it started. That is why the internal energy net change must be zero, leaving the net heat transfer to be equal to the net work done by the cycle introducing power. Notice that this does not mean that the efficiency of the cycle is 100%. That is, a heat engine cycle's efficiency is not defined as the net work out divided by the net heat. Efficiency is defined as the net work out of the system divided by the heat INPUT into the system.

1.5.1   *Example 1.5: Heat transfer from a cold glass of water to make it hotter*

Consider the glass of water shown in Figure 1.12. The temperature of the water is 100°F. The ambient temperature is given as 80°F. Let us assume

Figure 1.12.   Water exchanging energy in a warmer environment.

that 100 Btu of thermal energy is taken from the environment and added to the glass of warmer water. The First Law Equation for this Closed Mass System is given in Equation (1.7).

$$\sum Q = \sum W + \Delta E_{internal} \tag{1.7}$$

$$\sum Q = \Delta P.E. + \Delta K.E. + \Delta Et; \quad where: \Delta Et = C_v \times \Delta T.$$

But for the glass of water in this example: $\sum Q = 100\,\text{Btu}$, $\Delta P.E. = 0$ and $\Delta K.E. = 0$.

Therefore: $100\,\text{Btu} = C_v \times \Delta T$, provided that $\Delta T > 0$.

It is also true that the environment is providing the 100 Btu of energy and thus the first law equation applied to the environment results in the equation:

$$-100\,\text{Btu} = C_v \times \Delta T, \text{ provided that } \Delta T < 0.$$

The First Law permits the gain of the 100 Btu of energy from the colder environment into the warmer glass of water. However, the reality from this example very quickly indicates that never has it been observed that a warmer body gets hotter when exposed to a colder temperature. This is a significant dilemma that has plagued thermodynamics since its conception as an independent engineering science. This resulted in the development of the Second Law of Thermodynamics and the definition of entropy, which is another thermodynamics property. With the use of this property, a principle could be used for the first time to determine the direction and extent of a thermodynamics process. Entropy is defined as the amount of heat that can be transferred divided by the absolute temperature at the boundary of the exchangers. This can be thermodynamically represented by Equation (1.8a) and (1.8b).

$$\Delta S_{surroundings} = \frac{-Q}{T_{surroundings}}, \qquad (1.8a)$$

where the temperature, T, must be in absolute temperature units R (Rankine) or K (Kelvin), and the negative sign on the heat transfer, Q, indicates that the heat transfer is leaving the surroundings (the environment).

Similarly:

$$\Delta S_{system=water} = \frac{+Q}{T_{water}}, \qquad (1.8b)$$

where the heat transfer, Q, now has a positive sign to indicate that heat energy is received by the glass of water.

The entropy change for the system (the glass of water) and its surroundings (the environment) can now be determined by adding these two terms. The result is as shown to be a negative value.

$$\Delta S_{system=water} = \frac{+Q}{T_{water}}; \ \Delta S_{system=water} = +\frac{100}{460 + 100};$$

$$\Delta S_{system=water} = +.1786$$

$$\Delta S_{surroundings=environment} = \frac{-Q}{T_{water}}; \ \Delta S_{system=water} = \frac{-100}{460 + 80};$$

$$\Delta S_{system=water} = -.1852$$

$$\Delta S_{system=water} + \Delta S_{system=water} = -.00661$$

However, the Second Law of Thermodynamics states clearly that this sum must be greater than or equal to zero, in perhaps its simplest form. That the sum for this example is not greater than zero reveals that this process of a warm glass of water becoming hotter cannot be possible.

In fact, the reverse is always observed. That is, that the glass of water must always lose energy via heat transfer to the environment, if and when the glass is at a higher temperature than its surroundings. The universality of the Second (and First) Law(s) of Thermodynamics requires that this must be true. The calculation below finds that the sum is greater than zero when the environment gains the heat transfer that is lost from the glass of water.

$$\Delta S_{system=water} = \frac{-Q}{T_{water}}; \ \Delta S_{system=water} = -\frac{100}{460 + 100};$$

$$\Delta S_{system=water} = -.1786$$

$$\Delta S_{surroundings=environment} = \frac{+Q}{T_{water}}; \ \Delta S_{system=water} = +\frac{100}{460 + 80};$$

$$\Delta S_{system=water} = +.1852$$

$$\Delta S_{system=water} + \Delta S_{system=water} = +.00661.$$

This simple and straightforward example will be expanded in the next chapters to help solve many more complicated problems in thermodynamics. However, this example provides the essence of the Second Law of Thermodynamics that is applicable in its simplicity to the understanding of more difficult problems.

The requirement that a heat engine does not have a net cycle efficiency greater than the Carnot efficiency is only one face of the Second Law concept. Another face of second law efficiency is that no heat engine cycle can have 100% efficiency, unless the cold temperature reservoir is at 0 K or 0°R. We know this will not happen. In fact, according to physicists, the cosmic background from the Big Bang is still 3 K. There have been experiments that have led to Nobel Prizes, where only close to absolute 0 K or 0°R have been achieved — literally 0.0000000005 K in 2003 by Wolfgang Kettle — but never 0 K. Suffice to say in this first Chapter that for all practical purposes the Second Law of Thermodynamics indicates that no heat engine cycle can have as a sole effect, the absorption of heat energy to produce a new amount of useful work, while rejecting ZERO heat energy to a colder heat sink.

But why even discuss this when you are reading a textbook on the practical applications of engineering thermodynamics, when the closest an engineer can get to zero absolute is perhaps 300°F below zero? In fact, to achieve these temperatures, and given Earth's environmental temperature being a nominal 520°R, it would be necessary to use the principles of thermodynamics to transfer heat energy from space that needs to be cooler than the environment. The man-made systems that are developed for this heat exchange would require mechanical work to power these devices. These devices were labeled "refrigerators" long before the kitchen appliance with the same name became a necessity in most modern kitchens. It will be shown that these refrigerators (as understood in engineering) are actually heat engine cycles that operate in reverse. It is therefore true that the Second Law can also be stated to require that heat cannot be transferred from a cold to a hot reservoir without some external effort or work being applied to the system.

Thus we return to the original discussion of how the Second Law came to be regarded by the initial researchers in thermodynamics as being necessary. Heat cannot simply jump into a hotter reservoir than from which it came. The entropy change would be less than zero and the heat transfer would also occur without expending physical work, rotating a shaft or doing anything equivalent.

With all of this now presented, it is imperative to also understand that the engineer is strictly forced to adhere to using only absolute temperature scales in any thermodynamics analysis. Absolute temperature scales are needed by the physical reality of the Universe, rather than the more locally observable and hence measurable events in the real world, using recurring events of one or more material that describe a high and a low temperature. For example, the development of a thermometer — the first researchers would select a convenient high and low temperature that corresponded to any easily reproducible earthly effect. Certainly, the freezing and boiling points of water serve this definition very nicely. If we were to find inhabitants on other planets in the solar system, this material could be ammonia or methane, etc. there, but on Earth the freezing point of water is assigned by definition to be 0°C and the boiling point to be 100°C.

The person who invented this scale could have chosen 0 and their day of birth in the year as an example of the arbitrariness of the Celsius scale. But now, what about a very true scale that defines a temperature that cannot go below 0°C? For this, one will need the absolute scales of either the Kelvin (K) or the Rankin (R), both named after researchers in the 18th century who contributed considerably to the birth of the field of thermodynamics. The high level has no limit, except by understanding that a given high Kelvin reading, of say 273 K, will be observed to freeze water and 373 K to boil it.

The interior of the Sun is about 10,000,000 K as a result of the fusion reaction between hydrogen and helium. This can be considered as the top temperature on the Kelvin scale, not that it matters to any thermodynamics analysis. What matters is that the use of the Carnot equation for determining the maximum cycle efficiency that can operate between two temperature reservoirs is restricted to 100%, only if a continuous cold temperature of 0 K can be achieved. But then again, if that cold temperature of 0 K is achieved and held steady (i.e., the temperature does not change with time), in that perfect utopic world where there are only perfect heat engines and perfect refrigerators, the perfect heat engine would not have any heat to be rejected into the 0 K heat sink (or cold reservoir) because all of the heat input to the perfect heat engine would be converted to useful work. Thus, in that utopic world of perfect heat engines, perfect refrigerators would never be needed to remove heat energy from the 0 K cold space because then, with no heat ever entering the cold space, the cold space could only get colder and this is not possible unless you define this lower temperature as the new absolute zero temperature scale.[2]

1.6 Some real-world examples

The concept and the reality of the Second Law has been used by physicists and engineers to predict the outcome of an experiment, if only to put a boundary around the upper or lower limits of what is desired. It also enables the analyst to envision a world of thermodynamics utopia (assuming that this envisioning helps to solve thermo problems). In this world there are only perfect heat engines and refrigerators. In this world, if there is a hot temperature heat source and a cold heat sink available, and some heat (Q) to be rejected from the hot heat source, then we can calculate how much useful work can be generated using the perfect heat engine with Q as an input. The result is $Q \times (1 - Tc/Th)$. But if this perfect heat engine is compared to a real world engine that is forced to live in the real world, where there is friction, and the entropy change of the Universe is greater than but not equal to zero, then the work that can be generated is less than this amount. The difference in the work that could have been is called exergy loss (or irreversibility, or also loss in availability) because the difference is forever lost. A major "could have been". But where did this difference in work or power go? The reality is that it hasn't gone anywhere except to heat the Universe a little more, and this increases its entropy.

In the worst-case scenario, the real world doesn't recover any of this heat to make power. For example, if you fill your bathtub with reasonably tempered bath water (313 K) and finish your bath and drain the tub for the next person, then the heat energy that was consumed to make the 313 K tub water hot with no other work done (i.e., no rotating shaft to produce useful power for some useful purpose), then the loss of power can be enormous if that hot water was made from 313.0001 K. Then 1,000 units of heat multiplies $(1 - 300/313)$, where 300 is the ambient temperature around the tub,

---

[2]It is interesting to consider the moments immediately after the Big Bang, when nuclear and quantum physics indicate that the Universe began with an infinitesimal small space at 10^6 K + and that this space cooled and got infinitely large (and is continuing to expand as I write this). But then as the universe space cooled from 10^6 K where did the heat go? What space outside this Universe's Control Volume got hotter?

and results in 41.5 units of power or energy that could have been generated for each 1,000 units of heat, and this energy is now irreversibly, irretrievably lost and the entropy of the Universe has increased by (41.5 units of energy/300 K). Where did this equation come from? It is from the future. (Note that the reduction is stated to be of "available energy" and not the energy that most are familiar with.)

This available energy is a somewhat utopian entity that cannot be destroyed. It is only transformed from one form to another, but even in a utopic universe its availability can be destroyed. This is simply the recognition that depending on the transformation process, the conversion of one form of energy to another cannot be 100% effective and the amount of energy that is not converted to a form that could be best used by humanity in the real world is lost forever. For example, rubbing two sticks together will require mechanical energy from your arms or a mechanical linkage of any device whose function is to rub two sticks together. This will cause the sticks to become hot. In fact, if you are a boy scout you will know that the temperature can get as high as the auto ignition temperature of paper, or even better, dry leaves, and thus help in your chore of starting a fire. However, the transformation of some of that mechanical energy is lost due to heat transfer to the environment even as the sticks are getting hotter. That lost energy could have been put to good use in a utopic world where all engineers deal with perfect heat engines and refrigerators. The lost heat could have had some work or power generated if that heat was used as heat input to a perfect heat engine. Given the efficiency of that heat engine, the amount of heat input would have generated:

$$W = \eta_{Carnot} \times Q_h;$$ an amount of work that is no longer available or irreversibly lost.

1.7 Open flow systems

The presentation of the First and Second Law of Thermodynamics to a closed system must now be continued by looking at the analysis of a system that has mass flowing into or out of it. Once again it is important that the definitions of "open" and "closed" be understood and made clear to the reader. It is simply this: mass flowing into the system is simply part of the wider universe that the engineer for this problem thinks is necessary to identify and solve. The engineer is again prepared to analyze the parts of the Universe that are relevant to solving the problem by tracing a line around the relevant parts, which will go through a process that starts at a defined time and undergoes some sort of change until another prescribed period of time at the end of the process. All the events and paths are under the control of the engineer or are usually well defined by a very careful review of the problem statement, which is either clearly written or communicated orally to the analyst engineer.

The open flow method of applying the First and Second Laws of Thermodynamics is probably the most useful as most problems in thermodynamics involve some fluid that is flowing into or out of the system's CV. For example, turbines, compressors, valves, heat exchangers, pumps, tanks, nozzles and diffusers are mechanical systems that involve flow into and/or out of them. Well, what about things such as the piston cycle system? Yes, that system can certainly be modeled with the intake valve open to take in air and fuel. It is then closed and a process (actually, four) occurs and the exhaust valve is opened to dispel the exhaust products of combustion. Define your CV as the imaginary boundary around the drawing of an engine (or cylinders), and there is clearly a breaching of the system with mass flow, thus showing that the system here is clearly an open one. Then, there is heat input and there is work out — even if this does not happen simultaneously in real time, in that the heat input occurs before work is produced and more heat transfers out after the work produced — the Open Flow System Analysis method simply states that the heat in minus the heat out must be equal to the work done, because the internal energy of the engine cycle for the single cycle process is zero.

Please note that the engine cycle that we have just described is indeed just a cycle. Thus, it can be completely analyzed during one of the cycles to force the internal energy to be zero or n cycles, but this just increases the time duration of the entire event. Also, consider this: each piston-cylinder engine that the engineer is analyzing is actually a collection of one or more of the seven machines that were identified here. That is, the engine can be seen as a valve (carburetor or fuel injector), the heat transfer across the walls of the piston can be modeled as a heat exchanger, and the compression and expansion stroke can be modeled by the adiabatic (no heat transfer) compression and expansion process mentioned above. OR CAN IT?

Here is where it gets very interesting (and perhaps tricky) until you better appreciate what actually goes on during an Open Flow System. The Open Flow Analysis must consider not only the internal energies that enter or leave the system, but also what is conventionally called "flow work", which is a rather appropriate name for the energy that must be given up or gained by having the working fluid push its way into or out from a system. This flow work is easily described as $P \times V$ and has units of ft-lbf or ft-lbf/lbm of fluid flow. In the case of a closed system that is undergoing compression or expansion, there is no flow work because the fluid is assumed to be stored and never liberated from the closed system. How the fluid actually got into the system in the first place is rather esoteric, but it is assumed that at some time in the past it needed to have its flow rate expended (or gained). However, after that singular event of filling the system, it is closed and does not require the flow work to be tallied.

Returning to the piston-cylinder reciprocating engine, the fact that the system is closed for a small fraction of time means that the Closed Systems Analysis will only be of use to the analyst during this duration. For a more realistic picture of the process, the analyst may be better off choosing the Open Flow Analysis case, where a CV is placed around the entire piston engine, thus there being the need to only keep track of the air and fuel that enters the CV (i.e. the engine) and exhaust gas products that leave the engine, along with the rotating shaft power that represents the work produced. For this scenario, the flow work into and out from the system is added to the internal energy and the sum of these two parameters is called the enthalpy of the inlet and exhaust streams. The enthalpy of the system gives definition to one other specific heat capacity, which is defined for vapor fluids as:

$$C_p = \frac{\Delta H}{\Delta T},$$

which is not to be confused with the specific heat capacity with respect to volume, $C_v = \frac{\Delta U}{\Delta T}$. These two parameters are used frequently and are tabulated as functions of temperature (or also the pressure) in most textbooks.

A not so subtle takeaway from this is the need for the analyst to begin the solution of a thermodynamics problem by first asking him- or herself if the closed mass or open flow methodology is more appropriate for that particular problem. The more problems the analyst solves the easier it gets to know which one to start the problem solution with. In any event, there is only a 50/50 chance that the analysis is stated incorrectly, and even the stubborn analyst will stop and retrace his/her steps and begin again if the initial attempt was unsuccessful.

The equations for the First Law applied to an Open Flow System may be summarized as follows:

$$\sum Q + \sum \dot{M} \times (h + K.E. + P.E.)_{in}$$

$$= \sum W + \sum \dot{M} \times (h + K.E. + P.E.)_{out} + \frac{\partial U}{\partial t}_{cv}. \qquad (1.9a)$$

It always helps when starting something new to look for the familiar. In the case of the first law equation for energy in an Open Flow System, the familiar terms should be heat transfer, work and internal energy. However, the open flow equation is rather unique in its use of a time rate of these terms. Thus, the heat transfer, Q, is really a rate of heat transfer in units of energy per unit of time or power. Similarly, the rate at which work is done is called power. The rate at which the energy of the system, U, is changing per unit of time is also a power term, but it is useful to think of it as simply the rate at which the internal energy of the system is changing with time. As always, the "system" is identified by the engineer-practitioner to encompass only those elements of the real world that are of interest to the engineer-practitioner. The internal energy, U may include all of the other forms of energy, shown in Table 1.1, that have not been included in the equation that the system may have and that could change during a process. The most unfamiliar term in the equation is the term, H, and this identifies the enthalpy of the inlet and outlet streams.

Enthalpy is a "man-made" parameter, in that it really is simply the sum and product of three familiar fluid properties: pressure, specific volume and internal thermal energy such that:

$$h = u + P \times v \tag{1.9b}$$

It is normally collected as shown in Equation (1.9) because it is very commonly the "go to" parameter for an Open Flow System. Just as internal energy is one of the major terms in the analysis of a Closed Mass System, there is a fairly clear reason why enthalpy is associated with fluids that flow into and out of a system, where internal energy is used for a closed system. Once again, start with the familiar: internal thermal energy, here represented by U_T, which defines it as a fluid with thermal energy coming into a system and then leaving a system. Note: a system is considered an open one, as long as there is mass moving either in or out of it, but both conditions need not exist concurrently. For example, a balloon that is being filled with mass has no mass flow rate out. The energy of the person blowing up the balloon is stored in the pressure and the skin surface tension of the balloon, and thus the first law equation would have zero outlet enthalpy, a non-zero inlet enthalpy and a non-zero stored internal energy of the system.

Continuing with the definition of enthalpy leaves us explaining the "p×v" term shown in Equation (1.9b). The "P×V" term is called flow work because it determines how much energy is needed to push the fluid into the system or how much energy is "returned" when the fluid leaves the defined system and needs to "push" its way out of the system. The multiplication of p and v is simply taking the pressure at the inlet or outlet state and then multiplying it by the specific volume. Care must be taken about the units if working with the English or imperial system. The p product will have units of ft^3-lbf/lbm/in^2 if the pressure is in psia, and therefore the term must be multiplied by 144 and divided by 778 to convert that term to Btu/lbm, the typical unit for energy per unit of mass in the imperial system.

The sign convention used in Equation (1.9a) is the same as the Closed Mass System equation of the First Law. Heat transfer rates into a system are considered to be positive, while heat transfer rates out of the system are negative. Power into a system is considered negative and power out is positive. This is simply convention and the signs in front of each term would change if the opposite convention were to be adopted. However, the sign convention with power output being considered positive seems to make the most sense from a standpoint of wanting power out from a system and thus that power is a positive — a good thing to have.

The simplicity of the First Law as it appears for the Open Flow System must never deflect from the importance that this simple and elegant equation has in being relevant to every event that can occur within our Universe. The analyst often needs to apply this equation properly and this starts with the very important realization that the First Law is only able to be properly applied to solve a thermodynamics problem if one first identifies the system to be analyzed (i.e. whether the system is closed or open) and then keep track of the heat transfers and power going into and out of that system, respectively. It must cross the boundaries that the engineer analyst has constructed to define the system of interest. If the heat transfer or work does not breach the CV boundary, then that term is zero. Thus, any heat transfers or power exchanges within the boundary or out of the system is not used in the equation. It must cross the boundary. It is clear then that the judicious choice of a CV that properly defines the system of interest is essential (or critical) to the ease of how the thermodynamics of the problem can be solved using the First Law. This is where the experience (or luck) of the analyst engineer comes into play. However, luck, as it has been said, is when opportunity meets preparation. The preparation part comes from the study of thermodynamics in similar books as this, while the opportunity comes in the form of an open mind and interest to pursue experiences in this universe of options.

Several examples of how the First Law can be used are critical to its understanding and these examples are offered in the next section. However, the reader is reminded that this chapter is intended for a quick review of the basic principles that will be described in much more detail in the later chapters.

1.8 Enthalpy rules

The First Law for an Open Flow System is among the most useful equations that may be used by an engineer, particularly if that engineer works in the power industry or other related industries. It is also the easiest of the universal laws to use, assuming that the user applies it correctly. Fortunately, the application of the First Law becomes more routine and accurate with experience. The first means of applying this law is to assume certain physical properties of the system that may more often than not be applicable to that particular system. In addition, most systems are typically common enough systems that the First Law in its long form, as shown in Equation (1.9a), may be reduced to a shorter and simpler equation, but careful: it "...[can]not [be made] too simple..." as Albert Einstein once cautioned regarding difficult physics problems that can often be solved quickly if one looks for the simple, elegant solution. For example, in the case of thermodynamics, all mechanical systems in this Universe may be observed to be constructed of one or more of seven or maybe eight "building blocks". These building blocks are analogous to software icons or parts of a puzzle that can be put into an infinite number of arrangements to construct more and more sophisticated systems. But when taken separately, they are no more complicated than a very friendly and useful equation — a "lite version" of the more complicated but still elegant First law, which is stated in Equation (1.9a).

These basic building blocks may be analogous to the L, C, R or: inductors, resistors, capacitors and integrated circuits (ICs) that are used in electrical engineering. They are the DNA and RNA of the mechanical engineering world. Although these are limited to seven building blocks, it could be said that mechanical engineering is more versatile than electrical or biological nature systems. These basic blocks are: Turbines, compressors, pumps, nozzles, diffusers and heat exchangers, and some may request that tanks be added to the collection. The first law equations of each of these devices all have one thing in common: the "mantra" that Enthalpy Rules. That is, determining the enthalpy at the inlet and the exit state of each of these basic systems is the goal of the exercise. The enthalpy difference then leads to power in or out of the turbine (or compressor or pump), or the increase in velocity of the nozzle, the pressure of the diffuser or the heat exchange for the heat exchanger. Table 1.1 summarizes the equations for each of these basic systems, along with the assumptions that were made in order to derive these equations.

Again, a straightforward application of the First Law to a turbine, a heat exchanger and a nozzle will help demonstrate the utility of these equations.

1.9   Thermodynamics cycles — The chromosome of mechanical engineering

The seven basic mechanical engineering components can be used to derive an infinite number of cycles. Recall that cycles are the processes that begin and end at the same state points. Thus, at least two processes are needed to connect the end of one process to the start of the second process as shown in Figure 1.9. It will be made clear in Chapter 3 that this cycle cannot happen unless the processes are perfect and if there is no occurrence of heat transfer. It will be learned that the Second Law of Thermodynamics will require that the entropy change for these two processes must be equal to or greater than zero and that would most likely require the two processes to not overlay each other, as shown in this non-perfect, non-utopic universe that we inhabit. Thus, there are likely to be at least three processes that will start and end at the same state point after being able to exchange heat transfer or work, or utilize or steer internal energy.

It is more common in thermodynamics to have the four processes shown in Figure 1.9, able to consume heat transfer and produce net power, or for net power to be input to the cycle somewhere in the process and then inject and/or reject heat transfer from the environment. These cycles would be identified as heat transfers or a refrigeration cycle. Of course, a cycle may have more than four processes. Similarly, any industrial process may have many more components that proceed through a cycle in order to produce a product as an output.

One of the most common elements of any mechanical cycle is the need for a working fluid to be used with all of the components. The working fluid can either be any man-made chemical or naturally occurring fluid. It will not be of any surprise to the reader that the most common working fluids that are used in the most cycles are water and air. The reader may also be familiar with refrigerant fluids due to their common use in air conditioning systems. Recent news also mention that certain refrigerant chemicals have an impact on the environment. The reader may be familiar with the research (on using hydrogen as the working fluid and/or fuel in cycles) that is being sponsored by the US Department of Energy, as well as private auto manufacturers. However, the use of supercritical carbon dioxide as a working fluid may not be as familiar to the general public, though it has been getting increasing attention in recent times.

The two cycles shown in Figures 1.13 and 1.14 are the most common cycles developed for generating power and for air conditioning or freezing. These are the Rankine Cycle and the reverse Rankine Cycle (or vapor recompression cycle). This Cycle was named after a famous thermodynamics

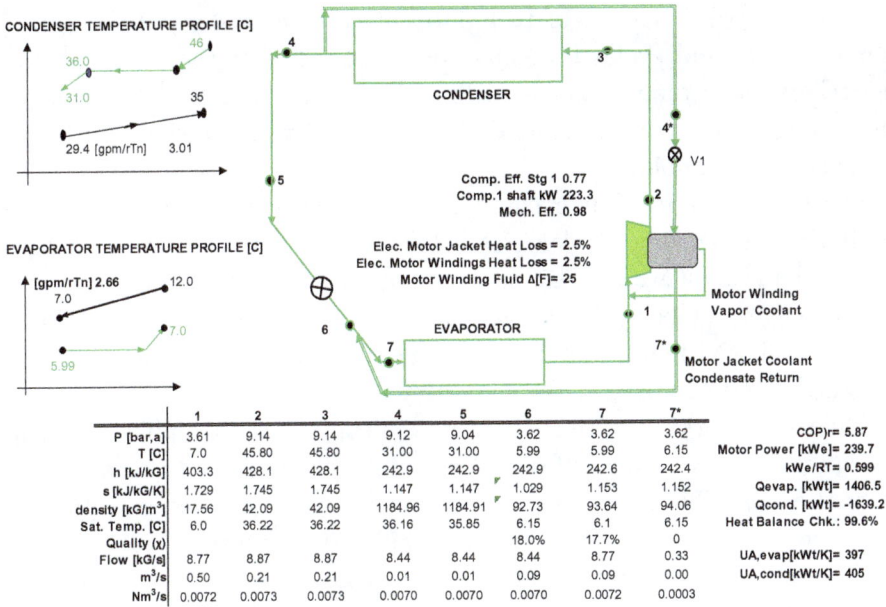

CONDENSER TEMPERATURE PROFILE [C]

46
36.0
31.0   35
29.4 [gpm/rTn]   3.01

CONDENSER

4*
⊗ V1

Comp. Eff. Stg 1 0.77
Comp.1 shaft kW 223.3
Mech. Eff. 0.98

Elec. Motor Jacket Heat Loss = 2.5%
Elec. Motor Windings Heat Loss = 2.5%
Motor Winding Fluid Δ[F]= 25

EVAPORATOR TEMPERATURE PROFILE [C]

[gpm/rTn] 2.66   12.0
7.0
7.0
6
7
5.99

Motor Winding
Vapor Coolant

EVAPORATOR

7*

Motor Jacket Coolant
Condensate Return

	1	2	3	4	5	6	7	7*
P [bar,a]	3.61	9.14	9.14	9.12	9.04	3.62	3.62	3.62
T [C]	7.0	45.80	45.80	31.00	31.00	5.99	5.99	6.15
h [kJ/kG]	403.3	428.1	428.1	242.9	242.9	242.6	242.6	242.4
s [kJ/kG/K]	1.729	1.745	1.745	1.147	1.147	1.029	1.153	1.152
density [kG/m³]	17.56	42.09	42.09	1184.96	1184.91	92.73	93.64	94.06
Sat. Temp. [C]	6.0	36.22	36.22	36.16	35.85	6.15	6.1	6.15
Quality (x)						18.0%	17.7%	0
Flow [kG/s]	8.77	8.87	8.87	8.44	8.44	8.44	8.77	0.33
m³/s	0.50	0.21	0.21	0.01	0.01	0.09	0.09	0.00
Nm³/s	0.0072	0.0073	0.0073	0.0070	0.0070	0.0070	0.0072	0.0003

COP)r= 5.87
Motor Power [kWe]= 239.7
kWe/RT= 0.599
Qevap. [kWt]= 1406.5
Qcond. [kWt]= -1639.2
Heat Balance Chk.: 99.6%

UA,evap[kWt/K]= 397
UA,cond[kWt/K]= 405

Figure 1.13. A refrigeration cycle including the cooling of the electric motor using refrigerant working fluid.

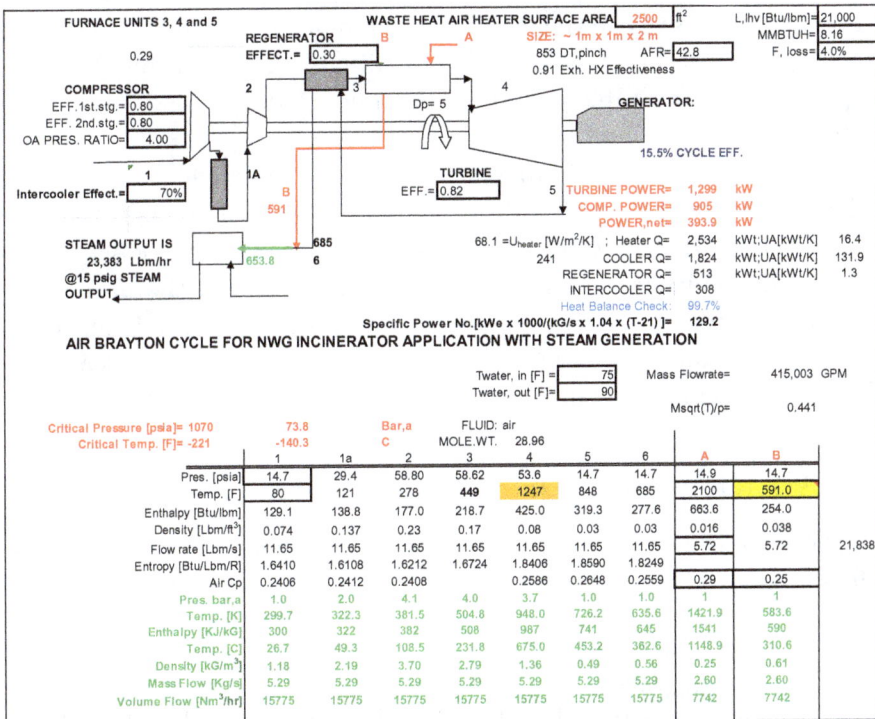

FURNACE UNITS 3, 4 and 5

WASTE HEAT AIR HEATER SURFACE AREA 2500 ft²
L,lhv [Btu/lbm]= 21,000
MMBTUH= 8.16
F, loss= 4.0%

REGENERATOR
EFFECT.= 0.30
0.29
SIZE: ~ 1m x 1m x 2 m
853 DT,pinch   AFR= 42.8
0.91 Exh. HX Effectiveness

COMPRESSOR
EFF.1st.stg.= 0.80
EFF. 2nd.stg.= 0.80
OA PRES. RATIO= 4.00

Dp= 5

GENERATOR:

15.5% CYCLE EFF.

Intercooler Effect.= 70%

TURBINE
EFF.= 0.82

TURBINE POWER= 1,299 kW
COMP. POWER= 905 kW
POWER,net= 393.9 kW

STEAM OUTPUT IS
23,383 Lbm/hr
@15 psig STEAM
OUTPUT

68.1 =U_heater [W/m²/K] ; Heater Q= 2,534 kWt;UA[kWt/K] 16.4
241   COOLER Q= 1,824 kWt;UA[kWt/K] 131.9
REGENERATOR Q= 513 kWt;UA[kWt/K] 1.3
INTERCOOLER Q= 308

Heat Balance Check: 99.7%
Specific Power No.[kWe x 1000/(kG/s x 1.04 x (T-21)]= 129.2

AIR BRAYTON CYCLE FOR NWG INCINERATOR APPLICATION WITH STEAM GENERATION

Twater, in [F] = 75   Mass Flowrate= 415,003 GPM
Twater, out [F]= 90
Msqrt(T)/p= 0.441

	1	1a	2	3	4	5	6	A	B	
Critical Pressure [psia]= 1070	73.8		Bar,a	FLUID: air						
Critical Temp. [F]= -221	-140.3		C	MOLE.WT. 28.96						
Pres. [psia]	14.7	29.4	58.80	58.62	53.6	14.7	14.7	14.9	14.7	
Temp. [F]	80	121	278	449	1247	848	685	2100	591.0	
Enthalpy [Btu/lbm]	129.1	138.8	177.0	218.7	425.0	319.3	277.6	663.6	254.0	
Density [Lbm/ft³]	0.074	0.137	0.23	0.17	0.08	0.03	0.03	0.016	0.038	
Flow rate [Lbm/s]	11.65	11.65	11.65	11.65	11.65	11.65	11.65	5.72	5.72	21,838
Entropy [Btu/Lbm/R]	1.6410	1.6108	1.6212	1.6724	1.8406	1.8590	1.8249			
Air Cp	0.2406	0.2412	0.2408		0.2586	0.2648	0.2559	0.29	0.25	
Pres. bar,a	1.0	2.0	4.1	4.0	3.7	1.0	1.0	1	1	
Temp. [K]	299.7	322.3	381.5	504.8	948.0	726.2	635.6	1421.9	583.6	
Enthalpy [KJ/kG]	300	322	382	508	987	741	645	1541	590	
Temp. [C]	26.7	49.3	108.5	231.8	675.0	453.2	362.6	1148.9	310.6	
Density [kG/m³]	1.18	2.19	3.70	2.79	1.36	0.49	0.56	0.25	0.61	
Mass Flow [Kg/s]	5.29	5.29	5.29	5.29	5.29	5.29	5.29	2.60	2.60	
Volume Flow [Nm³/hr]	15775	15775	15775	15775	15775	15775	15775	7742	7742	

Figure 1.14. An open air-Brayton Cycle used for waste heat recovery.

researcher, Lord Rankine. In this publication, the reader will be exposed to many units and cycles that have been named after their inventors and researchers. For example, names such as Watt, Joule, Kelvin, Rankine, Stefan Boltzman, Brayton, Lenoir, Otto and Diesel will become very familiar as the chapters unfold.

The next most familiar cycle is called the Brayton Cycle, known commonly the Gas Turbine Cycle (GTC). It is this cycle that has revolutionized air travel by allowing aircraft to travel at 500 mph while carrying up to 400 passengers. The reader learns that this is possible because the Cycle (that uses air as the working fluid) produces considerable power while being lightweight and compact at the same time.

The use of working fluids in a cycle requires that the variables in the first law equation be determined for that chemical, particularly the enthalpy and internal energy. These properties have been well established and published by many research institutions and are available for thermodynamics analysis. The advent of computers since the 1970s has made tabulated properties even more easy to use via lookup routines that have been programmed into software platforms such as the Excel spreadsheet. It is this platform that this textbook and thermodynamics rendition will be making use of most often to complete even the most difficult of thermodynamics analysis. To help the reader be most effective in the use of working fluids in thermodynamics analysis, Chapter 2 will begin with a thorough discussion of how fluid properties are "looked-up" in a table of thermodynamic properties and incorporated into thermodynamics analysis.

For the reader to be able to answer any thermodynamics problem, this detailed discussion will be essential. Thus, it is offered as the first significant study in this thermodynamics text, seconded only by the necessary discussion of metric units in thermodynamics. Taken together, the properties of the working fluids used in a thermodynamics problem and the metric units of the pressure and temperature energy are analogous to the learning of grammar rules in any new language.

1.10　A solution process for engineering problems

There is no one perfect pathway for solving engineering problems. Each engineer tends to develop a solution methodology that works for him or her; a process that is finely honed over many years of engineering practice. This author has found the following steps to be most effective in solving thermodynamics problems and, perhaps with some simple modifications, finding solutions for other engineering disciplines.

STEP 1. Translate the written specifications and/or oral presentations from whatever language is being used in thermodynamics. The meaning of this will become clearer in the next chapters, but as an example, the written specification may indicate that the system is closed and fixed. This is translated in thermodynamics to mean that the system does not have mass entering or leaving the system and that the volume of the system does not change, regardless of what the process may be from the beginning of an event to the end.

STEP 2. Draw a free hand sketch, approximate to scale if possible, of the written specification and label the drawing. For example, a representation of a piston-cylinder, a turbine or compressor that clearly shows the inlet or outlet flows, the type of fluid in the system, the operating pressure or temperature or mass, etc.

STEP 3. Very clearly, identify and print on the solution page, not only the given information but also the unknowns. This should be followed by writing down any and all equations that immediately come to mind as being relevant to the solution of the problem. For thermodynamics, the analyst cannot go wrong by listing the First and Second Laws of Thermodynamics, the conservation of mass.

STEP 4. Draw a process diagram using a T–s or P–v or T–v chart. More on these charts will be found in the next two chapters. Immediately follow up by drawing a property chart with at least seven columns labeled: Pressure, temperature, specific volume, quality, enthalpy, internal thermal energy and entropy. Some analysts insist on adding an eighth column for the mass flow rate. However, mass, mass flowrate, and volume flowrate must never be confused with a thermodynamics intrinsic property. This Property Table must have the same number of rows as there are state points that identify the start and the end of the process-paths that describe the problem at hand.

STEP 5. The Property Table established in Step 4 must most often be completed with information that is supplied or derived from the principles of thermodynamics. Usually, enthalpy at the very least must be determined before the components that make up the system process can be individually analyzed. With the enthalpy known, the thermodynamics solution is imminent by usually applying the First and Second Laws of Thermodynamics. Readers are provided with additional assistance in understanding and mastering this in the form of the case studies presented in this textbook. While they are not inclusive of all possible thermodynamics problems that the analyst might encounter, these case studies have been selected because of their wide applicability to most common thermodynamics problems.

STEP 6. The most important part of the solution methodology is to check the "sanity" or "reality" of the solution; never accept your answer and submit it as a solution until you have tried to identify any inconsistencies after a complex calculation is performed. In many ways, the analyst is using his or her intuition, borne from experience, to verify the correctness of the solution to the problem. For example, in Example 1.1 of this chapter, if the ball is determined to be moving at 100 mph, the solution should be suspect. If the power output from a turbine is too large or too small for the "scale" of the system (i.e., the flow rate and/or the nature of the application, or if the answer has the wrong sign), then the solution may be suspect. Perhaps the best way of determining whether a solution is correct is when the solution is "too good to be true", thus a second look is warranted. Also, a second look at the solution is needed if the result is either nowhere near what the analyst expects or even disproves the original idea that analysis was to verify. In either case, any extra time spent on reviewing the first answer is always time well-spent.

1.10.1   *Applying the solution process*

Let us apply this methodology to the solution of a thermodynamics problem that has become a classic problem in thermodynamics, because it derives rather non-intuitive results using the First Law of Thermodynamics. The closed vessel shown in Figure 1.15 has two chambers. The left chamber is filled with a vapor fluid with known pressure, P, and temperature, T. The right-side chamber is a vacuum. Thus, pressure of the right-side chamber is zero and the temperature is basically not defined. A wall that contains a valve separates these two chambers. When the valve is closed it does not allow the pressurized fluid to leak into the vacuum. Once it is opened, it allows the almost instantaneous expansion of the vapor into the right-side chamber.

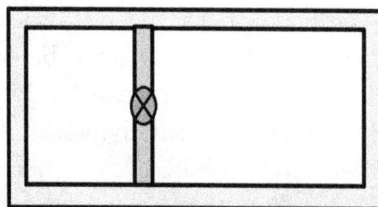

Length= L,total = 1 m
Diameter= D= .5 m
Left Chamber Length= 0.3 m
Temperature =T,right side= 500K
Pressure =Pleft side= 2 bar,a
Fluid = AIR

Figure 1.15.   Closed, rigid vessel with two chambers. The left side contains a known gas at a known pressure and temperature.

Assuming that (1) the dimensions of the vessel and chambers are known and that the vessels are rigid, and (2) the vessel is perfectly insulated and that the fluid is uniform throughout both spaces, what is the final temperature and pressure of the vessel when steady state and thermal equilibrium occur?

STEP 1. Carefully re-word the problem statement's use of the following words:

a. "Perfectly insulated" — no heat transfer is occurring with the outside environment.
b. "Rigid vessel" — the vessel walls do not expand or contract and thus volume of the vessel and chambers do not change.
c. "Thermal equilibrium" — the fluid temperature in both sides of the chamber are the same after the valve opens.
d. "Steady state" — the pressure and temperature do not change values with time.
e. "Closed system" — no mass enters or leaves the vessel during the time of interest.
f. "Uniform throughout" — the fluid is thoroughly mixed after the expansion.

STEPS 2 and 3. Re-draw the system with labels for all knowns and unknowns.

STEP 4. Construct a Property Diagram and Property Table (see Figure 1.16).

STEP 5. Solve for the unknowns using the First and Second Laws of Thermodynamics.

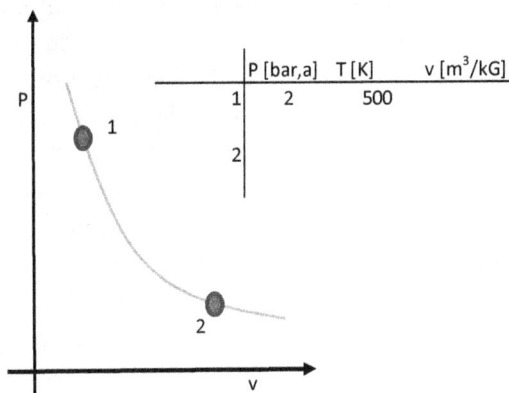

	P [bar,a]	T [K]	v [m^3/kG]
1	2	500	
2			

Figure 1.16. Property Diagram and Property Table for fluid in Figure 1.13.

In this problem the Property Table clearly shows that there are two properties known for state point 1. Thus, the third property, v, can be determined from the Perfect Gas Law. The properties of the second state point are missing. From the data given in the Problem Statement it can be determined that the specific volume, v2, can be found by knowing the new volume, and knowing that the mass of the system, M2, is the same as the initial mass of the system, M1. That leaves only one other property to be determined. But which one? Pressure, P2, or temperature, T2? To answer this question, we proceed to apply the First Law of Thermodynamics for a Closed Mass System:

$$\sum Q = \sum W + \Delta E_{internal}.$$

In this problem the internal energy resolves into only one internal energy type: internal thermal energy or

$$\Delta E = C_v \times \Delta T.$$

From an inspection of the system's diagram in Figure 1.13, there is no heat transfer and no work being done by the system of air, as it expands from the left into the right chamber. Thus, the change in the internal thermal energy is zero. But for the internal thermal energy to be zero, the temperature change for the air system must be zero. Thus, T2 = T1! If you are somewhat surprised by this result, you are not alone. Most initiates into the world of thermodynamics expect the pressure to decrease as the small left chamber opens to fill the now larger total volume. However, many would also expect that the sudden expansion of the (air) vapor would also decrease the temperature, leaving two unknowns with only one equation to use. In fact, if thermocouples or thermometers were to be placed everywhere within the two chambers and the temperature instantaneously monitored during the expansion process, there would be some decrease in the air temperature in the left chamber and an increase in the temperature of the contents in the right chamber as it is being filled. However, the problem states that the final equilibrium and steady state temperature of the fluid is to be determined after the valve is opened. The steady state and thermal equilibrium will cool the right side and heat the left side as the total fluid is thoroughly mixed. The net result is for the fluid temperature to not change. With this now understood, the second state point has two properties known: T2 and v2, and thus the pressure can be determined using the perfect gas law equation.

Case Study 16 will provide a more detailed, transient solution to this problem, wherein the temperatures of the left and right side are determined as a function of time.

STEP 6. Sanity check. The sanity check in this problem is to recognize that the many assumptions made in the Problem Statement restricts this problem from a very ideal situation. For example, there is no such thing as a perfectly insulated system or a system that does not have internal reversibilities caused by friction and energy loss. The expansion of the fluid would always result in friction energy lost to heating the walls of the vessel. The sudden expansion of the fluid into the right chamber would deflect the walls of the vessel. Even a very small amount would result in strain energy being adsorbed and lost through the hysteresis of the material. If one considers the amount of energy lost or energy absorbed, then the application of the First Law will still be useful along with the equations that account for these energy changes and thus enable the determination of the final equilibrium temperature once again.

Chapter 2

Thermodynamics Units, Energy, the Property of Fluids and the Thermodynamics Process

2.1 Thermodynamics units: Energy, power, temperature and pressure

Chapter 1 provided a very quick review of the concepts that will be covered in this textbook. Among them was the concept of energy. It was made clear that thermodynamics is truly the study of how different types of energy are transformed from heat transfer to work to internal energy. It was also noted in Chapter 1 that internal energy can be in the form of any number of energy types, which include the very familiar kinetic and potential energies, as well as spring, nuclear, chemical, electrical and magnetic energies. These transformations are restricted by the First and Second Laws of Thermodynamics and thus may not be reversible in the real world, where friction and entropy provide pathways for reducing available energy, which in turn does not allow for reversible processes.

Note the reduction that is stated is for the "available energy" and not "energy" (in the sense of the term most are familiar with). Available energy is a somewhat utopian entity: we cannot destroy energy in both the real and fictional utopic universes, as it is transformed from one form of energy (whether heat or work or internal or thermal) to another. However, we can destroy the availability of energy in our Universe.

This is simply the recognition that, depending on the transformation process, the conversion of one form of energy to another cannot be 100% effective, and the amount of energy not transferred in the course of conversion may no longer be available for future use. An example of the loss of available energy while conserving energy was given in Chapter 1 in the form of rubbing two sticks together to start a fire. Friction in that example is often taken as a small percentage of the total power energy per unit of time, i.e. power that is transmitted by the two gears moving at different speeds. This loss of friction energy results in the heating of the gears and requires that heat to be removed by an adequate flow of cooling oil. The friction energy is converted to heat, which in turn results in an increase in the temperature of the coolant oil stream. However, that heat could have resulted in useful power had it been used in a perfect heat engine in the form of a rotating shaft connected to a generator, fan or a mechanical transmission and the wheels of a vehicle. Given the efficiency of that perfect heat engine and the amount of heat input that could have been used in an ideal Carnot heat engine:

$$\text{Work} = \text{efficiency } (\eta) \times \text{heat input } (Q).$$

$$\text{Where: the heat engine efficiency is: } 1 - \frac{T_{cold}}{T_{hot}}$$

Because this mechanical power was not generated it is lost forever to the universe and thus unavailable. In thermodynamics, the words "available" and "availability" have a unique meaning. These words indicate an energy that can be "lost", but only in the sense that real-world engineers were not able to efficiently utilize all of the heat energy available for use in a heat engine, because real world heat engines cannot be as efficient as a Carnot (ideal or perfect) engine. The concept of available energy and unavailable work is a critical concept in thermodynamics. An understanding of this concept readies the student for the Second Law of Thermodynamics. If the reader is keeping count, then this is the second time he or she has encountered some aspect of the Second Law of Thermodynamics. If the reader understands these descriptions, then he or she has actually proceeded quite far in understanding the Second Law of Thermodynamics and is thus prepared for Chapter 3.

In Chapter 1, energy was simply defined as the ability to do work, which itself was defined as the ability to raise a weight in a gravity field. Heat transfer was simply defined as the energy exchanged due to a temperature difference. The "language" of thermodynamics must include the units of these three different forms of energy, as well as all of the other forms that were described in Chapter 1. It may be recalled that energy has many faces, such as thermal, internal, kinetic, potential, nuclear, strain, magnetic, surface tension energy, chemical and electrical, to name the most common forms. Despite this long list, the definition of energy or work or heat transfer must be common to all of these forms of energy. That definition indicates that raising a weight (i.e., a force) in a gravity field over a distance (length) is simply the product of force multiplied by distance. The units in the English system or imperial system for this is pound-force feet (lbf-ft). The units in the metric system are Newton-meters (N-m). Every field of study may have a different label for energy, but the label is often simply to give honor to a renowned inventor or researcher such as Watt or Joule. Nevertheless, the names are superfluous, in that the energy that needs to be inventoried via the First and Second Laws of Thermodynamics can be elegantly and simply stated as force multiplied by the distance through which that force moves.

The units of force or length in SI units are conventionally accepted to be pound-force (lbf) and Newtons (N) for force, and feet (ft) and meters (m) for distance. Thus, in the absence of any direction in a thermodynamics problem, the default units for the imperial and metric system are as indicated in Table 2.1(a) and (b). These units are always the "default" position in the event that the origin of an equation is unknown, and the units must be assumed in order to apply the equation properly. The conversion from the metric to the imperial unit system is not uncommon, particularly as the United States stands almost alone in its use of the imperial system. Having Table 2.1 at hand and being comfortable in its use provides thermodynamics analysts with the ability to move between either unit systems.

Another important consideration is the need to carefully account for the proportionality constant, also known as the gravitational constant, g_c. The gravitational constant has different values depending on its use in the imperial system or in the metric system, as shown here:

$$g_c = 32.2 \frac{lbm - ft}{lbf - s^2} = 1 \frac{kg - m}{N - s^2}.$$

Table 2.1.   Engineering imperial and metric units.

	Imperial	Metric
length	Feet (ft)	Meter (m)
mass	Pound-mass (lbm)	Kilogram (kg)
force	Pound-force (lbf)	Newton (N)
gravity (g_g)	32.2 ft/s^2	9.81 m/s^2
gravitational constant (g_c)	32.2 lbm-ft/lbf-s^2	1 kg-m/N-s^2
time	Second (s)	Second (s)
Btu/h =	1 hp ×	2545 Btu/h/hp
hp =	1 kW ×	1.341 hp/kW
kW =	1 watt ×	0.001 kW/W
watt =	1 J/s ×	1 watt/(J/s)
watt =	1 N-m/s ×	1 watt/(N-m/s)
N-m =	1 ft-lbf ×	1.3567 N-m/ft-lbf
ft-lbf =	1 Btu ×	778 ft-lbf/Btu
ft-lbf/s =	1 hp = ×	550 ft = lbf/s/hp

This gravitational constant is significant in its use in Newton's Second Law of Motion:

$$\sum F = \frac{M}{g_c} \times \frac{dV}{dt}. \tag{2.1}$$

The proportionality constant is essential for the units of force in Newton's Law to be derivable from the product of mass and acceleration — or more accurately stated, the force is proportional to the rate of momentum change. Similarly, the use of the gravitational constant, g_c, in kinetic and potential energies is essential to keep them defined as simply the product of force and distance.

For example, the kinetic energy equations for linear and angular kinetic energy are shown in Equations (2.2a) and (2.2b), respectively. The potential energy equation is shown in Equation (2.2c):

$$\text{Linear Kinetic Energy } (KE) = \frac{M}{g_c} \times \frac{V^2}{2} \tag{2.2a}$$

$$\text{Angular Kinetic Energy } (KE) = \frac{I}{g_c} \times \frac{\omega^2}{2} \tag{2.2b}$$

$$\text{Potential Energy } (PE) = \frac{M}{g_c} \times g_g \times \Delta H \tag{2.2c}$$

where M is the mass of the object, V is the linear velocity in ft/s or m/s, ω is the angular velocity in radians/s and ΔH, measured in either ft or m, is the height above the datum reference.

Note the units for these energies are as follows:

$$\text{Linear kinetic energy: } lbf - ft = \frac{lbm}{32.2 \frac{lb, - ft}{lbf - s^2}} \times \frac{ft^2}{s^2},$$

$$\text{Linear kinetic energy: } N - m = \frac{kg}{1 \frac{kg - m}{N - s^2}} \times \frac{m^2}{s^2},$$

$$\text{Angular kinetic energy: } N - m = \frac{kg - m^2}{1 \frac{kg - m}{N - s^2}} \times \frac{radian^2}{s^2},$$

$$\text{Potential energy: } lbf - ft = \frac{lbm}{32.2 \frac{lb, - ft}{lbf - s^2}} \times \frac{32.2 ft}{s^2} \times ft.$$

As can be observed, the cancellation of units with the gravitational constant in use provides a unit of energy, as it should. Similar use and check of units should always have energy in terms of a distance multiplied by a force unit.

As noted in Chapter 1, the energy parameter in thermodynamics has many "faces". Energy is elegantly defined as the "... ability to do work", while work is the "... raising of a weight in a gravity field" and heat transfer is the "... energy exchange due to a temperature difference". Each of these energy or work terms have units of lbf-ft or N-m. There are many forms of energy as described in Chapter 1 and summarized in Table 2.1.

Work also has many "faces". The most common work terms used in engineering thermodynamics involve moving a piston in a cylinder with a known volume through a distance or stroke. The amount of work produced is then the integral of the pressure with respect to the volume displaced. It may be expressed as shown in Equation (2.3a).

$$W = \int_{v1}^{v2} P(v)dv. \tag{2.3a}$$

During the very early days of steam engines, the integral of the P(v) with dv was performed using a P-v indicator diagram. The pressure and displaced volume of the steam piston were traced on the same chart paper and the engineer would determine the work performed per stroke of the piston by "simply" determining the area within the P-v closed curve. The advent of computers made this job much easier.

Apart from the terms already introduced, several other interesting work terms are used in mechanical engineering thermodynamics, such as stress-strain, surface tension and thermal expansion.

The stress (σ)–strain (ϵ) relationship is unique for any material. A typical graph of the stress–strain relationship is shown in Figure 2.1. The units for stress are force per unit area or lbf/in^2 or N/m^2. The units for strain are in/in or m/m. It is easily shown that the area under the stress–strain curve shown in Figure 2.1 is an energy per unit volume as may be derived from the work relationship shown below. The multiplication of the stress and strain with each of the units clearly reveals the energy per unit volume result.

$$W = \int_{v1}^{v2} \sigma(\epsilon)d\epsilon \tag{2.3b}$$

$$Energy\ per\ Volume = \frac{N}{m^2} \times \frac{m}{m}; \frac{Energy}{volume} = \frac{N-m}{m^3}.$$

Thus, if a mechanical element is subjected to a stress that causes it to elongate (strain), then the amount of work performed on the mechanical element is given by the product of the stress–strain multiplied by the volume of the mechanical element.

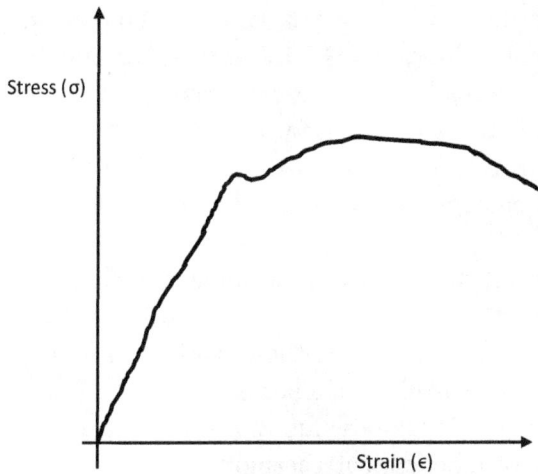

Figure 2.1. Stress-Strain diagram for structural material.

Surface tension is defined as the amount of energy per unit of surface area. The units are of surface tension: $N - m/m^2$. Thus, the surface of a liquid may be able to support a mass, M, if the surface tension of the liquid is strong enough or the mass is small enough to not break the surface. The classic example of the usefulness of surface tension is the image of a small insect that can walk on water because of its small mass, which is small enough to be supported by the impenetrable water surface. The use of detergents in water is the attempt to reduce the surface tension of the water and thus attempt to reduce the forces between the cloth fiber and the dirt or oil that is spotting the fabric.

The work of thermal expansion can be very large, and if it is ignored in the mechanical design of a system, it can be catastrophic for the element. The temperature change of a mechanical system will cause the length of the mechanical element to increase. If the thermal expansion is constrained and not allowed to freely extend, the stress in the element increases from zero to a very high level, depending on the size of the temperature change that causes the elongation and the amount of free movement that is allowed. Any increase in the stress that is not relieved by allowing the element to expand will cause work to be performed on the element according to this equation:

Thermal Expansion Work

$$= ((\varepsilon_T \times L \times \Delta T) - \Delta\epsilon) \times Y_{young's\ modulus} \times (\varepsilon_T \times \Delta T)$$
$$\times Volume\ of\ Element.$$

The Young's Modulus can be a very large value and depends on the material. For example, Young's Modulus is 10×10^6 psi for aluminum and 30×10^6 psi for steel. The thermal expansion coefficient (ε_T) is very small: 6×10^{-6} in/in for steel and 12×10^{-6} in/in for aluminum.

Other work terms most useful in the fields of electricity and magnetism involve moving a charge in a magnetic field and a charge through a voltage potential.

These work terms are important to know and understand because they may contribute to the work output or the work required for a thermodynamics process to occur. However, these work modes are also important to understand the State Postulate of Thermodynamics, a postulate that is very important in solving a thermodynamics problem. The state postulate identifies the number of independent thermodynamics properties that must be known in order to completely solve the thermodynamics problem at hand. It is stated as follows: the number of independent thermodynamics properties that are required to be known to completely identify the state of a thermodynamics system is equal to the number of work modes that are significant for the problem plus 1.

Let us explore a simple example using what the reader has learned from Chapter 1. Consider a closed vessel that contains air at an initial pressure of P = 100 psia and an initial temperature of 200°F (660°R). What is the density of the air that is contained at these initial conditions? The reader is quick to point out that the solution for the density can be found by using the Perfect Gas Law, as stated in Chapter 1, and repeated here:

$$P \times V = M \times R_u \times \frac{T}{MoleWt},$$

$$P \times v = R_u \times \frac{T}{MoleWt}.$$

It is noted that there are three thermodynamics properties identified in this equation: P, T and specific volume, v, or $1/v$ equal to the density, ρ. With this simple thermodynamic relationship, it is clear that the properties are well defined with respect to each other and any one of these properties can be calculated if the other two properties are known. This rather obvious mathematical fact is stated in a more thermodynamic manner by first recognizing that the single reversible work mode as identified in Equation (2.1) helps to determine how many independent properties are needed to be known before all other properties can be found. If only one reversible work mode is needed to complete a thermodynamic analysis, then two independent properties are sufficient to find all of the others. If a second reversible work mode, such as the stress-strain relationship shown above is also needed, the three independent properties are needed. In general, for "N" necessary reversible work modes there must be N + 1 independent properties known before all the other relevant properties can be found. This principle is called the State Postulate. Thus, given the reversible mechanical work mode shown in Equation (2.3a), the use of the Perfect Gas Law with two of the properties given is sufficient to calculate the single unknown. This simple reality leads to a far-ranging opportunity to complete significant thermodynamic analysis for gases, as will be demonstrated in this chapter.

2.2   Temperature in Rankine or Kelvin units

Thermodynamics requires the temperature to be on the absolute temperature scale. The two absolute temperatures are Rankine, °R, and Kelvin, K. The reader is likely more familiar with the more common temperature scales expressed as Centigrade, and Fahrenheit, °F; the latter is used almost exclusively in the United States. Unfortunately, these temperature scales were adopted before thermodynamics became an important part of civilization development. The Fahrenheit and Centigrade scales were defined based on the ability for anyone to reproduce the temperature scale in order to have a common reference with regard to temperature. The easiest references were the temperatures at which water, at sea level, freezes and starts to boil. Drawing a straight line connecting these two points on a temperature graph provides a geometric slope to the line that can now extend the line above and below the freezing and boiling references. But now the distance between these two points must be divided into equal parts. The enthusiasts for the Centigrade scale chose 100 and the enthusiasts for the Fahrenheit scale chose 180. The conversion of one scale to the other can be done with the equation:

$$°C = (°F - 32) \times 5/9.$$

Closely related to energy is the rate at which energy is used, or power. Power is often the default unit of the terms that appear in the First Law of Thermodynamics for open systems, whereas the energy term is most often used when applying the First Law to closed systems. As initially described in Chapter 1, the First Law applied to open systems considers the rate at which mass is flowing into or out of a system boundary. The rate of mass flow is given in units of pounds-mass per second (lbm/s) or kilograms per second (kg/s). Notice here that the statement of mass flow rate in the imperial unit system is given as pounds-mass, lbm, and not pounds-force, lbf. These units are similar in spelling but dissimilar in the quantity that they describe. Pounds-mass is a mass unit and pounds-force (used in calculating energy) is a force unit. There is a distinct difference between mass versus weight and both are often expressed in abbreviations that are used in calculations or technical reports; However, the mass and the force subscripts are often left out, leaving the exact same word behind: pounds. This is a critical error that can be catastrophic if not corrected. Simply stated, the pounds-mass of some substance will have the same mass content whether it is measured on Earth or any other celestial body in the Universe. The same cannot be said of something that has a quantity of pounds-force. A material has a weight of pounds-force due to the gravitational attraction between that object and the planetary object or any enormously larger mass. More precisely stated using Newton's Law of Gravity: $g_g = G\,M/r^2$, where $G = 6.68 \times 10^{-8}$ lbf-ft^2/lbm^2. If a material has a quantity of X pounds-force on Earth, then that same material mass will have a greater or lesser pounds-force quantity if it is brought to another place in the Universe. On the Moon, the weight of the object would be one-sixth of the weight measured by a weight scale on Earth. But the object has the same quantity of mass, whether it is on Earth or on the Moon.

Power, or the rate of energy, has units of foot pound-force per second (ft-lbf/s) or Newton-meter per second (N-m/s). In the metric system, a N-m/s is also called a watt (w) to give honor to James Watt, who is renowned for his work in developing the first steam engine. The general public will be less familiar with the unit ft-lbf/s, as well as the engineer, until a relationship between ft-lbf/s and horsepower (hp) is provided. The conversion is 550 ft-lbf/s = 1 hp. The relationship between hp and kilowatt (kW) is equally straightforward: 1 kW = 1.341 hp.

It is commonly expected that the units of hp or kW be used to identify mechanical or electrical work per unit of time. However, it is also acceptable to use kW to represent the amount of heat transfer that occurs between two temperatures. This is typically done in countries that use the metric system. However, to make it clearer that the quantity is a heat transfer rather than a work term, the kW often uses a subscript "t" for thermal when the power exchange is to denote heat transfer and not mechanical or electrical power. This is most acceptable in the metric system. The more conventional manner to indicate heat transfer in the imperial unit system is to use the British thermal unit per unit of time, h, or Btu/h. A Btu is defined as the amount of heat transfer that must occur to change the temperature of a pound-mass of water by 1 Fahrenheit, °F, or Rankine, °R.

It is thus also useful to consider the relationship between hp and Btu/h. Two very useful unit conversions are:

$$1\,\text{hp} = 2545\,\text{Btu/h} \text{ and } 1.341\,\text{hp} = 1\,\text{kW}$$

The names given for energy and power terms are chosen to honor researchers in science and engineering, particularly, when the field of thermodynamics was first being developed in the 1600s. The exception to that rule is the unit for heat transfer (see previous paragraph). The measure or sense of a Btu can be determined with a bucket full of water, a thermometer and a heat source. The researcher would be able to establish how much power, Btu/h, or energy, Btu, is being added to the system by heating a known amount of water up to any temperature (but not too close to the boiling point) and monitoring the time it took to get to that temperature. In today's modern engineering research facilities, this measurement can be done in any number of ways. However, the simplest and most elegant is still taking a known amount of water in a vessel that is insulated and heating the vessel using one exposed surface while measuring the time it takes to achieve any temperature increase that is prescribed and that can be accurately measured. Of course, it is also necessary to consider the material type and the mass of the container in the calculation; the container should be insulated as it is being heated, so as to minimize the amount of heat transfer lost to the environment. Notice that the word "minimize", and not "eliminate", is used with respect to heat transfer, which is the energy exchanged. This is correct because there can be no definite heat transfer process unless the temperature difference between the hot and the cold objects is zero. It is also interesting to note that the use of water is not arbitrary. Choosing water as the medium to heat is particularly attractive because it has a property that is unique in its magnitude. That property is called specific heat. While all materials have a specific heat value, the value of the specific heat for water is 1 Btu/lbm/R. Thus, its use in the equation (presented here for the first time) for heat transfer due to a temperature difference is more easily calculated.

$$Q = M_{\text{liquid}} \times C \times \Delta T. \tag{2.4}$$

The property of specific heat, C_p, must not be confused with the specific heat with respect to volume or C_v. The units are the same but the equation that uses the C_v determines the amount of thermal internal energy, E_t, that is added or removed from a system by changing its temperature from T_1 to T_2. This equation proves to be extremely important because it determines one of the major parameters in the first law equation for Closed Mass Systems.

Worked examples will show that E_t is very often essential to the modeling of heat engines.

$$E_t = M_{\text{liquid}} \times C_v \times \Delta T.$$

With the introduction of E_t, we will begin a more detailed discussion of fluid properties and how they are used to determine the heat transfer, work performed and internal energy change, which are values that are at the heart of every thermodynamics problem. It will be easily stated and demonstrated that the properties of a fluid are the essence ("the blood and bones") of the thermodynamics description of any natural or man-made material, liquid or vapor, or a mixture of the two latter forms. Properties such as specific heat and entropy are in fact established on a micro-scale in statistical thermodynamics. Their magnitude in the nano-world can be demonstrated to be well-founded, even as they are frequently used for real-world engineering problems in the macro-world. There are other similar properties that need to be identified. These properties include isobaric (i.e., constant pressure) compressibility, $\beta = \left(\frac{1}{v} \times \frac{\partial v}{\partial T}\right)_p$, and isothermal (constant temperature) compressibility, $K = \left(\frac{1}{v} \times \frac{\partial v}{\partial P}\right)_T$. These properties help determine how the volume of a substance is changed due to temperature and pressure changes as a result of the addition of heat or work on the substance.[1] The most common of these is specific heat, more specifically, specific heat with respect to either a constant pressure or constant volume, depending on how they are measured in the experimental measurements of such properties in the macro-world. Certainly, their magnitudes at different pressures and temperatures have been determined by the founders of thermodynamics. This textbook will not delve in the statistical exploration of the definition of specific heat or entropy, or any of the other properties mentioned above, because though such knowledge is important, we should first understand the common properties that can be used to help solve the most common thermodynamics problems. This is the intent and purpose of this textbook. Justification for this methodology is most evident by the availability of software programs designed to remove the burden of looking up such properties, thus leaving the user needing to only understand what these properties are, and how they characterize a material, particularly its liquids and vapors. The following section will provide that necessary understanding, starting with specific heat.

---

[1]The compressibilities are used in the following equation that determines the volume of a substance as a function of temperature and pressure changes: $dv = \beta \times v \times dT - k \times v \times dP$.

2.3 Thermodynamic property

In thermodynamics, the word "property" has a different meaning than that in the world of real estate. However, the adage "... location, location, location" may still apply to indicate the importance of the parameter. A thermodynamics property identifies the precise condition of a material. While it is true that fluid in the most interesting and viable engineering thermodynamics applications tend to be in a liquid or vapor state (so as to provide the necessary energy exchange in the most often used applications), the thermodynamics properties of a material can be defined for all four phases of a material: solid, liquid, vapor and plasma state.

Some properties of a material are very familiar, even to non-engineering students. For example, density is probably the most common property and it can pertain to a solid, liquid or vapor state of a material. Pressure and temperature are also two of the other very common properties of a material. Other common material properties are: thermo conductivity, thermal expansion coefficient and viscosity. These properties are all intrinsic to the material, meaning that if a specific quantity of that material (by volume or mass) were to be divided into "n" pieces, each of these pieces would still have the same density, pressure and temperature, assuming that there has been no energy exchange with that material during the division process. Interesting enough, if that specific quantity of material were to be split in any number of pieces, then the mass or volume of each of these pieces would obviously be reduced. However, the ratio of each piece's mass and volume would still define the same density (mass/volume), or the inverse of density, which is called specific volume, which is equal to volume/mass. Thus, a pressure gauge and a thermometer placed inside the material pieces would still display the same values of pressure and temperature before and after the division. The energy of each of these pieces of material would be reduced, but the energy per unit of mass would be the same before and after the division. Thus, energy per unit mass, also called specific energy, Btu/lbm, or kJ/kg, is an intrinsic thermodynamics property of the material, whereas the total energy of the piece and the mass and volume of each part of the divide material are extrinsic properties of the material.

There are other intrinsic properties that will be useful to the application of thermodynamics principles in solving engineering thermodynamics problems, for example, entropy, Btu/lbm/R, enthalpy and availability, A. The properties of a material are related in such a manner that knowing two of them can usually precisely determine its other properties. This "rule" is known as the State Principle and it is strictly true if there is only one mode of work from among the several identified earlier. Engineering thermodynamics deals most often with the Pressure × Dv work mode and not simultaneously with stress-strain or surface tension, etc. Thus, it is true for this textbook that two properties among the other important ones, such as pressure, temperature, density (or specific volume), entropy, enthalpy and availability, define the other intrinsic properties.

As noted earlier, the best and most useful example of the State Rule or Principle is easily observed by applying the Perfect Gas Law, which from a strictly chemistry point of view, is given as follows:

$$P \times V = M \times R \times T/\text{MoleWt}, \qquad (2.5)$$

where pressure is P, lbf/ft^2 or N/m^2, temperature is T, °R or K, mass is M, lbm or kg, MoleWt is molecular weight, and universal gas constant is R, 1545 lbf-ft/lb-mol/R or 8314 N-m/kg-mol/K.

The Perfect Gas Law is clearly only applicable for perfect gases, which do not exist in the real world. It is only a good substitute for some non-ideal gases and gases under certain conditions, such as relatively low pressure and high temperature. More will be stated about what is meant by relative low pressure and relative high temperature later in this chapter. The basic first lesson that demonstrates the State Principle, as noted earlier in the last section, is that the Perfect Gas Law has three of the most basic properties that can well define a fluid. These properties are: the pressure and temperature of the fluid, as seen in Equation (2.5), and density or specific volume. This last property is not as apparent until the reader performs a simple algebraic step of dividing both sides of Equation (2.5) by the mass, M. The left side then has a V/M term that was defined in the last section as specific volume, v. This becomes essential for the maximum use of the Perfect Gas Law in thermodynamics problem-solving because it links the one simple basic equation with the three most distinctive and easily measurable properties. Thus, if two of these three properties are known, then the third can be determined using the Perfect Gas Law. It is also noted that this Perfect Gas Law has a wide range of applications to almost any fluid that is a vapor (i.e., a gas) by changing its molecular weight. The universal gas constant, R_u, never changes for a particular unit system.

Computer programs or spreadsheets are able to integrate the Perfect Gas Law to solve significant thermodynamics problems. Certainly, the ubiquitous presence of properties for use with computer software, including spreadsheets, means that engineers no longer need the Perfect Gas Law in favor of the more accurate fluid properties that are easily programmed into a software platform or a spreadsheet. The author believes that the more versatile of these property routines is the one offered by the National Institute of Standards and Technology (NIST), USA, called the Reference Fluid Thermodynamic and Transport Properties Database (REFPROP), which is available through the NIST website. This author can personally attest to its usefulness, having used it professionally. However, for students who are just starting out, the Perfect Gas Law is very important and also useful for easily programming properties.

Perhaps one slight adjustment to the Perfect Gas Law would be the use of a compressibility factor, Z, that is routinely used to increase the accuracy of the Perfect Gas Law in representing the relationship of the three properties: P, T and v. The Perfect Gas Law becomes:

$$P \times v = Z \times R_u \times T/MoleWt.$$

The compressibility factor, Z, is found using compressibility charts as shown in Figure 2.2. To use these charts, it is first necessary to define what the critical pressure and temperature of a fluid are. This step will also help identify what it means to be "almost" a perfect gas when the actual pressure and temperature is known. These properties of critical pressure and temperature will be discussed in the next section, but only after more is learned about

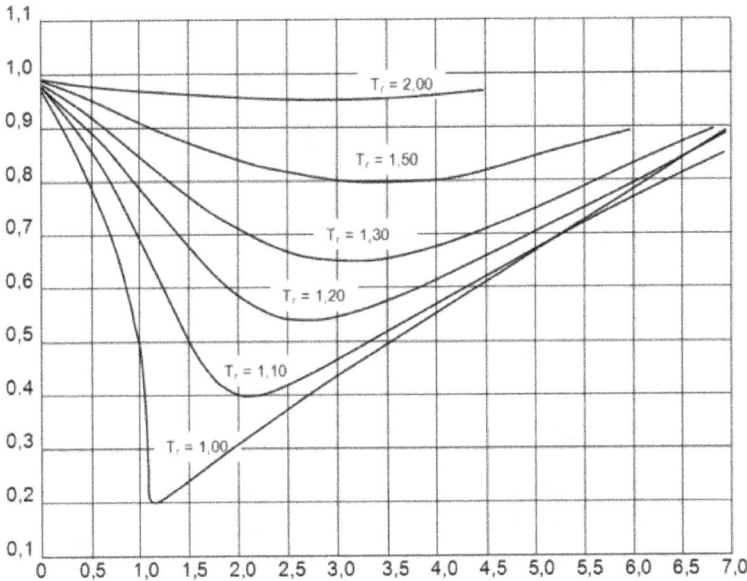

Figure 2.2. Compressibility, Z, as a function of P/Pc (along the x-axis) and T/Tc.

how we can determine whether a fluid is in fact a liquid, solid or gas (or vapor), or a mixture of 2–3 forms.

One should first try working out an example of the utility of the Perfect Gas Law, as it may be worthwhile getting a worked example with all its units consistent, in particular, the units of temperature that must be in absolute temperature scales (i.e. measured in Rankine or Kelvin). The easiest calculation to make is to determine the density of air in atmospheric pressure and temperature or 14.7 psia and 70°F, respectively. Example problem No. 2.1 will provide this solution using both the imperial and SI unit systems.

2.3.1   *Example 2.1: Determining the density of atmospheric pressure*

Use the perfect gas law to determine the density of atmospheric air at a pressure of 14.7 psia and 60F.

Imperial Unit system:

$$P\left[\frac{lbf}{in^2}\right] \times 144 \times v\left[\frac{ft^3}{lbm}\right] = 1545\left[\frac{ft-lbf}{lb-mole-R}\right] \times \frac{T[R]}{MoleWt.\left[\frac{lbm}{lb-mole}\right]}$$

$$14.7\left[\frac{lbf}{in^2}\right] \times 144 \times v\left[\frac{ft^3}{lbm}\right] = 1545\left[\frac{ft-lbf}{lb-mole-R}\right] \times \frac{(460+60)[R]}{28.966};$$

$$v = 13.1\frac{ft^3}{lbm}$$

Metric unit system:

$$101{,}300\left[\frac{N}{m^2}\right] \times v\left[\frac{m^3}{kg}\right] = 8314\left[\frac{N-m}{kg-mole-K}\right]$$

$$\times \frac{(273+15.33)[K]}{28.966\left[\frac{kg}{kg-mole}\right]}; \quad v = 0.831\frac{m^3}{kg}$$

2.4   Relationships between liquid, vapor and two-phase mixtures of liquid and vapor phases

Every substance can be described graphically as having a functional relationship between pressure, temperature and specific volume. This graphical representation is very useful in solving thermodynamics problems, just as how a road map is important to the traveler who must get from point A to point B. This analogy should be taken very seriously, for the graphical representation of a fluid is the first step in a multistep process that is almost

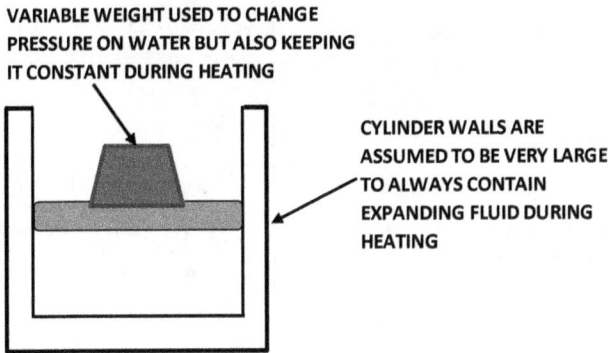

Figure 2.3. A simple piston-cylinder containing water and variable weight.

guaranteed to speed up the solution of a thermodynamics problem, because it helps an engineer to visualize solving an engineering problem. There are two other graphical tools provided in the next sections of this chapter. The first graphical representation is called the property chart. This chart is not to be confused with the Property Table or the Process Table that will be defined and used later in this chapter.

To best understand the construction of this fluid map of property relationships, a thought experiment is offered to the reader. In this thought experiment, the reader is asked to consider a piston and cylinder that is entirely filled with liquid water (see Figure 2.3). Note that the words "liquid" and "water" are not redundant, for water can be a vapor (sometimes called steam) or a solid (ice), or in a two-phase (liquid and vapor) or three-phase (liquid, vapor and solid; a triple-point) state. The length of the cylinder can be as long as needed to contain the expanding water, as heat is added to the cylinder.

The experiment continues with the reader observing that three measurements — for pressure, temperature and specific volume — are constantly being made and recorded. The last parameter will change due to the change in the volume of the cylinder. In Figure 2.3, the reader can see that the weight on the piston can be easily changed, thus causing a different pressure to be exerted on the contained, trapped and closed mass system of water. The reader may find it easier if values are added; if the cross-section of the cylinder is $1\,in^2$ and 100 lbf of weights are added to the cylinder, $100\,lbf/in^2$ is thus created on the water.

Water is added to the closed system only at the start of the experiment and no additional water is added or removed during the experiment. The experiment is straightforward: it starts by recording the initial specific

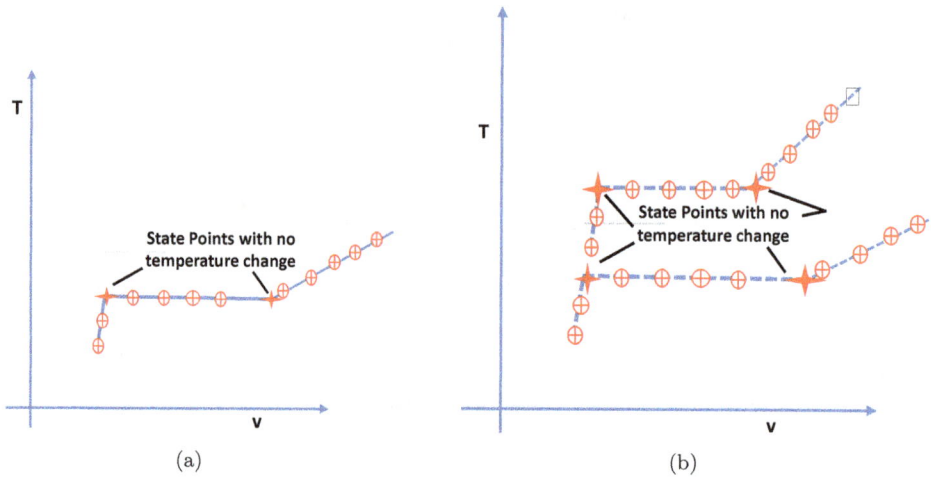

Figure 2.4.   The temperature-specific volume traces from the thought experiment; (a) single data line as observed in the first experiment with a certain set of weights applying constant pressure on the water; (b) second data line of the second experiment observed, this time conducted using a heavier set of weights than used in (a) and that are also held constant during the experiment, plotted on the same graph.

volume and pressure of the system on a chart measuring temperature versus specific volume, as shown in Figure 2.4(a). Heat is added to this Closed Mass System and the pressure and temperature is recorded along with the specific volume. During the experiment, the weight, and hence the pressure on the water, is kept constant. Also, saying that "... heat is added" is a simple way of saying that something hotter (i.e. of a higher temperature) than the cylinder is placed near (or is in contact with) the cylinder wall. This is based on the fact that heat transfer is energy exchanged due to a temperature difference.

The experiment starts at atmospheric pressure and temperature with 1 lbm of water in the cylinder. The temperature starts to rise and there is a very, very small change in the specific volume. The temperature and the specific volume are faithfully recorded as time moves forward. Taking reference from Figure 2.4(a): (1) as heat is added to the system, a trace of almost a vertical straight line is observed from the instruments; (2) as time elapses, the temperature seems to hold constant even as the specific volume increases; (3) as more time goes by, there is a point where the temperature starts to increase again, with a noticeable increase in specific volume. The experiment proceeds further for a long period of time, but the relatively straight curve as shown in Figure 2.4(a) continues unabated.

In our second experiment, once all the equipment used in the first experiment is back to room temperature, we add a heavier weight to the cylinder

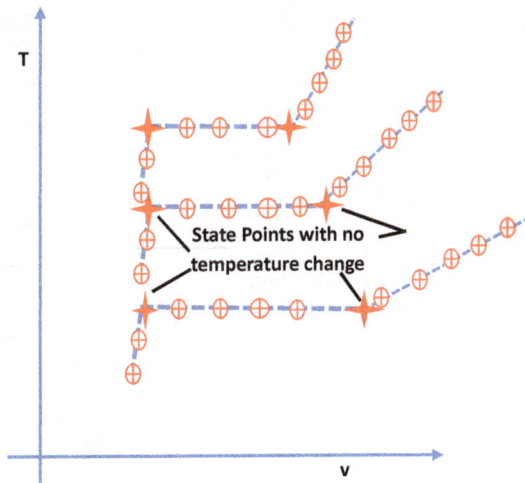

Figure 2.5.   A third experiment with a higher constant pressure.

that increases the pressure on the cooled liquid (water) to, for example, 100 psia, and not 14.7 psia, as was tested moments earlier. The experiment is repeated, and a new data line is drawn as shown in Figure 2.4(b). The low specific volume recorded at low temperatures traces a curve that is almost identical to the first line at 14.7 psia. As the experiment continues, there is a point in time when the temperature of the fluid stops changing even as the fluid's specific volume continues to increase, as shown in Figure 2.4(b). As time elapses, the temperature starts changing as more heat is added to the system. After a suitable period of time, both graphs look as shown in Figure 2.4(a) and (b).

Once again, the cylinder and its contents are allowed to cool, which means that the heat transfer is stopped, and the thermal energy of the cylinder and water lose heat to the environment. Of course, it is assumed that the environment is cooler than the cylinder, otherwise heat transfer out of the cylinder-water system would not occur, and thus the cylinder is cooling. The experiment is restarted after the same amount of water has reached atmospheric temperature, at which time another weight is added to the piston so that the pressure is now 200 psia. The experiments results are shown in Figure 2.5. This experiment can be repeated for any number of times, with increments of pressure until 3,200 psia. The pressure threshold is interesting because when the experiment is allowed to go to completion, there is now no noticeable point where the temperature stays constant at any time. The curve of temperature versus specific volume is not linear. It may be somewhat curved, but no constant temperature is observed.

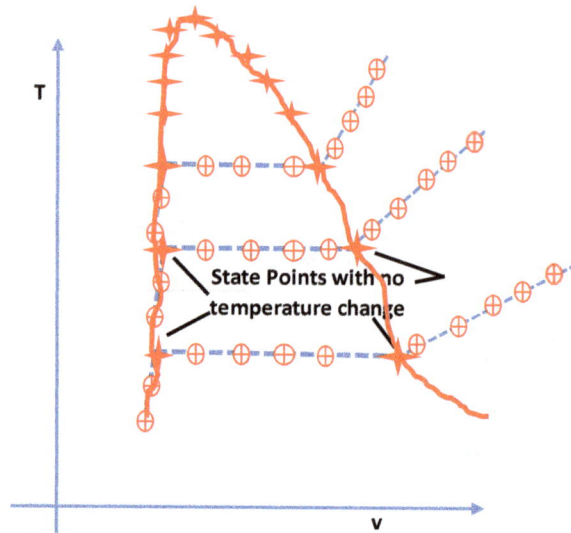

Figure 2.6.   Connecting the loci of points denoting where the temperature first stops changing during heating and then restarts changing as heating continues.

Now, we will connect all the points where the temperature started to stay constant on the left side and then do the same on the right side. This is shown in Figure 2.6. The somewhat bell-shaped curve is often called "the dome" and it becomes our landmark on this graphical representation of the fluid and the relationship of pressure, temperature and specific volume. This landmark is essential for a thorough understanding of the fluid: whether the fluid is liquid, vapor or somewhere in between, or existing as a two-phase, non-reacting mixture. A particularly noticeable feature of the T-v diagram using actual water properties, as shown in Figure 2.7, is the almost vertical line that makes up the left side of the dome. This is a very common feature and indicates that the liquid phase of most fluids does not change specific volume or density with respect to temperature or pressure. There is some change, but not much, and the change is not noticeable on a typical T-v diagram. An inspection of the right side of the dome indicates the very opposite of a vertical line. In fact, the very large slope of the curve is evidence of a very large change in the specific volume, with respect to pressure and temperature. Water has probably the most severe change in specific volume (or density) as a function of temperature and pressure.

The use of property charts, such as those shown in Figure 2.7, provide a graphical picture of the processes that are being studied. Before the digital era, property charts served as a very quick way of determining all of the necessary thermodynamics properties that were required to solve a problem.

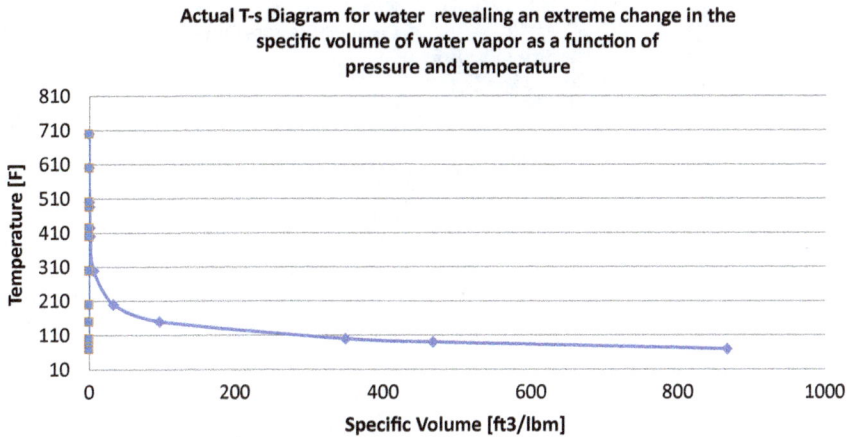

Figure 2.7. An actual T-v diagram for water.

The values for the properties were taken from these charts. With the advent of laptops, the use of charts has been diminished because the values of the thermodynamics properties can be stored and quickly retrieved from computer programs, such as Excel, MatLab or Fortran. The most common of these software property routines is REFPROP. REFPROP is easily obtained from NIST (www.nist.gov) for a very reasonable fee. This software combined with the readers' favorite software platform is strongly recommended.

However, the graphical display of a fluid's properties, as illustrated in Figures 2.4 and 2.6, are still extremely useful. The graphical depiction of a fluid property's state point at the beginning and the end of a particular process can even be hand sketched on the analyst's calculation sheet, providing him or her with a very useful visualization of the process that is being analyzed. This is particularly true of a heat engine or any closed process. It is also is true for any thermodynamics process. The choice of using a T-v, P-v or T-s diagram, or even an enthalpy-specific volume (h-v) diagram, depends on the nature of the problem that is being analyzed.

A study of Figure 2.8 can discern points 1, 2 and 3. Point 1, at the left of the dome, is in a liquid region. If this graph were to scale and the temperature and specific volume are known from the x and y axes, then those properties would indeed represent what we identify as liquid water. Point 3 is a vapor or gas state that gains the same temperature according to the graph, but at an extremely high specific volume that would indicate that this water state is a vapor or steam. Point 2 is inside the dome and it can be shown to be a system of water that is partially liquid and partially vapor. In fact, if the cylinder was transparent and an aquatic animal (say, a fish)

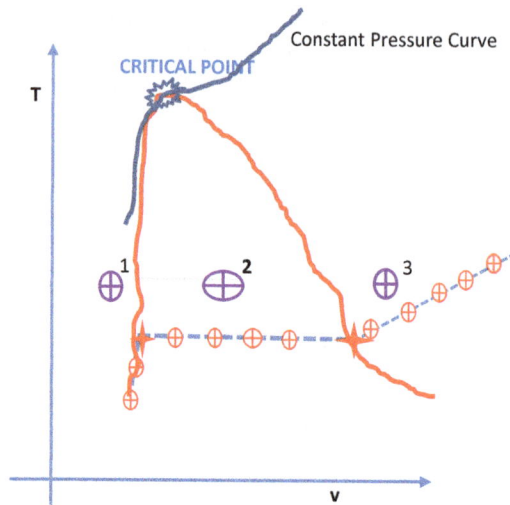

Figure 2.8.   T-v diagram showing state points 1, 2 and 3.

was placed inside this cylinder — assuming it can survive any pressure and temperature — it can then be observed that the fish is swimming and/or floating on the interface between the liquid and vapor water. As point 2 moves to the right, the interface surface gets smaller and smaller until the poor aquatic animal is left stranded in only water vapor. The only way to rescue the animal then is for the experimenter to quickly cool the cylinder, to allow the water vapor to condense into liquid state again, until it reaches the initial cold and liquid water state that we began with in this experiment.

The curve shown in Figure 2.8 is known as the constant pressure curve, and in this case the aquatic animal that was happily floating in the liquid water is now stranded because the liquid at extreme pressure (at a point that is called the critical point) has changed into vapor, and thus lost its buoyancy potential. This region of transition is very much in doubt as to the accuracy of any fluid properties, whether they be density, viscosity or thermal-conductivity. Thus, this region of the critical point and some area around it is usually abandoned with respect to any useful engineering, or at least not until substantial work is done to quantify the properties of any fluid in that area, so as to reduce the risk of damaging equipment that had been designed with one property but then found to have an entirely different value, and thus be unable to operate properly.

Every fluid property shares this dome-like structure although the dome's shape can vary widely. In fact, the range of different shapes proves that though fluid is good for some thermodynamics processes, it is perhaps poor

T-s Diagram for Water

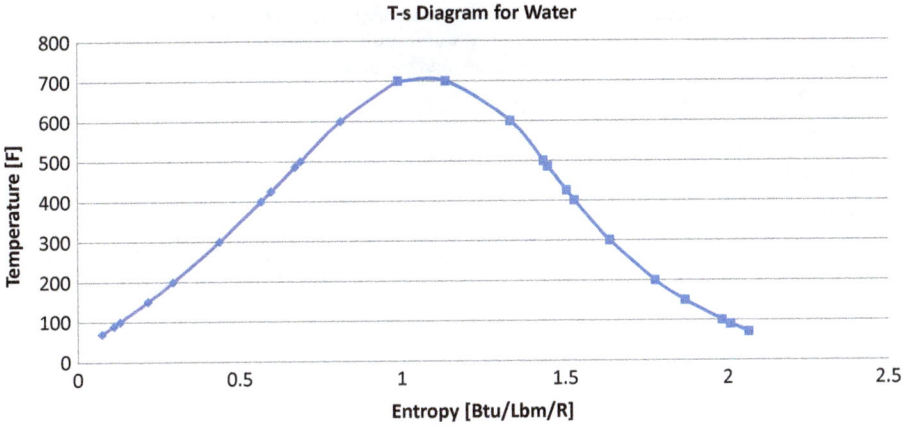

Figure 2.9. Actual T-s diagram for water.

for other desirable process. In Chapter 3 it will be shown that a dome shape, as shown in Figure 2.8, is good for heat engines producing useful mechanical power, while fluids that are identified with dome shapes that have the right side, saturate vapor curve, with a positive slope are better suited for perhaps refrigeration or air conditioning, i.e., better for making things cooler instead of making power. This will be further discussed in Chapter 3. Figure 2.9 displays the "dome", but on a graph that uses T and s (an entropy property) as the coordinate axes. The shape is different from those shown previously. In fact, the dome is even more "dome- or bell-" shaped. It will be shown that the T-s diagram for a fluid is very helpful in analyzing a thermodynamics problem, particularly if the problem involves a heat engine. Any vertical line on a T-s diagram connects two or more points that have the same entropy at the start and at the end of a process. The Second Law of Thermodynamics states that a constant entropy process is a thermodynamically perfect process. Thus, the T-s diagram can be readily used to indicate a perfect process, if only to then gauge how well a real process is compared to the perfect process.

It is appropriate to label the most distinctive parts of this dome and thus become more familiar with the thermodynamics jargon used by experienced engineers. Referring to Figure 2.9, the left side of the dome is called the saturated liquid curve because any fluid property that "lands" on this side is a liquid. Any fluid property resting exactly on the line is a saturated liquid, in the sense that any small amount of heat energy added to that liquid will start to vaporize that liquid. The amount of liquid converted to vapor is dependent on the amount of heat energy added to the liquid. If sufficient heat

Actual T-s Diagram for water revealing an extreme change in the specific volume of water vapor as a function of pressure and temperature

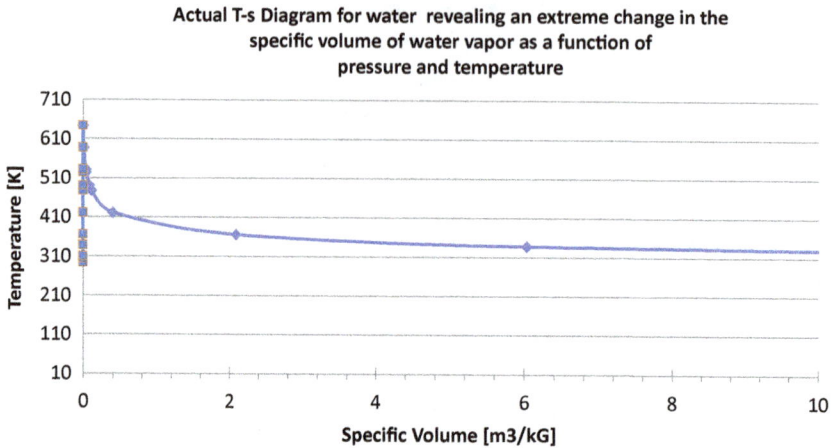

Figure 2.10.   T-s diagram for water in metric units.

is added to convert all the liquid into vapor, then the evaporation process at constant pressure would cause the end state to migrate from the left side to the right on a constant temperature line and terminate along the right side of the curve. The right side of the dome is called the saturated vapor curve because any additional amount of heat will increase the temperature of the vapor again. If the vapor remains on the right side of the curve, then it is saturated vapor. These two points on the left- and right-hand sides of the dome can also be identified by a parameter that is also a true property of the fluid: quality (x). Quality is defined as the ratio of the amount of mass vapor to the amount of fluid mass (vapor and liquid) that it has in any system volume. That is, if we had started with 1 lbm of liquid water at the start of our experiment, we should still have 1 lbm of water at its end. It would have traversed the dome at constant pressure and temperature during its transformation from liquid to vapor state. If the cylinder had been transparent, the engineer would have been able to observe a partitioning of that same amount of water into M_{liquid}, on the bottom of the vessel and M_{vapor}, in the head-space above the liquid interface. Quality is defined simply as $x = M_{vapor}/(M_{vapor} + M_{liquid})$.

This thought experiment has been conducted with no dependence on the pressures, temperatures and specific volumes used. However, Figure 2.10 shows the water temperature versus specific volume chart in the metric system. The reader is directed to turn his or her attention to the virtually vertical, saturated liquid "curve" on the left side of the dome. This line gives evidence of the almost independency of the water's specific volume (or density) to the temperature or pressure of the water. It is now time to use it to determine the state of water — liquid, vapor or two-phase — that it exhibits.

It is imperative that the engineer be able to "read" any fluid property chart and locate the start and end of a process. This is necessary because the application of the thermodynamics laws is usually facilitated by knowing a starting and an end point in time, just as it is important to always use a Control Volume (CV) around the system, i.e., a collection of components that is of the most interest to the problem at hand. A later requirement of the solution process provides the boundary to which heat transfer, mass flow rate and work terms must be observed to breach or cross. If this boundary has not been crossed by either of these, then that particular parameter should not be used in the First or Second Law of Thermodynamics. This is much like applying a CV around a mechanical object in order to construct a Free Body Diagram of forces that are acting on the mechanical part. Without such a diagram, the static equilibrium of the mechanical part cannot be assumed and thus the unknown forces cannot be determined using the basic principles of mechanics.

All fluid properties of a material will start with a table of saturated liquid and vapor properties of pressure, temperature and specific volume. Added to this will be several other very important properties such as internal thermal energy, entropy and enthalpy. These last two properties will be used in Chapter 3. Here, it is only necessary to know that these properties are among the many useful properties that can be simply "looked up" in the Property Table, such as that shown in Table 2.2(a) and (b). Table 2.2 is the saturated Property Table for water. It is very discernible by the pairs of columns that are given as a function of temperature or pressure. The two left-most columns show the temperatures and pressures that define a quality of zero or one, as shown by the figures in Table 2.2. These points are the left and right points of the dome that were generated earlier in our thought experiment. As you will see, from the top to the bottom of the saturation Property Table, the temperature is rising along with an increase in pressure. Now look at the specific volume designated saturated liquid, v_f, and specific volume of saturated vapor, v_g. If these two coordinates are plotted on a graph and the same process is repeated throughout the entire Property Table, the result is the dome that was generated in our thought experiment. Figure 2.10 is one such plot. Once again, the region to the left of the loci of points, known as the saturated liquid line, will be a liquid region. The region to the right of the saturated vapor line will indicate the region in

Table 2.2. Saturated property table for water (a) in imperial units; (b) in metric units.

(a)

T [F]	P [psia]	ft^3/kg		Btu/lbm		Btu/lbm/R		Btu/lbm	
		v, liq.	v, vapor	h, liq	h, vapor	s, liq	s, vapor	u, liq.	u, vapor
40	0.12	0.01601969	2443.32	8.04	1079.4	0.016	2.160	8.04	1024.3
50	0.18	0.016024	1702.81	18.08	1083.8	0.036	2.127	18.08	1027.6
60	0.26	0.01603497	1205.98	28.10	1088.2	0.056	2.095	28.10	1030.9
70	0.36	0.01605175	867.11	38.10	1092.5	0.075	2.065	38.10	1034.2
80	0.51	0.01607371	632.38	48.10	1096.8	0.093	2.037	48.10	1037.4
90	0.70	0.01610035	467.40	58.09	1101.1	0.112	2.009	58.09	1040.6
100	0.95	0.01613129	349.83	68.08	1105.4	0.130	1.983	68.08	1043.9
110	1.28	0.01616622	264.97	78.07	1109.7	0.147	1.958	78.07	1047.1
120	1.70	0.01620491	202.95	88.06	1114.0	0.165	1.935	88.06	1050.3
130	2.23	0.01624717	157.09	98.06	1118.2	0.182	1.912	98.05	1053.4
140	2.89	0.01629283	122.82	108.06	1122.3	0.199	1.890	108.05	1056.6
160	4.75	0.01639394	77.18	128.08	1130.6	0.232	1.849	128.07	1062.7
180	7.52	0.01650757	50.17	148.14	1138.6	0.263	1.812	148.12	1068.8
200	11.54	0.01663337	33.61	168.24	1146.5	0.294	1.777	168.21	1074.7
220	17.20	0.01677124	23.13	188.40	1154.1	0.324	1.745	188.35	1080.4
212	14.71	0.01671464	26.78	180.33	1151.1	0.312	1.758	180.28	1078.1
240	24.99	0.01692129	16.31	208.63	1161.3	0.354	1.715	208.55	1085.8
260	35.45	0.01708383	11.76	228.95	1168.2	0.382	1.687	228.83	1091.0
280	49.22	0.01725933	8.64	249.37	1174.7	0.410	1.661	249.21	1095.9
300	67.03	0.01744849	6.47	269.91	1180.7	0.437	1.636	269.69	1100.5
320	89.67	0.01765218	4.91	290.60	1186.3	0.464	1.613	290.30	1104.7
340	118.02	0.01787152	3.79	311.45	1191.3	0.491	1.591	311.06	1108.5
360	153.03	0.01810788	2.96	332.50	1195.6	0.516	1.569	331.99	1111.8
380	195.74	0.01836293	2.34	353.77	1199.3	0.542	1.549	353.11	1114.6
400	247.26	0.01863874	1.86	375.30	1202.2	0.567	1.529	374.45	1116.9
420	308.76	0.01893781	1.50	397.12	1204.4	0.592	1.510	396.04	1118.6
440	381.48	0.01926325	1.22	419.27	1205.6	0.616	1.491	417.91	1119.6
460	466.75	0.01961889	1.00	441.81	1205.9	0.641	1.472	440.11	1119.9
480	565.95	0.02000957	0.82	464.78	1205.1	0.665	1.453	462.69	1119.4
500	680.55	0.02044147	0.68	488.27	1203.1	0.690	1.434	485.69	1118.0
520	812.10	0.02092263	0.56	512.35	1199.8	0.714	1.415	509.21	1115.5
540	962.24	0.02146379	0.47	537.14	1194.8	0.738	1.396	533.32	1111.9
560	1132.73	0.02207971	0.39	562.77	1188.0	0.763	1.376	558.14	1106.8
580	1325.46	0.02279149	0.32	589.44	1179.0	0.788	1.355	583.85	1099.9
600	1542.51	0.02363066	0.27	617.42	1167.4	0.814	1.333	610.67	1091.0
620	1786.21	0.02464744	0.22	647.11	1152.3	0.841	1.309	638.96	1079.3
640	2059.25	0.02592972	0.18	679.19	1132.6	0.869	1.281	669.30	1063.9
660	2364.88	0.02766067	0.14	714.96	1106.1	0.900	1.249	702.84	1042.8
680	2707.28	0.03036148	0.11	757.89	1067.6	0.936	1.208	742.67	1011.8
700	3093.00	0.03665249	0.07	823.00	991.7	0.991	1.136	802.01	948.8
705.1025	3200.11	0.04941365	0.05	895.12	898.2	1.052	1.055	865.84	868.6

(*Continued*)

Table 2.2. (*Continued*)

(b)

T [K]	P [kPa]	m³/kg		kJ/kg		kJ/kg/K		kJ/kg	
		v, liq.	v, vapor	h, liq	h, vapor	s, liq	s, vapor	u, liq.	u, vapor
300	3.54	0.001003	39.08	112.6	2549.9	0.393	8.517	112.6	2411.6
320	10.55	0.001011	13.95	196.2	2585.7	0.663	8.130	196.2	2438.5
340	27.19	0.001021	5.73	279.9	2620.7	0.917	7.801	279.8	2464.8
360	62.19	0.001034	2.64	363.8	2654.4	1.156	7.519	363.7	2490.1
380	128.85	0.001049	1.34	448.1	2686.2	1.384	7.274	448.0	2514.1
400	245.77	0.001067	0.73	533.0	2715.7	1.601	7.058	532.7	2536.2
420	437.30	0.001087	0.43	618.6	2742.1	1.810	6.866	618.1	2556.2
440	733.67	0.00111	0.26	705.3	2764.7	2.011	6.691	704.5	2573.3
460	1170.89	0.001137	0.17	793.4	2782.9	2.205	6.530	792.1	2587.2
480	1790.47	0.001167	0.11	883.3	2795.8	2.395	6.379	881.2	2597.1
500	2639.20	0.001203	0.08	975.4	2802.5	2.581	6.235	972.3	2602.5
520	3768.95	0.001245	0.05	1070.5	2801.8	2.765	6.094	1065.8	2602.4
540	5236.91	0.001294	0.04	1169.3	2792.2	2.948	5.953	1162.5	2595.5
560	7106.25	0.001355	0.03	1273.1	2771.2	3.132	5.807	1263.5	2579.9
580	9447.97	0.001433	0.02	1383.9	2735.3	3.321	5.651	1370.4	2552.7
600	12344.82	0.00154	0.01	1505.4	2677.8	3.519	5.473	1486.4	2508.3
620	15900.58	0.001704	0.01	1645.7	2583.9	3.740	5.253	1618.6	2434.3
640	20265.21	0.002077	0.01	1841.8	2395.5	4.038	4.903	1799.7	2281.1
645	21515.21	0.002353	0.00	1931.1	2281.0	4.172	4.715	1880.5	2185.2
647	22038.41	0.002798	0.00	2029.4	2148.6	4.322	4.506	1967.8	2071.6

which the fluid exists in a vapor state. The region inside the dome indicates the region in which the fluid exists in a two-phase state, or as a combination of vapor and liquid. The quality property defined earlier is the proportion of the pathway across the dome at a constant pressure and temperature. Thus, a 75% quality means that a CV of material (in this case, water) is 75% water vapor and 25% water liquid. The specific volume of the state point with a quality of 75% can be found through:

$$v = \chi \times v_g + (1 - \chi) \times v_f,$$

where χ is the quality of the two-phase fluid.

The reader's attention is now drawn to the remaining columns of enthalpy, entropy and internal thermal energy. These pairs are the saturated liquid and saturated vapor values of these thermodynamics properties. The same equation can be used to determine the enthalpy or entropy inside the dome if the quality is known along with the pressure or the temperature.

From the State Principle, the reader will recall that knowing only two properties is sufficient to determine all of the others. The only exception is that if the property happens to lie inside the dome, then both of these

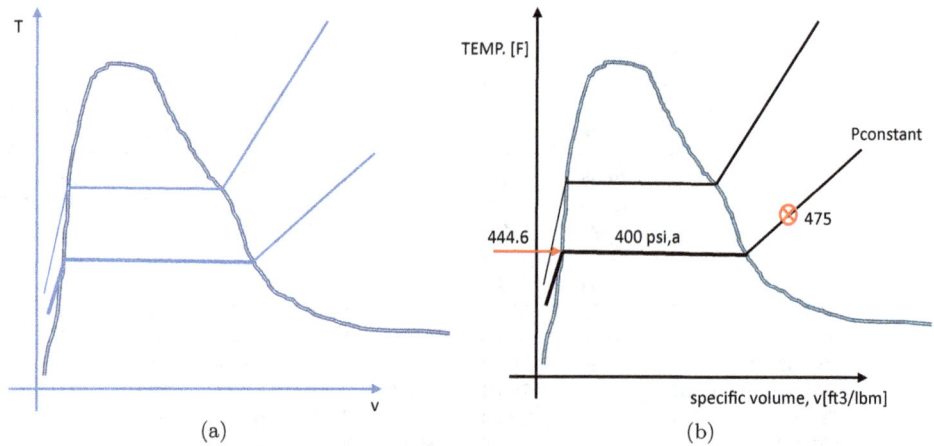

Figure 2.11.   A manual sketch of the T-v property diagram not drawn to scale. (a) Unlabeled diagram providing several recognizable features: the saturated liquid and vapor loci and two constant pressure curves; (b) Diagram labeled with the relevant problem information.

two known properties cannot only be temperature or pressure. It must be temperature and another property, such as specific volume or quality, or enthalpy or entropy.

In order to gain some experience in the quickest and most accurate way of locating the starting and end points of a thermodynamics process (and hence obtaining the solution to a thermodynamics problem quickly), here are a few examples on how the reader should use the property chart. A general step-by-step procedure is worth considering and it is useful to conduct these steps by first drawing a dome (not to scale) on a T-v axis and then drawing a constant pressure line (again, not to scale), see Figure 2.10(a).

Step 1. Draw the T-v diagram and include a constant pressure curve, as shown in Figure 2.11(a).

Step 2. Assume that either pressure or temperature is one of the known properties, and then label the saturation liquid and vapor points with the corresponding pressure or temperature and the specific volume as shown in Figure 2.11(b). In this example, assume the desired point is 475°F and of a pressure of 400 psia. Is it in a liquid, vapor or two-phase state? Assume that the constant pressure line drawn during Step 1 is 400 psia. Next, label the saturated temperature that corresponds to this pressure.

Step 3. Use the results from Step 2 and determine if the known temperature 350°F is above or below the saturated vapor and liquid temperature that you labeled the diagram with. In this example, the temperature is higher.

Then, with this now known, find the intersection of the temperature and the pressure curves. You will find this intersection to be to the right of the saturated vapor curve of the dome. Therefore, the property state of the water fluid is a vapor or steam.

Step 4. If you find that the fluid is a vapor, as in this example, then the saturation property table has done its job and you now need to use another table: the superheated property table of the fluid — which will be discussed in the next section.

For now, let us try applying the methodology to three more examples. In the first example, the property state is of 300 psia and 100°F. The sequence of Steps 1 to 4 is shown in Figure 2.12(a) to (c).

In this example, the fluid's property is located in the liquid region. The values of the temperature and pressure are given as inputs to the problem, but what are the specific volume, enthalpy, and entropy of the fluid property state of the fluid, which is obviously not on the saturated liquid line but in the region to the left of the line? The fact is that the saturated fluid property table can only provide the properties on the saturated liquid line or the saturated vapor line, or within the 2-phase region, assuming that the two given properties were not only temperature and pressure, but rather temperature or pressure and one other property. In the case where the fluid property is to the left of the saturated liquid line, the saturated temperature determines the properties of that liquid. The reason for this is because while temperature is a very strong factor in the properties of the enthalpy, internal energy and specific volume, pressure is not a strong factor. Thus, for the example given, where the temperature equals to 100°F, the entropy, internal energy and enthalpy can be found from the saturated property table using the saturated temperature. These values are shown in the table below. Notice that the pressure corresponding to 300 psia is equal to 2,069 kPa. However, though the properties shown in the table correspond to the desired temperature of 100°F or 310.8 K, only the liquid properties are relevant for a liquid phase.

Let us try another example. Find the property states of 300 psia and 500°F. The sequence of Steps 1 to 3, including the table, is shown in Figures 2.13(a) to (c).

In this example, the fluid properties fall in the vapor region. The properties must be found using another property table called the Superheated Steam Table, which is available for every fluid. This table is distinguished by a "NY city block" look, as shown in Table 2.3, that is, the Superheated

(a)

(b)

(c)

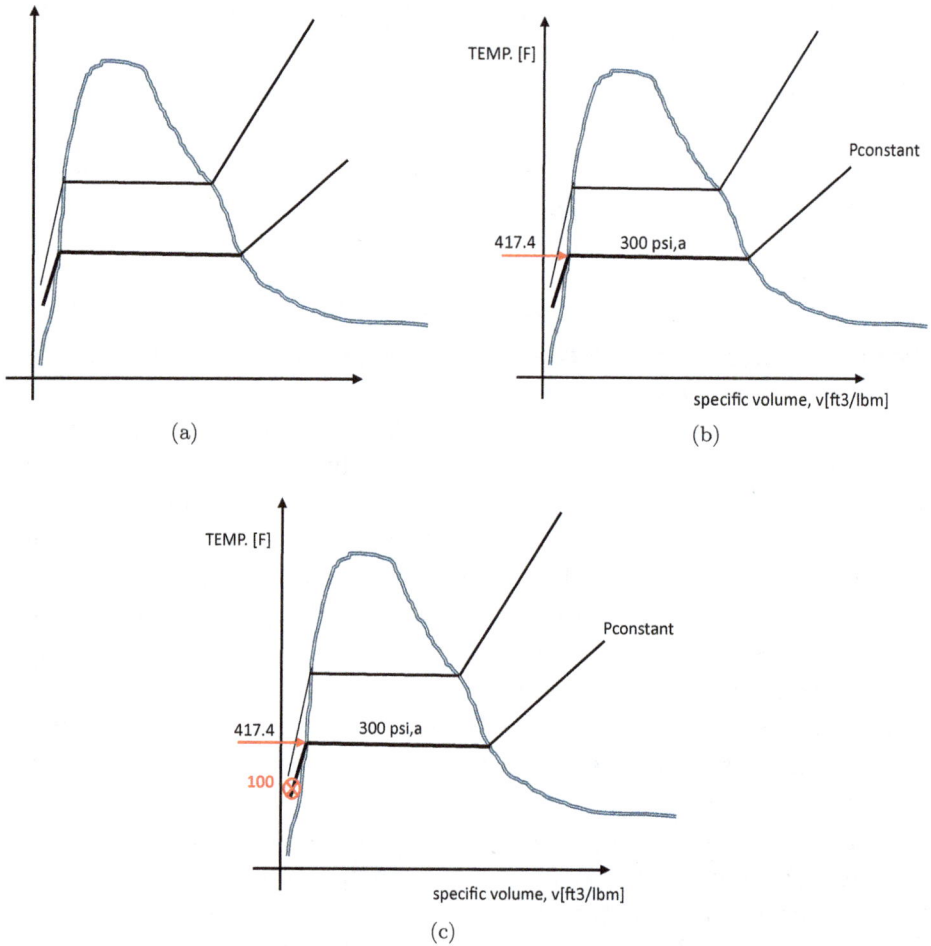

Figure 2.12.   Progression of the solution methodology. (a) A manual sketch of the T-v diagram not drawn to scale; (b) Labeling of the constant pressure line with the values of saturated pressure and temperature; (c) Locating the desired temperature (100°F) on the T-v diagram and recognizing that the state point falls to the left of the saturated temperature line, identifying the fluid as a liquid.

Table 2.2.   (c). A single row of Table 2.2(b) identifying the correct temperature and pressure solution.

		m^3/kg		kJ/kg		kJ/kg/K		kJ/kg	
T [K]	P [kPa]	v, liq.	v, vapor	h, liq	h, vapor	s, liq	s, vapor	u, liq.	u, vapor
310.78	2068.97	0.001007	22.01	157.6	2569.3	0.541	8.301	157.6	2426.2

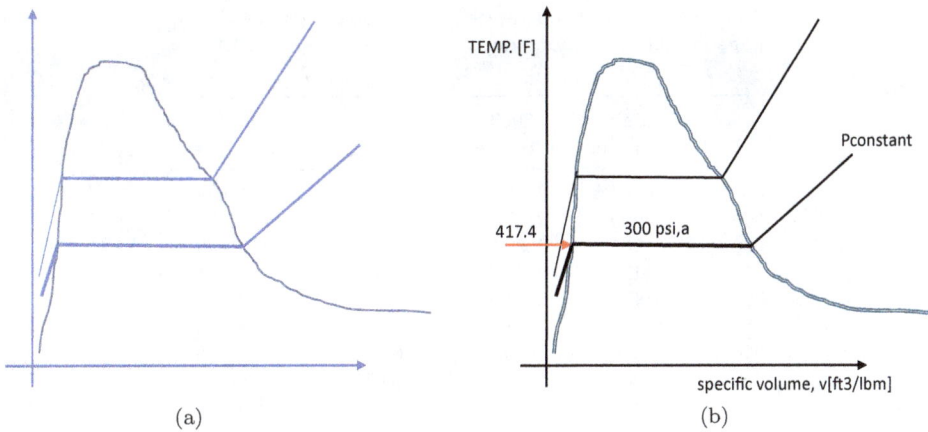

(a)

(b)

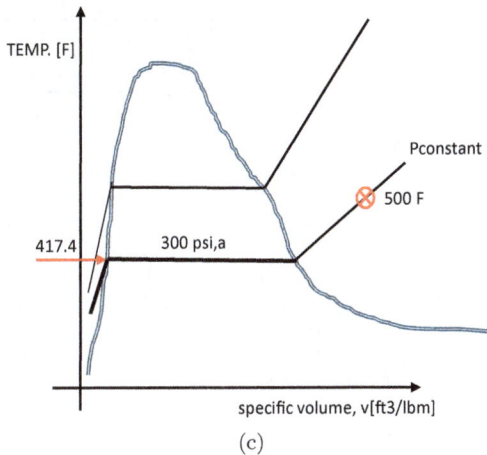

(c)

		m³/kg	kJ/kg	kJ/kg/K	kJ/kg
T [K]	P [kPa]	v, vapor	h, vapor	s, vapor	u,vapor
533	2069	0.11	2925.37	6.57	2697.30

(d)

Figure 2.13.   Progression of the solution methodology to find the property states of 300 psia and 500°F. (a) Unlabeled manual sketch of the T-v diagram not drawn to scale; (b) Labeling of the constant pressure line with the values of saturated pressure and temperature; (c) Locating the desired temperature (500°F) on the T-v diagram; (d) A line from the Superheat Property Table identifying the correct temperature and pressure solution.

Steam Property Table Compressibility Chart consists of "blocks" of constant pressure with a range of temperatures displayed along with the properties of specific volume, enthalpy and entropy. For the given conditions of 300 psia (2069 kPa) and 500°F (533 K), the process consists of finding the

Table 2.3. Partial superheated steam table for water (a) in imperial units; (b) in metric units.

(a)

T [K]	P [kPa] = $\sqrt{v}$ [m³/kg]	2069 h [kJ/kg]	P [psia] = s [kJ/kg/K]	300 u [kJ/kg]
487.2	0.0964	2799.12	6.33	2599.78
500	0.1005	2837.51	6.40	2629.59
520	0.1065	2892.07	6.51	2671.68
540	0.1122	2942.82	6.61	2710.71
560	0.1176	2991.24	6.70	2747.90
580	0.1229	3038.13	6.78	2783.92
600	0.1280	3083.98	6.86	2819.18
650	0.1404	3196.02	7.03	2905.51
750	0.1643	3416.42	7.35	3076.46

(b)

T [K]	P [kPa] = v [ft³/lbm]	2069 h [btu/lbm]	P [psia] = s [btu/lbm/R]	300 u [Btu/lbm]
417.4	1.5435	1204.15	1.5121	1118.4
450	1.6369	1227.27	1.54	1136.3
500	1.7670	1258.70	1.57	1160.5
550	1.8884	1287.82	1.60	1182.9
600	2.0046	1315.67	1.63	1204.3
650	2.1172	1342.79	1.65	1225.2
700	2.2273	1369.48	1.68	1245.8
750	2.3356	1395.93	1.70	1266.2
800	2.4424	1422.26	1.72	1286.6
850	2.5481	1448.56	1.74	1307.0
900	2.6529	1474.87	1.76	1327.5
950	2.7570	1501.26	1.78	1348.1
1000	2.8605	1527.75	1.80	1368.8

"300 psia block" and then the 500°F temperature row, before simply reading the enthalpy, specific volume and entropy values. In this example, the values are 2925.37 kJ/kg, 0.11 m³/kg, and 6.57 kJ/kg/K, as highlighted in Figure 2.13(d).

Here's one more example with the complete sequence from Steps 1 to 4: What are the fluid properties for a fluid that is 550°F and has a specific volume of 1.884 ft³/lbm?

Step 1. Draw a sketch of the T-v diagram, as shown in Figures 2.14(a) to (c).

Step 2. Insert the values of the specific volume for the saturated liquid and vapor on the graph on the x-axis, as shown in Figure 2.14(b) using Table 2.4.

Step 3. Observe that the desired value for the specific volume is much greater than the saturated vapor's specific volume at the saturated temperature of

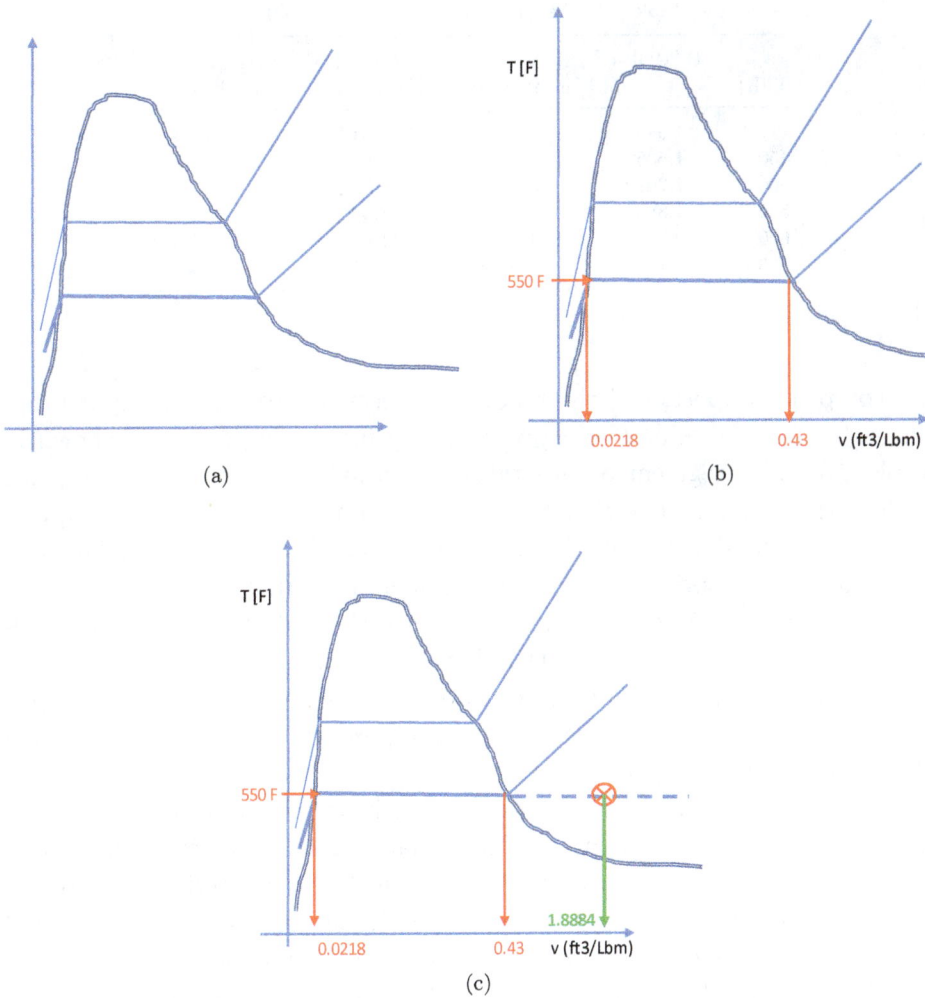

Figure 2.14.   Manual sketch (not drawn to scale) of the T-v diagram for a fluid that is 550°F and has a specific volume of 1.884 ft³/lbm. (a) Unlabeled diagram; (b) Labeling the constant pressure line with the values of saturated pressure and temperature; (c) Locating the desired temperature (100°F) on the T-v diagram and recognizing that the state point falls to the left of the saturated temperature line; hence identifying the fluid as a liquid.

550°F. This means that the state point with the given thermodynamics properties must lie to the right of the saturated vapor dome, as shown in Figure 2.14(c).

Step 4. With this knowledge from Step 3, the reader will know that the fluid is in a vapor state. This also means that the saturated liquid and vapor property tables have done their job but cannot be used to determine

Table 2.4.   Segment of vapor property table.

T [K]	P [kPa] = v [ft³/lbm]	2069 h [btu/lbm]	P [psia] = s [btu/lbm/R]	300 u [Btu/lbm]
417.4	1.5435	1204.15	1.5121	1118.4
450	1.6369	1227.27	1.54	1136.3
500	1.7670	1258.70	1.57	1160.5
550	1.8884	1287.82	1.60	1182.9
600	2.0046	1315.67	1.63	1204.3
650	2.1172	1342.79	1.65	1225.2

the complete properties of the fluid. It is necessary to use the vapor property table for the fluid. The vapor property table was previously shown in
Table 2.3 and a segment of this table is repeated in Table 2.4. Using this
table with the knowledge that the specific volume, v = 1.8884 ft³/lbm, and
pressure = 300 psia, the row that starts with the 550°F is selected. This row
also contains the specific entropy, enthalpy and internal energy.

The presentation of the "dome" and its various identified labeled "landmarks" misses one very important landmark: the critical point, which is
located at the apex of the dome. The pressure and temperature at this apex
point are called the critical pressure and temperature and are usually labeled
T_c and P_c. These properties are very important because the region in the
immediate vicinity above the critical point is where the properties of the
fluid can be very unstable and are not reliably known. For this reason, any
engineering work done in this region — called the supercritical region —
must be very carefully done to ensure that the mechanical components will
operate as intended. For example, a pump that is intended to pressurize a
liquid may suddenly find the density and viscosity of the fluid it is pressurizing to be closer to the vapor state than a liquid state if the inlet to the
pump is fluid above the critical point. In that case, the pump may cavitate;
that is, try to pump a vapor. The better choice is to design a compressor to
handle the compression of the fluid that is above the critical point.

The critical pressure and temperature are also used to determine the
compressibility constant, Z. The compressibility constant, Z, is used in the
perfect gas equation, as shown in:

$$P \times v = Z \times R_u \times \frac{T}{MoleWt}.$$

The compressibility factor, Z, corrects the Perfect Gas Law to be more accurate for many real gases. A chart called the Compressibility Chart is shown
in Figure 2.15. The chart shows the relationship between the compressibility

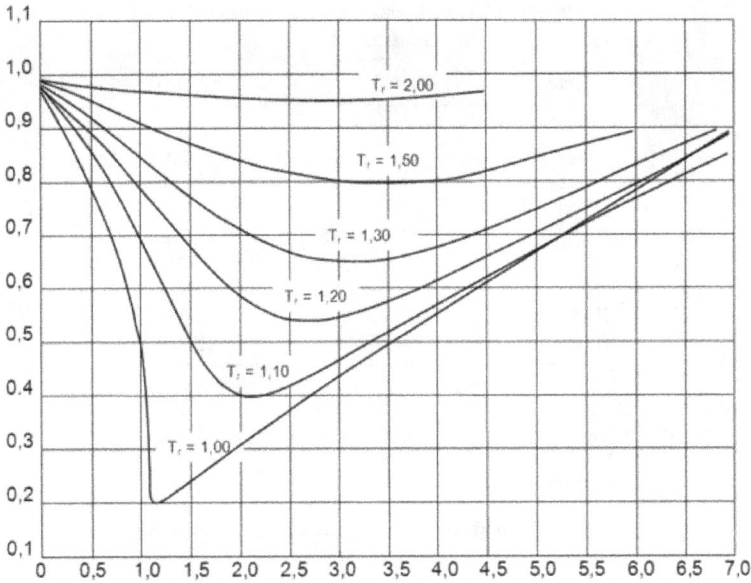

Figure 2.15. Compressibility Chart to determine, Z. $P_R = \frac{P}{P_c}$ along x-axis.

factor, critical pressure and temperature for the real gas. The chart is also helpful in proving that a real gas becomes more and more like an ideal gas when the state point pressure is low with respect to the fluid's critical pressure and when the fluid is very hot compared to the critical temperature.

As an example of how this equation is used, consider Example 2.2.

2.4.1   *Example 2.2: How two properties determine all the others*

What is the density of nitrogen at a temperature of 300°F and pressure of 28 psia? The Compressibility Chart shown in Figure 2.15 for nitrogen requires the critical pressure and temperature for nitrogen. The critical pressure and temperature of nitrogen may be found to be 33.9 atm or 498.3 psia (34.36 bar = 3.436 MPa and 126.1 K–146.9°C = −232.4°F). An example of the fluid properties is shown in Table 2.5. These critical pressure and temperature are then used to determine their respective ratio with the given pressure and temperature. These ratios were found to be: Pr = 0.056 and Tr = 3.34. It is important to always remember to use the absolute temperature scale when the temperature, T, appears alone in an equation such as T/Tc. Using Figure 2.15, the value of Z is found to be 0.99.

Table 2.5. Critical pressures and temperatures
for some common substances.

Substance	$T_c(°C)$	$P_c(atm)$
NH_3	132.4	113.5
CO_2	31.0	73.8
CH_3CH_2OH (ethanol)	240.9	61.4
He	−267.96	2.27
Hg	1477	1587
CH_4	−82.6	46.0
N_2	−146.9	33.9
H_2O	374.0	217.7

Using this compressibility, Z, in the Perfect Gas Law the specific volume
is:

$$P \times v = 0.99 \times R_u \times \frac{T}{MoleWt},$$

$$v = 0.99 \times 1545 \times \frac{\frac{460+300}{28.0}}{28 \times 144},$$

$$v = 10.30\frac{ft^3}{lbm}; \rho = \frac{1}{v}; \rho = 0.093\frac{lbm}{ft^3}.$$

The result of the calculation is a value of a specific volume, which is equal
to 10.3. Recalling that the density is the reciprocal of this specific volume
value, the density equals to 0.097 lbm/ft^3.

It is necessary to note that the compressibility factor, Z, is used in a
linear, first-order equation that is in its lowest form, which can be used to
adjust the Perfect Gas Law in order to have a relationship between pressure,
temperature and specific volume. Another more complex form is shown in
Equation (2.7).

$$P \times v = Z \times R_u \times \frac{T}{MoleWt} + A \times T + B \times T^2, \qquad (2.6)$$

where the constants A, B, C, etc., are empirically determined to be the best
fit of the measured properties with the equation. All such equations that
relate the three properties — P, T and v — are called Equations of State.

However, with the advent and ubiquitous availability of computers and
laptops, the compressibility factor and the various Equations of State have
been computerized and used in computer program models of thermodynamics systems. Among the most common of these computer programs are
the REFPROP properties from NIST. The NIST information is given in
Figure 2.16. These properties are relatively inexpensive and easy to embed

NIST Standard Reference Database 23

NIST Reference Fluid Thermodynamic and Transport Properties—REFPROP
Version 8.0

User's Guide

Eric W. Lemmon
Marcia L. Huber
Mark O. McLinden

Physical and Chemical Properties Division
National Institute of Standards and Technology
Boulder, Colorado 80305

April, 2007

U.S. Department of Commerce
Technology Administration
National Institute of Standards and Technology
Standard Reference Data Program
Gaithersburg, Maryland 20899

Figure 2.16. Reference for NIST's REFPROP "Add-In" for Excel and other compatible software platforms.

in any number of computer platforms, such as MatCad or MATLAB, or even Fortran compilers. However, the simple spreadsheet computer platform is among the easiest to use for solving thermodynamics problems. Thus, this textbook will focus on the solutions of thermodynamics problems in this textbook by using spreadsheets as the solution platform.

As an example of the ease of using REFPROP, consider the solution of the same problem that was stated earlier in Example 2.2, after it has been added to the spreadsheet using the "Add-In" feature of the Excel spreadsheet. What is the density of nitrogen at 28 psia and 300°F?

2.4.2 *Example 2.3: Solving Example 2.2. using REFPROP software "ADD-IN" in a spreadsheet to calculate density*

The cell statement for density is as follows:

= density ("nitrogen", "PT", "E", 28,300) where the units of P and T are psia and F

= 0.096 lbm/ft^3;

or in metric units

= density ("nitrogen", "PT", "S", 3.436,126.1) where the units of P and T are MPa and K

= 1.543 kg/m^3

2.5 Thermodynamics process paths

Although computer "look-up" tables are a very quick way of determining the desired properties, the reader is still encouraged to use free hand diagrams, as shown in Figures 2.11 to 2.14, to initiate and guide the solution of the thermodynamics problem. The property diagrams can use any two properties — "x" and "y" pairs — on a two-dimensional Cartesian coordinate display. For example, the most common thermodynamics property pairs are: T-v, P-v, P-h (pressure enthalpy) or T-s (temperature–entropy). These diagrams are particularly useful when two or more thermodynamics properties are connected to form a "path" to and from a thermodynamics state that has a specific purpose. Each connection of a pair of state points is called a process or process path. For example, Figure 2.17 is a P-v diagram (shown for the first time in Chapter 1) on which has been placed six thermodynamics states. Thus, each of the state points is completely and uniquely defined by the temperature and specific volume shown. Once the state point has been identified with the temperature and specific volume, all the other thermodynamics properties can be determined using REFPROP or by using either the property tables or property charts for the fluid. The ability to find all of the thermodynamics properties at a state point with only two properties is based on the State Principle, which states that the number of independent properties that are needed to perfectly define a thermodynamics state of a fluid is equal to one plus the number of work modes involved. In most of the engineering thermodynamics that this textbook is concerned with, the single work mode is the mechanical work mode being considered, as given by the equation:

$$Work = \int_{v1}^{v2} P(v) \, dv.$$

Example 2.2 is an example of how two properties determine all the others.

Why is it important that two properties can be presented graphically on a two-dimensional coordinate system? The simple answer is that all thermodynamics problems that are encountered in engineering must begin at some known condition of a fluid (i.e., a thermodynamics state point for the fluid) and the thermodynamics problem must end at a known thermodynamics

AB Constant Pressure
AC Constant Volume
AD Constant Temperature
AE Polytropic Process
AF Adiabatic Process

P [psia]; [Pa]

START (A)

v [ft³/Lbm], [m³/kg]

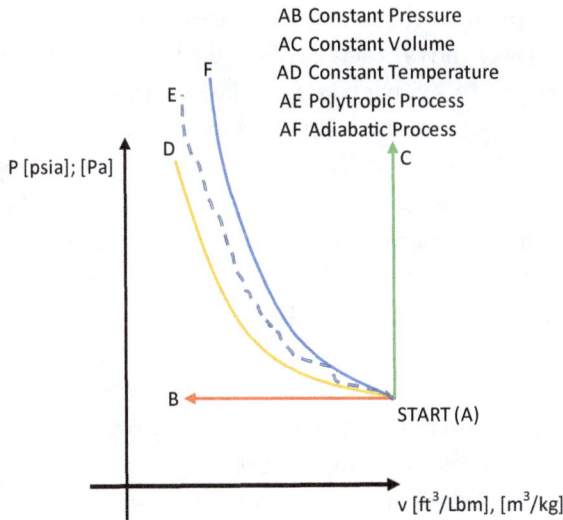

Figure 2.17.   P-v diagram illustrating the five viable "real-world" processes.

state point. What makes a thermodynamics problem unique is how these two state points are connected.

Given two or more state points and by applying the thermodynamics laws to the process or path that connects these state points, there can be an infinite number of ways that these state points can be connected. An analogy to a process path is an app providing street map directions from points A to B. These apps are excellent in being able to determine the fastest route or the route that uses the most or least number of freeways. They even know where one-way streets are located and can direct users on how to avoid them. Thermodynamics has its own way of determining one-way streets. These directions are imposed by the thermodynamics principles, in particular, the Second Law of Thermodynamics. Many other pathways are constrained by the fact that there may not be a mechanical method for producing the desired pathway. It can be shown that of all of the possible pathways that can connect points A to B, there are five basic paths that can model the operation of the world's most often used engines, whose sole purposes are to transform heat energy into a rotating shaft that can provide torque to a useful cause. These five pathways are defined as: constant pressure, constant volume, no heat transfer (i.e., the adiabatic process), constant temperature and almost zero heat transfer (i.e., the polytropic process). These five processes are shown in Figure 2.17, where they are superimposed on the same T-v diagram for air. Chapter 1 provides a more detailed description of each process, including how to calculate the heat transfer, internal energy and work. Air is a fluid very commonly applied in these five processes because it is ubiquitous, and thermodynamics takes advantage of this by injecting it into engines and then using its oxygen to oxidize the carbon-hydrogen based fuel. Oxidation is another name for the rapid combustion of fuel with oxygen that releases chemical energy that is chemically part of the carbon-hydrogen bond. It is also useful to have air used in these cycles because the Perfect Gas Law is extremely simple to program into a computer model with something as simple as a spreadsheet. More will be said about these in Chapter 3 when these processes are used with the first law equation as it is applied to a Closed Mass System. The presentation will also be associated with the construction of a computer model of both a Diesel and Otto engine, with the expectation that applying these processes to real engines provides more incentive to actually learn and understand the processes.

To help handle all the thermodynamics properties that may be of use in the solution of a thermodynamics problem, it is important to identify a thermodynamics "bookkeeping" tool, which is called a Property Table. This tool is very useful, particularly when there are many thermodynamics processes and thus many thermodynamics state points.

An example to show how all of the above can be used to solve a basic thermodynamics problem is shown in Example 2.4. This example also provides the reader with the sequence of steps that are used to solve a thermodynamics problem.

2.5.1   *Example 2.4: Finding the state points of water heated in a closed vessel and the net heat transfer and work for the processes*

A closed, insulated container of water is known to contain 5 lbm of water in a fixed volume of $2\,\text{ft}^3$. The pressure gauge on the closed vessel measures a pressure of 30 psia. The vessel is heated until the pressure reaches 300 psia. Determine the initial and final state points of the fluid. Determine the work done, heat transfer and internal energy change.

Step 1. Draw a diagram that can translate the written problem into an engineering diagram and that clearly identifies the known and the unknown parameters. This is shown in the drawing in Figure 2.18, which clearly identifies the closed vessel.

Step 2. The Property Table is the next useful engineering tool, apart from the engineering diagram. The beginning of the Property Table is shown in Table 2.6. It is a very useful "bookkeeping" tool to identify the known and

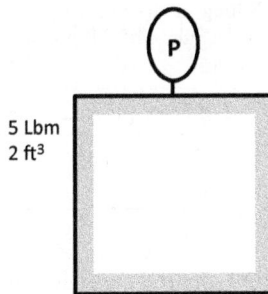

Figure 2.18.   An engineering diagram describing the written problem.

Table 2.6.   The beginning of the property table
used in solving Example 2.4.

P [psia]	T [F]	v [ft³/lbm]	u [Btu/lbm]
1	30		
2	300		

unknown properties at each of the state points that are under consideration. For this example, there are only two state points.

In this case, the known properties from the problem statement are the pressures at state points 1 and 2. In order to solve this and most of the other thermodynamics problems, it is necessary to fill this table with the values of all the other properties. Based on the State Principle, in order to fill each row of this table it is necessary to know at least two properties. Clearly, the wording of the problem only precisely gives the pressure of the closed vessel at the start and end of the process. There must be information one can infer from the context of the worded problem that will provide a thermodynamics "clue" as to the value of one of the other unknown properties.

A careful reading of the problem reveals that the 2-ft³ vessel is closed and that it contains 5 lbm of water. Thus, the specific volume, v, is the ratio of the volume of the vessel and the mass, or:

$$V = 2\,\text{ft}^3/5\,\text{lbm}; v = 0.4\,\text{ft}^3/\text{lbm}.$$

Thus, for state point 1 the specific volume is $0.4\,\text{ft}^3/\text{lbm}$ and it can be placed into the table, as shown in Table 2.7. An even more careful translation of the phase "... closed vessel" in the problem statement indicates that the end of the process has the same vessel volume and mass, thus the specific volume is the same as the specific volume at the start of the process. Thus, $v_1 = v_2 = 0.4\,\text{ft}^3/\text{lbm}$, as shown in Table 2.7.

Table 2.7.   Filling in the property table with the specific volume calculated for the system in Example 2.4.

	P [psia]	T [F]	v [ft³/lbm]	u [Btu/lbm]
1	30		0.4	
2	300		0.4	

This table now has sufficient properties identified at each state point (i.e., in each row) to determine all the other state properties. This may be done using look-up Property Tables or REFPROP software.

However, before proceeding to determining these properties, it is useful to demonstrate how to use a property chart.

Step 3. A property chart should be hand drawn on the analyst's worksheet (see Table 2.7) and labeled with the known properties. As seen in Figure 2.19(a), the 30 psia constant pressure curve must be labeled with the saturation temperature, in this case, 250°F, and the saturated liquid and vapor properties, v, $_{\text{sat.liq}}$ = 0.017 ft³/lbm and v, $_{\text{sat.vapor}}$ = 13.75 ft³/lbm, respectively. With Figure 2.19(a) as the starting point, the same chart can be used to identify where the specific volume v1 = 0.4 ft³/lbm is located with respect to the property "dome". For this problem, v1 is located in the dome and thus it is recognized as a two-phase fluid.

It is noted that at first glance, the reader may be puzzled because the fluid stored in the closed vessel is in a two-phase state rather than as simple, liquid water. But it is important to note that the problem statement identifies the fluid as water, and while most readers may believe this to imply that the fluid is liquid, the term water is simply the common name for H_2O, a very common fluid that most people experience as a liquid. In fact, water can be liquid, two-phase or vapor. In this problem the fluid is in a two-phase state.

The construction of the fluid chart can now be utilized to graphically determine where the end state point lies on the property chart. As shown in Figure 2.19(b), a vertical line from state point 1, representing a constant specific volume, v, to state point 2 should terminate at the 300-psia constant pressure curve. Using property tables or REFPROP, the saturation liquid and vapor points on the "dome" provide values for the specific volume. A quality of 25% can be calculated from:

$$\chi = \frac{(0.4 - .0189)}{(1.54 - .0189)}.$$

Returning to Step 2 the properties of state points 1 and 2 can now be filled in by going to the saturation property table for water, previously shown

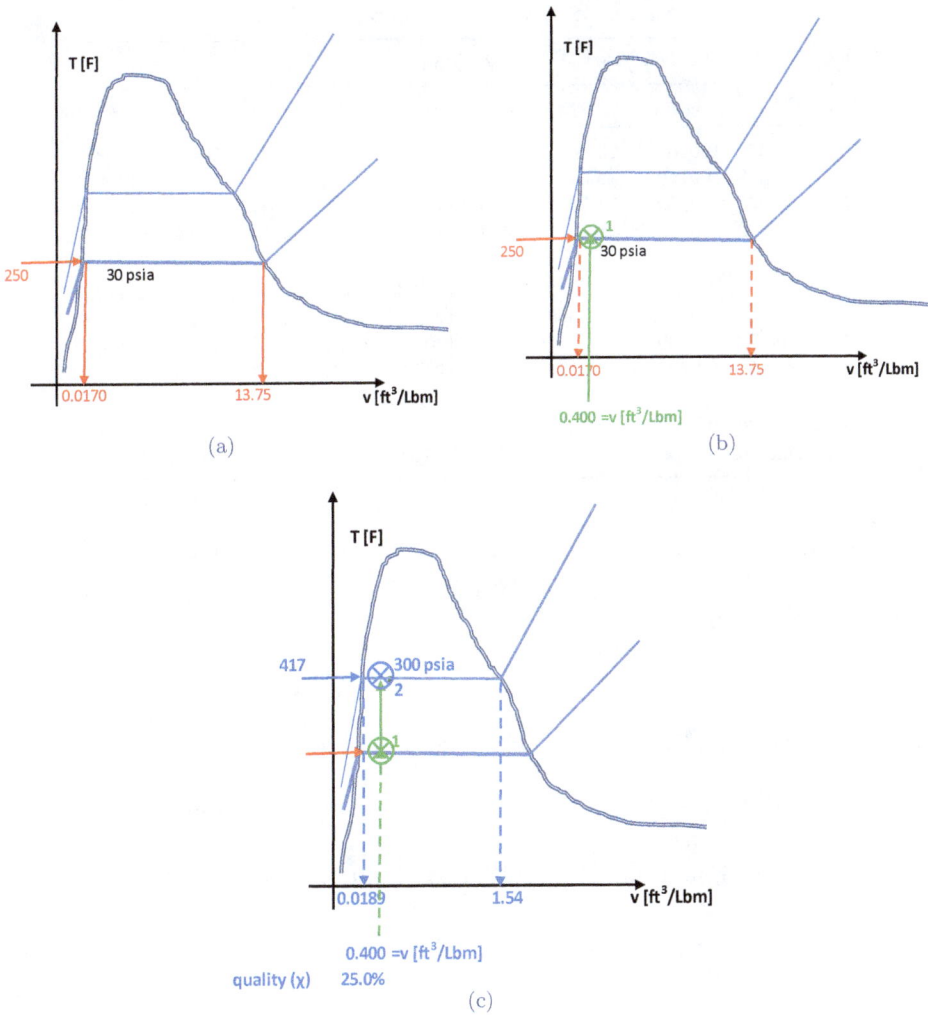

Figure 2.19. The three hand-drawn thermodynamic drawings presenting the three-step methodology useful for solving most thermodynamic problems and applied to solving Example 2.4. (a) Saturated pressure is labeled, as is saturated temperature; (b) Saturated liquid and vapor specific volume are identified using info from (a); (c) constant volume process locates the intersection of the 300 psia (one of the second given information from the problem) and the specific volume.

in Table 2.2. An excerpt from this large table, is shown in Table 2.8. The extracted rows specifically for the pressures of state points 1 and 2, which are 30 psia and 300 psia, respectively, are shown in two rows at the bottom of the table.

The exact properties at 30 psia and 300 psia must be determined from linear interpolation.

Table 2.8.   Saturation property table for water.

		ft³/kg		Btu/lbm		Btu/lbm/R		Btu/lbm	
T [F]	P [psia]	v, liq.	v, vapor	h, liq	h, vapor	s, liq	s, vapor	u, liq.	u, vapor
212	14.71	0.016715	26.78	180.3	1151.1	0.312	1.758	180.3	1078.1
240	24.99	0.016921	16.31	208.6	1161.3	0.354	1.715	208.6	1085.8
260	35.45	0.017084	11.76	228.9	1168.2	0.382	1.687	228.8	1091.0
280	49.22	0.017259	8.64	249.4	1174.7	0.410	1.661	249.2	1095.9
380	195.7	0.0184	2.3361	353.8	1199.3	0.542	1.549	353.1	1114.6
400	247.3	0.0186	1.8638	375.3	1202.2	0.567	1.529	374.4	1116.9
420	308.8	0.0189	1.5006	397.1	1204.4	0.592	1.510	396.0	1118.6
440	381.5	0.0193	1.2177	419.3	1205.6	0.616	1.491	417.9	1119.6
250.3	30.00	0.0170	13.75	219.1	1164.9	0.368	1.701	218.98	1088.5
415.4	300.00	0.0189	1.54	394.2	1204.1	0.589	1.512	393.16	1118.4

The values of internal energy are found at each state point by using the calculated quality (χ) at each state point, or $\chi_1 = 2.8\%$ and $\chi_2 = 25\%$ by using the equation:

$u_1 = \chi_1 \times u_v + (1 - \chi_1) \times u_l$; with the saturated properties

$u_l = 219\,\text{Btu/lbm}; u_v = 1088.5\,\text{Btu/lbm}$; and $u1 = 243.2\dfrac{Btu}{lbm}$.

$u_2 = \chi_2 \times u_v + (1 - \chi_2) \times u_l$; with the saturated properties

$u_l = 393.2\,\text{Btu/lbm}; u_v = 1118.4\,\text{Btu/lbm}$; and $u1 = 574.4\dfrac{Btu}{lbm}$.

Step 4: The calculations have determined the values of the internal energy at state points 1 and 2. These should now be entered into the Property Table that we used in Step 2 of the solution process (i.e. Table 2.6) and continued in Table 2.7, and that is now completed as Table 2.9. Table 2.9 provides a neat and clear identification of the properties that will be used to solve the last part of the problem, namely, what are the heat transfer, work and internal energy change for the process that begins at state point 1 and ends at state point 2?

The change in internal energy is easily observed to be simply the numerical difference between the internal energies between state points 1 and 2. Care must always be taken to take the difference as the property value after the thermodynamics event (i.e., state point 2) is subtracted from the value of the thermodynamics property before the event, as given in the equation:

$$\Delta u = u_2 - u_1; \Delta u = 574.4 - 243.2; \Delta u = 331.2\frac{Btu}{lbm}.$$

Table 2.9. Property table for the example problem now completed with the other relevant properties.

	P [psia]	T [F]	v [ft³/lbm]	u [Btu/lbm]
1	30	250.3	0.4	243.2
2	300	417.4	0.4	574.4

The problem statement also requires calculating the heat transfer and the work done between the two state points. How can the heat transfer and work between the start and stop of the thermodynamics process be determined? The First Law of Thermodynamics is useful here, because it is a perfect equation that connects the work, heat transfer and internal energy changes. However, as described in Chapter 1, the First Law of Thermodynamics can be expressed by two similar equations, depending on whether the process is known to be a constant mass process or a continuous flow process. Determining this now becomes essential in order to write the correct form of the First Law of Thermodynamics. For this problem the wording of the problem clearly states that the vessel is fixed, or in other words, its volume does not change. Thus, the First Law of Thermodynamics is repeated here as:

$$\Sigma Q = \Sigma Work + \Delta U. \tag{2.7}$$

With the volume fixed throughout the process, the change in volume between the two state points is zero. As indicated in Chapter 1, the work performed during this process must therefore be zero. This leaves the first law equation to answer the question of how much heat transfer must occur in order for the process to occur as stated.

In summary, the internal energy change is 331.1 Btu/lbm or 331.1 × 5 lbm = 1,656 Btu; the work done is zero, and the heat transfer required for the process is also 1,656 Btu.

Step 5. The last step of any thermodynamics problem solution is the need to perform a sanity or reality check. For this problem it may help to recognize that the amount of energy used (1,656 Btu) is small compared to the amount of chemical energy contained in a pound of liquid or gaseous fuel. For example, gasoline has a heat content of approximately 21,000 Btu/lbm. Thus, the pressurization of a closed mass, fixed volume container from 30 psia to 300 psia requires approximately 0.079 lbm or 2.5 ounces of fuel. It will be described in Chapter 4 that during a very common engine cycle called the Otto Cycle, the detonation of the fuel in the piston-cylinder is assumed to occur at "fixed volume"; that is, fast enough that the piston may be considered to not have moved. It will also be shown that the change in the pressure can be very high, occur very quickly (it is assumed to be essentially instantaneous) and reach magnitudes as high as shown in this worked example.

Another example of a sanity or reality check is considering the heating of water in a pressure cooker on a common kitchen stove. Five pounds of water is approximately 63% of a gallon. Of course, the water in a pot is truly a liquid, is not two-phase and is filled with atmospheric pressure (14.7 psia). However, after the lid is put on and secured tightly, the additional heat quickly starts to vaporize the water and, depending on the rate of heat transfer into the fixed pressure cooker pot, the heat input of 1,656 Btu can take less than 20 minutes. Of course, long before the 20 minutes has elapsed, the pressure cooker's pressure relief valve would have "popped" to vent the sealed chamber because the pressure would have reached 30 psia within minutes starting from its initial atmospheric pressure of 14.7 psia.

An additional important consideration is that all of the analysis above has been made by assuming that all of the required heat transfer occurred with no heat loss to the atmosphere, and is thus not available for the actual heating of the vessel. In reality, and despite the great effort to reduce the amount of heat loss from the vessel to the atmosphere as the vessel gets hotter from the heating, there is no such thing as a 100% effective insulation. Thus, some of the heat input will be lost to the atmosphere. In order to accomplish the heating that is required, more than the heat transfer value shown in this answer will need to be added to the vessel. For the sake of argument, let us say that 500 Btu of heat energy is lost to the atmosphere during the heat transfer between state points 1 and 2. Thus, one can say that the heating efficiency or heating effectiveness is thermodynamically:

$$\zeta = \frac{1,656}{1,656 + 500}; \zeta = 0.76 \ or \ 76\%.$$

Lastly, it is important for the reader to always consider another reality of many thermodynamics problems: many seemingly different types of thermodynamics problems are actually very similar in their solutions; for example, they use the same thermodynamics equations (and certainly the same solution methodology as shown in this example). This analogy is very important for the analyst to understand, appreciate and use when appropriate. The last two examples of an engine's combustion process and the heating of a pressure cooker pot are examples of the types of analogies that can be applied to the initial problem statement.

Chapter 3

Thermodynamics First Law Applications to Closed Mass and Open Flow Analyses

Chapters 1 and 2 are intended to prepare the reader with the tools to effectively use the First Law of Thermodynamics. The concepts of energy, work and heat transfer have been defined in ways that will help the reader understand and apply these concepts in real world of thermodynamics applications. Similarly, the possibly infinite number of potential thermodynamics processes was reduced to only five that the author deems the most used by and/or useful for the thermodynamics analyst solving real-world processes. These five processes (or pathways for a Closed Mass System) are: constant pressure, constant temperature, constant specific volume, no heat transfer and the approximately no heat transfer processes. The author then proceeded to make an even bolder statement: that every real-world engineered system may be seen as composed of one or more of only seven basic mechanical devices, and that these are the bases of thermodynamics system designs, much like how A, C, G and T — the four bases of the DNA molecule in biology — can have almost infinite arrangements, and how they make each of us unique yet similar to each other. The seven basic mechanical devices are: turbines, compressors, pumps, heat exchangers, valves, nozzles and diffusers. It may also be successfully argued that a reservoir or mixing tank should be considered as an eighth mechanical device, despite the convincing argument that a missing tank could be regarded as a wall-less, heat exchanger where fluids can be mixed. This chapter will review these concepts in much more detail.

Chapter 1 introduced a solution methodology that begins with determining whether the problem is best solved by constructing either a Closed Mass or an Open Flow system analysis Control Volume (CV) around the system. A Closed Mass System does not allow mass to cross the CV border and thus the mass is constant from the beginning to the end of the process.

Conversely, an Open Flow System allows mass flow to transfer across the CV border in at least one place. The flow rate may come in and/or out of the CV; i.e. it may come into CV and leave at the end, into the CV without leaving, or leave from one or more openings in the CV without an inlet. Of course, it is assumed that in the last scenario the CV was assigned with a finite amount of mass in the system at the beginning of the process.

Recall that a thermodynamics system is any collection of elements that is of particular interest to the analyst and has been selected to help solve the thermodynamics problem. The last chapter ended with a discussion on the definition of processes — pathways (occurring to a system) that start from an initial state point and conclude at a final state point. The application of the First Law of Thermodynamics is directed to each individual process or to a collection of processes. It is only necessary to define a specific starting point and a specific ending point, during which time the First Law is applied to determine how much of the system's Internal Energy (E) has been changed but not destroyed or how much work or heat transfer is performed during the process path in question; perhaps converting some the initial Internal Energy into heat or work.

3.1   The first law of thermodynamics using closed mass analysis

The first step in solving any thermodynamics problem is to determine if it is best solved using either a Closed Mass or an Open Flow Analysis. Your choice can make the difference between an easy, straightforward solution or a difficult and/or approximate solution. As indicated above and in Chapter 1, the Closed Mass Analysis can be made easier by recognizing the most common processes that can be readily performed in the real world of engineering. It is a fact that all major or minor engines that have been developed since the 1700s and the advent of thermodynamics analysis can be built from only five processes. The reader is reminded that while there are an infinite number of processes that can be identified as the pathway from an initial state to a final terminal state, there are only five processes that can actually be built with the technology that is available today. Nonetheless, these five processes have produced remarkable engines that can transform thermal energy into rotary shaft energy that is able to raise a weight in a gravity field, and that is equivalent to performing work required of a rotating shaft.

It is important for the analyst to remember that a process path starts at some point in time and with the Control Volume system at an equilibrium state, only to be terminated at a different state point at a later point in time. The time duration for the process can be as short or long as it takes to bring the process from a starting point state to a termination. For example, a Closed Mass Analysis is almost the perfect method for analyzing the heat engines prevalent in today's society. It can be completely described by the five processes. The Closed Mass Analysis for a heat engine focuses on the typical piston-cylinder, internal combustion engine. The Closed Mass is only true during the time when the intake and exhaust valves are both simultaneously closed after air is inducted into the engine and before the exhaust valve opens only after the power stroke. The time for this process is very short. For example, a piston engine typically has to fulfil five processes to complete in one revolution: the intake of air at constant pressure, compression of the air charge (quasi-adiabatic or almost no heat transfer), detonation of the fuel (may be treated as a combination of constant volume and constant pressure), the power stroke (quasi-adiabatic), and the expelling of the products of combustion (constant pressure). A piston-cylinder engine operating at 3,600 rpm completes 60 revolutions in one second, or each revolution takes 1/60 seconds. During this time the combustion of the fuel will have occurred along with the intake, power stroke and discharge processes. With these considered, the time taken for the intake and exhaust valves to completely close can be 1/120 of a second. And yet the First Law of Thermodynamics works perfectly fine during this period, and for a time duration that is orders of magnitude larger.

The five processes that "rule the heat engine world" are:

1. Constant Pressure or $\Delta P = 0$
2. Constant volume process or $\Delta v = 0$
3. Constant Temperature process or $\Delta T = 0$
4. No heat transfer or $\Delta Q = 0$
5. Almost no heat transfer $\Delta Q \sim 0$

Each of these processes should be graphically drawn on a P-v or T-v coordinate system, such as the one shown in Figure 3.1, to facilitate the solution of the problem required in Step 5. The use of a P-v chart is recommended because there is an added benefit to using the graphic to describe the energy exchange that occurs during any process. That benefit is seen by taking the work expression: Work = Force x Distance traveled, and substituting the expression for Pressure = Force/Area into the force term to get Work = Pressure × Area × Distance moved. But the reader quickly recognizes that the Area × Distance product is a swept volume change during the process and hence the significance of using a P-v chart is the observation that the

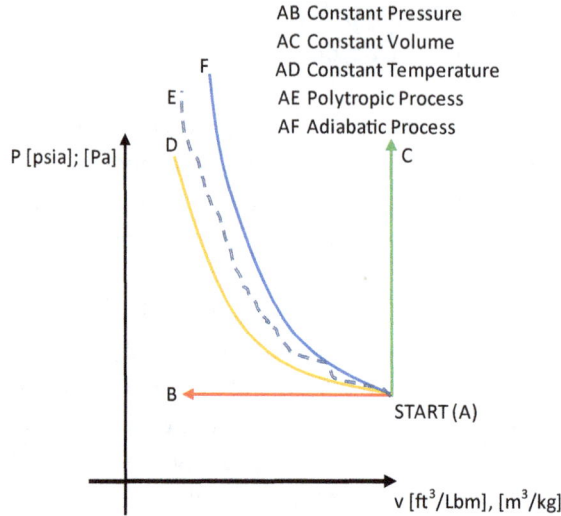

Figure 3.1.   The five basic processes for a Closed Mass System.

area under the process curve (path) is actually the work that is either input to the process or work that is removed from the process.

More precisely stated, the work occurring during a process is perfectly represented by the integral:

$$Work(W) = \int_{v_1}^{v_2} P(v)dv, \tag{3.1}$$

where P(v) is identified as the pressure at a state point as a function of specific volume.

This equation is almost iconic in its definition of mechanical work as a function of pressure and volume.

3.1.1   *Process 1: Constant pressure process or* $\Delta P = 0$

The constant pressure process is easily observed to be simply a straight, horizontal line across the swept volume of the process as shown in Figure 3.1. The integral of work is again very easily performed and is simply the product of the constant pressure multiplied by the swept volume or Work $= P \times (\Delta v)$. Care is taken to respect the sign that results from this equation, because in engineering the Δ delta symbol means the "difference" of the parameter that is given. For example, Δv means the difference between the specific volume at the end of the process minus the specific volume of the system at the beginning of the process, or $\Delta v = v_2 - v_1$. Once again, it is clear that the process path must have a beginning and an ending, during which the First

Law of Thermodynamics is applied. The reader must also be careful with the units: if psia and ft^3/lbm are the units used for P and v, then the expression must also be multiplied by 144/778 in order to convert the units to Btu/lbm. If the SI unit system is used, then the expression is multiplied by 1/1000 in order to convert it to the common units of kJ/kg.

The amount of internal thermal energy change or Δu (with units of Btu/lbm or kJ/kg) is always $C_v \times \Delta T$, where C_v is the specific heat of a gaseous substance with respect to volume and the ΔT is the temperature change during the process, but again expressed as the difference in the temperature after the process has concluded, compared to the temperature of the state point of the system when the process was initiated. C_v must not be confused with C_p, the latter being the specific heat with respect to the constant pressure process. It is also true that both the specific heats are only defined for fluids that are in their gaseous state. It can be shown that the $C_p = C_v + R_u/MoleWt$. Thus, $C_p > C_v$ and the ratio of C_p to C_v is recognized as an important property: $k = C_p/C_v$, known as the specific heat ratio. Substances that are either solid or liquid, i.e., not a gas or plasma, will also have a specific heat property, but the specific heat for a solid or liquid will not be distinguished as either C_v or C_p; it is denoted simply as C. The reason why liquid or solid substances need not distinguish between C_v and C_p is entirely due to the way C_v and C_p are experimentally determined. The value of C_v for a substance is determined by fixing that substance so that its volume does not change, hence the use of a subscript "v" to mathematically denote a constant volume heating process. Similarly, the value of C_p for a substance is determined by keeping the pressure on the surface of the substance at a constant pressure during the heating process. In both cases the heating process will change the temperature of the substance. The amount of heat transfer, ΔQ, that results in a change in the temperature, ΔT, is thus known and the ratio: $\Delta Q/\Delta T$ determines the specific heat, C. However, for liquids and solids that are incompressible, i.e. where the density of the liquid or solid does not change when exposed to high pressure, the ratio of $\Delta Q/\Delta T$ does not change whether the liquid or solid is held with a constant volume of constant pressure during the heating process. This ratio does change if the heating process is done on a gas. In any case, specific heat, whether denoted as C, C_v or C_p, is a property that is dependent on pressure and temperature. The values for C, C_v or C_p can be found in thermodynamics handbooks or the internet, as well as by using the REFPROP properties software identified in Chapter 2. The amount of heat transfer that will occur as a result of this constant pressure process is simply determined by adding the work done $(+ \text{ or } -)$ to the ΔU $(+ \text{ or } -)$ during the process. Why? Because that is the correct application of the First Law of Thermodynamics to a system that has just undergone a constant pressure process: $\Sigma Q = \Sigma W + \Delta u$.

3.1.2   *Process 2: Constant volume process or* $\Delta v = 0$

A constant volume process is also precisely represented on a P-v chart by a vertical line starting from the initial state point, which moves either vertically up or down, depending on whether the process is heating or cooling the system. The process path moves vertically upward if the system is being heated. Conversely, the process path moves vertically downward if the system is being cooled down. The amount of work done in a constant volume process is the simplest calculation of all, considering that if the position does not move, thus there is no chance of moving a weight in a gravity field, then the work must be zero during the process. The internal thermal energy change is still as simple as before: $\Delta u = Cv \times \Delta T$. And certainly, the heat transfer is the addition of these two terms due to the first law equation.

3.1.3   *Process 3: Constant temperature process or* $\Delta T = 0$

A constant temperature process is slightly more difficult to visualize until one realizes that the constant temperature process for a Closed Mass System can be represented by the perfect gas law equation at both the start and the end of the process. Thus:

$$(P \times V)_{start} = R_u \times \frac{T}{MoleWt}; (P \times V)_{end} = R_u \times \frac{T}{MoleWt}.$$

Take the ratio of these two equations and you will find that:

$$(P \times V)_{start} = (P \times V)_{end},$$

only to realize that this is the expression for the shape of a hyperbola on a Cartesian coordinate system of P versus v, as shown in Figure 3.2. The process path indicates that for the given initial and constant temperature throughout the process, the state points must "land" on this hyperbola path between the initial and final state points. The calculation of work (area under this hyperbola curve) is a little trickier, but will be relatively straight-forward assuming the thermodynamics analyst were to use the perfect gas law to represent how the pressure of the system varies with changes in the specific volume of the system. The integration of Equation (3.1) results in

Pressure-specific volume (v) with Constant Temperatures [R] as Shown

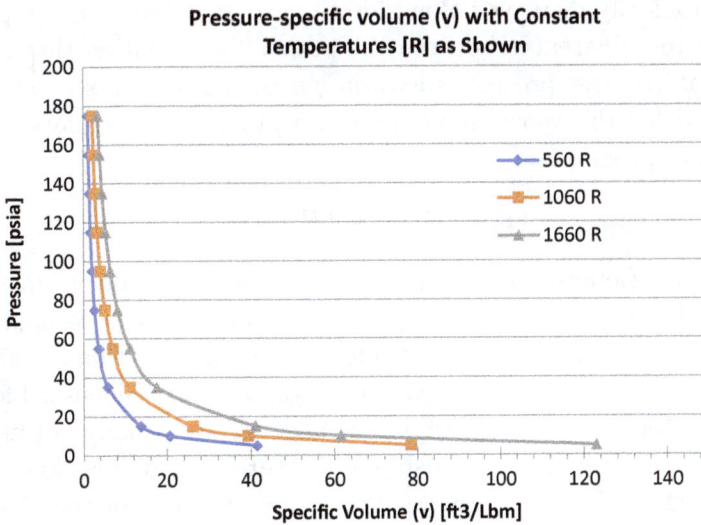

Figure 3.2. Constant temperature process paths or isotherms using the relationship $Pv = $ Constant.

the expression for work during the process (and hence area under the curve):

$$\text{Work} \left[\frac{\text{ft} - \text{lbf}}{\text{lbm}} \right] = \text{P} \times \text{v} \times \ln \left(\frac{v_2}{v_1} \right). \tag{3.2}$$

The product of P V used in Equation (3.2) can be either the PV product at the start or at the end of the process. That is:

$(PV)_{\text{start}} \times \ln(V_{\text{end}}/V_{\text{start}})$ or equally valid: $PV_{\text{end}} \times \ln(V_{\text{end}}/V_{\text{start}})$

or even $(PV)_{\text{end}} \times \ln(P_{\text{end}}/P_{\text{start}})$.

In all cases, the positive or negative sign of the resultant calculation must not be ignored. According to the convention stated in Chapter 1, the positive sign (+) means that work is produced (output) by the system and the negative sign (−) means that work is input into the system. Once again, the internal thermal energy expression, $\Delta u = Cv \times \Delta T$, remains unchanged, and because the process under consideration is a constant temperature process, $\Delta T = 0$, the internal thermal energy change is 0. Once again, the heat transfer is the sum of these two energies, as stated by the First Law of Thermodynamics for a Closed Mass System.

3.1.4 Process 4: No heat transfer or $\Delta Q = 0$

The no heat transfer process path is of particular interest because it is often used in the construction of a heat engine. The no heat transfer process path is best graphically displayed alongside the constant temperature process to emphasize the differences between the paths. The equation that represents the no heat transfer process is a variation of the hyperbola equation that was derived for the constant temperature process. The equation for a no heat transfer process is:

$$(P \times v^k)_{start} = (P \times v^k)_{end},\tag{3.3}$$

where the parameter used in the power of the specific volume term is actually the ratio of the specific heats, C_p and C_v,, and hence it is appropriately enough called the specific heat ratio for the vapor or gas fluid state. Recall that C_p and C_v are different for vapors or gases, but are identical for liquids and solids. However, the subscripts "v" and "p" are usually not used.

This equation for the relationship of pressure and volume along a no heat transfer path is derived from the First Law of Thermodynamics coupled with the perfect gas law. This textbook in thermodynamics is intended to be different than most in its expressed purpose of not presenting derivations of thermodynamics equations, unless the derivation is vital for better overall understanding of the subject. The derivation of the relationship between P and v in an adiabatic process is one such very important derivation. The results are used throughout thermodynamics, both at an introductory level and in the more detailed expositions of the subject. The derivation also provides the reader with an introduction to the use of the derivative representation of the internal thermal energy and work, as they may be expressed mathematically. The derivative is as follows.

$$dQ = du + Work; \ but \ for \ adiabatic \ processes \ Dq = 0;$$

$$du = C_v \times dT \ and \ Work = Pdv.$$

But also:
$$P = R_u \times \frac{T}{v}/MoleWt,$$

$$-C_v\frac{dT}{T} = \frac{R_u}{MoleWt} \times dv/v,$$

$$-C_v \int \frac{dT}{T} = \frac{R_u}{MoleWt} \times \int \frac{dv}{v},\tag{3.4a}$$

$$-C_v \ln\left(\frac{T_2}{T_1}\right) = \frac{R_u}{MoleWt} \times \ln\left(\frac{v_2}{v_1}\right).$$

It is also true, and the reader should be very aware, that: $C_p = C_v + \frac{R_u}{MoleWt}$.

However, care must be made to make sure that the units are consistent: whether one is using Btu/lbm/R or kJ/kg/K.

Dividing both sides of the equation $C_p = C_v + \frac{R_u}{MoleWt}$ by C_v, and recalling that the specific heat ratio, k, is equal to $\frac{C_p}{C_v}$, then the equation resolves to: $k - 1 = \frac{R_u}{MoleWt} \times \frac{1}{C_v}$.

Combining this last equation with Equation (3.4a), we find that: $\frac{T_2}{T_1} = \left(\frac{v_2}{v_1}\right)^{(1-k)}$.

Recalling the perfect gas law and applying it to state points (1) and (2) we find that:

$$\frac{P_2 v_2}{P_1 v_1} = \left(\frac{v_2}{v_1}\right)^{(1-k)}.$$

And from this equation, with some algebra, we find that:

$$\frac{P_2}{P_1} = \left(\frac{v_1}{v_2}\right)^k \text{ or } P_2 v_2^k = P_1 v_1^k. \tag{3.4b}$$

The work expression for the adiabatic (or later the polytropic) process can now be found from integrating the expression for work:

$$Work\,(W) = \int_{v_1}^{v_2} P(v)dv.$$

Using the mathematical relationship given in Equation (3.4b) for P and v, the integration is found to be:

$$Work = [(P \times v)_2 - (P \times v)_1]/(1 - k). \tag{3.4c}$$

Figure 3.3 illustrates three "no heat transfer" (also called adiabatic) process paths that start at the temperatures shown. It is important to note that unlike the isothermal process paths, the temperatures in an adiabatic process path do not stay constant but in fact increase during compression.

The adiabatic and isothermal process paths are thermodynamically very different from each other despite their similar appearances. It is worthwhile to present both process paths on the same curve, as shown in Figure 3.4, in order to distinguish the two process paths. It is also very important to note that these two process paths are theoretical. That is, it is rare to experience a 100% constant temperature or no heat transfer process path. However, these perfect processes provide two very important thermodynamics concepts that at least provide the ideal potentials for a process path.

As always, the internal thermal energy uses the same expression: $\Delta u = C_v \times \Delta T$. Lastly, the heat transfer is the sum of these by using the First Law of Thermodynamics. It is important to note that for the adiabatic process path this heat transfer for the process must be zero and thus the

P-v Diagram with Adiabatic Relationship [Pv^k = constant]

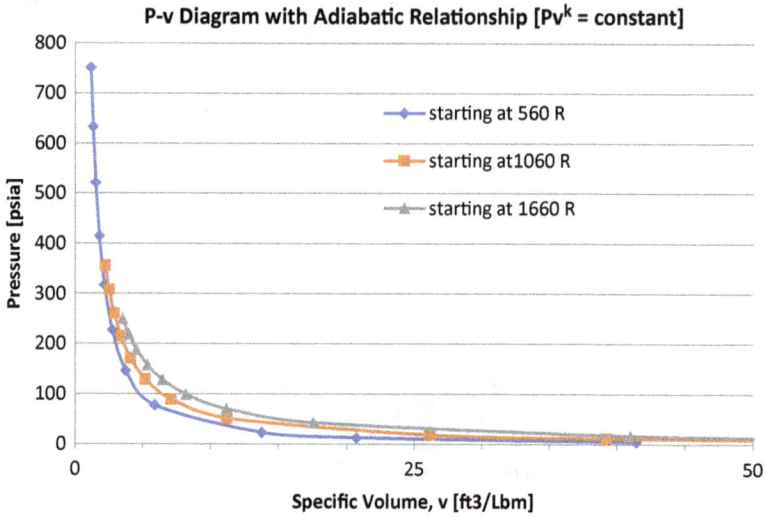

Figure 3.3.   No heat transfer (also known as adiabatic) process paths that follow the mathematical relationship: Pv^k = constant.

P-v Diagram for air showing Adiabatic Curve and Isothermal Curve

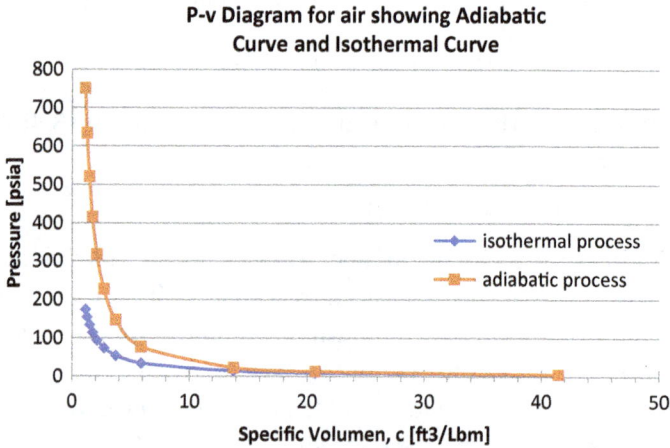

Figure 3.4.   Comparison of adiabatic and isothermal process paths that start from the same temperature. Note that the adiabatic path achieves a higher outlet temperature as the process of compression proceeds from right to left.

internal energy change will be equal to the work performed (or required) during the process.

3.1.5   *Process 5: ALMOST no heat transfer or* $\Delta Q \sim 0$

The last common process path is the "almost no heat transfer" process path. This is necessary because in this "real world" of engineering application, there is never an instance when a process can truly be said to be an absolutely no heat transfer process. Heat transfer is the energy exchange due to a temperature difference. If there is even the slightest temperature difference between the system in question and the local ambient, then even the best and thickest insulation will not prevent some heat transfer loss from the system (or to the system). In fact, there is a critical limit on the thickness that a system can be insulated with before the larger surface area actually enhances the loss of heat energy to the surroundings, so that heat is gained by the system from the surroundings. Thus, the process path of almost no heat transfer is too common and needs to be empirically measured to determine a parameter that is labeled the polytrophic constant (n). The polytropic index may be determined empirically to be different for an expansion or a compression process. The reader may have already guessed that the expression for the curve that represents an almost no heat transfer process is very similar to the no heat transfer process equation. It differs only in the use of the polytrophic constant, which is in place of the specific heat ratio, k, as shown in the equation below:

$$(P \times v^n)_{start} = (P \times v^n)_{end}. \tag{3.5a}$$

The similarity does not stop there, because the equation for the work performed during the process is simply:

$$Work = [(P \times v)_2 - (P \times v)_1]/(1 - n). \tag{3.5b}$$

Care must be taken to watch the units for pressure and specific volume. In the imperial (i.e. English or non-metric) system, the typical units for measurement are as follows: psia for pressure, ft^3/lbm for specific volume, and Btu/lbm for Work. Therefore, the equation for work shown in Equation (3.5) should be multiplied by $144\,in^2/ft^2$ and divided by $778\,ft - lbf/Btu$. In the metric system, the units of measurement are typically N/m^2 for pressure, m^3/kg for specific volume, and kJ/kg for work. Thus, Equation (3.2) should be divided by 1,000.

Given that the internal thermal energy expression for a perfect gas is still dependent on the temperature difference and the specific heat capacity, C_v, it is still:

$$\Delta u = C_v \times (T_2 - T_1). \tag{3.6}$$

Figure 3.5 displays the differences in the process paths for an adiabatic process and a polytropic process.

That leaves only the first law equation again to determine the heat transfer (albeit very small by the definition of almost no heat transfer) to be the sum of the work and the internal thermal energy during the process.

P-v Diagram of Adiaqbatic Processes with Different Specific Heat Ratio, k and polytropic indexes n1and n2

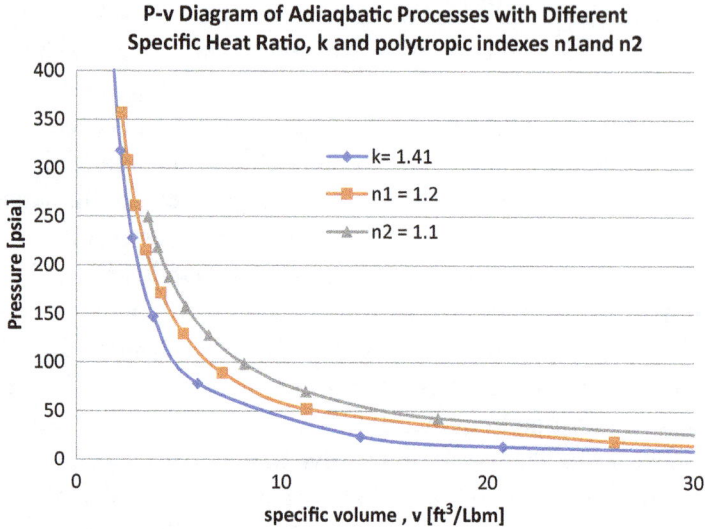

Figure 3.5.   Comparison of P-v curves for different indices.

Table 3.1.   The first law equations for a Closed Mass System for each of the five basic processes.

Process	Heat transfer	Internal thermal energy	Work
Constant temperature	$\sum U + W$	$Cv \times \Delta T$	$Pv \times \ln\left(\dfrac{v2}{v1}\right)$
Constant pressure	$\sum U + W$	$Cv \times \Delta T$	$P \times \Delta V$
Constant volume	$\sum U + W$	$Cv \times \Delta T$	0
No heat transfer (adiabatic)	$\sum U + W$	$Cv \times \Delta T$	$(p2V2 - P1V1)/(1 - k)$
Small heat transfer (polytropic process)	$\sum U + W$	$Cv \times \Delta T$	$(p2V2 - P1V1)/(1 - n)$

In these five processes and the resultant work, internal thermal energy and heat transfer equations are used frequently enough to deserve a special look-up table for easy reference, as shown in Table 3.1.

The polytropic equation, which represents the "almost no heat transfer" process, is very useful in thermodynamics. This is particularly true when one considers that there is no one process that is purely adiabatic in real world engineering applications. There is always some heat transfer loss from the process, even when a large effort is made to insulate the device. The work expression for the polytropic work process provides a closed equation, wherein the start and end state points pressure, P, and specific volume, v, are inputs into the equation, and the work for the process is easily solved. It is interesting to also point out that the ability to plot a thermodynamics

Table 3.2. Volume and
Pressure data.

V	Pbar
300	15
361	12
459	9
644	6
903	4
1608	2

process on a P-v diagram provides an opportunity to determine the work done during the process by a straightforward numerical integration. Numerical integration of a mathematical curve has never been easier given the availability of spreadsheets and other similar computation software. As an example, consider Example 3.1.

3.1.5.1   *Example 3.1: Determining the work done for a polytropic expansion process*

A polytropic expansion process has been experimentally determined to have a pressure-volume relationship, as seen in Table 3.2. Determine the work for the process using numerical integration and then check this result using the theoretical equation for work done during a polytropic process (see Equation (3.5)).

The P-V process as determined by the measured data in Table 3.2 can be plotted on a P-V diagram as shown in Figure 3.6. Using a spreadsheet or any equivalent mathematical software platform, it is easy enough to determine a curve fit equation for the P-V diagram. That equation is presented in Figure 3.6.

That equation can be used in the integration Equation (3.2). It is also possible to use the columns of volume and pressure to determine the work done in the process via a straightforward numerical integration. The integration proceeds by determining the area under the P-V process curve by calculating the area along the process curve at each step of the volume displacement. The result of such a calculation is shown in Table 3.3. The process proceeds as follows:

STEP 1. Determine the average pressure between each volume, V, step. For example, for the volume step between 300 and 361, the average pressure between 15 and 12 is 13.5. Proceeding to the next volume step, the pressure between volumes 361 and 459 varies from 12 to 9. Thus, the

**Pressure-Volume Diagram
of Expansion Process {Curve fit shown}**

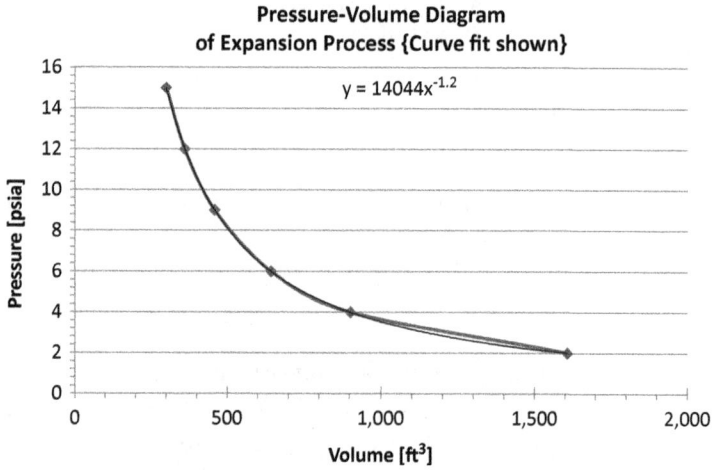

$y = 14044x^{-1.2}$

Figure 3.6.   P-V Diagram for Table 3.2.

Table 3.3.   Table with Example 3.1's solution for work by using
numerical integration.

V	Pbar	Pavg	P × DV
300	15		
361	12	13.5	824
459	9	10.5	1029
644	6	7.5	1388
903	4	5	1295
1608	2	3	2115
$Work = [(P \times v)_2 - (P \times v)_1]/(1 - n)$ -6420		$\sum Work =$	6650 3.5%

average pressure for that step is 10.5. Proceed in a similar manner until the
entire column is calculated.

STEP 2. Using the average pressure that was calculated in Step 1, multiply
this average pressure by the difference in the volume for that volume step.
This is the area under the curve in units of ft-lbf once the volume and
pressure units are converted to the imperial unit. Repeat this calculation for
each volume step.

STEP 3. Add the column of incremental areas that were calculated in Step 2.
The total is the total area under the P-V curve.

STEP 4. Perform a "sanity or reality" check by comparing the numerical
calculation with the work calculated from the equation for work, assuming

that the process is a polytropic process. The curve fit is clearly a power equation with a coefficient that was determined to be 1.2. Using this value as the polytropic index and by using Equation (3.5), the calculation indicates a value that is within 3% of the value obtained in the numerical integration. The accuracy can be improved upon by simply dividing the overall volume change into many more equal divisions. This should be of no surprise as it is based on sound mathematical principles from one of Newton's most famous inventions — the calculus.

This example of how a numerical integration can be completed will prove to be invaluable to the reader, as it will be very useful in determining one of the three terms in the first law equation. When combined with the internal thermal energy change, which can be very simply calculated by using Equation (3.3) (or as found in Table 3.3) and by knowing only the terminal temperatures, the third term: heat transfer can then be determined.

3.2   Process, process and process: Cycles

Whether one is buying or selling real estate, housing agents usually tell prospective buyers and sellers that there are three key factors to profitable and valuable real estate: location, location and location. In the engineering thermodynamics world, this refrain can be changed to: "process, process and process". The very early presentation and emphasis of the five basic thermodynamics processes that can be used to model all real-world applications of a Closed Mass Analysis demonstrates this author's respect for the need to understand processes in solving thermodynamics problems.

A thermodynamics application starts by identifying a process path that the system is designed to experience, as it starts from an initial condition and ends where the analyst has designed it to end. A thermodynamics process is analogous to a street map that determines the easiest, quickest or safest path from one point to another. Step 5 in the step-by-step procedure to solving any thermodynamics problem identifies the plotting of a process on a Cartesian coordinate system as being vital to the solution of the problem. The process path is commonly drawn on a P-v or a T-v diagram. This author finds the P-v diagram useful in solving thermodynamics problems and will use it in this section as a means of explaining the utility of diagramming the process paths.

Just as there are many routes to a destination on a map, one or more of the five most useful processes appear 100% of the time because they are replicable by mechanical elements that may constitute the system studied. The mechanical systems are typically engines in the very generic sense that

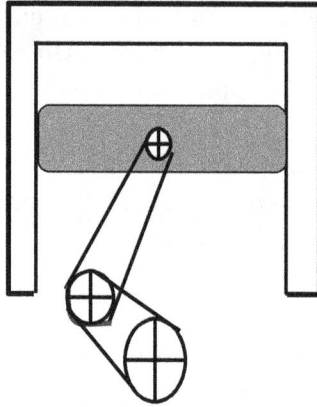

Figure 3.7.   Piston-Cylinder System.

they are commonly used to transform thermal energy into useful rotary shaft energy, which in turn can be connected to generators, etc. Thus, it seems reasonable to discuss the utility of process path diagrams by simultaneously presenting a prime example of how such process paths are used to produce useful results in engineering thermodynamics.

For this demonstration the reader is offered the development of a computer spreadsheet model of the two most common heat engines that are used throughout the world today: the modified or dual-pressure Diesel or Otto engine. The reader may be alarmed at being introduced to two major heat engines so early on in the study of thermodynamics, but they need not be, as the thermodynamics of these engine cycles are easy to understand given that they use only the five processes that are shown in Figure 3.1.

The diesel engine is an engine that does not use a spark plug to ignite the combustible air-fuel charge in the piston-cylinder, while the Otto engine cycle is the precursor engine developed in the late 1880s that uses one to ignite the fuel–air charge. However, both engines use the very familiar piston-cylinder arrangement, which can be easily drawn as shown in Figure 3.7.

A Closed Mass Analysis is suitable for this thermodynamics analysis only during the period when the intake and exhaust valves of the engine are closed. This occurs immediately after air is inducted from the open atmosphere into the cylinder and the valve shuts the flow path. The four-stroke engine is the most common engine in use today in transportation, as it is amenable to cleaner combustion and will be used in the engine model. The first stroke of four-stroke engine may be begin when the piston is exactly at the center at the top of the cylinder, achieving the minimum volume between the piston and the cylinder. The intake valve then opens to enable the air to be drawn into the engine using the piston's downward movement toward the bottom of the cylinder. This intake of air operates in a manner similar to respiration in the human body, which is performed by the joint actions of the diaphragm and the lungs: where the lungs fill with air drawn from the atmosphere when the diaphragm moves downward. Once the engine cylinder is filled, with the piston now at its bottom extension, the intake valve closes, and the second stroke, the compression stroke, begins. The compression stroke compresses the trapped air, and in so doing also heats the air to extremely high temperatures.

The third stroke, the combustion stroke, is achieved differently in diesel engines vs. Otto engines. In the case of the diesel engine, the high temperature created by the compression stroke is sufficient to ignite the special oil — diesel fuel — that has a relatively low auto ignition temperature. An Otto engine must use higher-octane fuel to avoid this pre-explosion until its spark plug is controlled to ignite the fuel-air charge at the correct moment. Upon combustion, the exploded high-pressure air is expanded, pushing on the piston to produce the desired rotary power for the engine. This power can then be used to properly charge a vehicle, drive a generator, turn a propeller or fulfil any suitable power demand. The fourth stroke, the expansion stroke (also known as the exhaust or discharge stroke), concludes when the piston reaches the maximum stroke or travel, at which time the exhaust valve opens in order to discharge the products of combustion — accomplishing their purpose of providing engine power. The sound of the exhaust valve opening gives the engine its typical unruffled noise at every other revolution (in a four-stroke engine), because though the expansion cannot reduce atmospheric pressure, it can have the exhaust valve opening at pressures as high as 50 to 60 psia. Once the discharge stroke is completed, the piston returns to the top of the cylinder once again and the cycle begins anew. The entire time duration for these four strokes may only be 0.01 to 0.02 seconds, depending on the speed (rpm) of the engine.

Two very important facts must be noted from several points made in the last sentence: (1) the Closed Mass Analysis can be performed for one or many processes that take either a very long or short duration to complete. The decision to use a Closed Mass Analysis does not depend on the time duration of the process, (2) the use of the word "cycle" in the last sentence is very important. A cycle is the collection of process paths (you need at least two) that will have the process start and end at the same point. All real-world engines share this common feature where they operate very quickly and in cycles, as noted, taking as little as 0.01 seconds to compete a cycle. However, the reader will also learn in Chapter 4 that certain important engines can operate in semi-cyclic operations. For example, the steam engine, which was used during the 1800s all over the United States and Europe, produced high pressure (250 psia) and high-pressure steam (900 psia). The high pressure produced imposed on pistons that were in turn linked to the train's drive wheels. The expanded steam was then discharged directly into the environment and did not return to the initial state that it was in. The return of the water vapor after the piston expander removed the motive energy from the high-pressure steam would have competed the cycle. However, such a device (called a condenser) would be large, heavy and expensive, and so it was deemed more suitable (and more economical) to simply discharge the expanded steam into the atmosphere. This meant water tanks were required along the train routes, to enable trains to replenish their water supply which was otherwise constantly being discharged into the atmosphere in the form of vapor. However, this commodity is no longer inexpensive to waste, plus environmental controls would not allow this form of wastage today. The steam process, while not precisely qualifying as a true cycle, nonetheless can be treated as if the system competed the cycle. A similar condition is true of the modern day gas turbine engine, in that air is inducted into the engine and discharged back into the atmosphere without actually returning to the starting point, except that the atmosphere is the common starting point for all thermodynamics activities that involve air as the working fluid in the cycle.

Let us proceed to the actual model construction of a modified dual-pressure diesel engine. The five familiar processes that constitute this important thermodynamics cycle are shown in Figure 3.8. These are:

- Process 0-1: Induction of air into the piston-cylinder.
- Process 1-2: No heat transfer (adiabatic) or almost no heat transfer (polytrophic) compression.
- Process 2-3: Constant volume pressurization.
- Process 3-4: Constant pressure heat input or temperature increasing.
- Process 4-5: No heat transfer or polytrophic expansion.
- Process 5-1: Constant volume exhaust valve opening.
- Process 5-0: Discharge of combustion products.

The Process Diagram shown in Figure 3.8 completes the first of the three graphics that are required to implement a thermodynamics solution. The two remaining graphics include the Property Table and the Process Table. The completed Property and Process Tables are shown in Table 3.4(a) and (b). The construction of these tables is straightforward. However, it is strongly

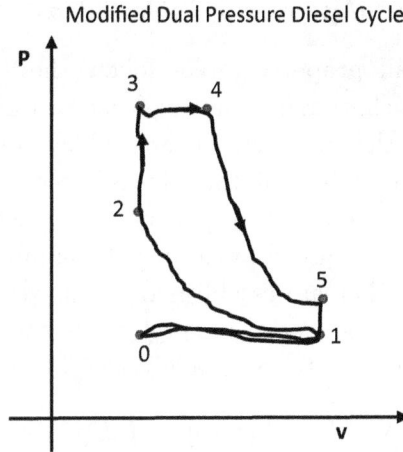

Figure 3.8.   Dual-pressure, modified diesel cycle.

Table 3.4.   (a) Property Table and (b) Process Tables for dual–pressure, modified diesel cycle.

PROPERTY TABLE FOR DIESEL HEAT ENGINE

	P (psia)	T(R)	v (ft^3/LBm)
1	14.7	534	13.46
2	344.6	1252	1.35
3	825.6	3000	1.35
4	825.6	4500	2.02
5	59.1	2147	13.46

(a)

PROCESS TABLE (units: BTU/LBm)

	HEAT	INTERN. ENERGY	WORK
1-2	-11	122.1	-133.0
2-3	297	297.1	0.0
3-4	358	255.0	102.8
4-5	14	-400.0	413.6
5-1	-274	-274.2	0.0
SUM	383.4	0.0	383.4

(b)

recommended to complete the Property Table after the Process Diagram has been constructed. The calculation of the Process Table is completed using the equations shown in Table 3.1, but only after the properties of P, T and the specific volume are first determined.

3.2.1   *Construction of property Table 3.4(a)*

The construction of all property tables follow the same rule: each state point on the property chart must have at least two known or easily determined properties from the statement of the problem and/or from reasonable assumptions. The reader will recall that this is based on the State Postulate, which states that the number of properties that must be known to determine all the others is determined by adding the number 1 to the number of reversible work modes that are established by the problem. In engineering thermodynamics, the work mode is expressed as the "area under the P-v curve on a P-v property chart", or in analytical form:

$$\text{Work, mechanical} = \int P(v) \, dv.$$

As indicated in Chapter 1, there are many such work modes, such as "stress–strain", surface tension, electric field and charge, and magnetic field and charge. If any combination of these work modes is present, then the number of properties that are needed to completely define the state of a system is again the number of these work modes present plus 1.

In this textbook, thermodynamics is considering a reversible mechanical work mode. Thus, there is a need for $1+1 = 2$ properties to be known before all the others are fixed and can be determined. With these two properties known, all the other properties can be determined from either property tables from textbooks, the Perfect Gas Law or the Equations of State Properties that have been theoretically or empirically determined, and are most likely either stored in hard copy or soft copy, such as those available via REFPROP (by NIST, as discussed in Chapter 2). Regardless the source of the fluid properties, knowing two properties of a state point is sufficient to look up all the other properties and thus prepare for the final calculation of the processes that involve heat transfer, work or thermal internal energy.

The completion of the Property Table is essential before attempting to "solve" the thermodynamics problem in any other manner. The need for this table to be completed is clear, for it contains all the properties that are needed to calculate the internal thermal energy and work for the individual processes. With these two energies determined, the First Law of Thermodynamics is all that is required to verify that the sum of these is equal to the heat transfer that has occurred in the process.

The best way of learning thermodynamics is to apply equations for several very common heat engine cycles. These equations are based on thermodynamics principles that have been presented thus far in this text. The next section proceeds to guide the reader to apply these equations but the reader must assist in this guide by actually preparing a spreadsheet (or equivalent computer platform that the reader is comfortable using) with the equations that are presented.

The reader may find that this author's application of thermodynamics principles and equations to heat engine cycles is done much earlier than in other thermodynamics textbooks. This is because the five basic thermodynamics processes have direct application in these heat engine cycles, and it is the expressed intention of this author to have the reader apply these principles as quickly as possible in order to witness firsthand how quickly the thermodynamics principles can be applied to what may seemingly be a difficult engineering analysis.

3.3 Heat engine cycle

The modeling of a modified dual-pressure Diesel Cycle begins with a print-out for the dual-pressure (modified) diesel engine cycle display shown in Figure 3.9. This heat engine cycle was first explained in Chapter 2 and is one of many successfully developed engine cycles since the advent of thermodynamics analysis. However, the dual-pressure Diesel Cycle, along with the Otto Cycle and the Brayton Cycle, are the dominant heat engines used throughout the world today. Most automobiles, isolated power generation systems, emergency generator sets, and engines in fire trucks, locomotives and lawn mowers, use heat engines that are based on either the Diesel or Otto Cycles. It is noted that in Figure 3.9, the inputs to the spreadsheet are identified in the cells with black borders. This should help the reader discern the input parameters of the engine from the output calculated by the cycle analysis. It is noted that many of the parameters are inputs to the analysis. These inputs would be available from the design specifications of the engine. For example, the piston-cylinder size must be identified, as must the bore, stroke and the number of pistons used in the engine. The speed is input, but so too is the operating temperature (T_3) for the engine.

The application of the First Law of Thermodynamics is best performed in a methodical manner aided by a Process Table. The Process Table for the diesel engine cycle is shown in Figure 3.9. It is a table that contains a row for each process that occurs with the problem at hand and contains enough columns to list all the properties that are important and needed to solve the problem. The most common and a more complete list of properties include: P, T, v, u, h and s.

The best way of demonstrating this solution technique that includes all of the steps itemized above is to solve a relevant thermodynamics problem.

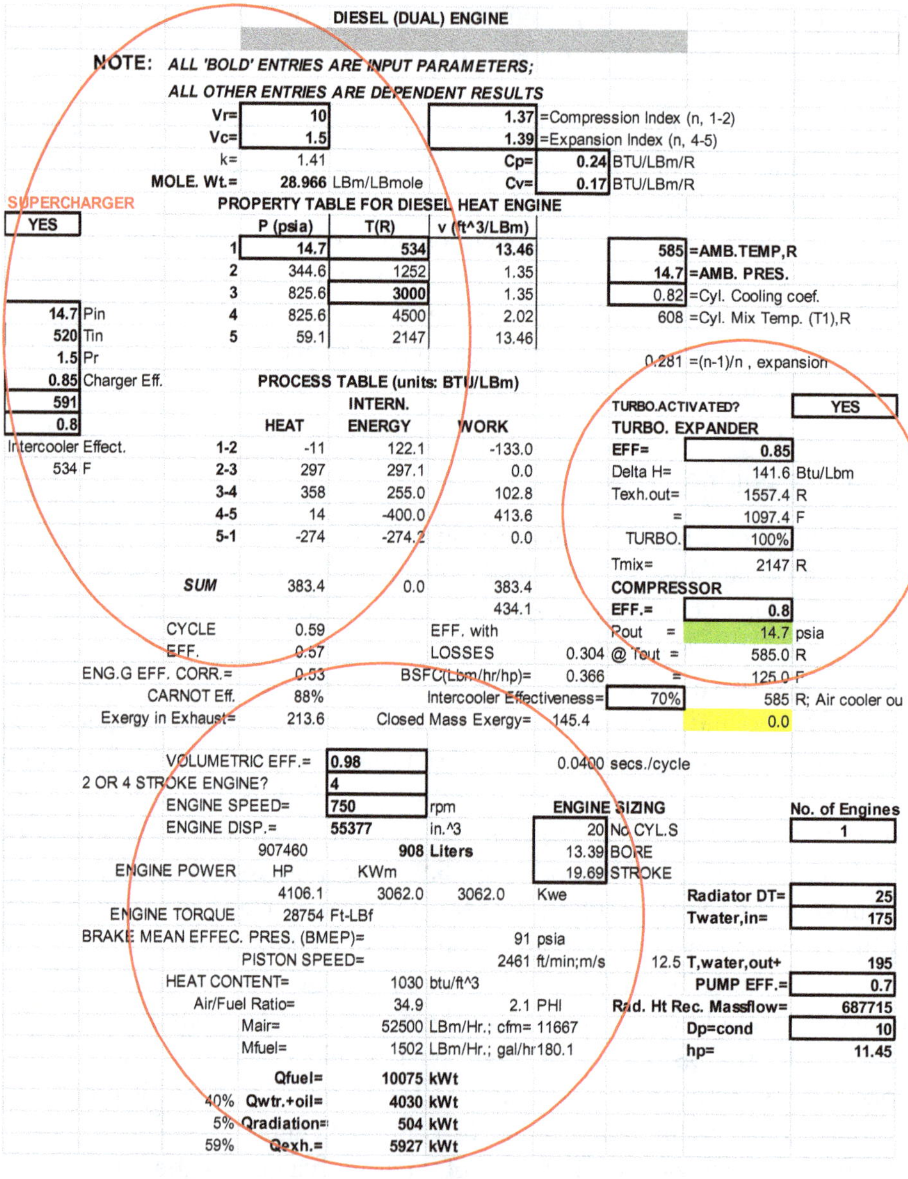

DIESEL (DUAL) ENGINE

NOTE: ALL 'BOLD' ENTRIES ARE INPUT PARAMETERS;
ALL OTHER ENTRIES ARE DEPENDENT RESULTS

Vr=	10	1.37	=Compression Index (n, 1-2)
Vc=	1.5	1.39	=Expansion Index (n, 4-5)
k=	1.41	Cp=	0.24 BTU/LBm/R
MOLE. Wt.=	28.966 LBm/LBmole	Cv=	0.17 BTU/LBm/R

SUPERCHARGER
YES

PROPERTY TABLE FOR DIESEL HEAT ENGINE

	P (psia)	T(R)	v (ft^3/LBm)
1	14.7	534	13.46
2	344.6	1252	1.35
3	825.6	3000	1.35
4	825.6	4500	2.02
5	59.1	2147	13.46

585	=AMB.TEMP,R
14.7	=AMB. PRES.
0.82	=Cyl. Cooling coef.
608	=Cyl. Mix Temp. (T1),R

14.7 Pin
520 Tin
1.5 Pr
0.85 Charger Eff.
591
0.8
Intercooler Effect.
534 F

0.281 =(n-1)/n , expansion

PROCESS TABLE (units: BTU/LBm)

	HEAT	INTERN. ENERGY	WORK
1-2	-11	122.1	-133.0
2-3	297	297.1	0.0
3-4	358	255.0	102.8
4-5	14	-400.0	413.6
5-1	-274	-274.2	0.0
SUM	383.4	0.0	383.4
			434.1

TURBO.ACTIVATED?		YES
TURBO. EXPANDER		
EFF=	0.85	
Delta H=	141.6	Btu/Lbm
Texh.out=	1557.4	R
=	1097.4	F
TURBO.	100%	
Tmix=	2147	R
COMPRESSOR		
EFF.=	0.8	
Rout =	14.7	psia
@ Tout =	585.0	R
=	125.0	F
	585 R; Air cooler ou	
	0.0	

CYCLE	0.59	EFF. with	
EFF.	0.57	LOSSES	
ENG.G EFF. CORR.=	0.53	BSFC(Lbm/hr/hp)=	0.304 @ Tout =
CARNOT Eff.	88%	Intercooler Effectiveness=	0.366
Exergy in Exhaust=	213.6	Closed Mass Exergy=	70%

VOLUMETRIC EFF.= 0.98
2 OR 4 STROKE ENGINE? 4
ENGINE SPEED= 750 rpm
ENGINE DISP.= 55377 in.^3
907460 908 Liters

ENGINE POWER	HP	KWm
	4106.1	3062.0 3062.0 Kwe

0.0400 secs./cycle

ENGINE SIZING
20 No CYL.S
13.39 BORE
19.69 STROKE
No. of Engines
1

ENGINE TORQUE 28754 Ft-LBf
BRAKE MEAN EFFEC. PRES. (BMEP)= 91 psia
PISTON SPEED= 2461 ft/min;m/s
HEAT CONTENT= 1030 btu/ft^3
Air/Fuel Ratio= 34.9 2.1 PHI
Mair= 52500 LBm/Hr.; cfm= 11667
Mfuel= 1502 LBm/Hr.; gal/hr180.1

Radiator DT= 25
Twater,in= 175
12.5 T,water,out+ 195
PUMP EFF.= 0.7
Rad. Ht Rec. Massflow= 687715
Dp=cond 10
hp= 11.45

	Qfuel=	10075 kWt
40%	Qwtr.+oil=	4030 kWt
5%	Qradiation=	504 kWt
59%	Qexh.=	5927 kWt

Figure 3.9. A screenshot from the Excel spreadsheet model of the modified, dual-pressure diesel engine cycle.

Among the most relevant of engineering thermodynamics problems is the modeling of a reciprocating engine that can be modeled using a Closed Mass Analysis. Recalling that a Closed Mass Analysis is one that applies the laws of thermodynamics to a system that does not change mass content by having neither an inlet nor an outlet stream of mass breaching the CV, which had been selected by the analyst to identify the relevant system in question. In the case of modeling a reciprocating engine, the system is the fixed mass of air that is trapped in the piston-cylinder engine once the intake valve is closed and before the exhaust valve is opened, even though the time duration for this is barely 0.02 seconds. As indicated previously, the time duration of the "before and after" is irrelevant.

In Figure 3.9, the spreadsheet display is shown in three distinct parts. Section 1 displays the basic cycle calculation using the five processes discussed in Chapters 1 and 2, and which will also be demonstrated in this chapter. Section 2 completes the engine cycle analysis by providing the opportunity to input the engine size (including bore, stroke, number of cylinders and speed). Outputs such as power and the brake mean effective pressure will also be given. Section 3 shows the addition of a turbocharger that is added to the engine cycle. This subject will be covered in the last part of this chapter during the presentation of open flow thermodynamics. Each of these parts will be discussed in turn, as the subject matter that is the basis for the equations used to construct each section is presented. At the conclusion of each section, the student will be able to change certain inputs and see the effect on the engine performance, primarily power output.

This particular engine cycle is the modified or dual-pressure diesel engine cycle. As noted earlier, this cycle can be ideally represented by five processes, shown in Figure 3.8 and reproduced again below, starting with the piston at the bottom of its stroke (position 1) when the air charge has been inducted into the engine and the intake valve is closed, thus effectively trapping the air charge for a fraction of a second while the five processes are started and completed to form a complete cycle.

3.3.1 *Section 1. Construction of the property table*

The processes proceed with the compression stroke from state point 1 to 2, which is essentially a polytrophic process. That is, it is almost a no heat transfer process and it also recognizes that no process in the real world of engineering can ever be truly "adiabatic". The second process (which happens from state point 2 to 3) is a constant volume process and represents the instantaneous combustion of the first amounts of fuel injected into the cylinder. Again, in the real world of engineering, this fuel injection begins at state point 2 and continues until state point 4, and is only idealized as shown in process 2–3 and then followed by 3–4, as the injection process continues until the fuel injection is terminated. Real-world engineering can never have the fuel injection at constant volume. However, for this idealized cycle analysis this is a reasonable assumption in order to proceed quickly to a theoretical solution that can later be understood to only be the ideal solution, and thus have it adjusted by several cycle correction factors. The ability to quickly program the constant volume, followed by the constant pressure process as shown for processes 2–3 and 3–4, has the advantage of being able to be done using the Perfect Gas Law and in two steps, as compared to what may otherwise have required a hundred or more micro steps along the actual P-v path from 2 to 4. The process path continues with the expansion stroke (that occurs from points 4–5), with the exhaust valve popping open at 5, which essentially suggests that the pressure in the cylinder at the end of the stroke essentially collapses to atmospheric pressure at state point 1. Thus, completing the cycle.

Each of the process steps are taken one step at a time, but only after the property table is first completed. The property table uses the Perfect Gas Law, and the compressibility factor, Z, is taken to be 1 to expedite the calculation. Note: Remember that the property table can be filled when only two properties for each state point are known, while using the Perfect Gas Law can get the third property. For example, starting with state point 1. The pressure and temperature are known from the ambient conditions. Therefore, the temperature is found with the Perfect Gas Law, using the equation:

$$T = \frac{(P \times v) \times Mole_{wt}}{R_u}.$$

The second state must have the specific volume known and then use the fact that the process is polytrophic in order to get the pressure. The specific volume is easily obtainable because the compression ratio for the engine is selected by the engine designer (you, the reader) to be 10:1 or 15:1, etc. The spreadsheet in Figure 3.10 shows a compression ratio, $V_r = 10:1$. This means that v_2 is v_1/v_r.

The pressure at state point 2 can then be determined by noting that the process is a polytrophic process, which means that the equation that couples the state points 1 and 2 is as follows:

$$(P \times V^n)_{start} = (P \times V^n)_{end}.$$

Thus, P_2 can be calculated.

$$P_2 = P_1 \times Vr^n.$$

State point 3 has the same specific volume: $v_3 = v_2$ and the only reasonable second property is to select the maximum temperature that the engineer designer would need to select in order to classify the material of the engine.

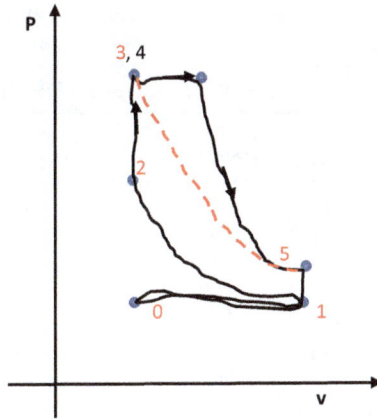

Figure 3.10.   The Otto Cycle transformed from the Diesel Cycle by setting $V_c = 1$ and thus $v_3 = v_4$.

For this case study, t_3 is selected to be 3,000 °R. With these two properties the pressure can be found from, you guessed it, the Perfect Gas Law.

The pressure at state point 4 is clearly the same as 3; thus, $P_4 = P_3$. The specific volume at state point 4 can be known again because the engine designer would be in control of how much fuel should be injected into the engine between 2–3–4. This quantity is identified as the cut-off ratio, V_c, and this parameter is simply defined as $V_c = v_4/v_3$. The larger this value the more fuel is injected into the engine, and the greater the power generated by the engine. In this case study the cut-off ratio is 3:1. Therefore $v_4 = v_3 \times V_c$. Thus, the pressure, P_4, and the specific volume, v_4, are known and the temperature can be calculated. The process from 4–5 is once again a polytrophic expansion and the relationship given for the process from 1–2 is used here, except perhaps the polytrophic expansion constant n_e may be different. The result is the equation shown below:

$$P_5 = P_4 \times V_r^{1/n}.$$

The final process from 5–1 is very simply determined to see that the process is a constant volume.

With all the properties determined, the property table is complete. The next step is to construct the process table. Each row of the process table is simply the application of the First Law of Thermodynamics for a Closed Mass System, but with a caveat: The Closed Mass System is worth repeating because an Open Flow System has a different equation for each of the processes. The reader is also reminded that the energy terms of heat transfer, work and internal thermal energy are all with units of energy per unit mass.

Table 3.5.   Process table for a modified dual-pressure diesel cycle.

	PROCESS TABLE (units: Btu/lbm)		
	HEAT	INTERN. ENERGY	WORK
1-2	−11	122.1	−133.0
2-3	297	297.1	0.0
3-4	358	255.0	102.8
4-5	14	−400.0	413.6
5-1	−274	−274.2	0.0
SUM	383.4	0.0	383.4

The process table is most easily completed by first calculating the thermal internal energy change. This is very straightforward for a perfect gas because the internal thermal energy is the product of the specific heat with respect to volume, C_V, multiplied by the temperature difference for each process. This is shown in the following equation:

$$C_V \times (T_{n+1} - T_n).$$

The work for each of the processes will use the equations for work, as summarized in Chapter 2 and repeated in Table 3.1 for easy reference.

Certainly, the constant volume process from state point 2 to 3 and 5 to 1 has work equal to zero.

The heat transfer for each of the processes follows from the application of the First Law, where:

$$\Delta Q = W + \Delta u.$$

The result of the completed process table is shown in Table 3.5.

It is noted that the sum of the thermal internal energies is zero, as it must be for the cycle. The net amount of work performed per cycle is determined by adding the column of the work terms. The heat terms must sum up to be equal to this work if the internal energy is zero, but the most important observation is to identify where the heat transfers are positive. At these processes, the heat transfer is positive because the heat is input into the cycle. The cycle efficiency is defined as the ratio of the net amount of work for the cycle divided by the sum of the positive heat transfer terms. For this case study the efficiency is:

$$\text{EFF} = W_{net}/\text{heat input}.$$

This efficiency term is put into the spreadsheet as shown in Table 3.6. The Carnot efficiency of the cycle is also added to the spreadsheet. The Carnot

Table 3.6. Engine performance parameters.

CYCLE	0.59	EFF. with				
EFF.	0.57	LOSSES	0.304			
ENG.G EFF. CORR.=	0.53	BSFC(Lbm/hr/hp)=	0.366			
CARNOT Eff.	88%	Intercooler Effectiveness=	70%		585 R; Air cooler ou	
Exergy in Exhaust=	213.6	Closed Mass Exergy=	145.4		0.0	

VOLUMETRIC EFF.=	0.98		0.0400 secs./cycle			
2 OR 4 STROKE ENGINE?	4					
ENGINE SPEED=	750	rpm	**ENGINE SIZING**		**No. of Engines**	
ENGINE DISP.=	55377	in.^3	20 No.CYL.S		1	
907460		908 Liters	13.39 BORE			
ENGINE POWER HP		KWm	19.69 STROKE			
4106.1		3062.0 3062.0	Kwe		Radiator DT=	25
ENGINE TORQUE	28754 Ft-LBf				Twater,in=	175
BRAKE MEAN EFFEC. PRES. (BMEP)=			91 psia			
PISTON SPEED=			2461 ft/min;m/s		12.5 T,water,out+	195
HEAT CONTENT=		1030 btu/ft^3			PUMP EFF.=	0.7
Air/Fuel Ratio=		34.9	2.1 PHI		Rad. Ht Rec. Massflow=	687715
Mair=		52500 LBm/Hr.; cfm= 11667			Dp=cond	10
Mfuel=		1502 LBm/Hr.; gal/hr 180.1			hp=	11.45

	Qfuel=	10075 kWt
40%	Qwtr.+oil=	4030 kWt
5%	Qradiation=	504 kWt
59%	Qexh.=	5927 kWt

Cycle is defined as the maximum efficiency that can be achieved by any heat engine cycle operating between two temperatures of the hot and cold reservoirs. For this case study the Carnot efficiency uses the cold ambient temperature and the hottest temperature in the cycle, T_4, in the equation:

$$\eta_{carnot} = 1 - \frac{T_{cold}}{T_{hot}}.$$

Once again, the reader is reminded that the temperatures must be in the absolute temperature scale of either Rankine, °R, or Kelvin, K.

3.3.2   Section 2. Engine performance parameters

The lower part of the spreadsheet in Figure 3.9 (with respect to the part noted as Section 1 in the previous review) provides more detail of the inputs and outputs from the engine cycle analysis that are common to the definition of an engine's performance.

The amount of power produced by the engine and the torque is the most common engine performance description. Not as familiar to the novice but equally important and relevant to the engine designer are the engine's Brake Specific Fuel Consumption (BSFC) and the Brake Mean Effective Pressure (BMEP).

Each of these will be defined and, with equations, presented in the next section.

3.3.2.1   *Engine power*

The power developed by the engine is by definition the energy generated over a period of time. The sum of the work energy per power stroke is shown at the end of the Work column. However, the unit of measurement for this specific energy is clearly Btu/lbm. In order to determine the power generated by the engine, this term must be multiplied by the amount of air and fuel mass in the cycle during the power stroke. The total displacement of the engine that can be filled with air from each power stroke is precisely known from the physical geometry of the engine. This geometry is typically given by identifying the bore, stroke and number of cylinders in the engine. But it is also important that the number of power strokes per unit of time be determined, in order for the continuous flow rate of air and fuel through the engine to be determined with the units of lbm/h or kg/s.

This can be very simply done by recognizing that the engine cycle shown in Figure 3.8 continually repeats itself for as long as the engine is operating. Each power stroke of the engine occurs in a time period that is measurable by knowing the revolution speed, rpm (rev.s/min), of the engine.

All of these variables can be combined to determine the flow rate of air and fuel through the engine by using the equation shown here:

$$\textbf{Power [kW]} = \textbf{W}_{\textbf{net}}\textbf{[Btu/lbm]} \times \textbf{Flow rate [lbm/h]}$$

$$\times \textbf{1/2545} \times \textbf{1/1.341,}$$

where:

$$\textbf{Flow rate} = \textbf{(Engine Displacement (in}^3\textbf{)/1728)} \times \textbf{(1/v}_1\textbf{)}$$

$$\times \textbf{RPM} \times \textbf{\{2/(2 for 4 stroke engine OR 1 for}$$

$$\textbf{2 stroke engine)\}}$$

$$\times \textbf{60 min/h} \times \textbf{Vol. Eff.} \times \textbf{Eff. Correction.}$$

The parameter, **Vol. Eff.,** is the volumetric efficiency of the engine. It is a variable that is a function of the engine speed, operating temperature and size, and must otherwise be determined by experiments on an actual piston-cylinder system. It may also be theoretically determined using computational fluid dynamics (CFD) coupled with heat transfer analysis. The volumetric efficiency may be seen to be a correction of the air density (rho $= 1/v_1$) that is used in the equation.

The air density is typically the ambient air density for a naturally aspirated engine and not a turbocharged engine. Naturally aspirated engines are engines that breathe in air from the ambient with the assistance of a blower or air compressor. The ideal situation is for the ambient air to be directly ingested into the engine at the ambient air temperature. Unfortunately, the air will actually be heated as it travels through the air filter and all of the closed piping that directs the air from outside of the vehicle to the engine cylinders before it is injected into the cylinder. Thus, with the air slightly heated, the density of the air is decreased and therefore the actual amount of air that can fill the geometry of the engine displacement is decreased.

With the decrease in air mass comes a decrease in the amount of fuel that can be burned with that air mass, and with a decrease in the amount of fuel that can be burned comes a decrease in the maximum power that the engine can produce. Thus, the lower the volumetric efficiency, the lower the power output from the engine. The volumetric efficiency is also a measure of how fast the air can be naturally inducted into the cylinder before the intake valve closes. For very fast running engines with high rpm, the time that the intake valves are open is very small. During this short period of time, the flow rate of air must be high enough to allow the engine cylinder to fill. Clearly, the faster the engine the less time is available for the air to be naturally aspirated into the engine. A relevant analogy for this would be a room full of people who need to leave the room via a single door as quickly as possible. If this door is held open, a maximum evacuation rate can be observed. If the door is replaced by a revolving door and the door turns very slowly, the speed of egress is reduced, though the room can still be evacuated. However, if the revolving door is spun very fast, the number of people that can fit through the door may be reduced to zero. Of course, this is not the best analogy because people have a relatively large volume with respect to the size of the door opening, as opposed to the size of an air molecule compared to the opening of the intact valve. However, it is clear that the speed of the engine can affect the volumetric efficiency of the engine.

Another very strong effect on the lowering of the air density is the heating of the incoming charge of air by the remnants of the products of combustion after the exhaust valve was opened and the piston properly pushed out most of the combustion products from the cylinder. The correct term is "most" and not "all" because the cylinder length is a little longer than the piston stroke, so as to enable the valve to have some space to lower into the cylinder without damaging the piston, and the piston must have a finite compression ratio, which is defined as the ratio of the volume at the point where the piston is at its bottom-most position, to the space remaining when the piston is at its top position.

For the spreadsheet solution the volumetric efficiency is taken to be 0.98, unless more relevant empirical data is available.

The last term, **Eff. Corr.**, in the equation is also an arbitrary value taken as 0.53 in this spreadsheet. This parameter accounts for the fact that the engine cycle analysis that is being performed is truly very ideal. The process paths are very clean and crisp, with constant volumes and pressures that would not actually occur in real engine performance. It is used here simply because these processes are easily modeled and allow the reader to experience a significant and relevant application of thermodynamics via a brief introduction of the First Law for a Closed Mass System. The actual value of the Eff. Corr. is made available by referring to the engine manufacturer's "cut sheet" (or engine specification sheet) and calculating the engine's cycle efficiency. Then, by using that engine's bore, stroke, number of cylinders, fuel heat content and estimates for the highest temperature in the cycle, a comparison is made between the ideal efficiency for the cycle, as calculated by the model, and the actual engine efficiency, as published by the engine manufacturer. That ratio is used to correct the ideal efficiency in order to more accurately model the engine.

The Brake Specific Fuel Consumption (BSFC) of the engine is another very common way of expressing the engine efficiency. However, it is defined by the actual physical quantities of an engine that has the most relevance: the power generated, Hp or kW, and the amount of fuel burned, lbm/h or kg/s, along with the heat content, Btu/lbm, of the fuel. The BSFC is calculated using the formula:

$$\text{BSFC} = \text{Fuel flow rate}/\text{Hp} \ [\text{lbm/h/hp}].$$

The efficiency can be determined from knowing the BSFC by using the formula:

$$\text{Efficiency} = 1/\text{BSFC} \times 2,545/21,000,$$

where the 2,545 Btu/hr per hp is a unit conversion and the 21,000 Btu/lbm is the high heat content of the fuel that is burned in the engine.

The high heat content refers to the fact that if all the products of combustion were to be cooled to ambient temperature and all of the heat was liberated, then the heat of condensing the water vapor as a product of the combustion process will have released its latent heat energy. A low heating value assumes that this condensation heat is not accounted for because it has little value. Unfortunately, the choice of using either of these terms for the heat content of the fuel is sometimes misleading with respect to the efficiency. In the United States, it is convention to use the high heating value when defining the engine efficiency. In Europe, engine, furnace and combustion system manufacturers may often use the lower heating value. As can be seen in the equation of efficiency, the lower heating value will calculate a higher efficiency, perhaps an unfair higher efficiency, unless the astute engineer recognizes the potential conflict in the definitions. The difference in the quoted engine efficiency when using one or the other could be as much as 8–10%.

There are several other important engine parameters that are also used in the engine design industry that can precisely distinguish one engine design from the other. These parameters are piston speed and Brake Mean Effective Pressure (BMEP). The piston speed is determined from the stroke and the rpm using the formula:

$$\text{PistonSpeed[ft./min]} = 2 \times \text{stroke[inch/rev]}/12 \times (\text{rev./min.})$$

The BMEP of the engine is the average pressure that the engine develops during the cycle to produce an amount of energy per revolution. The BMEP is basically the average pressure determined from knowing the energy, W_{net}, and the total swept volume of the piston.

The formula for the BMEP is:

$$\text{BMEP} = W_{net}[\text{Btu/lbm}]/(v_2 - v_1)[\text{ft}^3/\text{lbm}]*778[\text{lbf} - \text{ft/Btu}]/144[\text{in}^2/\text{ft}^2].$$

The worked example of a dual-pressure diesel engine provides a very convincing example of the First Law of Thermodynamics at work by using the closed mass methodology. It provides a straightforward method for modeling a common engine, with the ability to change the operating parameters via simple keystrokes of the computer model. It is recommended that the reader perform the following exercise.

3.3.3 *Example 3.2: Developing a digital engine model and using the model for engine analysis*

The purpose in having the student develop an engine model using Excel (or any other computer calculating platform) is to quickly see the effect of changes in the operating conditions of the engine without being burdened by repetitive and possibly error-prone calculations. With the engine model prepared, the reader can use it to provide the following calculations with only a minimum number of keystrokes to provide inputs to the model:

(a) Replicate the cycle condition shown below and submit your output page. This is your Base Case.
(b) Change P_1 to 12 psia, 14.7 psia, 24 psia and 30 psia, and for each of these pressures, copy the engine, W_{net}, per revolution output for each

calculation, so as to be able to plot W_{net} versus P_1. Be sure to properly label this graph.

(c) Restore P_1 to 14.7, but now change the T3 to 2000 °R, 3000 °R and 4000 °R, and again store the power for each case so that you can plot W_{net} versus T_3.

(d) Restore the model to the Base Case and now change the V_r to 7.0, 9.0, 12.0 and 15, and again store the output power for each case, and plot W_{net} versus V_r.

Once the student has performed this analysis, the student may be surprised to realize that the network per cycle, Btu/lbm, does not change with respect to inlet pressure, P_1. To be more specific, the temperature, T_3, V_r and V_c are not changed, and the processes from 1 to 2 and 4 to 5 are adiabatic processes. If these processes are polytropic (i.e., a process that the heat transfer is almost zero), then there is a small change in W_{net} for the cycle as a function of the change in P_1. This can be shown by deriving the expression for W_{net} by considering each of the work terms for each process. These equations are as follows:

Starting with the equation for work for an adiabatic (no heat transfer) process: $W = \frac{P_2 \times v_2 - P_1 \times v_1}{1 - k}$

For a reciprocating engine the Compression Ratio (Vr) can also be used to determine P2: $P_2 = P_1 \times V_r^k$

By substituting the equation for P_2 into the Work Equation and with some algebra, one will find:

$$\text{Work} \equiv P_1 \times v_1 \times (v_R^{(1-k)} - 1)/(1 - k)$$

But it is also true that: $\frac{T_1}{T_2} = (\frac{v_1}{v_2})^{(1-k)}$

From this relationship between T_1 and T_2 one can see that it also depends only on the compression ratio, Vr.

Lastly, it is understood that for the diesel engine the only other contribution to the work per cycle is the work output during the constant pressure fuel injection process from (3) to (4). But this work output is dependent on the volume difference which in turn is due to the cut-off ratio (Vc) such that:

$$v_4 = v_3 \times V_c \text{ and } Work_{3-4} = P_3 \times (v_4 - v_3)$$

But for the diesel engine it is true that: $v_3 \equiv v_2 = V_r \times v_1$

Thus, the adiabatic work equation is only a function of the compression ratio and not the inlet pressure, P_1.

The equation for the net work done per cycle is based on the work produced or absorbed by the five processes; two of which are constant volume processes with therefore no work (0) generated or consumed. The total net work done is based on:

$$Total\ Net\ Work = W_{4\,to\,5} + W_{3\,to\,4} + W_{1\,to\,2}$$

With some algebra and collecting of terms is determined to be:

$$Total\ Net\ Work = \left(\frac{R_u \times T_3}{Mole.Wt.}\right) \times \left\{\frac{V_c \times \left(\frac{V_c}{V_r} - 1\right)}{1-k} + (V_c - 1)\right\}$$

$$+ \left(\frac{R_u \times T_1}{Mole.Wt. \times (1-k)}\right) \times [V_r^{k-1} - 1]$$

It is important to note that although the Net Work per lbm of air charge and per revolution does not vary with inlet pressure, the amount of power output will change. This seemingly ambiguous and non-intuitive statement is made clearer when it is understood that the density of the air charge does change when the inlet pressure changes and thus the mass of the air charge per revolution does change. Thus, the amount of fuel that can be burned per unit of air changes, and therefore so will also the amount of power output; the latter is measured in hp or kW units and not power per unit mass of air charge.

From the equation for Total Net Work shown above, it can be observed that the net work is a function of three engine design variables that can be changed by the engineer to affect the engine's performance. These three variables are: Compression ratio, v_r, Temperature, T_3, and the cut-off fuel ratio, V_c. The equation also shows that the ambient temperature and the molecular weight of the gas, hence the type of fluid used in the engine, affects the W_{net}, which is not the same as the power output of the engine. The power output is determined by calculating how much air is passed through the engine at any time. The flow rate is dependent on engine revolutions per minute, rpm, the inlet pressure and temperature, as well as engine size or displacement. The inlet pressure and temperature will affect the engine power only because the air density will increase if the pressure is increased or the temperature, T_1, is decreased. This effect is the reason why turbochargers and aftercoolers are used to increase the power of the engine without changing the displacement.

The dual-pressure diesel engine model can be easily changed to another common engine: the Otto Cycle. The Otto Cycle is similar to the dual-pressure diesel engine in its use of adiabatic (or close to adiabatic) processes connected by two constant volume processes, as shown in Figure 3.10. The diesel computer model can be converted to the Otto Cycle model by simply using a fuel cut-off ratio, V_c, equal to 1. This essentially has $v_3 = v_4$ and thus the cycle avoids the constant pressure process. The result is the Otto Cycle. The student can now compare the engine cycle efficiencies and the net power for these two common engines. It is important to note that the compression ratio, V_r, for the diesel engine must be much higher than the V_r for the Otto Cycle. This is necessary because the temperature of the compressed air in a diesel engine must exceed the auto ignition temperature of the diesel fuel, and this can only occur when the compression ratio is typically greater than 12:1. The fuel used in an Otto Cycle must never auto ignite due to the temperature achieved after the compression process. The ignition must be caused and controlled by the spark ignition system integrated into the Otto Cycle.

Among the first observations from such a comparison will be the higher efficiency for the diesel engine compared to the Otto Cycle. Thus, the work output is higher, but so too is the Brake Mean Effective Pressure (BMEP) and the maximum operating temperature. The reader should now be able to confirm these two major observations.

3.4   Engine part-loading

All the thermodynamics cycle equations presented in this section have been used to model the design point engine specification. That is, the engine speci-fication that identifies the maximum power at the maximum speed. However, the common automobile engine is rarely used to its design point design spec-ification. That is, the engine is part-loaded for most of its operating hours. For example, a 454 V8 engine (used in Chevrolet's Corvette automobile series) can attain 450 hp at 6,000 rpm; but common driving requirements rarely require engine speeds to be above 2,500 rpm and operating at 200 hp. While this power and speed may be momentarily reached when the vehicle properly and safely accelerates into traffic or avoids an accident, the actual engine speed and power output while at cruising speed in "city traffic" are very low with respect to the design point.

How does the vehicle achieve this part-load condition? If the reader has successfully completed the Diesel and Otto engine models, he/she can demonstrate how the performance of the engine at part-load is obtained. The reader should use the engine model and input a new inlet pressure, P_1, of only 5 psia. That is, instead of the engine "breathing in" air at atmo-spheric pressure, the engine is forced to receive an air pressure that is sub-atmospheric; in this case, 5 psia. The reader should also reduce the operating speed to 2,000 rpm. The reader immediately observes that the engine power has been drastically reduced; from 450 hp to 47 hp. Clearly, reducing the speed and intake pressure has the effect of de-rating the engine's power, but not necessarily the engine's efficiency. In reality the engine efficiency will be reduced due to the presence of the almost constant friction forces that "rob" the engine's fuel of producing effective power.

But how is the atmospheric pressure intake of the engine reduced to only 5 psia? The answer is straightforward: a valve placed between the ambient air and the engine intake manifold causes this pressure reduction. This valve is often called the "throttle valve", "throttle" or even the "gas pedal" (which is connected to either to the fuel injector or carburetor of the engine, depending on the era during which the automobile was manufactured). The actual function of a valve and how it reduces the pressure of the fluid flow stream that enters will be much better understood once the student completes Chapter 4. For now, it suffices to say that the purpose of a valve is to reduce the pressure of the flow stream that enters it, which it does very well. By forcing air to pass through the valve before it enters the engine, and by connecting the engine to the gas pedal, the amount of power that the engine develops is completely controlled to match the power required at any instance of operation. This control includes the vehicle's engine speed as well. In fact, it would be accurate to call the driver of the vehicle the "engine governor". An engine governor is a term usually given to the mechanical and electrical device that controls the speed of the engine. Here, the driver performs this governing action by recognizing when the vehicle needs to operate linearly faster (with respect to the inertial road reference). If the linear speed of the vehicle needs to be faster, the driver increases pressure on the accelerator pedal. Conversely, he eases up on the pedal to go slower. The vehicle dynamics that are affected by the air fluid drag, mechanical friction of the engine, and friction of the tires on the road determine how much power is needed by the engine at any time to maintain that speed. The matching of the engine power and speed to the vehicle's linear speed is thus accomplished.

3.5   The Carnot cycle analysis

The reader will note from the previous development of the diesel engine that an expression for the Carnot efficiency was given using the ambient temperature and the maximum temperature, T_4, for the diesel engine and $T_3 = T_4$ for the Otto Cycle engine.

The development of the equations for the work energy, internal thermal energy and the heat transfer for the adiabatic and constant temperature processes provides the necessary information to actually construct the classical Carnot Cycle; it is after all a thermodynamics cycle (like the Diesel Cycle

Compression Ratio =	10			psia	R	ft³/Lbm
				P	T	v
3			1	20	600	11.1
			2	200	600	1.11
	4		3	9353	1800	0.1
			4	935.3	1800	0.7
2		1				
				Q	Δu	W	BTU/Lbm
			1 to 2	-94.72	0	-94.72
Process 1-2 is Constant Temp			2 to 3	-1.68	204	-205.675
Process 2-3 is No Heat Transfer			3 to 4	284.15	0	284.15
Process 3-4 is Constant Temp			4 to 1	1.68	-204	205.68
Process 4-1 is No Heat Transfer						
				189.43	0.00	189.43
eff.=	0.666667					
Carnot	0.666667					

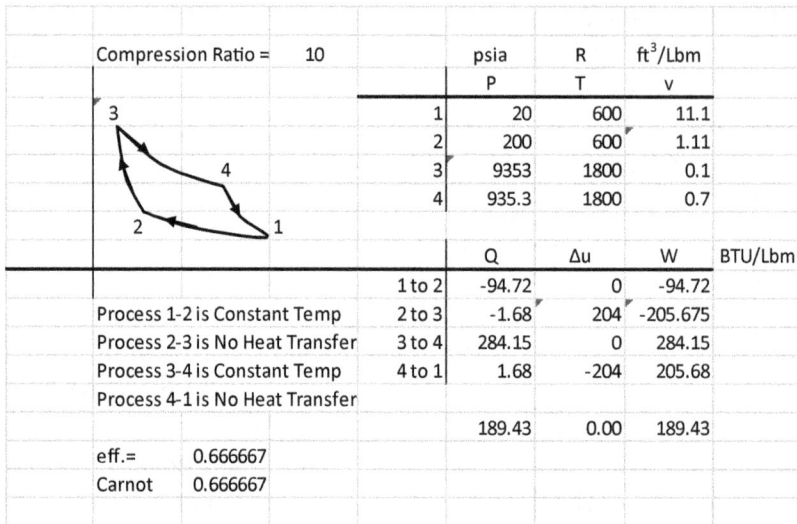

Figure 3.11.   The Carnot Cycle analysis example.

in its use of processes) that closes on itself, but it is unlike any other type of cycle because it is the cycle that provides the maximum efficiency for a heat engine between two temperatures: the high and the low temperature reservoirs.

The classical Carnot Cycle is emblematic of maximum efficiency. The magnitude of that efficiency can be determined by using the equations shown previously, starting with the work performed for each of the constant temperature and adiabatic processes, as given in Table 3.1 and repeated here:

$$Work\ at\ constant\ temp.(isothermal) = P \times v \times \ln\left(\frac{v_2}{v_1}\right),$$

$$Work\ with\ no\ heat\ transfer(adiabatic) = [(P \times v)_2 - (P \times v)_1]/(1 - k),$$

where k = Cp/Cv.

Figure 3.11 shows the result of the calculation that has been completed for the temperatures shown. For this cycle two efficiencies are shown. The first is a check of the maximum possible cycle efficiency, the Carnot efficiency given by the equation, $\eta_{\text{carnot}} = 1 - T_{\text{cold}}/T_{\text{hot}}$, The second cycle efficiency is determined by the equation of heat engine efficiency that is defined as the net useful work output of the cycle divided by the heat input to the cycle. As may be readily observed, these efficiencies are the same, as they must be, because the cycle shown is the Carnot Cycle.

$$Heat\ engine\ efficiency\ (\eta) = W_{\text{net}}/Q_{3\ to\ 4}.$$

The Carnot Cycle is constructed with two adiabatic processes and two isothermal processes. The Second Law of Thermodynamics that will be discussed in Chapter 5 will be used to derive the requirement that the efficiency of any cycle that is equivalent to a Carnot Cycle's efficiency must only have heat addition and heat rejection during the constant temperature processes. Only in this manner can the entropy of the two processes and the external source and sink of the heat transfer be summed to be equal to zero. The two isothermal processes may be connected by either the constant pressure or constant volume process. These two cycles are the Ericson (with two constant pressure processes) and Sterling Cycles (with two constant volume processes), respectively. These two cycles are shown in Figures 3.12(a) and (b). Included in Figure 3.12(a–b) are the respective Property and Process Tables for the cycles. The reader should refer to the process equations that are summarized in Table 3.1 and validate the work, internal thermal energy and heat transfer that are shown in the figure. These calculations are straightforward, particularly when for each of these perfect cycles, the constant temperature process that they all must share if they are to be perfect heat engines is that the process's internal thermal energy must be zero based on the equation for the internal energy changes, $C_v \times (T_2 - T_1)$ and $C_v \times (T_4 - T_3)$. The work process for the Sterling Cycle's constant volume process must also be zero due to the constant volume process. The adiabatic processes in the Carnot Cycle must obviously have zero for the heat transfer.

A Sterling Cycle engine has been built and tested by a number of manufacturers. However, the cycle efficiency is measured to be less than the maximum possible Carnot Cycle efficiency due to the friction and the need for heat transfer between the two constant volume processes to be 100% effective, an impossibility. It is an impossibility because the heat transfer between these two constant volume processes must be done at zero temperature difference. The best that can be done at an effectiveness that is less than 100%, is for some of the heat to be transferred between the two constant volume processes.

An example of perhaps the simplest of the four heat engines presented in this chapter is called the Lenoir Cycle, which is defined by only three processes (i.e., the minimum number of processes needed in order to have a true thermodynamics cycle) and is shown in Figure 3.13. Also shown are the Property and Process Tables for these processes. The Lenoir Cycle is most commonly used as the theoretical model of a pulsed-jet engine, that is, the periodic but rapid combustion of fuel (detonation) instantaneously produces a high pressure that can be expanded in a turbine or piston to produce useful work or high gas velocities. In a different scenario, the high

ERICSON CYCLE

Compression Ratio = 20

	psia P	R T	ft³/Lbm v
1	20	600	11.1
2	400	600	0.56
3	400	1200	1.1
4	20.0	1200	22.2

	Q	Δu	W
1 to 2	-123.23	0	-123.23
2 to 3	143.14	102	41.14
3 to 4	246.46	0	246.46
4 to 1	-143.14	-102	-41.14
	123.23	0.00	123.23

Process 1-2 is Constant Temp
Process 2-3 is Constant Pressure
Process 3-4 is Constant Temp
Process 4-1 is Constant Pressure

Cycle eff. from table=　0.5
Carnot Cycle Calc.=　0.5

(a)

STERLING CYCLE

Compression Ratio = 20

PROPERTY TABLE

	psia P	R T	ft³/Lbm v
1	20.0	600	11.1
2	400.0	600	0.56
3	800.0	1200	0.56
4	40.0	1200	11.1

PROCESS TABLE

	Q	Δu	W
1 to 2	-123.23	0	-123.23
2 to 3	102.00	102	0
3 to 4	246.46	0	246.46
4 to 1	-102.00	-102	0.00
	123.23	0.00	123.23

Process 1-2 is Constant Temp
Process 2-3 is Constant Volume
Process 3-4 is Constant Temp
Process 4-1 is Constant volume

Cycle eff. from table=　0.5
Carnot Cycle Calc.=　0.5

(b)

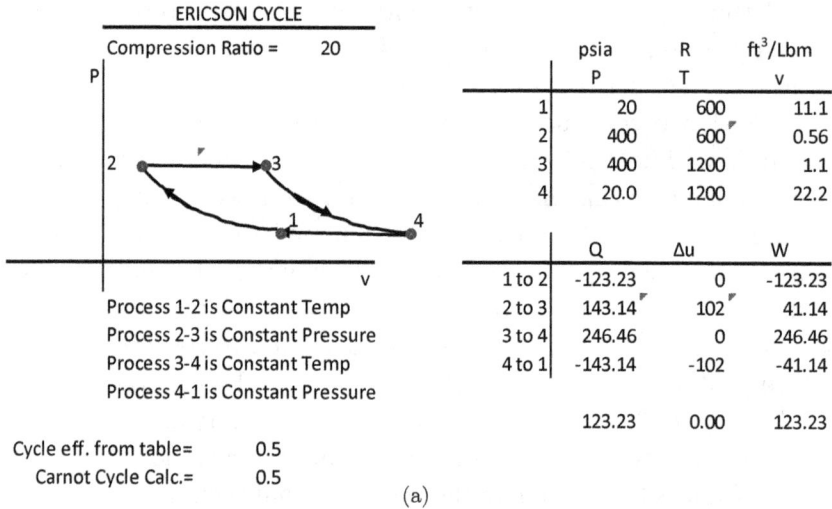

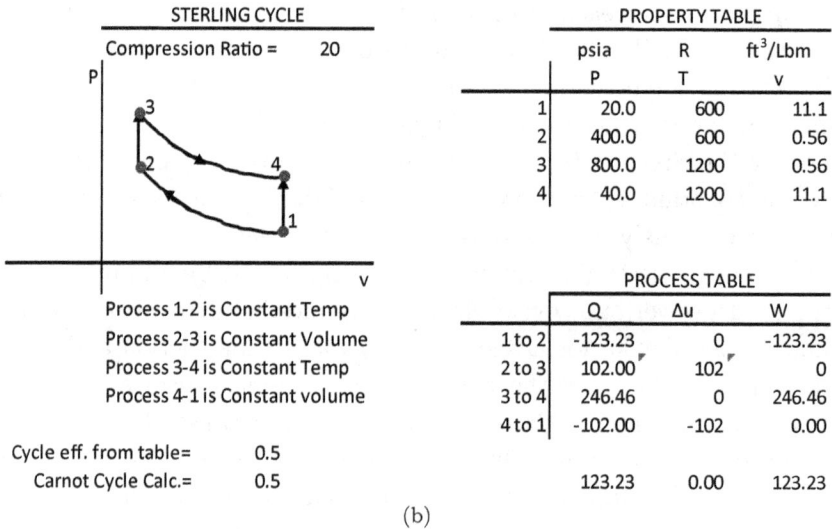

Figure 3.12. (a) The Ericson Cycle, where units for Q, Δu and Work are Btu/lbm and (b) the Sterling Cycle, where the units for Q, Δu and Work are Btu/lbm.

velocities are used to propel an aircraft or perhaps even a land-based rocket sled.

As may be observed in Figure 3.13, the Lenoir engine consists of a constant volume, constant pressure and a no heat transfer process. Once again, the Property Table is constructed with the necessary information that defines the state points in the cycle. The Process Table can then be constructed using the equations presented in this chapter for work, internal thermal

Compression Ratio = 10					psia	R	ft³/Lbm
					P	T	V
			1		14.7	1000	25.2
			2		14.7	100.0	2.52
			3		369	2512	2.52

	Q	Δu	W	BTU/Lbm
1 to 2	-214.70	-153	-61.70	
2 to 3	410.02	410.0207	0	
3 to 1	2.11	-257.021	259.13	
	197.43	0.00	197.43	

Process 1-2 is Constant Pressure
Process 2-3 is constant volume
Process 3-1 is No Heat Transfer

Cycle Eff.= 0.48
Carnot Cycle Eff.= 0.60

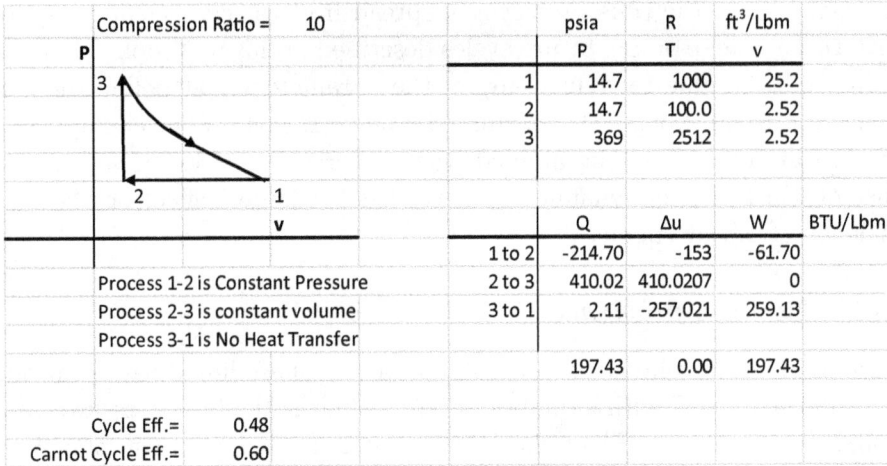

Figure 3.13.   The Lenoir engine cycle.

energy and heat transfer. Taking each process separately, the changes of the above can be determined. The sum total of the work (taking care to keep the sign convention resulting from the equation) is the net output work for the cycle. As a check for the accuracy of the Process Table, the sum of the three internal thermal energy changes must be zero, since the cycle begins at any of the three state points and must end at the starting point. The heat input or output from each of the processes can then be easily determined by applying the First Law of Thermodynamics for a Closed Mass System, that is, summing the internal energy change and the work for each process.

The Lenoir Cycle is much simpler than the Carnot (and its close "cousins", the Erickson and Sterling Cycles), Otto and Diesel Cycles. However, this simplicity comes at a cost in the lower cycle efficiency. As may be observed in Figure 3.13, the cycle efficiency for the Lenoir engine is theoretically lower than the Carnot Cycle's. The reader should expect the Lenoir Cycle to have a lower efficiency when one considers that the heat input to the cycle is during the constant volume process. That is, when the fuel in the engine is burnt, and stored chemical energy is released. For any heat engine to be equivalent to the Carnot Cycle's efficiency, all heat transfer into or out from the cycle must occur at the constant temperature process.

The reader is strongly encouraged to reproduce this cycle. Once the reader's cycle is validated by giving the same results as shown in the Property and Process Tables, the reader is encouraged to change the state point temperature and/or specific volumes to witness how the cycle efficiency and network output change. This last "chore" is in fact the goal and the reward for having spent the time preparing a spreadsheet (or equivalent computer platform) for these or any future cycles described in this textbook. That goal is simply to be able to change any of the parameters that define the heat engine and to quickly see the calculated results. If this is done enough times for each of the cycles programmed, the reader will develop a very strong sense of which thermodynamics property has the largest effect on the cycle efficiency of a heat engine.

3.5.1 *Example problems*

1. Determine the volume (ft^3) of 2 lbm of a two-phase liquid-vapor mixture of R134a at 50°F with a quality of 50%. What is the final pressure (psia) and Temperature (°F) of the R134a if it is heated at constant volume until all the fluid is at a saturated vapor state?
2. The pressure in a piston-cylinder contains 2 lbm of R134a and is kept constant at 50 psia. The initial temperature is 100°F. Heat is removed from the cylinder until the quality is 50%. Determine the heat transfer from the system. Plot this process on the following T versus v diagram.

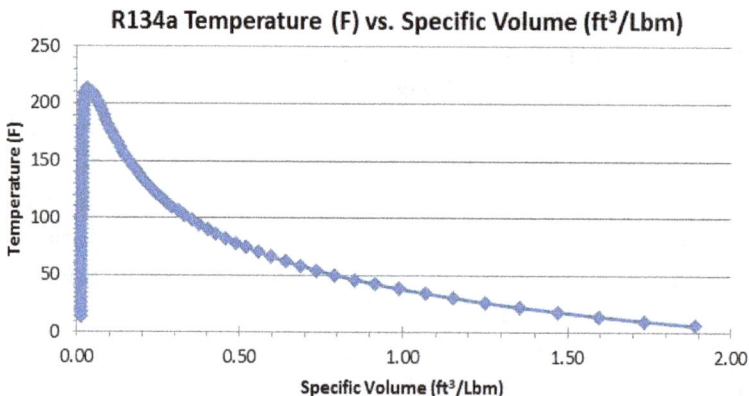

R134a Temperature (F) vs. Specific Volume (ft³/Lbm)

3.6   Transient analysis using the closed mass First Law of Thermodynamics

The First Law of Thermodynamics is typically applied during steady state conditions. Steady state in thermodynamics means that the system under study does not change its energy state, that is, its temperature does not change as a function of time. Certainly the First Law applies any time, whether it is applied to a system that is at ambient temperature (and is destined to achieve a higher or lower temperature as energy) or work is added or removed from it.

The First Law will account for the amount of energy that is being stored into the system, even as the system is starting to achieve what it was designed to do. For example, an automobile engine that is installed in your automobile that is parked on your driveway for many hours is at ambient temperature. All the metal in the system, the lube oil and grease, the electrical system, and the fuel, are all at ambient temperature state. When the ignition is turned on for your daily drive to work in the morning, the engine responds by having the engine cranked and the fuel fed to the engine, as air is inducted into the engine. The engine cycle processes are quickly established, as noted in the case study detailed above. However, all the state points used in the case study assume that the system is at the steady state and that the engine is operating at its full load design point. As fuel is burned in the piston-cylinder, the metal temperature of the piston and the cylinder, as well as the temperature of the cooling water in the engine block jacket, start to increase. The temperatures increase until there is a sufficient temperature difference between the walls of the cylinder and the cooling water in the jacket coolant passages, so as to enable the correct amount of heat transfer from the walls to the water and eventually from the hot water to the colder environment. These "sufficient temperature" differences are dictated by the surface area between the engine walls and the coolant, as well as the heat transfer coefficient that is produced by the chaotic high combustion gas swirls and the somewhat more streamlined water coolant. The heat transfer coefficients are determined primarily by empirical experiments by the designers of the engine block and/or heat exchangers. Suffice to say, you will be able to derive the heat transfer coefficient by using either analytical analysis (the subject of another course of study labeled "Heat Transfer") or empirical measurements on either full scale or similar scaled prototypes of the full-scale system. The heat transfer coefficient is defined as the measure of the ability of a fluid to be able to conduct, convect or to radiate heat from a warm to a colder body. If the heat transfer coefficient is low, then the surface area or the temperature difference must be larger to enable a fixed amount of heat transfer to occur. There are three forms of heat transfer: conduction, convection and radiation, as discussed in Chapter 1. In each of these heat transfer types, the amount of heat transfer is proportionate to the temperature difference, surface area and the heat transfer coefficient. Recalling the three basic heat transfer equations:

$$\dot{Q}_{convection} = h_{ht.convection\ coef.} \times A_{surface} \times \Delta(T_{hot} - T_{cold})$$

$$\dot{Q}_{conduction} = k_{thermal\ conductivity} \times A_{surface} \times \frac{\Delta(T_{hot} - T_{cold})}{\Delta X}$$

$$\dot{Q}_{radiation} = \sigma_{tstefan-Z\,Boltazman\,Constant} \times A_{surface}$$
$$\times F_{view\,factor} \times \epsilon \times (T_{hot}^4 - T_{cold}^4)$$

These equations become integral to the transient behavior of the system that is being heated or cooled during the initial transient period. Of course, it is entirely possible that the transient period is the duration of the entire period in which the application of the First Law is of the most interest. That is, it is certainly possible that a real-world engineering problem will never operate at a steady state, but only have a transient behavior from the beginning to the end of the event. For example, a rocket engine may never achieve steady state during the duration when its engines are fired. It is also possible that this is anticipated and hoped for in the event that if the steady state were to be achieved, the temperature at which it happens could be much higher than what the metal of the rocket engine could withstand.

In more routine applications of the First Law, however, the steady state temperature is achieved within a very small time period with respect to the operating time of the system. With reference to the example concerning the engine in the automobile that is transporting you to work, it likely reached steady state temperature within 10–20 minutes. If your drive to work is over 20 minutes long, the engine system will then be operating at temperatures that are not changing with time, assuming that the output of the engine is relatively steady. Of course, traffic conditions will dictate whether that assumption is correct as the effect of traffic plays an important role in the steady conditions of the engine. For example, an automobile moving slowly in traffic (or not at all) still needs to burn fuel to keep the engine idling. The heat from the cylinder at idling state is not as hot as its normal running temperature, but it is still much hotter than the ambient temperature, thus that heat must be rejected into the atmosphere. In fact, all of the fuel burned while the engine was idling and when the vehicle was stuck in traffic must be rejected into the atmosphere. If that ambient is too hot, the engine's wall temperatures may start to increase in order to help overcome what is likely to be low heat transfer coefficients when the ambient air through the radiator is not moving as fast as it should have when the vehicle was moving at 20–40 mph. The area of the piston wall certainly cannot change and if the heat transfer coefficient is compromised (made lower), then the only consequence according to the equations above is to have the temperature increase. But what "tells" the temperature to increase, so as to compensate for the lower heat transfer coefficient and the need to reject more heat to the ambient? The answer: The First Law of Thermodynamics. And its effect can easily be seen and measured by applying the closed mass method for the First Law of Thermodynamics.

Restating the First Law as follows, with the internal thermal energy of the system replacing all of the possible internal energies that the system may have:

$$\sum \frac{Q}{\Delta time} = C_v \times \frac{\Delta T}{\Delta time} + \sum \frac{W}{\Delta time}. \tag{3.7}$$

This equation should be familiar to the reader by now. However, there is a subtle difference where the time element is imposed on both sides of the equation. This changes the heat transfer to become a rate of heat transfer, and likewise, the temperature change of the system becomes a rate of temperature change. Now consider what happens when the equation of any of the three forms of heat transfer (or all three) are used on the left side of the equation. During the first moments of starting the system (think of the engine's piston-cylinder as an example), the ability of the walls of the cylinder to be cold enough to reject the heat from the engine is virtually zero or close enough to zero to be reliable. The recourse for that is for "fuel energy input" to be stored in the walls of the engine or the water in the engine. As the next time increment occurs, the walls of the cylinder start to heat up and will be able to reject more heat into the environment or to a nearby colder medium, to enable the heat exchanger. This increase in wall engine temperature continues until the amount of energy being stored by the wall is equal to the ability of the heated wall to reject that heat into the environment or coolant that surrounds that wall. At that moment, the heat rejection by one or all of the heat transfer modes equals the heat input, and the wall temperature stops increasing.

All this can best be demonstrated by our engine cylinder example from earlier, or even an espresso coffee maker; the worked example of the transient temperatures that occur to the coffee maker is in Case Study 1. The common elements in these examples are: the presence of metal that is heated by a form of heat transfer, and that if the heat transfer coefficients are long and high enough, then the system of the metal train track, engine cylinder or espresso coffee maker will achieve a steady state. The solution method is the same, thus the First Law of Thermodynamics applied to the Closed Mass System is also the same. For further analysis, the reader can either rely on Isaac Newton's calculus or use a spreadsheet solution in the event that the system of differential equations is nonlinear and therefore somewhat difficult to achieve a closed form answer. The spreadsheet platform also enables an immediate transition to the plotting of the results, that is temperature versus time, for the mass of the system. Another worked example for this chapter consists of a train track and an iron rail that is subject to solar energy flux throughout the day. The energy flux from the Sun continues to change with the time of day and with the ambient temperature as well as shading and location above the equator.

The problem to be solved in this section is determining the transient temperature of a train rail. The analysis proceeds by using the spreadsheet output shown in Table 3.7. The parameters that are to be either held constant or easily changed are shown in the cells that have a bold perimeter. As seen in Figure 3.15, there are five columns, starting with the time column on the left. The next columns include the rail temperature, the heat transfer into and out of the rail, and the rate of temperature change, $\Delta F/s$. The initial time period, $T = 0$, has the rail at the ambient temperature, shown in this example to be $100°F$. The heat transfer into the rail is due to the solar energy flux that is prevalent at that location. The solar heat flux is given as $+200 \text{Btu/h/ft}^2$. The heat transfer out of the rail is determined from the radiation equation given here as:

$$Q_{radiation} = -\sigma \times F_v \times \epsilon \times (T_{rail}^4 - T_{ambient}^4),$$

where $\sigma = 0.1718 \ e{-}7 \, \text{Btu/h/ft}^2/R^4$, $F_v = $ view factor $= 0.75$ (because three-quarters of the rail surface is exposed to the ambient to be cooled), and ϵ is the emissivity of the rail that is assumed to be 1 because the rail can be considered a highly polished surface.

The temperature change with time ($\Delta T_{rail}/\Delta \text{time}$) is solved using the First Law of Thermodynamics for this Closed Mass System with the equation:

$$\sum \frac{Q_{radiation} + Q_{solar\ flux}}{\Delta time} = M \times C_v \times \frac{\Delta T}{\Delta time} + \sum \frac{W}{\Delta time},$$

where Work $= 0$.

Then, solving for ($\Delta T_{rail}/\Delta \text{time}$):

$$\frac{\Delta T}{\Delta time} = \frac{Q_{radiation} + Q_{solar\ flux}}{\Delta time \times M \times C_v \times 3600s/hr}.$$

A more traditional presentation is shown in the following equation, wherein the rate of heat transfer is represented by: $\dot{Q}$

$$\frac{\Delta T}{\Delta time} = \frac{\dot{Q}_{radiation} + \dot{Q}_{solar\ flux}}{M \times C_v \times 3600s/hr}.$$

The solution for the rate of temperature change is then added to the temperature at the start of the heat transfer process to get to the next temperature at time, t_{n+1}.

Table 3.7.   Input and output for the transient train rail analysis example.

Width=	4							
Length=	12	inch						
Height=	4	inch						
Wall Thickness=	0	inch						
Metal Density=	489	lbm/ft^3						
Water mass=	0.00	lbm						
Vol.=	0.2618	ft^3						
Rail Mass=	54.333	lbm						
Specific Heat metal=	0.12	Btu/lbm/F						
Water Specific Heat	1.0	Btu/lbm/F						
Heat Transfer Coef.=	5	Btu/h/F/ft^2						
Time Increment=	50	sec.s						
Q,conv. & Qrad. heat loss=	5	Btu/h/F	1.54E-09	Btu/h/R^4				
Tambient=	100	F						
Solar Heat Flux,input=	200	Btu/h/ft^2				DT/Dtime		
232.5 minutes		Time	Trail	Qin	Qout	F/h		
	0.00	0	100	200	0.00	30.7		
	0.83	50	100.4	200	2.59	30.3		
	1.67	100	100.8	200	5.15	29.9		
	2.50	150	101.3	200	7.68	29.5		
	3.33	200	101.7	200	10.18	29.1		

The next time increment row is then repeated in order to recalculate the rate of temperature change for that time step. This is repeated for N steps until the transient temperature is observed to become zero, as evidence of having achieved a constant temperature for the rail. The reader is encouraged to duplicate this analysis and use the results shown in Figure 3.14 to validate the program.

The reader is also encouraged to consider the large number of analogies that can be constructed for the transient train rail analysis example and other exciting engineering examples, such as the piston in the piston-cylinder example, which has been analyzed per revolution for its work using the Closed Mass Analysis method. The piston, like the rail, proceeded through a temperature transient until it reached a thermal equilibrium thermal state. The same Closed Mass Analysis is used in the same manner except for the heat transfer terms and perhaps the need to break up the piston into smaller pieces. However, each of these pieces is treated the same way, except that with the many pieces came many equations with the same number of unknowns. Ultimately, this results in a matrix algebra solution technique that has an nXV matrix to solve an n number of unknown temperatures.

Train Rail Temperature Transient in an Ambient of 100 °F and Solar Flux= 200 Btu/h/ft^2

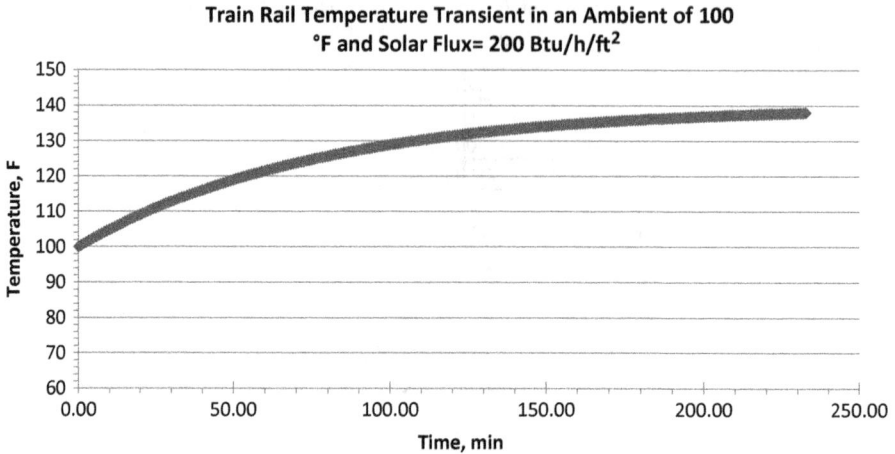

Figure 3.14.   Results of the transient train rail analysis example.

Similar types of analyses are treated in typical Heat Transfer textbooks. In the same manner, the engine cylinder (or crankcase) can be treated as a transient mechanical system, as can each gear in the transmission or any other mechanical parts of the engine. The difficulty in performing an accurate analysis is the need to break the mechanical elements into many parts and being able to accurately model the heat transfer into or from the mechanical element. Often, the heat transfer involves all three of the heat transfer modes.

3.7   Open flow analysis using the First Law equation

The Closed Mass Analysis methodology and application technique of the First Law of Thermodynamics provides insight into the common applications of thermodynamics, for example, the piston-cylinder system. Of course, most readers will likely expect to see the First Law applied to an engine system by broadening the CV, which is the imaginary boundary that is required to identify the system of interest — usually to include the entire engine and not just the air-fuel that is closed within the piston-cylinder when the intake and exhaust valves are closed. The expectation is that fuel and air is entering the engine via the fuel tank and pump, with the air intake coming through the grille of the vehicle and the air filter. This very common expectation cannot be handled by using a Closed Mass Analysis, because there is clearly fluid flow (fuel and air as well as the exhaust products of combustion) through the engine. To apply the First Law of Thermodynamics to this application, the analyst must consider using the Open Flow Analysis methodology. In fact, in the world of thermodynamics problems, the reality is that more

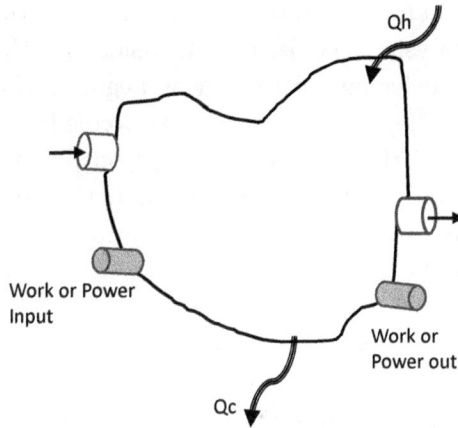

Figure 3.15. Open Flow System representation.

than 80% of the problems are best not handled by using only an Open Flow Analysis technique. Certainly the First Law is the same, but its expression is somewhat different in order to take into account the need to literally push a fluid stream into a system environment, and then to have it push out or be pushed out from the environment.

This "push" or "being pushed" effort is called the Flow Work and is identified, as shown in Figure 3.15, as the need to overcome the P-v product of the pressure and the specific volume (inverse of density): $P \times v = \text{Flow} - \text{Work}$. As always, watch the units!

The other most obvious observation is the use of the rate of change of mass, heat and work, which is expressed as: M dot ($\dot{M}$), Q dot ($\dot{Q}$) or Work dot ($\dot{W}$), which is also known as power. The units of the specific energy rate per unit mass are therefore: Btu/h and Btu/h/lbm, or kJ/s, kW or kJ/S/kg.

The full expression of the First Law of Open Flow Systems is then observed to be:

$$\sum \dot{Q} + \sum_{in} \dot{M} \times (h + KE + PE)$$

$$= \sum \dot{W} + \sum_{out} \dot{M} \times (h + KE + PE)_{cv} + \frac{\partial U}{\partial t}. \qquad (3.8)$$

The summation sign, $\sum$, is the sum of the parameters shown within the parenthesis. However, the analyst must ensure that the heat transfer into or out from the CV is given a positive or negative sign. The same is true of the power into or out from the CV and the rate of internal energy change of the CV $(\Delta E/\Delta t)_{cv}$.

The sign convention for heat transfer, power, etc., is the same as it was for the Closed Mass Analysis. That is, heat transfer into the CV is considered positive and heat transfer out of the CV is negative. Power delivered into the CV is considered negative, and the convention is to assign a positive sign if power is developed by the CV and is thus an output of the CV. As before, power can be best thought of as a rotating shaft that represents the rate of energy or power that can raise a weight in a gravity field. A rotating shaft represents a turbine of a pulley that can easily be visualized as the raising of weight in a gravity field.

The reader will also note that kinetic and potential energy terms have taken a central place in the equation of the First Law applied via an Open Flow Methodology as compared to the Closed Mass Analysis, which uses only three main energy parameters: heat, internal thermal and work energies. The kinetic and potential energy parameters are much more useful in an Open Flow Analysis due to the inlets and outlets, which with respect to the CV boundaries may be at different heights to a chosen datum, and the open flow assumes that the fluid streams have a velocity and hence a kinetic energy. Of course, it will also be evident after some thought (or some experience of applying the open flow equation) that the kinetic and potential energies for most applications are either small quantities compared to the internal thermal energy and flow work (i.e., the product of pressure and volume, or Pv) and power or heat transfer rates, or that they are going to cancel each other out because the inlet and outlet streams are similar in magnitude.

This last observation should help the reader realize that the left side of the first law equation is a "bookkeeping" of all the streams that flow into the CV, while its right side keeps track of the streams that exit the CV. The term "bookkeeping" is used because once the CV is selected, the analyst must carefully and clearly label the streams that are coming in and out of the CV. This places an emphasis on the need for a CV to properly use the First Law with either the Closed Mass or Open Flow Methodology. The reader is also reminded that the mass flow rate is expressed outside the sum of U + PV + KE + PE, thus these terms must all be measured in Btu/lbm. Also, when the mass flow rate is multiplied by these terms the correct resulting units are Btu/h, or energy per unit of time as required.

The last term in the equation is the "catch all" for any rates of energy of the CV system that have not been included in the other terms used in the equation, such as nuclear, electrical, chemical, magnetic, spring energy, etc. The reader will recall that the Closed Mass Analysis also included a "file Cabinet' catch-all that included all these energies. This was later reduced to only the internal thermal energy, which turned out to be the most common of the energies employed to solve most thermodynamics problems — the most common being the cylinder-piston system. It must also be noted that this last term can cause havoc during the application of the First Law for problems, because the rate at which the energy changes with time is very difficult to determine or measure for extensive systems. However, it needs to be used in cases where there is only a stream (or multiple streams) coming in to but not out of the CV, or vice versa. It is because the CV is likely to be changing with time and the last term in the equation will account for that energy change per unit of time. An example is given in Example 3.3. This solution technique is very similar to the transient analysis that was demonstrated for the train rail.

This difficulty with the last term can be eliminated for most thermodynamics problems, simply by requiring the process to be performed at steady state conditions. Steady state is defined as having the energy within the CV not changing with respect to time. That is not to say that the energy is zero or of any magnitude for that matter; it means that the internal energy of whatever type is not changing throughout the duration of the thermodynamics analysis, for which the First Law of Thermodynamics is in its open flow expression. The reader may also recall that in his/her capacity as an analyst, he/she can determine when the thermodynamics analysis begins and ends. However, it is important to pick a time where the internal energy of the CV is not changing with time. This is akin to starting and finishing the analysis on an engine, only after it has been heated from a cold start and then kept running at a constant fuel flow rate, thus having a constant power output. This process may take one to ten hours, but once the steady state is achieved, the last term can be set to zero and the First Law of Thermodynamics will become much easier to handle.

The reader may have noticed that the transient analysis presented in the last section using Figure 3.14 is an example of the steady state that was discussed in the last paragraph. Figure 3.14 presents the rail temperature as a function of time. The temperature is observed to achieve a steady state temperature, i.e., the temperature of the rail does not change with time after 150 minutes. This has occurred because the heat transfer into the rail is exactly equal to the amount of heat energy leaving the rail once it has reached a temperature of 130°F. This is therefore the steady state temperature.

Continuing with the description of the first law equation for an open flow process, the final observation to be made is the collection of the two terms, $U + PV$ (or $u + Pv$), for specific energy and specific volume units, or the internal thermal energy, u, plus the flow work, pV. This sum appears every time an Open Flow Analysis is performed, so much so that this collection of terms has been given its own name: enthalpy.

For the Open Flow Analysis, the thermodynamics property of enthalpy is what internal thermal energy is for the Closed Mass Analysis. That is, the enthalpy is a basic parameter that is used most often to calculate the power, heat transfer or internal energy changes of the CV. Enthalpy can be represented by the equation, $\Delta h = C_p \times \Delta T$, just as how internal energy can be represented by $\Delta u = C_v \times \Delta T$ for ideal gases.

There is another very important similarity between the Closed Mass and the Open Flow Methodology for applying the First Law of Thermodynamics. For the Closed Mass Analysis, the reader may recall that there are five basic process paths — constant pressure, constant volume, constant temperature, no heat transfer and almost no heat transfer — that need to be learned, understood and applied. Similarly, in using an Open Flow Analysis there is an infinite number of ways that a thermodynamics process in an open flow CV can proceed from a point 1 state point to a point 2 or 3 or even n state points. However, there are only seven basic mechanical systems that can actually be made to perform as expected in this "real world" of engineering. These seven systems, along with their icons and block diagram that represent them in a thermodynamics diagram, and the assumptions that must be valid in order to use these equations are shown in Table 3.8. The author can also be convinced that an eighth device: a mixing tank should be added to this list. However, a mixing tank can be thought of as a heat exchanger without a wall that separates the hot fluid from the colder fluid. The author has a standing offer for the reader: examine any mechanical system and let me know if this mechanical system has a mechanical element that is not one of the seven elements shown in Table 3.8.

Regardless of whether there are seven or eight mechanical systems, all simple or complex systems that are used throughout the world today are constructed using one or more of these special mechanical systems, to perform single or multiple processes. These processes can still be closed onto themselves and thus form a continuous cycle, or they can be started and then stopped at a different state point. The cycle is defined as a process that begins and ends at the same state point. For this reason, the cycle must have at least two processes that can be produced by two of the components shown in Table 3.8. Figures 3.16 to 3.23 illustrate many of the most significant thermodynamics cycles that utilize one or more of these seven mechanical devices. These cycles are named as shown in the figures. There can be many more such systems that generate power, provide cooling, produce chemical products, or in general, transform one form of energy into another or into a

Table 3.8.   The seven mechanical devices that can be
used to construct all engineering systems.

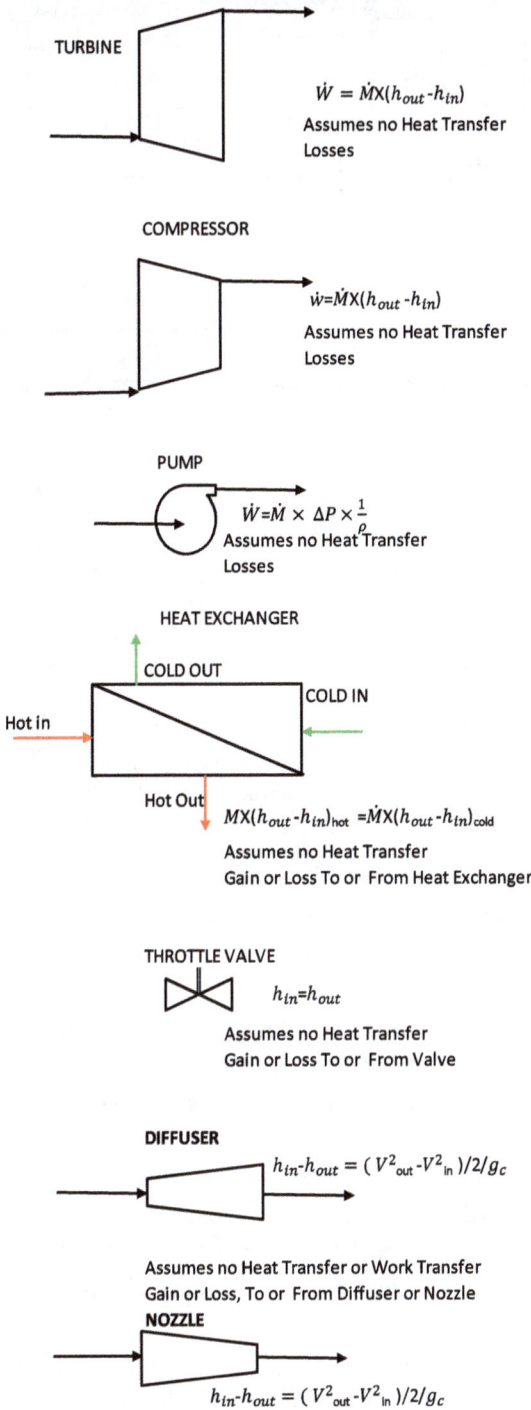

TURBINE

$$\dot{W} = \dot{M}X(h_{out} - h_{in})$$

Assumes no Heat Transfer
Losses

COMPRESSOR

$$\dot{w} = \dot{M}X(h_{out} - h_{in})$$

Assumes no Heat Transfer
Losses

PUMP

$$\dot{W} = \dot{M} \times \Delta P \times \frac{1}{\rho}$$

Assumes no Heat Transfer
Losses

HEAT EXCHANGER

COLD OUT

COLD IN

Hot in

Hot Out

$$\dot{M}X(h_{out} - h_{in})_{hot} = \dot{M}X(h_{out} - h_{in})_{cold}$$

Assumes no Heat Transfer
Gain or Loss To or From Heat Exchanger

THROTTLE VALVE

$$h_{in} = h_{out}$$

Assumes no Heat Transfer
Gain or Loss To or From Valve

DIFFUSER

$$h_{in} - h_{out} = (V^2_{out} - V^2_{in})/2/g_c$$

Assumes no Heat Transfer or Work Transfer
Gain or Loss, To or From Diffuser or Nozzle

NOZZLE

$$h_{in} - h_{out} = (V^2_{out} - V^2_{in})/2/g_c$$

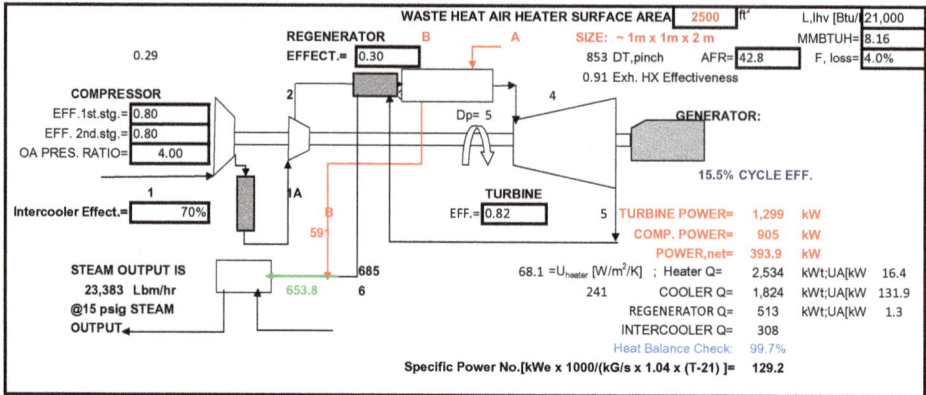

Figure 3.16.   Gas turbine engine operating on the Brayton Cycle (see Case Study 4).

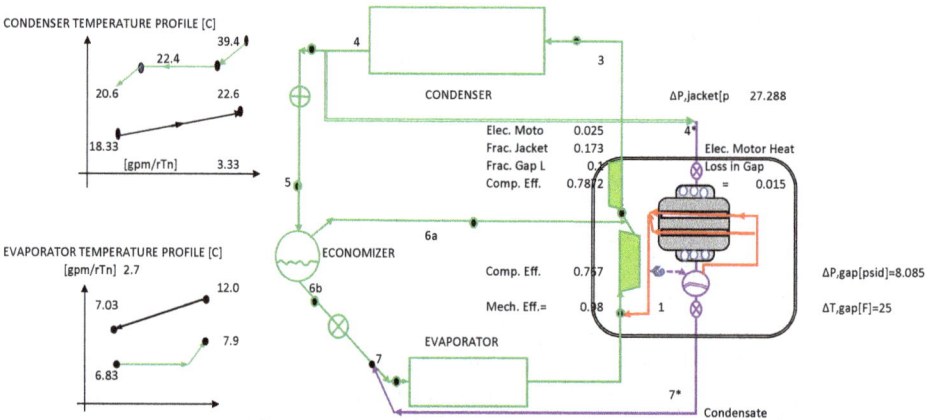

Figure 3.17.   Vapor Recompression (Refrigeration) Cycle using a Cascade or Economizer System (see Case Study 8).

desired product. The purpose of showing these cycles here is two-fold: (1) to emphasize that though such systems may look complicated, they are actually relatively straightforward to follow once it is understood that the First Law applies to a CV around one (or more) of the devices, and (2) that the reader will see these cycles again in the case studies that accompany this textbook, wherein each cycle will be explained in more detail.

The work, heat transfer, enthalpy or energy state of the process or the CV can be well defined by the open flow equation of the First Law of Thermodynamic. The First Law can even be made simpler by allowing and accepting a few reasonable assumptions, as long as it is understood that these assumptions may not be strictly true in the real world of engineering applications,

Figure 3.18 (Rankine Cycle System) — diagram labels:

Critical Pressure [Mpa]	3.7
Critical Temperature [C]	154

TURBINE-GENERATOR

Control Valve   R 101   Shut-OFF Valve

CYCLE EFF.S

0.8	Eff.t-s	Thermal 15.0%
0.95	Mech.xGear Eff.	Mechanical 14.2%
296	Power (kWm)	
0.96	Eff. Elec.	Electrical 13.7%
284.4	Power (kWe)	Net 12.1%
252.3	Net Cycle Power	
0.18	Dia (m)	
15000	RPM	0.75 Ns
3	No. of Stg.s	7.303 Ds

ΔP valve 35 kPa   R201

ORC VAPORIZER   By Pass Valve

(Hot Side)   ΔP[kPa] 1.5

HEAT SOURCE   H 01

7.1	MMBtu/hr
2,082	kWt

Evap. Pinch Pt. Temp. [C] 34

R 301   Regen. Effectiveness= 65%

C 201   ΔP[kPa] 10

ΔP heat source= 0.001 kPa
Cp,heat sour 1.08 kJ/

ΔP ORC Evap. Fluid 45 kPa

REGENERATOR (Cold Side) ΔP[kPa] 1.5

C 101
ΔP [kPa] 1.40   WATER COOLED CONDENSER
6.15 MMBtu/hr
-1,803 kWt
(AIR COOLED CONDENSER
Est.d Fan kWe) 81

H 02

R 302

R501

Subcool Liq. 3 C

R 401 PUMP

Pump Eff. 55%

-32.0 kWm

	UA [kWt/K]	kWt	HX Dia (m)	HX Length (m) & Weight (kG)	
Evaporator :	14	2,082	0.37	3.66	1,364
Regenerator :	14	665	0.32	5.18	1,364
Condenser :	54	1,803	0.45	3.66	1,818

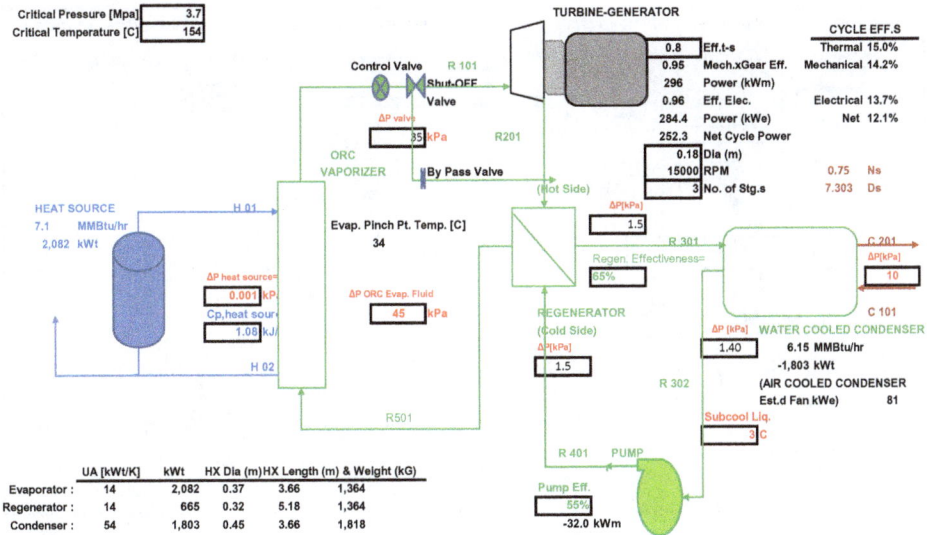

Figure 3.18.   Rankine Cycle System using a refrigerant as the working fluid (see Case Study 2).

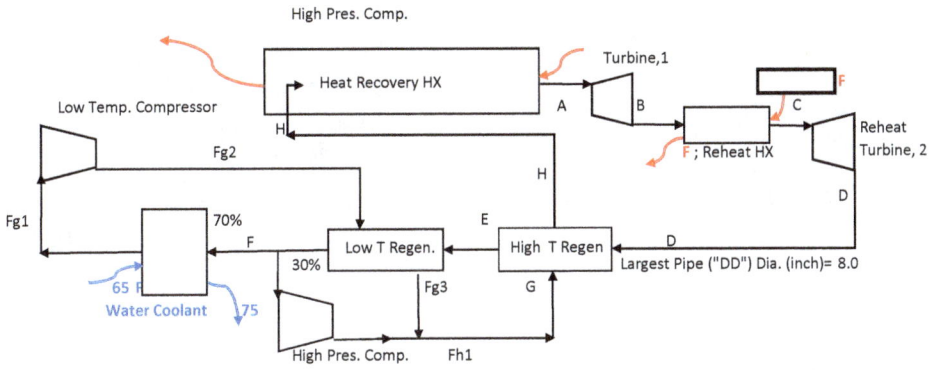

Figure 3.19 (Supercritical CO$_2$ Cycle) — diagram labels:

High Pres. Comp.
Heat Recovery HX
Turbine,1
Low Temp. Compressor
Fg2
A   B   C   F
Reheat Turbine, 2
F; Reheat HX
H   D
Fg1
70%   F
30%
Low T Regen.   High T Regen
E
65 F
Water Coolant 75
Fg3   G   D
Largest Pipe ("DD") Dia. (inch)= 8.0
High Pres. Comp.   Fh1

Figure 3.19.   Supercritical CO$_2$ Cycle using an Open Brayton Cycle (see Case Study 5).

but allow at the very least a "best possible" performance for the process in question. These assumptions are summarized for each of the seven systems in Table 3.8, along with a final equation that defines the relationship between enthalpy, work or heat transfer, and the internal energy of the CV.

The reader's attention is also brought to the fact that an Open Flow System, as the name implies, means that a mass flow stream, $\dot{M}$, must breach the CV. The flow rate measured can be either into or out from the CV; the system will be treated as an Open Flow System, even if there is only a single inlet and/or outlet. This observation is linked to another important

S-CO₂ OPEN BRAYTON CYCLE-SEQUESTRATION SYSTEM FOR 335 Mwe POWER PLANT WITH REGEN., INTERCOOLED COMPRESSORS AND REHEAT

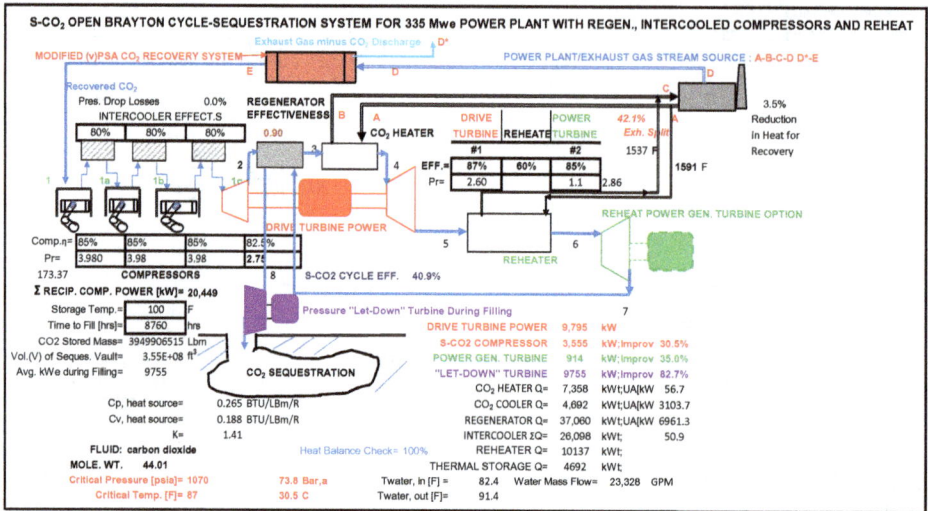

Figure 3.20. A SCO$_2$ system with CO$_2$ recovery and sequestration (see Case Study 12).

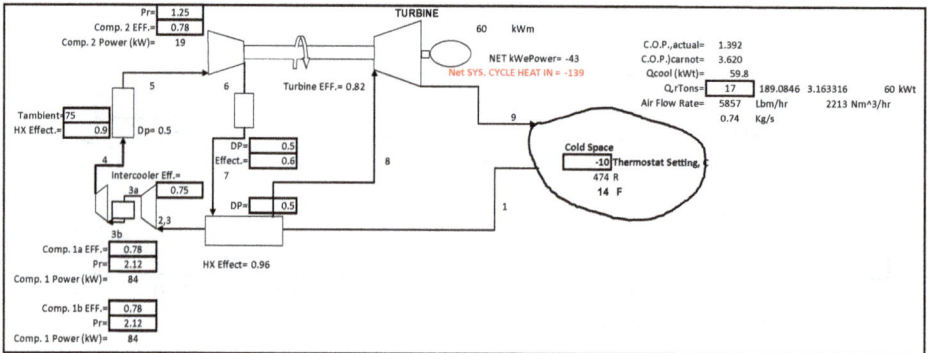

Figure 3.21. Refrigeration system using air as the refrigerant in a Reverse Brayton Cycle.

universal principle: the conservation of mass (or mass flow rate). That is, the mass flow rate into a CV must be accounted for as being either stored in the CV or observed to exit the CV. Conversely, the mass flow into a system equals the mass flow stored per unit of time, plus the mass flow rate exiting the system. In most cases, the mixing tank, perhaps being the exception of the seven mechanical systems identified in Table 3.8, will not have the mass increase or decrease in the CV, but rather the mass flow rate into the CV will equal the mass flow rate out from the CV.

It is also important to understand that mass is one of those parameters that defy the adage that you cannot mix "apples and oranges". The same

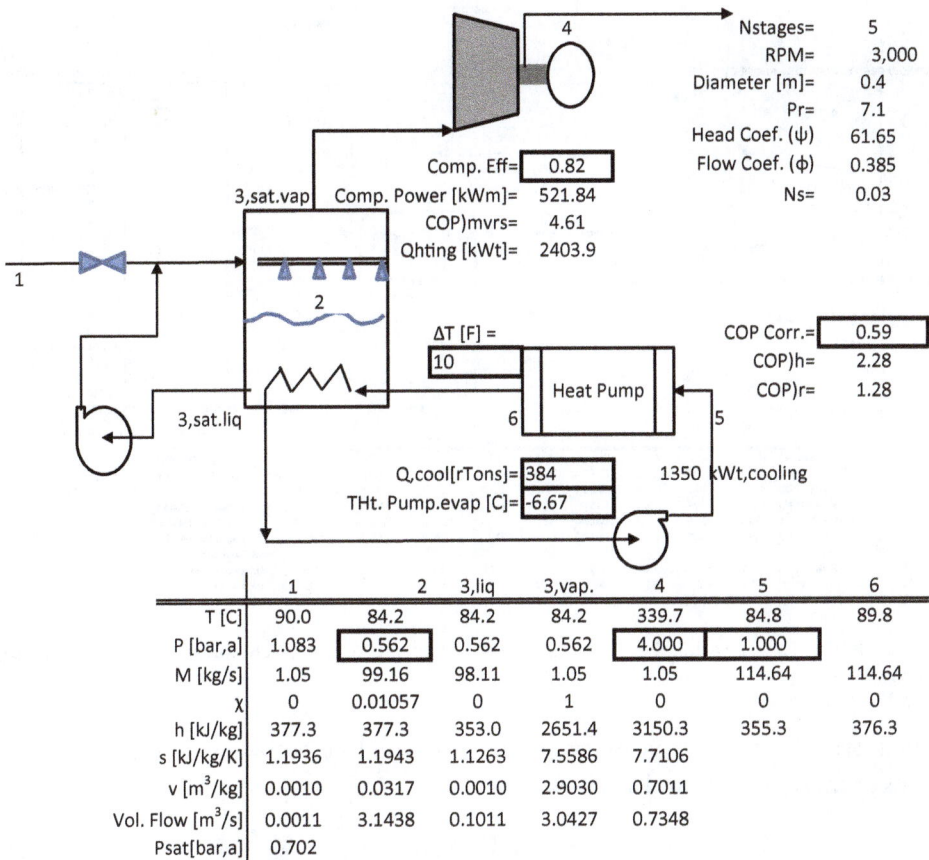

Figure 3.22.   Mechanical Vapor Recompression System (MVRS) steam heat pump (see Case Study 10).

can be said for energy, as was described in Chapter 1. For example, if there are 10 1-lbm apples entering a CV along with 20 1/2-lbm oranges into a second entrance, then the output from the CV will be $(10 + 10)$ lbm of fruit, assuming that the CV is not storing any of the fruit. However, if the CV is storing some fraction of the fruit mass that is entering the CV, then it is easy to determine the amount of mass leaving the CV by using the conservation of mass equation:

$$M_{in} = M_{out} + \partial M_{stored} \tag{3.8a}$$

or:

$$\dot{M}_{in} = \dot{M}_{out} + \frac{\partial \dot{M}_{stored}}{\partial t}. \tag{3.8b}$$

There is also a very important relationship that relates the mass flow rate with density, the cross-section of the area through which the mass

Super Heat= 25
Elec. Motor Heat Loss, jacket= 2.9%
Frac. Jacket Liq. Flashed to evap. (f1,evap)= 6%
Frac. Gap Liq. Flashed to evap. (f2,evap)= 10%

Elec. Motor Heat Loss in Gap & Quality
5.0% 11.4%
422.3

CONDENSER TEMPERATURE PROFILE [C]

36.0 47.9
33.0 35
30.0 [gpm/rTn] 3.33

CONDENSER
Economiser Comp. Eff.= 0.706
Comp.1 shaft kW= 6.3

239.94716
3338.7016
3578.6488
3578.6488

ECONOMIZER

EVAPORATOR TEMPERATURE PROFILE [C]
[gpm/rTn] 2.66 12.0
7.0
6.9
6.0

Heater
Q [kWt]= 1539
26.0 C

EJECTOR

Pump Power [kWm]= 14.6

EVAPORATOR

3338.74
3338.70
-264.52
1747.77
1746.13
264.52

7.10
6.54
4.99

7.87% 99.598396

COMPRESSOR SPECIFICATION SUMMARY

75 rTons	1st. Stage Inlet	
P,in [bar,a]=	3.59	
T,in/T,out [C]	7.10	30.2
M,total [kg/s];Vol. [m³/s]=	1.54	0.088
(Mgap [kg/s])=	0.0327	
Comp. Eff.=	0.741	
Economiser Into Comp. 2nd Stage		
P,in [bar,a]=	6.24	1.735
T,in [C]	22.8	
M,economizer,6a [kg/s]=	0.126	
1 stg.Comp.Temp.out [C]=	30.2	
2nd. Stage Mix Inlet		
P,in [bar,a]=	6.24	
T,in [C]	29.84	
Comp. Eff.=	0.706	
2nd. Stage Mix Outlet		
P,in [bar,a]=	9.14	1.465
T,out [C]	47.3	
M at 3 & M at 7* [kg/s]=	1.700	0.097
4*, M,jacket cool.[kg/s]=	0.129	

	1	2	3	4	5	6a	6b	6c	7
P [bar,a]	3.59	9.13	9.14	9.11	6.26	6.26	6.26	31	3.62
T [C]	7.1	125.68	111.38	32.97	23.13	23.13	23.13	171	6.00
h [kJ/kG]	403.4	510.9	503.8	245.8	245.8	411.3	231.7	544.0	231.8
s [kJ/kG/K]	1.730	1.977	1.939	1.157	1.158	1.717	1.110	1.967	1.114
density [kG/m³]	17.48	30.08	31.56	1176.84	299.00	30.44	1214.30	101.18	131.84
Sat. Temp. [C]	5.9	36.20	36.22	36.13	23.13	23.13	23.13	88.0	6.2
Quality (x)					8.0%				12.1%
Flow [kG/s]	1.54	6.54	7.10	7.10	7.10	0.57	1.54	4.99	1.54
m³/s	0.09	0.22	0.23	0.01	0.02	0.02	0.00	0.05	0.01
Nm³/s	0.0013	0.0054	0.0059	0.0059	0.0059	0.0005	0.0013	0.0041	0.0013

COP]r= 0.17
Motor Power [kWe]= 20.8
Qevap. [kWt]= 263.7
Qcond. [kWt]= -1832.5
Qheater [kWt]= 1538.9
Heat Balance Chk.: 99.9%
UA,evap[kWt/K]= 439
UA,cond[kWt/K]= 82

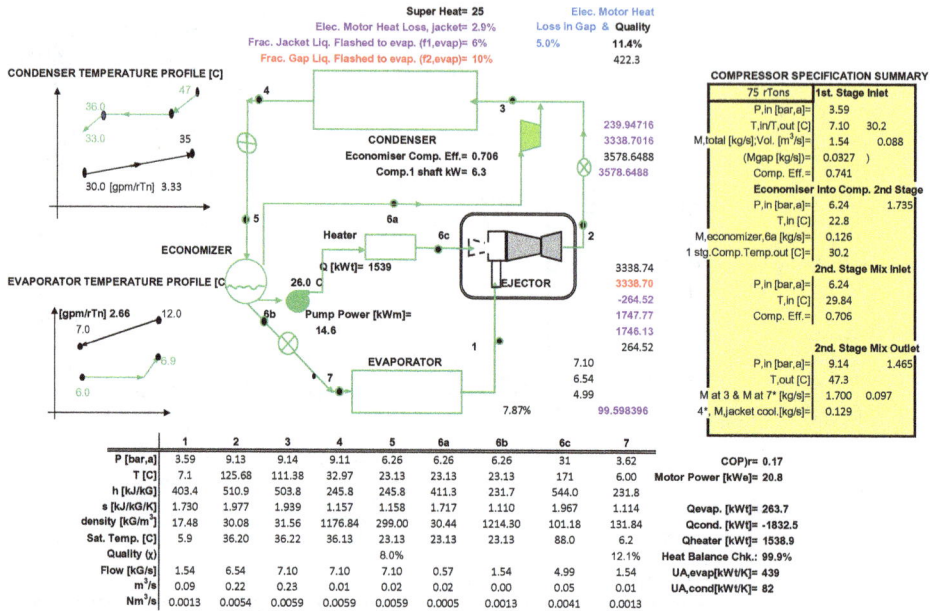

Figure 3.23. Refrigeration Cycle using an ejector in place of one compressor stage (see Case Study 14).

must travel, and the velocity of the mass. That equation is elegant as it is extremely useful, and is shown as:

$$\dot{M} = \rho \times V \times Area_{flow}, \tag{3.9}$$

where $\dot{M}$ is the mass flow rate usually in lbm/s or kg/s, velocity, V, is in ft/s or m/s, and Area is the cross-sectional area of the conduit carrying the fluid in ft^2 or m^2.

This equation may be straightforward, but yet powerful in its application to thermodynamics problems. Of particular note is how it is used along with the principle of perfect expansion to design a converging–diverging nozzle also known as a supersonic nozzle. This particular device is made up of a nozzle that is in series with a diffuser, hence its name. But when a gas such as air or superheated steam is at a pressure higher than approximately two times the ambient pressure, then the flow rate of the air (or steam) will actually reach velocities that are greater than the sonic velocity of that air or steam. This ability to convert thermal energy into very high-speed velocity is what makes a jet engine provide the thrust forces that can propel an aircraft to very high speeds.

An example of how the open flow equation of the First Law of Thermodynamics is applied to these special "real world" mechanical systems is useful

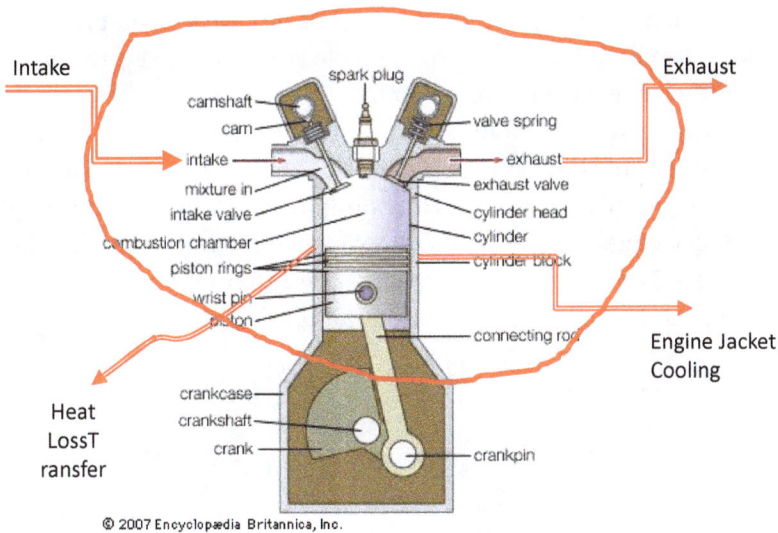

Figure 3.24.   Open Flow System used to analyze a piston-cylinder engine.

and given here, and is focused on the same piston-cylinder engine system that we have discussed in some detail in the previous section.

Figure 3.17 shows that system, but now identifies a CV that envelops the entire cylinder engine and not just the air fluid inside the cylinder. This makes a very large difference: the air flow, fuel flow and the exhaust products will now breach the CV and thus no longer define a Closed Mass System. In a similar way, the heat transfer from the cylinder that would have crossed the closed volume CV and entered the walls of the cylinder is now simply entering the atmosphere that surrounds the cylinder. Certainly, it is also possible that the CV in this Open Flow Analysis still borders the insides of the cylinder wall. In this case the heat transfer enters the wall and is removed by the coolant water that is flowing through the cylinder walls. But to make this analysis correct and complete, the inlet coolant water and outlet coolant water flow rate must also breach the CV. This would be considered one of the mass flow rates at their respective temperatures and therefore respective enthalpies. This heat transfer is correctly identified as both convection and a radiation heat transfer. Clearly, the work output is the crankshaft that breaches the CV.

Given the diagram of the cylinder-piston as detailed in Figure 3.24, the correct application of the First Law using the Open Flow System Analysis is given in Equation (3.10) using the labels for each flow stream.

In this equation the enthalpy that is associated with the fuel flow rate is a measure of the chemical heat content, also called the latent heat content or heat of formation of the fuel. In fact, every carbon-hydrogen based chemical has a latent heat content that can be released upon very rapid chemical oxidation, otherwise called combustion.

It is also necessary to point out how the exhaust gas flow rate out of the CV is the sum of the mass flow rate and the fuel flow rate that enters the cylinder. This is a perfect real world example of how the air mass, plus the fuel mass must equal the mass of the exhaust gas that exits the CV given, such that there is not any mass stored in the cylinder-piston system.

Recalling the First Law of Thermodynamics for an Open Flow System:

$$\sum \dot{Q} + \sum_{in} \dot{M} \times (h + KE + PE)$$

$$= \sum \dot{W} + \sum_{out} \dot{M} \times (h + KE + PE) + \frac{\partial U}{\partial t}\bigg|_{cv}. \quad (3.10)$$

For the piston-cylinder problem, the following terms in the first law equation are for the system, as shown in Figure 3.17, with the CV as drawn.

$$\sum \dot{Q} = \dot{Q}\,radiation + \dot{Q}\,jacket,$$

$$\sum \dot{W} = Engine\,shaft\,power \sum_{in}^{n,inlets} \dot{M} \times (h + KE + PE)$$

$$= \dot{M}air \times h_{air} + \dot{M}fuel \times h_{fuel}, \quad (3.11)$$

$$\sum_{out}^{n,\,outlets} \dot{M} \times (h + KE + PE) = \dot{M}exhaust \times h_{exhaust},$$

$$\frac{\partial U}{\partial t}\bigg|_{cv} = 0, \; because\,the\,system\,is\,assumed\,to\,be\,at\,steady\,state.$$

It is noted that the heat transfer from the engine jacket is given as a term in the summation of heat transfer in Equation (3.11). Instead of including it in Equation (3.4), the heat transfer could also have been included in Equation (3.10) where it would be added to the product of the exhaust gas flow rate and enthalpy. Also, the enthalpy of the fuel is also given the label, L_{hhv}, the latent heat content with respect to the high heat value. More details of this parameter will be given in Chapter 5, but in this chapter this parameter accounts for the amount of chemically stored energy of a substance, particularly a substance that has an arrangement of carbon and hydrogen atoms. A typical value for gasoline is 130,000 Btu/gal or 21,000 Btu/lbm. This may be compared to the heat content of hydrogen of 61,000 Btu/lbm or natural gas with 1,030 Btu/ft^3. It is also noted that the Kinetic and

Potential Energies (KE and PE) have been eliminated from the summation terms. This recognizes that the KE and PE are relatively small values and thus negligible comparted to the fuel energy and/or heat transfer that occurs with this engine.

The only unknown in this single equation is the Work term that can now be readily solved. Caution is given to keeping the units consistent. The Work unit will be Btu/h or kJ/s, which is also more commonly called a kilowatt (kW).

The reader should be able to clearly see how this equation passes the reality check. That is, the fuel entering the system is the only flow stream that provides energy to the system, with the radiation and jacket heat transfers being the energy releases from the CV. The exhaust, fuel and air intake flow rates are the only flow rates into or out of the CV. The sum of these energies and products of mass flow rate and their respective enthalpies provide a direct solution to the rate of work or power output from the CV.

It is not too early to identify that the Second Law of Thermodynamics requires that the cycle release heat energy to a low temperature reservoir — in this case, the ambient, if the system is exposed to a heat source and if there is work or power that is produced. This is basically one of the two statements of the Second Law of Thermodynamics, called the Kevin–Planck statement of the Second Law of Thermodynamics, which also defines the maximum cycle efficiency that a heat engine can attain when it receives heat and rejects heat from a hot and cold temperature reservoir, respectively. According to the Second Law of Thermodynamics, the Carnot efficiency of the cylinder-piston system is defined as:

$$\eta_{carnot} = 1 - \frac{T_{cold}}{T_{hot}}. \tag{3.12}$$

Assume that the highest temperature achieved in the piston-cylinder is 3,000°R and that the ambient is 520°R. Thus, the highest engine efficiency can be:

$$\eta_{carnot} = 1 - \frac{520}{3,000},$$

$$\eta_{carnot} = .827 = 82.7\%.$$

Table 3.9 provides a summary of the operating conditions for the piston-cylinder system illustrated in Figure 3.24 that the reader may use with the first law equation to practice its use. For example, the reader should verify that the net power output is 206 hp.

Using the first law expression, the efficiency is defined as $\eta = \frac{Work_{net}}{Q_{fuel}}$.

Table 3.9.   Engine specifications.

Air Flow Rate [Lbm/hr]= 2,000
Fuel Flow Rate [Lbm/hr]=100
Fuel Heat Content [Btu/Lbm]=21,000
Engine Efficiency[$\sim$]=?
Exhaust Gas Heat Rate [Btu/hr]=7,87,500 Exhaust gas Flow rate [Lbm/hr]=2,100 Exhaust Gas Temp. [F]=1,502
Engine Jacket Cooling [Btu/hr]=6,30,000
Radiation Heat Transfer [Btu/hr]=1,57,500
Work, net [Btu/hr]= 525,000 Work, net [hp]=?

Using the engine specifications given in Table 3.9, the efficiency is determined to be:

$$\eta = \frac{525,000}{100\ Lbm/hr \times 21,000\ Btu/Lbm},$$

$$\eta = .25.$$

Another short-handed way of describing the heat engine system with respect to the hot and cold temperature reservoirs is shown in Figure 3.25. The circle shown between the two heat reservoirs is left empty, but could just as easily include any thermodynamics cyclic process that produces power.

A similar representation that identifies a thermodynamics refrigerator is shown in Figure 3.26. It depicts a thermodynamics representation of a system that can withdraw heat from a cold temperature reservoir and emit (or reject) it into a hotter temperature reservoir. This may appear to be contrary to what has been stated as the direction of heat transfer: from hot to colder temperatures and thus complying with the Second Law of Thermodynamics. However, that statement is true IF there is no energy (or power) being put into the system. In the system shown in Figure 3.27, work or power is being used to drive the heat transfer from the cold reservoir to the hotter temperature reservoir. The First Law of Thermodynamics is still intact, for it is true that the energy into the cycle is equal to the energy from

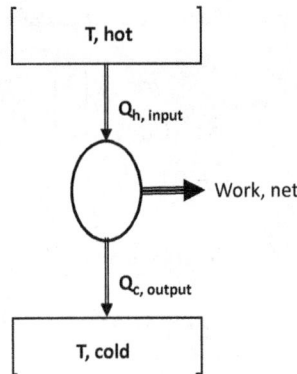

Figure 3.25.   Representation of a heat engine with heat input from a hot temperature reservoir.

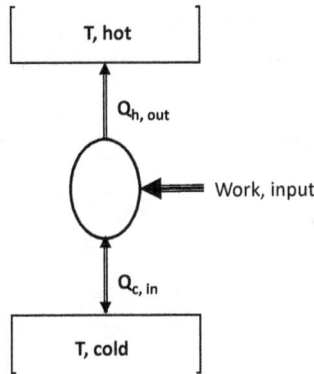

Figure 3.26.   Heat pump refrigeration system.

the cycle; or in equation form:

$$Q_c + Work_{input} = Q_h.$$

In this case the maximum efficiency of the refrigerator system is given another name — the Coefficient of Performance (COP). This is primarily because the COP can and is desired to be much greater than 1 for refrigerators, while the word "efficiency" usually denotes performance that is less than 1. The ideal refrigerator equivalent to a Carnot engine performance is a Carnot Refrigerator and is defined as:

$$\eta_{carnot} = \frac{T_{cold}}{T_{hot} - T_{cold}}. \tag{3.13a}$$

It is also possible to identify and focus on the ability to have the refrigeration system reject heat into a hotter temperature reservoir and thus provide

r245fa CONDENSING STEAM WASTE HEAT RECOVERY SYSTEM

Critical Pressure [Mpa]	3.7
Critical Temperature [C]	154

TURBINE-GENERATOR

0.82	Eff.t-s
0.94	Mech.xGear Eff.
295	Power (kWm)
0.96	Eff. Elec.
283.5	Power (kWe)
267.2	Net Cycle Power
0.2	Dia (m)
12000	RPM
1	No. of Stg.s

CYCLE EFF.S
Thermal 8.2%   Mechanical 7.7%
Electrical 7.4%   Net 7.0%
0.46 Ns   3.859 Ds

Control Valve — Shut-OFF Valve
R 101   ΔP valve 10 kPa   R201

ORC VAPORIZER

HEAT SOURCE
13.05 MMBtu/hr
3,825 kWt

H 01

Evap. Pinch Pt. Temp. [C] 10

ΔP heat source 2 kPa

ΔP ORC Evap. Fluid 20 kPa

H 02

(Hot Side) ΔP[kPa] 1.5   R 301   Regen Effectiveness 50%

REGENERATOR (Cold Side) ΔP[kPa] 1.5   R302

R501

ΔPcond. [kPa] 2   C 201   ΔP[Pa] 10   C 101

CONDENSER
12.04 MMBtu/hr
-3,527 kWt
(AIR COOLED CONDENSER
Est.d Fan kWe) 71
Subcool Liq. 3 C   Pinch Pt.= 2.8

R 401 PUMP
Pump Eff. 55%
-16.3 kWm

	UA [kWt/K]	kWt	HX Dia (m)	HX Length (m)	& Weight (kG)
Evaporator :	254	3,825	1.58	3.66	17,727
Regenerator :	17	183	0.35	5.18	1,364
Condenser :	282	3,527	1.03	3.66	7,727

	H 01	H 01	H 02
Pres. bar,a	1.10	1.10	1.08
Temp. (C)	104.3	104.3	100.8
Enthalpy (KJ/kG)	2680.7	2680.7	421.9
Super Heat Temp. Diff. (C)	2.0	2.0	
SubCooled Temp. Diff. (C)			1.0
Density (kG/m³)	1	1.12	1.12
Mass Flow (Kg/s)	1.69	1.69	1.69
Nm3/hr		5,433	5,433

	R 101	R 201	R 301	R 302	R 401	R 501	C101	C201
Pres. bar,a	9.00	3.06	3.02	3.00	9.22	9.30	1.38	1.28
Temp. (C)	90.8	62.6	53.0	42.8	43.3	50.5	23.0	40.0
Enthalpy (KJ/kG)	471.4	454.6	444.8	256.1	257.0	266.8		
Super Ht T. Diff. (C)	5.0		7.2	0.0				
Sub. T. Diff. (C)			0.0	2.8	43.0	36.2		
Density (kG/m³)	49	16	16	1290	1291	1270	999	994
Mass Flow (Kg/s)	18.7	18.7	18.7	18.7	18.7	18.7	49.67	49.67
Min. Pipe Diameter (mm)	152	279	838	101	101	101	177	177
Volume Flow Rate (m³/s)	0.383	1.175	1.142	0.014	0.014	0.015		

	H 01	H 01	H 02
Pres. psia	15.95	16.0	15.7
Temp. (F)	219.7	219.7	213.4
Enthalpy (Btu/Lbm)	1154.4	1154.4	181.7
Super Heat Temp. Diff. (F)	3.6	3.6	
SubCooled Temp. Diff. (F)	0	0	1.8
Density (Lbm/ft³)	0.07	0.070	0.070
Mass Flow (Lbm/s)	3.73	3.73	3.73
Volume Flow Rate ft³/s	92.55	92.55	0.06

	R 101	R 201	R 301	R 302	R 401	R 501	C101	C201
Pres. psia	130.5	44.3	43.8	43.5	133.7	135	20.01	18.54
Temp. (F)	195	145	127	109	110	123	73.4	104
Enthalpy (Btu/Lbm)	203.0	195.8	191.6	110.3	110.7	114.9		
Super Ht.T Diff. (F)	9		13					
Sub. T Diff. (F)				5.0	77	65		
Density (Lbm/ft³)	3.04	0.99	1.02	80.46	80.51	79.20	62.3	61.9
Mass Flow (Lbm/s)	41.1	41.1	41.1	41.1	41.1	41.1	109.3	109.3
Min. Pipe Diameter (Inch)	6	11	33	4	4	4	7	7
Volume Flow Rate ft³/s	13.52	41.47	40.29	0.51	0.51	0.52		

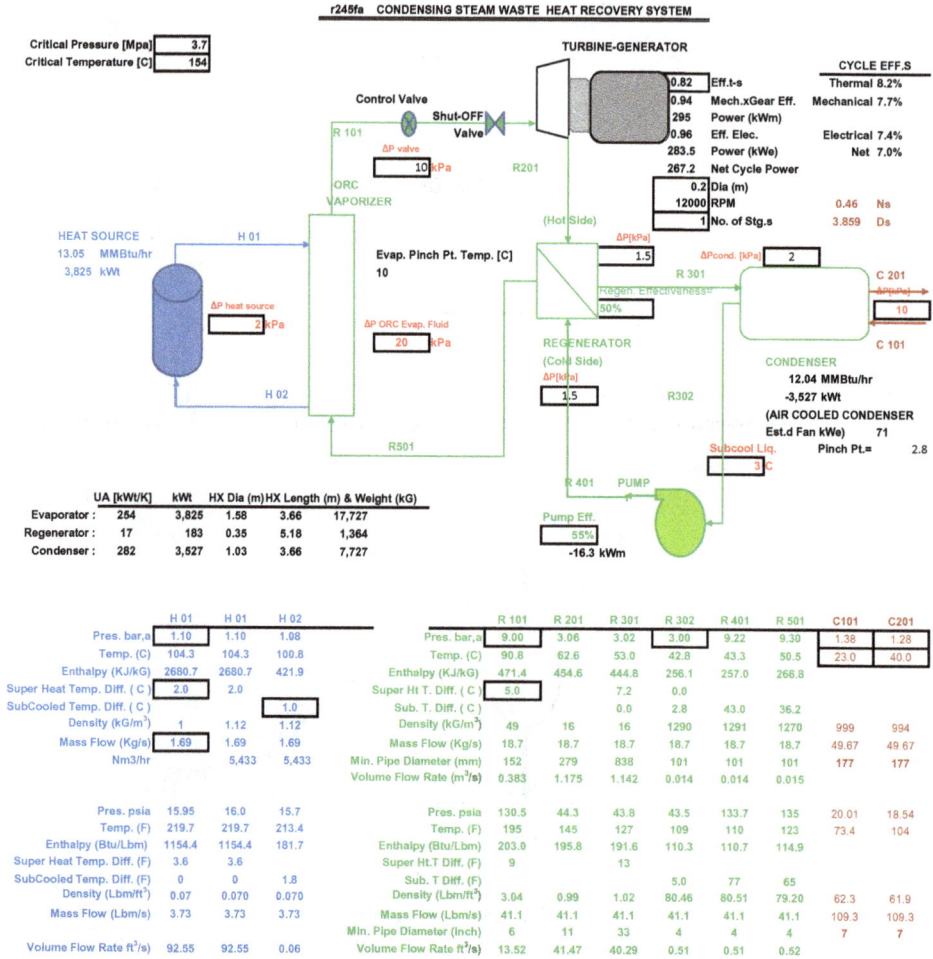

Figure 3.27.   Rankine Cycle System used to produce electric power; the turbine and condenser components are highlighted.

necessary heating. This system is traditionally called a heat pump and the COP is defined as:

$$\eta_{carnot} = \frac{T_{hot}}{T_{hot} - T_{cold}}. \tag{3.13b}$$

It can be easily shown that (COP)h = (COP)r + 1 by recalling that the cycle must conserve energy and thus $Q_{out} = Q_{in} + \text{Work}_{in}$.

Another example of the utility of the First Law of Thermodynamics applied to an arrangement of one or more of the seven mechanical systems that are routinely used in mechanical engineered systems is given here. Consider the block diagram in Figure 3.27 that uses some of the seven mechanical components previously mentioned in Table 3.8. The arrangement of the components shown in Figure 3.27 is deemed to be a cycle because the system can be observed to close on itself. Furthermore, if a CV is placed around all these elements, the system is also a Closed Mass System as no mass flow crosses the CV boundary. You may remember that the cycle may produce or absorb power from the net heat transfer that occurs, and the First Law for a Closed Mass System is available to determine this net power. The cycle shown in Figure 3.27 is a Rankine Cycle. As described in Chapter 1, the Rankine Cycle is one of the most useful and often-used cycle for electric power generation. However, in Figure 3.27, the working fluid contained in the piping of this Rankine Cycle is not the most common chemical used in most if not all Rankine Cycle systems used by electrical utilities (i.e. water); in this case, the working fluid is a refrigerant known as R134a, which is more commonly used in refrigerant cycles for cooling purposes. Thus, using what has come to be called an Organic Rankine Cycle (ORC) System, the fluid can be used for power generation also. More of this application will be given in the case studies portion of this textbook. There are two purposes for displaying the cycle in this section: (1) to begin describing how the Closed Mass System, as shown by the cycle, can also use the open flow method for solving the heat transfer of power into or out from the components shown, and (2) to show the reader that the seven basic mechanical systems that make up this cycle (and many more similar cycles) can very quickly be analyzed with the basic equations of the First Law of Thermodynamics in either long or short form — the latter as presented in Table 3.8. The most useful and productive power cycles do not get more difficult than the Rankine Cycle!

For example, given the heat transfer and the power input for the turbine, the net output work of the turbine can be solved by the first law equation (in long form):

$$\sum \dot{Q} + \sum_{in}^{n,\,inlets} \dot{M} \times (h + KE + PE)$$

$$= \sum \dot{W} + \sum_{out}^{n,\,outlets} \dot{M} \times (h + KE + PE) + \left(\frac{\partial U}{\partial t}\right)_{cv}.$$

Of course, as mentioned in the previous section, the basic assumption is that the turbine is well insulated, operates at steady state, and that the kinetic and potential energies into and out from the turbine cancel each other. The resultant equation for the work of the turbine is:

$$\dot{W}_{turbine} = \dot{m} \times (h_{out} - h_{in}). \tag{3.14}$$

This equation clearly solves for the power output of the turbine by knowing the mass flow rate, inlet and outlet enthalpies, and the amount of heat transfer loss from the turbine. The enthalpies are determined by knowing the state point properties, such as pressure and temperature at the inlet and outlet of the turbine.

However, the application of the CV around the entire collection of the mechanical elements cannot provide the details of any of the components. In order to be able to discern the more detailed behavior of any of the mechanical components, it is necessary to draw a CV around the component to essentially isolate it from the others. This has been done in Figure 3.27, where the turbine and condenser have been isolated with CVs. With these CVs in place, the First Law of Thermodynamics applied via an Open Flow Analysis will show that the mass flow rate breaches the CV, as does the shaft of the turbine and any heat transfer lost by the turbine — assuming that the no heat transfer (or negligible heat transfer) loss from the turbine enables the analyst to use Equation (3.14).

For the CV around the condenser, the First Law of Thermodynamics is applied, and the resulting equation is given in Equation (3.14) taken from Table 3.8.

$$\dot{Q}_{condenser} = \dot{m} \times (h_{out} - h_{in})_{r134a}$$

and
$$\dot{Q}_{condenser} = \dot{m} \times (h_{out} - h_{in})_{water}.$$

Two observations can be made: (1) these two equations can be set equal to each other and if either of the mass flow rates is not known, the equations can be used to solve for the unknown flow rate. In fact, this is how the mass flow rate of the coolant water for the condenser was determined; (2) the heat transfer equation is the same for a single or multi-phase (usually, two) fluid system. That is, the only important variable to be determined from the properties of pressure and temperature (or any other two properties) is the enthalpy, h, into the condenser and out from the condenser. In the case of the condenser, the fluid enters the heat exchanger as a superheated vapor and exits the heat exchanger as a subcooled liquid.

The reader should confirm the results shown for the turbine power and the heat transfer from the condenser by using these equations with the mass flow rate and enthalpies given in the Property Table that is also included in Figure 3.27, albeit with the properties now shown in the left most column and the values for the state points given in the rows labeled with the state point number.

The same procedure can be performed for each of the other mechanical elements. The result will be the heat transfer or the power that is input or output from the mechanical system. In each case, the equations based on the First Law of Thermodynamics and the reasonable assumptions of no heat transfer (or little to no kinetic or potential energy changes) will be the same as shown in Table 3.8, where the most relevant property is the enthalpy before and after each of the components. Assume that the state point for the components in the block diagram shown in Figure 3.27 is also given in the Property Table shown in Table 3.10. This table also has the enthalpy for each of these state points, including the entropy property (which is found from the same Property Tables for the fluid) in this steam. The reader will note that not much information has been given about the entropy of a property and its importance in the analysis of any thermodynamics process. This is because more details will be provided in the next chapter. Suffice to say, the entropy after each mechanical element that has no heat transfer must be greater at the exit of the mechanical element relative to its inlet state. The heat exchanger is the exception to this rule and more detail will be given in Chapter 4 about how the entropy of the Universe increases for a heat exchanger.

Using the Property Table in Table 3.10, the heat transfer, the work done or the velocity gained by each of the mechanical systems is given in the right most column of Table 3.8, which includes the multiplication algebra that is used to calculate this heat transfer or power. It is left to the reader to confirm these results using the same enthalpies shown in Figure 3.27. The reader can then use the new enthalpies in Figure 3.27 for the same cycle process, but with different pressures and temperatures before and after each component. He or she should first determine the enthalpies and entropies for each state point for the refrigerant R134a and calculate the heat transfer or the power done for each mechanical component.

The attentive reader may be looking for two of the seven mechanical elements that are not readily apparent in Figure 3.27. These two mechanical elements are the nozzle and the diffuser. In fact, both these mechanical elements can be shown to be part of the "inner" workings of the turbine. The nozzle is that part of the turbine that transforms the potential energy of a high enthalpy into high velocity. The high velocity fluid is then directed onto the turbine airfoils or blades at the correct angle, so as to maximize the change in the magnitude and direction of the momentum of the fluid. This change in momentum produces the force that acts through a moment arm that is equal to the radius of the turbine impeller (wheel). The vector product of the moment arm and force produces a torque on the turbine shaft. The diffuser may be positioned at the exit of the turbine to recover some of the pressure across the turbine and thus improve the

Table 3.10.   Detail of Property Table originally shown in Figure 3.27.

Metric units

	H 01	H 01	H 02	R 101	R 102	R 201	R 301	R 302	R 401	R 501	C 101	C 201
Pres. bar.a	7.00	7.00	6.98	23.00	22.275	9.06	9.02	9.00	23.47	23.51	1.38	1.28
Temp. (C)	165.9	165.9	163.8	84.2	82.9	46.6	38.5	32.6	34.2	40.1	20.0	30.0
Enthalpy (KJ/kG)	2762.5	2762.5	691.4	444.6	444.6	428.9	420.3	245.2	247.5	256.1		
Super Heat Temp. Diff. (C)	1.0	1.0	1.0	10.0	8.7			0.0				
SubCooled Temp. Diff. (C)						2.9	2.9	2.9	40.5	34.6		
Density (kG/m³)	1	1.12	1.12	114	109	41	43	1179	1183	1159	1000	997
Mass Flow (Kg/s)	13.89	13.89	13.89	152.6	152.6	152.6	152.6	152.6	152.6	152.6	635	635
Nm3/hr	44,532	44,532	44,532									
Min. Pipe Diameter (mm)				279	279	457	558	279	279	304	639.00	639.00
Volume Flow Rate (m³/s)				1.343	1.397	3.683	3.520	0.129	0.129	0.132		

Imperial units

	H 01	H 01	H 02	R 101	R 102	R 201	R 301	R 302	R 401	R 501	C 101	C 201
Pres. psia	101.50	101.5	101.2	333.5	323.0	131.3	130.8	130.5	340.3	341	20.01	18.54
Temp. (F)	330.7	330.7	326.9	184	181	116	101	91	94	104	68	86
Enthalpy (Btu/Lbm)	1189.6	1189.6	297.7	191.5	191.4	184.7	181.0	105.6	106.6	110.3		
Super Heat Temp. Diff. (F)	1.8	1.8	1.8	18	16			0				
SubCooled Temp. Diff. (F)	0	0				5	5	5.3	72.9	62.3		
Density (Lbm/ft³)	0.07	0.070	0.070	7.08	6.81	2.58	2.70	73.52	73.74	72.23	62.3	62.2
Mass Flow (Lbm/s)	30.56	30.56	30.56	335.7	335.7	335.7	335.7	335.7	335.7	335.7	1405.8	1405.8
Min. Pipe Diameter (inch)				11	11	18	22	11	11	12	25	25
Volume Flow Rate ft³/s	133.54	133.54	0.54	47.40	49.28	129.96	124.21	4.57	4.55	4.65		

turbine's power output and turbine efficiency. That is, the turbine nozzles can expand to a lower pressure than the pressure that is maintained by the condenser at the exit of the turbine, though the diffuser can increase that lower pressure and match it to the higher pressure at the exit of the turbine by slowing the exit velocity to a minimum needed to discharge the fluid flow in a reasonably sized pipe diameter.

Once the reader is comfortable with the process of using the First Law for each of the seven basic mechanical systems, he should consider the graphical representation of the process path that each of these mechanical components will exhibit on a P-v, T-s or P-h diagram. The choice of the x-y coordinate label is only dependent on which of the graphical representations of the process path is the most helpful in solving the thermodynamics problem at hand.

For example, the block diagram for the cycle of components, as shown in Figure 3.27, traces the pathways shown in Figure 3.28. It is also noted that the process diagram is plotted on a temperature–entropy (T-s) coordinate system rather than a P-v or T-v diagram, as has been done in earlier illustrations. The T-s diagram is often chosen because it enables the Second Law of Thermodynamics to be used as another "tool" that can assist the analysis to solve the problem quickly.

The Second Law of Thermodynamics states that the "...entropy of the Universe for a process must increase as the process proceeds". Entropy was introduced to the reader very briefly in Chapter 1 and more will be detailed in Chapter 4. It is sufficient to indicate here that entropy is a property of a system and can be tabulated in the tables, much as how temperature, internal energy, pressure, enthalpy and specific volume are tabulated as functions of each other to define a state point in a process. It will be shown in Chapter 4 that the Second Law of Thermodynamics provides a "direction" for the process, in that the end of a process must be to the right of its starting point. The only exception to this is for a heat exchanger process where the entropy might appear to allow the end of the process to move from the right to the left and thus violate the Second Law of Thermodynamics, but in fact a very precise application reveals that although the entropy via a heat exchanger at the end of a cooling process moves the entropy change of the *system* to the left, the entropy change of the *surroundings* that occurs because of the heat rejection from the heat exchanger is more positive in value than the negative value of the system, thus the *sum* of this system plus the surroundings' totals is greater than zero. This very brief explanation of the Second Law, plus one of its applications to solving a thermodynamics problem will be discussed in much more detail in Chapter 4. On a T-s

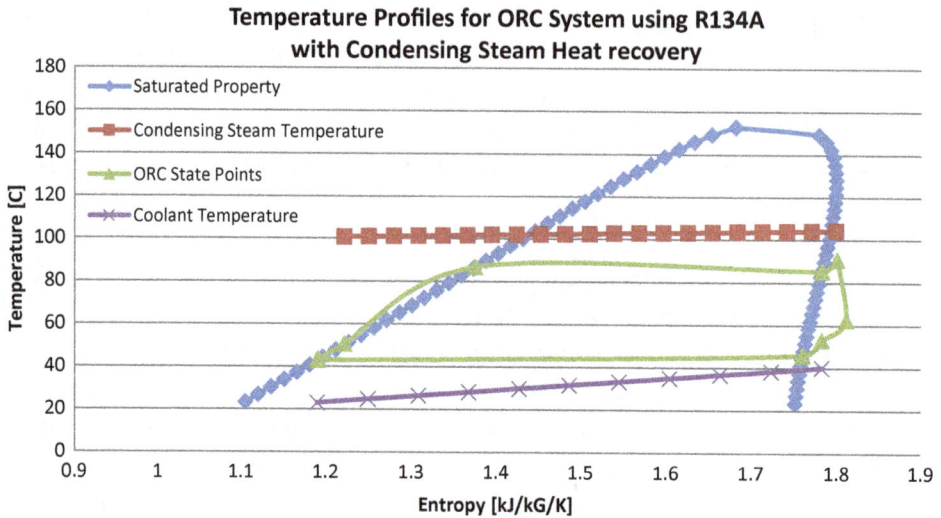

Figure 3.28.   T-s diagram for the Rankine Cycle in Figure 3.27.

process diagram, if the starting state point for any of the seven mechanical engineering devices is known anywhere on the diagram, then the outlet state can only be to the right of the starting point. This indicates that the outlet of the device is always greater than the entropy at the inlet. Of course, this is true only for the devices that have one basic assumption: that the heat loss or gain for that device is zero or very close to zero. For this condition, the system entropy change can only be positive with respect to the inlet state. The exception to this "rule of thumb" is the heat exchanger, where the purpose for the device during a heat transfer can therefore hardly be expected to be zero or approximately zero.

This cyclic process path from state point R101 to R102 and then R201, etc., as shown in Figure 3.27, is quickly discernible by the reader once the real-world constraints for each of the mechanical systems is understood. The turbine, for example, must always operate between two pressures: one high at the inlet and one lower at the exit. Thus, the analyst must understand a representation of a turbine expander in a block diagram, so as to be able to connect a high pressure with a lower pressure state point. The turbine cannot cause a lower pressure to become a higher pressure. For that to occur, the mechanical device must be a compressor. A compressor is a device that has as its primary function the need to compress the fluid, which must be in gaseous state, from a low pressure to a higher pressure. In both cases the constraints of the Second Law of Thermodynamics require the entropy at the exit of the device to be equal to or greater than the entropy at the inlet of the device. Thus, as may be observed in Figure 3.28, the trace of the pathways from the turbine or the compressor is towards the greater entropy. This is best done if the x-axis of the property diagram is entropy and not the specific volume as it had been in Chapters 1 and 2, and the first parts of Chapter 3.

The valve that is placed before the turbine has a very distinct process path based on the fact that a valve's sole functional responsibility is for the inlet pressure to the valve to be decreased without a subsequent loss in energy. Recalling that the valve is modeled with a constant enthalpy process, this requires a valve in the perfect world of Carnot engines to have the same temperature at its inlet as it has at its outlet but at a reduced pressure. Of course, in our real world engineering processes, the valve will have a slightly different temperature between the inlet and outlet, and most likely a decrease in its enthalpy to represent a small energy loss due to any number of real-world effects. Such effects are heat transfer loss and friction due to the fluid flow through the valve. Of course, the valve does not produce useful work in all cases. If it did produce even some useful work, the device would be performing more like a turbine expander than a true-to-its-design mechanical valve. That is, a valve and turbine will behave in the same way with respect to the inlet pressure being dropped to a lower pressure, but both cannot produce any useful work that a rotation of a shaft driving a generator or the wheels of a vehicle does, for instance.

3.7.1   *What is the process path for the valve?*

A valve's function is to reduce pressure with as little energy loss in the process as possible. As shown in Table 3.8 and illustrated in Figure 3.28, this translates into a process that has no change in enthalpy across the valve, assuming that the heat transfer and not just the friction energy is assumed to be zero. Thus, the enthalpy into the valve equals the enthalpy out of the valve. However, the entropy of the process will increase as required by the Second Law of Thermodynamics. Thus, the expected pathway is from left to right, as may be observed in Figure 3.28. For an ideal or perfect gas, the assumption of no change in enthalpy across the valve also implies that the change in temperature across the valve is zero. This is direct from the equation of enthalpy or:

$$\Delta h = C_p \times (T_{out} - T_{in}).$$

Of course, this holds true for a gas that is near perfect and obeys the Perfect Gas Law. A real gas will have some change in the temperature due to the small effect of pressure on the process path. Thus, the pathway for a valve as it may be displayed on a temperature–entropy process diagram is a horizontal path with respect to the temperature.

3.7.2   What is the process path for a heat exchanger?

A heat exchanger does not change pressure between its inlet and outlet. Thus, once the pressure of a heat exchanger is identified, the outlet path can either be to increase the energy or decrease the energy of the outlet, that is, to either increase or decrease the temperature of the outlet stream. That a heat exchanger does not change pressure may seem non-intuitive at first for the reader, especially if he or she is thinking that a boiler could blow up due to the high pressure built up in the heat exchanger. However, the build-up of pressure is not the "fault "of the boiler–heat exchanger, but rather the consequence of other actions in the process. For example, as may be observed in Figure 3.27, if the valve that is placed between the turbine and the boiler is accidentally closed and the feed pump shown in the figure is not stopped, the pressure from the feed pump can pressurize the boiler above its design point and thus cause it to fail. The valve's function of opening and closing is the cause of the boiler failure and not the unilateral result of poor design or unexpected function. Thus, the process trace is up or down along the constant pressure line as shown.

3.7.3   What is the process path for the compressor?

The compressor process path is the reverse of the turbine process, in that the compressor's function is to increase the pressure of the vapor. Thus, the process path for a compressor will always be from a low constant pressure process curve to a higher constant pressure process curve and directed to the right, relative from where it began on a T-s process diagram. The process path must have an entropy that is larger at the exit of the compressor, compared to the inlet as required by the Second Law of Thermodynamics. The entire process path, the inlet and the outlet state, must be entirely to the right of the saturated process curve of any fluid. This is necessary because the compressor is the mechanical device of choice for pressurizing a gas fluid and not a liquid. The pressurization of liquids is performed by the pump and will only be plotted to the left side of the saturated liquid property curve. The two-phase zone or the middle of the saturated property curve of a fluid will never have a compressor process or a pump process identified. The reason for this is because a state point is found to be inside the two-phase region, that is, the state point has a quality $0 < x < 1$. The two-phase fluid must be treated as separate fluids, typically with the liquid phase at the bottom of the container, pipe or reservoir, and the gas phase taken from the top (or the head space) of the tube or containment vessel that holds the two-phase mixture. This separation of the phases can either be done naturally by gravity pulling down the liquid at the bottom of the container or by the natural buoyancy of vapor at the top of the container. If necessary, there is a separator device that can help separate the two phases more efficiently to ensure that the headspace vapor does not carry any or at least a minimum amount of the fluid in the liquid phase. Given this separation of the phases into two distinct streams, the vapor phase will be directed to the compressor, and the liquid stream directed to the pump.

3.7.4 *What is the process path for the nozzle and diffuser?*

The nozzle's function is to also drop the nozzle inlet pressure and transform the high-pressure energy into higher velocity at a lower pressure at the exit of the nozzle. Similar to the turbine and the compressor, the exit state of the nozzle must have an increase in the entropy as required by the Second Law of Thermodynamics. Once again, the process path must therefore start from a high constant pressure curve and proceed to a lower pressure but also higher entropy. The larger velocity cannot be displayed on the T-s process chart. However, constant enthalpy curves can be displayed on the T-s diagram and these curves can be used to determine the higher velocity at the exit of the nozzle.

A diffuser is used to increase the pressure compared to the pressure at the inlet of the diffuser. The energy for this increase in pressure is derived from the high inlet velocity to the diffuser that is reduced to a much lower exit velocity. However, even the best diffusers will only increase the pressure by a relatively small amount by using the kinetic energy at the inlet of the diffuser. The process path is therefore from a lower pressure curve to a slightly higher-pressure curve, while also having the exit entropy larger in magnitude to the inlet of the diffuser.

To test the reader's understanding of how the process paths of any one of the seven mechanical devices can appear on a T-s diagram, consider the individual processes that are shown drawn in the same T-s diagram (see Figure 3.29). The reader should be able to identify the device that could have caused the process path that is shown. The answers are given at the end of this chapter, along with worked examples that apply the First and Second Laws of Thermodynamics for Open and Closed Mass Systems.

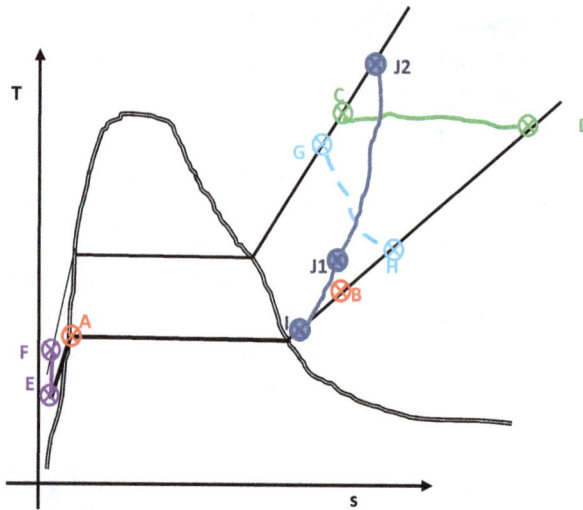

What mechanical machine is most likely to have a process path as shown?
There may be more than one answer per process path

A to B _____
C to D _____
E to F _____
G to H _____
I to J1 _____
I to J2 _____

Figure 3.29.   A thermodynamics problem for the reader to complete.

More significant examples of the use of the First Law of Thermodynamics as it is applied to Open Flow Systems are given in the following example problems. Try solving them.

1. An Inventor comes into your lab with a sealed enclosure that contains a new heat engine that can produce 10 kW with a heat input of 25 kW. The inventor will not allow you to examine the engine because it is not yet patented, but the temperature instrumentation on the control panel indicates a very steady, high temperature of 1,000°F, with a minimum temperature of 200°F. What can you conclude from your brief measurements and analysis?

2. What is the entropy change for the Inventor's system (mentioned in Question 1) at the constant HIGH temperature AND the constant low temperature? Comment on whether this agrees with the Second Law of Thermodynamics.

3. A steam turbine (see below) has an inlet pressure of 500 psia for the turbine and a temperature that is 1,000°F. The outlet temperature of the heat exchanger is 14.7 psia at a temperature that is 250°F. The flow rate through the turbine is 10 lbm/s and you can assume that the turbine has no heat losses.

    A. Determine the power output of the turbine.
    B. Determine the entropy change at the exit of the turbine.
    C. Does this turbine violate the Second Law of Thermodynamics? Why or Why not?

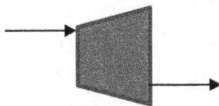

4. The heat exchanger (see below) operates with 10 lbm/s of water at 14.7 psia with an inlet state of 250°F. It rejects heat until the fluid is at a temperature of 100°F. Determine the heat transfer from the heat exchanger and the entropy change of the "Universe" if the ambient to which heat is rejected is a constant 60°F. What is the more common name for this heat exchanger

5. In a real world application, the heat exchanger mentioned in Question 4 will be cooled by water that has been taken from a river (or ocean) at 60°F, but the water will be allowed to be heated to 80°F before it is returned to the river or ocean. Assume that the heat source is slightly different than the one given in Question 4, in that it comes in as saturated water vapor (quality =1) at 14.7 psia and leaves as saturated water liquid (quality =0).

6. In your own words describe why a solar-heated hot water system has less exergy destruction (or less irreversibility) than an oil or natural gas fired furnace that is typically used to heat hot water?

7. You have been asked to develop a new power cycle using at least three of the following mechanical devices that are readily available from various vendors: turbine, compressor, pump, valve, nozzle, diffuser and heat exchanger. You may choose any cycle that you find in the literature or textbook. The efficiency of the

cycle will be the metric that your customer will use to decide whether to buy your cycle or your competitors'. You must use R134a as the working fluid, but you can choose any pressure or temperature for any of the components.

A. Devise a cycle and draw its block diagram.
B. Draw the process on a T-s diagram, labeling the state points 1 through n.
C. Fill in the property table that is associated with this cycle.
D. Calculate the cycle efficiency for as much or as little heat input that you care to use.
E. Compare this cycle efficiency with the Carnot Cycle efficiency, using the appropriate information from the state points.

8. The engineering partner of the Inventor in Question 1 returns with another hidden machine that she claims can produce 1 kWe for about 2 hours. She is only willing to reveal that the system uses air at a temperature of 500°F and a pressure of 100 psia, and has a volume, V, of 100 ft^3. Just as you thought you would have to do a complex analysis of her claims, you recall that a recent lecture on exergy (or availability) involved exactly the same problem. In fact, you will recall that the Instructor provided a convenient equation that simplified the calculation. This will make the validation of her claim much easier, BUT in order to make sure the solution is clear to the Inventor, you must state in your own words what the equation represents, along with your calculations showing whether her claims are or are not valid.

3.8   The thermodynamics solution procedure

The solution of any thermodynamics system follows the following step-by-step procedure. The reader was shown an example of this solution methodology in Example 1.3 in Chapter 1. That procedure is summarized for easy reference here:

1. Determine if the problem is best solved if the system is a Closed Mass or Open Flow System.
2. Construct the Control Volume around the system elements of interest. That is expected to help you solve the problem easily, or at least in a straightforward manner.

3. Label the Control Volume with ALL the known parameters or properties of the system; literally translating the problem statement that is in writing or that is offered to the analyst in oral communication.
4. Identify the unknown parameter(s).
5. Draw these three graphic tools: (a) a property chart that describes the process path as a function of at least two thermodynamics properties (such as T-v, P-v) connecting two or more state points of the system, (b) a Property Table that describes the properties at each state point, and (c) the energy process summary of each process path wherein the heat transfer, Q, the Internal Energy, U, and the Work done on or by the system, W, is described in detail.
6. The first objective is to fill in the Property Table with all of the information given in Steps 3 and 4. Remember that each process can be completely defined if and only if two properties are known for that state point, assuming that those two properties are not only P and T if the state point is in the two-phase region.
7. Write down the correct equations that are expected to help solve the problem. Like any algebra problem, the number of unknowns in Step 4 must match the number of available independent equations. The First Law of Thermodynamics is the first equation that must be written. The second, third, and n equations will come from basic physics, chemistry or fluid dynamics. Examples of these equations are the most useful way of demonstrating this step.
8. Solve for the unknowns.
9. Check the answer using the "sanity" check, which determines whether the answer is reasonable in quantity, magnitude or sign (positive or negative). Check the answer against the solution of a similar problem that has either been previously done by you or has been demonstrated in a textbook.

Chapter 4

Entropy and Thermodynamics Analysis using the Second Law of Thermodynamics

A thorough understanding of the Second Law of Thermodynamics is essential for any practicing engineer. The First Law of Thermodynamics that was presented for the first time in Chapter 1 and then discussed in detail in Chapters 2 and 3 is more easily understood than the Second Law of Thermodynamics. The reader will recall that the concept we label as "energy" is entirely defined as a force moved through a distance, and that this energy must never increase or decrease in magnitude between the start and the stop of an event, though it may not be in a useful form after the event has occurred. During the event, energy may be transformed from one form to another, and that form may be either useful or useless in terms of real-world applications of energy.

Let us not focus on the Big Bang because someone might ask: "If the Universe started from pure energy to form the masses — the universe of planets, stars and galaxies — that we now have, what put that energy there in the first place?" Let us proceed and agree that the laws of thermodynamics are useful here, even if it is difficult to imagine or understand the "beginning" of all things. Let us move further onto why engineers during the 1700s needed more than just the First Law of Thermodynamics to complete their understanding of how energy was transformed and whether there was an upper bound of converting one form of energy to another, more useful form of energy. Even though the analyst has a strong intuitive feel for the First Law of Thermodynamics and has some facility in applying it to the solution of energy-related physic problems, the First Law still falls short in identifying the direction of a process. The First Law, for example, could allow a glass of cold water on a table to get colder without being connected to a machine or without being otherwise influenced by an external effect. The First Law of Thermodynamics would permit energy to be removed and transferred to the open atmosphere. As long as the magnitudes of the energy removed from the glass of cold water are equal to the amount of energy that is put into the atmosphere, the First Law of Thermodynamics is satisfied. But of course, no one in the real world has ever observed a glass of cold water get colder in an environment that is hotter than it. This very clear and ubiquitous observation can be explained and understood by the Second Law of Thermodynamics. Using this law, the concept of entropy constrains the direction that a thermodynamics process must take. But the Second Law does not only limit the direction a process may take. Perhaps its most useful attribute is that it can determine the best possible outcome of a thermodynamics process. This ability can serve to gauge whether a real-world process is as good as it can be, and if not, it determines how close to the possible result can it achieve.

A good analogy would be during the dark ages when alchemists were struggling to turn common metals into sometime more precious, like gold. Most would agree that it would have been good for the alchemists to know that what they were trying to do was theoretically impossible based on a physical law of the Universe.

The Second Law of Thermodynamics can help the engineering analyst to determine the possible limit when designing a new engine, perfecting a new chemical reaction or defining how a living cell can process carbohydrates into sugars and thus ultimately into energy. Of course, we must not ignore the fact that even though alchemists had no guidelines to advise them on what was possible (or not), they uncovered many chemical reactions that benefited mankind through their efforts of sticking to the problem. It is interesting to ponder how much slower the accumulation of scientific knowledge would have been if they had been aware of the futility of their efforts.

With the Second Law entrenched in the foundation principles of the Universe, an engineer can be confident that the engine that he/she built is as good as it gets. Certainly, achieving the best engine as defined by the Second Law of Thermodynamics sets the mark of perfection. It is even more important than defining the limits to the real world's possible achievements when one considers the importance that the Second Law gives to temperature. Yes, the temperature that anyone can measure using the inexpensive pouch thermometer is finally given its throne to master the Universe. From the beginning of the Universe and its endless expansion, temperature continues to drop and will continue to do so until the temperature reaches zero or absolute °R or °K on the absolute temperature scale. Unless of course the gravitational attraction of all of the Universe's masses attract each other again to bring the Universe to a zero entropy (no chaos) state but at almost infinite temperatures.

While the future of the Universe is interesting to ponder from a thermodynamics philosophical vantage point afforded by the Second Law, it is much more practical to apply it to problems that affect the human condition, which may be remedied by the real-world engineer. Therefore, let us explore how the Second Law can be applied to real-world engineering problems, by starting to define entropy in thermodynamics terms — for example, how entropy increases due to chaos increasing — and not just in philosophical terms.

4.1　The definition of entropy

Leaving aside the entire philosophical quandary about the Universe descending into chaos and thus into increased entropy, the engineer's more useful equation for entropy has two useful equations that can be used to calculate the value of the property that engineers call entropy. One of these is for the macro-world of engineering while the other is more useful in the nano-world, where the probability of atomic events is the rule. Since the advent of the study of thermodynamics, the macro-world view of entropy has been more useful for devices that are larger than a biological cell — for example, steam engines, automobile engines and pumps. Then the arrival of electron microscopes and the continued exploration of nanophysics, as it may apply to biological systems as well as to electrical engineering systems, required a more probabilistic method for calculating entropy.

In the nanoworld the entropy change is: $S = k \times \sum_{i}^{n\,particles} p_i \ln(p)$, where k = Boltzman's constant = 1.38×10^{-23} J/K (Boltzman's = R_u/Avogadro's No. = 6.025×10^{22} molecules/g mol) and p is the probability of the state of a substance.

Note that the probability parameter is a measure of the chaotic motion of the nanoelements. As the motion becomes more chaotic and less predictable, then the entropy of the system of elemental particles becomes larger. This nanoscale reality becomes the source for the edict that entropy increases as the randomness of the system particles (also known as 'chaos') increases; in other words, as chaos increases so does entropy. It is also interesting to note that physicists as recently as 2012 have noted that though individual molecules can sometimes violate the Second Law of Thermodynamics on a nanoscale, but on a macro scale, the Second Law of Thermodynamics remains intact and predictable with the cumulative effects of very large numbers of nanoparticles all contributing to the net results of a process.

This reality may certainly be true, but then how does one explain that the entropy of the Universe has increased for a common activity, such as the writing of this textbook? After all, the textbook is the culmination of many pages of drafts of the text and derivations for the equations that have been used throughout, and that chaotic jumble of corrected pages, through a process, ends up as clean, neatly numbered, edited pages of a textbook. From chaos to neatness, from high entropy to low entropy. How is this possible if the entropy of the Universe must increase as required by the Second Law of Thermodynamics? The answer is straightforward if one understands that the "Universe", as defined in thermodynamics, is composed of two things: The System — a collection of items that is of most interest to the engineer/analyst — and its surroundings. Think of the System as those elements inside the Control Volume (CV) and the surrounding as everything outside the System. These, when taken together, are considered to constitute the "Universe" from a thermodynamics point of view. A sum of the changes in the entropy of the system plus the surroundings will always be greater than or equal to zero, taking into account the correct sign of the heat transfer to denote its direction. Heat transfer into a system is considered to be "positive". Heat transfer out of a system is considered to be "negative".

In the macro-world where the First and Second Laws of Thermodynamics have been the basis for the bulk of the thermodynamics processes that have benefited mankind, the equation to calculate entropy change is:

$$dS = \frac{\Delta Q}{T}. \tag{4.1}$$

The temperature must be in the absolute temperature scale and must be held constant during the exchange of heat transfer. That the temperature be held constant is very critical to the calculation of entropy change for a system. Chapter 2's introduction to the Carnot engine indicated that a heat engine that prescribes to be the most efficient must tend to receive and reject heat only during constant temperature processes. The reason for this may have been unclear as a basic principle, but is now made clearer by observing the link between the perfect heat engine and the need to have entropy preserved. That is, that the entropy does not increase. Instead, the total change of the entropy of the surroundings plus the entropy of the system must be equal to zero for a perfect, reversible process. However, the reality facing most heat transfer processes is that the temperature of a system fluid must change as heat is added or removed from the fluid. The exception to this is when the fluid is changing phase from liquid to vapor (evaporating) or from vapor to liquid (condensing).

When the process involves a change in temperature, the First Law with the help of calculus can be used to determine the entropy change of the system. The derivation of this change in system entropy is very straightforward and is as powerful in its utility as it is easy to derive.

4.1.1 *Derivation of change in system entropy*

Starting with the First Law, with Tds replacing Dq, the derivation can proceed as follows:

$$Tds = dU + pdV. \tag{4.2}$$

The reader should be able to easily identify the dU as the increment in the thermal internal energy, and the second term as the expression for an incremental change in the work that occurs during the process.

The internal thermal energy can be expressed as: $dU = C_v \times \Delta T$.

Substituting this equation into Equation (4.3a) and then dividing the right and left sides of the equation by T, establishes a differential equation that can be easily integrated, if the properties of specific heat are assumed to be constant, and the gas that is being analyzed for its change in entropy can be treated as an ideal gas. The result is:

$$\Delta S = C_v \times \ln \frac{T_2}{T_1} + \frac{R_u}{MoleWt.} \times \ln \frac{v_2}{v_1}. \tag{4.3a}$$

If the equation for enthalpy is:

$$h = u + Pv \, in \, its \, derivative \, form: dh = du + pdv + vdp$$

and the expression for du is substituted into Equation (4.3a), then it can be shown that the equation for entropy can also be expressed as:

$$\Delta S = C_p \times \ln \frac{T_2}{T_1} - \frac{R_u}{MoleWt.} \times \ln \frac{p_2}{p_1}. \tag{4.3b}$$

With the use of either equation, any process that changes temperature can now have its entropy determined. The choice of which one to use is perhaps best determined by knowing if there are changes in the volume or pressure during the process.

But there is a subtlety that needs to be emphasized when calculating the entropy change for a process, in addition to whether the temperature changes at constant temperature. The subtlety is that there are really two "parts" for any single thermodynamics process — one part undergoes the process, while the other involves its surroundings that may or may not have had heat transferred to or from it, in order for the process to occur. Consider

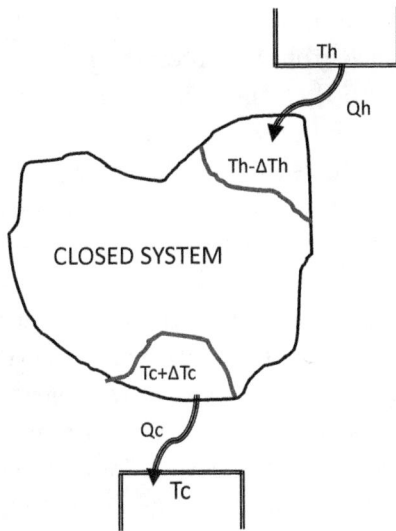

Figure 4.1.   Closed Mass System exchanging heat with two temperature reservoirs.

the CV shown in Figure 4.1. One can consider the inside of the CV as the system and the surroundings as the outside of this system. Given that for any CV, these two entities — the system and its surroundings — are all that the thermodynamics "Universe" contains, it is the sum of these two entities that the total entropy of the Universe must increase and be equal to or greater than zero.

For this to be calculated correctly, the sign convention of the heat transfer that may be occurring to or from the system must be considered properly. Heat transfer into a temperature reservoir is considered a positive value. Heat transfer out of the temperature reservoir is considered negative. It is vital that the direction is considered with respect to the temperature reservoir. In Figure 4.1, the heat transfer into the reservoir must be from a hotter heat reservoir than the surface/boundary of the CV. The heat transfer from the hot reservoir is negative because it is leaving the hot reservoir. The heat transfer must be a positive value from the vantage point of the CV boundary, where the temperature at the CV is less than the heat source's temperature. The heat transfer that leaves the hot reservoir resulting in a negative entropy change is the same heat transfer that re-enters the CV temperature, resulting in a positive entropy change. The sum of these two must always be greater than zero. The system's entropy change may be observed to be a "look up", in the sense that the entropy properties can be looked up in any collection of thermodynamics properties to determine how the entropy of the system has changed during the process. While these look-up Property Tables were traditionally part of the thermodynamics textbook, these days, such look-up properties can be obtained from the internet whenever they are needed.

It is important to note that the sum of these two entropy changes gets closer and closer to zero as the source and the CV temperatures approach one another. The sum of these two entropies is equal to zero when these temperatures are the same. However, if these temperatures are the same, then there cannot be a heat transfer exchange between the heat source and the system. In summary, the Second Law, that states that "...the entropy change of the Universe must be greater than or equal to zero", can never be violated. The heat transfer must always be zero between two temperature reservoirs for a perfect process to occur or that heat transfer is impossible across two temperatures that are at the same temperature. These observations lead to the conclusion that the Second Law of Thermodynamics cannot be violated, either in the real world or any alternate universes.

This revelation also leads to an interesting and acceptable analysis methodology. If it is true that the entropy between a high temperature reservoir and the CV boundary approaches zero as these two temperatures approach each other, then this can be used to heighten the understanding of entropy change as it applies to a heat engine or any thermodynamics process that requires a heat transfer exchange. That is, one can assume that the heat transfer that enters a system at one or more boundaries of a CV around that system is actually leaving the temperature reservoir at the same temperature as the temperature at the CV boundary. When this concept is used, whenever there is heat transfer entering at a system boundary, it can be assumed to be at the system's temperature. However, this heat transfer may be considered as a negative heat transfer because it is leaving the temperature reservoir. The same concept would be applied at the point in the system boundary where heat transfer is leaving to enter a cold reservoir. In that instance, the heat transfer is a positive value because the heat is entering the "cold" reservoir, which is actually at the same temperature as the cold boundary of the system that is losing its energy by heat exchange. This thought process maintains consistency that the entropy-change of the "Universe", as defined by the totality of the system and its surroundings, is greater than or equal to zero.

The alternative is to rely on another very well-known concept: the Clausius Inequality for heat engines. The typical mathematical formulation of the Clausius Inequality appears at first to violate the statement of the Second Law, as seen in the equation shown here.

$$\Delta S \leq \int_{T_1}^{T_2} \frac{dQ}{T}. \tag{4.4}$$

At first glance, the less than or equal sign looks erroneous for it seems to violate the stated principle of the Second Law. This is not the case, however. The "ambiguity" is easily dispelled when it is made clear that the Clausius Inequality requires the heat transfer into the system *at the CV boundary* to be a positive value because it is entering the system. In a similar manner the heat transfer leaving the CV somewhere at the system boundary must be negative because it is leaving that temperature reservoir. All heat transfers leaving a system are considered negative. If you were to apply values to the heat transfer and temperatures at the boundary and discover that unless the equation for the sum of the heat transfers around the system heat engine is allowed to be as given in the equation above, the entropy change of the Universe would be negative. Of course, the Clausius Inequality says nothing about the entropy change of the surroundings, which is part of the Second Law of Thermodynamics. However, even a quick example indicates that for any given high and low temperature at the system boundaries, the heat transfer entering the high temperature boundary may only be a few degrees higher in temperature, and that the heat transfer leaving the system's cold boundary and entering an even colder temperature reservoir may only be a few degrees colder, with the consequence that there will not be enough entropy change in the surroundings to overcome the entropy change of the system as used by the Clausius Inequality.

So, why bother with the Clausius Inequality? It is because it provides an easier way of looking at the heat transfer into or out of the system without resorting to the methodology described earlier in this section, when the heat leaving a high temperature reservoir is at the same temperature of the hot system boundary, and similarly for the cold boundary heat exchange loss. The entropy change between the hot temperature reservoir and the hot system boundary can be considered zero in a utopic world where heat transfer could perhaps still occur across zero temperature difference with zero entropy change. The system entropy can then continue to be determined by using the Property Tables for the fluid that is under consideration as the system.

4.2 Closed Mass and Open Flow System equations for entropy change

The CV shown in Figure 4.1 illustrates a Closed Mass System. The heat transfer designations shown indicate two heat transfers entering the closed mass and one heat transfer exiting the system. Thus, it is immediately understood that the temperature providing the heat transfer must be greater than the temperature at the surface of the closed mass, where that heat transfer is exchanged. Similarly, the heat loss from the system must be at a temperature that is higher than the heat sink it is in communication with. The entropy change of the Closed Mass System follows the description that was given in the previous section — the total entropy change of the Universe must consider the surroundings as well as the system. Thus, the total entropy change is the sum that follows:

$$\Delta S_{system} + \Delta S_{surroundings} = \sigma_{destruction} \qquad (4.5)$$

or as may be expressed using Clausius' Inequality:

$$\Delta S_{system} - \int_{T_{1b}}^{T_{2b}} \frac{dQ}{T} = \sigma_{destruction}, \qquad (4.6)$$

where $\sigma_{destruction}$ is the parameter whose magnitude is always greater than or equal to zero per the Second Law of Thermodynamics. The "b" subscript with the temperature parameters refers to the boundary of the CV.

Let us put some values to these parameters to verify the Clausius Inequality and then refer to Figure 4.1.

$$T_h = 1,000°R, T_c = 500°R, \Delta T_h = 50°R, \Delta T_c = 200°R,$$
$$Q_h = 100\,Btu, \text{ and } Q_c = 60.$$

4.3   Entropy change for an open flow, control volume system

An open flow, CV system is shown in Figure 4.2. Recalling that the entropy change of the "Universe" must be greater than zero, and continuing from where we left off with the Second Law for a Closed Mass System, the Second Law of Thermodynamics applied to the open flow CV system shown in Figure 4.2 can be expressed as:

$$\sigma_{destruction} = \sum \dot{M_e} s_e - \sum \dot{M_i} s_i + \sum \frac{\dot{Q_k}}{T_k} + \frac{\partial S_{cv}}{\partial t}, \qquad (4.7a)$$

where the subscripts for the heat source and heat sink term, $\sum \frac{\dot{Q_k}}{T_k}$, refer to the surrounds AND NOT the boundary of the CV. In many textbooks, these subscripts may be "b" and refer to the CV boundary. In those instances, Equation (4.7a) must be written as:

$$\sigma_{destruction} = \sum \dot{M_e} s_e - \sum \dot{M_i} s_i - \sum \frac{\dot{Q_b}}{T_b} + \frac{\partial S_{cv}}{\partial t}. \qquad (4.7b)$$

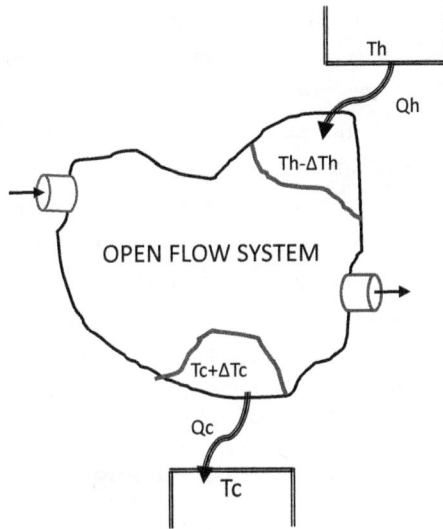

Figure 4.2.   Open Flow System exchanging heat with two temperature reservoirs.

4.4   The concept of irreversibility (or exergy destruction)

The determination of the entropy change of the "Universe" must be equal to or greater than zero, as expressed by the equation: $\Delta S_{universe} >= 0$, which leads to an interesting and important concept: irreversibility, or exergy destruction. Both these equivalent terms are presented in more detail in the following sections. Suffice to say, irreversibility (or exergy destruction) is a measure of the inability of "real world" engineers to accomplish the maximum amount of work (or power) from a source of energy that is available at a temperature, T_h, above a datum temperature, T_o, that is provided by nature. Irreversibility (or exergy destruction) can be most intuitively understood to be the numerical difference between two work or power parameters and expressed as:

$$\dot{I} = \dot{W}_{ideal,maximum,theoretical} - \dot{W}_{actual,\,real-world,\,measureable\,and\,useful}. \quad (4.8)$$

It can also be shown that:

$$\dot{I} = T_o \times \Delta S_{"universe"}. \quad (4.9)$$

Thus, it is possible to determine irreversibility by either of these equations. But what is the need and significance of knowing irreversibility or exergy destruction? Simply put, it provides an opportunity for the engineering analyst to know how close to perfection the cycle or process that is under consideration is. Referring again to the alchemists from the Middle Ages, the ability to predict the maximum work or power that can be achieved is akin to giving them the knowledge of whether their chemical experiments would have produced gold from lead or stone.

Many examples can demonstrate the strength of the concept of irreversibility (or exergy destruction) and a simple example would be enough to dramatize the significance of this concept. However, it is first necessary to describe the Carnot Cycle and the simple but elegant equation that determines the numerical value of the Carnot efficiency as a function of only two temperatures; the very same two temperatures that are used in the determination of the irreversibility: T_{hot} and T_o. Understanding the Carnot Cycle is the key to understanding the concept of irreversibility.

4.5   Carnot cycle efficiency for a heat engine

The Second Law of Thermodynamics leads to a very important and elegant simple equation that is also often stated as the alternate "face" to the Second Law of Thermodynamics. Consider the heat engine system shown in Figure 4.3. The hot temperature heat source and a cold temperature heat sink are clearly identified, and a cyclic heat engine is operating between these two reservoirs with the heat continually coming from the hot temperature reservoir. An amount of heat transfer that is less than the heat energy entering the heat engine is shown leaving the cycle and entering into the cold thermal reservoir. If one considers the heat transfer leaving the hot reservoir and the small amount of heat transfer entering the cold temperature reservoir equal to the difference between the heat input and the work performed by the heat engine (as determined by the First Law of Thermodynamics), the entropy change for the temperature reservoirs added to the entropy change for the cyclic internal processes (which is zero) will be the result shown in Equations (4.10a–c):

$$\Delta S_{hot} + \Delta S_{cold} + \Delta S_{system} = 0, \; for \; a \; reversible \; cycle, \qquad (4.10\text{a})$$

$$\Delta \frac{Q_h}{T_h} + \Delta \frac{Q_c}{T_c} + \Delta S_{system} = 0, \qquad (4.10\text{b})$$

$$\text{For a reversible cycle: } \Delta S_{system} = 0. \qquad (4.10\text{c})$$

By observing the positive (+) sign convention for heat into a temperature reservoir and the negative (−) sign convention for heat exiting a temperature reservoir, then

$$\frac{Q_h}{T_h} = \frac{Q_c}{T_c}. \qquad (4.11)$$

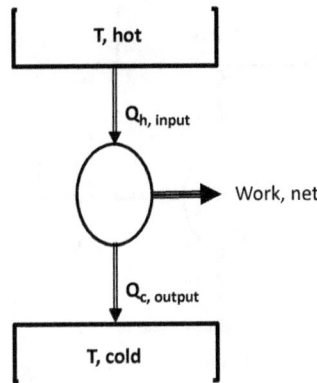

Figure 4.3.   Carnot Cycle heat engine.

Cycle efficiency is defined as the work output divided by heat input, as shown in Equation (4.11).

$$\eta = \frac{Work}{Heat\,Input}.$$
(4.12)

Using the equal sign to identify the best possible cycle for the heat engine and using Equation (4.12):

$$\eta = 1 - \frac{T_c}{T_h}.$$
(4.13)

This equation is the famous Carnot Equation, which is named after its developer Sadi Carnot.[1]

The equivalent to this is the thermodynamics refrigerator and heat pump, which is diagrammed in Figure 4.4. Using the definition of refrigerator efficiency, known as the Coefficient of Performance (COP), then the equation for the perfect refrigerator is shown in Equation (4.14a).

$$COP_R = \frac{Qcooling}{Work\,Input}; COP_{R,\,max.\,carnot} = \frac{T_c}{T_h - T_c}$$
(4.14a)

The perfect heat pump is similarly derived and is given in Equation (4.13b).

$$COP_H = \frac{Qheating}{Work\,Input}; COP_{H,\,max.\,carnot} = \frac{T_h}{T_h - T_c};$$
(4.14b)

---

[1] The equation developed by Nicolas Léonard Sadi Carnot is comparable in its importance to thermodynamics as Einstein's famous equation that equates mass and energy: $E = MC^2$. Thermodynamics lost a major scientific mind when Sadi Carnot died of cholera before he was 30 years old.

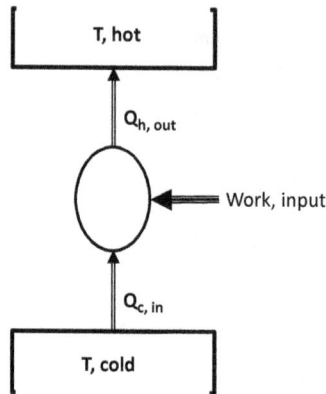

Figure 4.4.   Carnot refrigerator/heat pump.

The COP_R in Equation (4.14a) refers to the refrigeration cycle where cooling is the desired result of using the refrigerator, with the need to provide some effort from outside the system in the form of work or power (shown as Work Input). Similarly, for the heat pump, COP_H is used when there is a need for heat to be delivered to a space. As before with the refrigeration cycle, there must be an effort exerted in the form of work or power into the system.

In fact, there are two corollary statements of the Second Law of Thermodynamics that must be considered. These are the Kelvin–Planck and Clausius statements as they are applied to a heat engine cycle. These laws indicate that a heat engine working in a cyclic manner for continuous power generation cannot have the input of heat transfer at the high temperature part of the heat engine cycle converted all to useful work or power output. That would be equivalent to indicating that the engine has a 100% cycle efficiency, which would violate the second Law of Thermodynamics if the heat engine cycle is operating between a high temperature and, at the very least, the dead state or local ambient environment. The only time this could happen is when the cold local temperature is at zero K or zero R. However, neither of these temperatures have been established in laboratories or the physical universe. In fact, the temperature of the known universe is measured to be 2.5 K.

In a similar manner, the Clausius statement of the Second Law of Thermodynamics indicates that a refrigerator or heat pump (shown in Figure 4.4) cannot have the extraction of heat energy from a cold space as the sole effect on a cycle, without also providing some work or power to the cycle. Likewise, there cannot be heat input to a hot space that is only based on the work or power input to the cycle, because the heat pump (or refrigerator) cycle would be infinite in its COP. Once again, there has never been an instance when, for example, a cold glass of water sitting on a table in a warm room spontaneously gets colder until the process is stopped or 0 K or 0°R is attained. The First Law of Thermodynamics is not sufficient to provide a constraint from this happening. In the case cited, for example, the energy removed from the cold glass could be given to the warm environment and thus the energy extracted, being equal to the energy supplied, is in step with the First Law. However, the world would be a very cold place if all substances spontaneously get colder without any work or power being used to affect the heat extraction from the cold space. The Anthropic Principle can be used to cite that the Second Law is not valid, but you, the reader, would not be doing this textbook any justice if you think that the Second Law of Thermodynamics is not valid or universal in its application.

The ideal, perfect heat engine, refrigerator and heat pump are essential for putting engineering thermodynamics into perspective, with respect to what can be achieved in the real world of engineering. They present the engineer with a goal (though some would say a constraint) that can be strived for but never exceeded when contemplating the design and engineering of the best heat engine. In fact, the reader should understand the purpose of a perfect heat engine and refrigerator is to help readers better understand and have a non-intuitive reality of how the Universe works to increase entropy and make unavailable a magnitude of work or power due to an exchange of heat between two temperature reservoirs. This unavailable work or power is now given the relatively new name of "energy destruction". In older textbooks this term is known as "irreversibility". Both these terms have units of energy or power.

It is worth exploring several examples of how these concepts of a perfect heat engine and perfect refrigerator and/or heat pump in order to achieve a better understanding of the true effectiveness of the best possible engine and refrigerator/heat pump.

4.5.1   *Example 4.1*

What is the maximum cycle efficiency that can be experienced for a heat engine that operates between a high temperature reservoir of 1,000°R and a low temperature reservoir of 600°R? What is the power that can be generated with a heat energy input of 1,000 Btu/h?

Answer: The maximum cycle efficiency is equal to the Carnot Cycle efficiency as given by Equation (4.12). With the temperatures given, the Carnot

Cycle efficiency is:

$$\eta = 1 - \frac{T_c}{T_h}; \eta = 1 - \frac{600}{1000}; \eta = 40\%.$$

With a heat input of 1,000 Btu/he, the net useful "ideal" power is $1,000 \times .4 = 400\,Btu/h$.

4.5.2 Example 4.2

What is the minimum power required to cool a space if 12,000 Btu/h must be extracted from that cold space, which must be maintained at a temperature of 0°C and ambient temperature of 25°C?

Answer: The minimum required power is what can be achieved using a Carnot refrigeration cycle as modeled using Equation (4.13a).

$$COP_R = \frac{Qcooling}{Work\,Input}; COP_{R,max.\,carnot} = \frac{T_c}{T_h - T_c};$$

$$COP_{R,max.\,carnot} = \frac{0 + 273}{(25 + 273) - (0 + 273)}; COP_{R,max.\,carnot} = \frac{273}{25};$$

$$COP_{R,max.\,carnot} = 10.92.$$

Therefore, $Work, minimum = 12,000/10.92$ and $Work, minimum = 1,099\,Btu/h$.

It is interesting to note that the heat input quantity of 12,000 Btu/h is given a special name in thermodynamics. It is referred to as a Refrigerant Ton (RT). This is due to the practice of using blocks of ice for refrigeration when refrigeration systems were not available. This quantity assumes that a ton of ice can absorb 144 Btu/lbm of ice (known as the latent heat of fusion) to convert the solid water into liquid water during a 24-hour/day period.

Thus, $1RT = 1\,ton \times 2,000\frac{lbm}{ton} \times \frac{144\frac{Btu}{lbm}}{24}h$ and $1RT = 12,000\,Btu/h$.

4.6 The utopic world of perfect engineers, perfect engineering processes and determining irreversibility

The elegance and utility of the second law equations as expressed in the Carnot Cycle efficiency (efficiency $= 1T_{cold}/T_{hot}$) and $I = $ exergy destruction $(E_d) = T_{ambient} \times$ entropy change of the Universe $= W_{real} - W_{net}$ sets the stage for another helpful tool in understanding and using the Second Law to visualize the possibilities and limitations of a thermodynamics process. This author suggests considering two worlds. The first world being

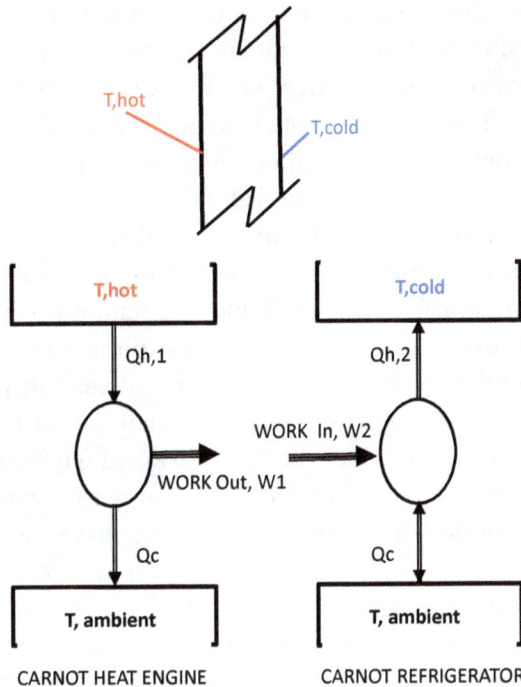

Figure 4.5.   Heat transfer through a tube wall used in a heat exchanger.

our own "real world" of possibilities steeped in the reality that the Second Law must require that the entropy of the Universe be greater than zero. The second world is what might be called the "utopic world", which would be populated by perfect engineers who deal with perfect heat engines, refrigerators and heat pumps. In fact, the engineers in this world would only need to utilize perfect heat engines and refrigerators with ratings that are as large or small as necessary to accomplish what is desired and asked of them by their real-world engineering counterparts. These engineers will be able to provide real-world problems with solutions that involve only the correct application of a Carnot Cycle heat engine and/or a Carnot Cycle refrigerator or heat pump. For example, in the real world, the heat transfer through a wall can only occur due to a temperature difference between the two surfaces of the wall, as shown in Figure 4.5. This heat energy transfer due to a temperature difference produces no useful, measurable power because the intent in the real-world application of a wall is to exchange heat between the two wall temperatures. This is the basis for the operation of a heat exchanger. The hot fluid on one side of the heat exchanger tube wall is exchanging heat with the cooler tube wall. This energy is then used to heat the cool fluid stream to a higher temperature. The goal is to cool something on one side of the heat exchanger while heating the coolant stream. When this "assignment" is given to the perfect engineers, they are able to see the problem somewhat differently. They will regard the heat transfer that needs to be done across the two wall temperatures as an opportunity to produce useful, measurable power or work. How will they accomplish this task?

Consider the graphic shown in Figure 4.5, which illustrates a wall through which heat is transferred with wall temperatures, T_{hot} and T_{cold}. Here, you will see that if heat transfer, Q, is available at a temperature, T_{hot}, then that energy could have first been passed through a perfect Carnot Heat Engine and produced useful power or work. But of course, the perfect engineers cannot simply stop there because the colder wall would then not have the incoming heat energy to do whatever the real-world engineers had originally intended to do. This is handled perfectly by the perfect engineers, who will simply construct a perfect heat pump and extract just enough of heat energy from the ambient heat reservoir (sometimes referred to as the dead state) to provide the same amount of heat transfer, Q, that was required by the real-world engineers to be presented or released at the cooler wall temperature. Of course, this heat pump requires power or work energy, W_c. The net gain in ideal perfect power or work energy is the difference between these two powers or work energies, i.e., $W_{ideal, \, maximum, \, theoretical} = W_t - W_c$.

This net gain of energy or power will never be accomplished because the real-world engineers did not intend to produce work or power, since they only needed heat re-transfer to occur to accomplish whatever the heat exchanger's function was intended to be. That they did not or could not achieve even a fraction of the possible work or power recovery identifies an irreversible or exergy destruction that is lost forever. Equation (4.15) defines irreversibility (or exergy destruction) as follows:

$$\dot{I} = \dot{W}_{ideal, maximum, theoretical} - \dot{W}_{actual, \, real-world, \, measureable \, and \, useful}. \quad (4.15)$$

The first term is the difference between the Carnot Cycle power produced and the Carnot Refrigerator power consumption. The second term in this example is zero because the real-world engineers did not intend nor therefore attempt to produce useful power from the energy exchange across the wall.

Using the equations for the Carnot Heat Engine cycle efficiency and the Carnot Refrigerator as presented previously in this chapter (see Equations (4.14a) and (4.14b)), an equation for the irreversibility, I, can be derived as a function of the amount of heat exchanged, Q_h, the two wall temperatures, T_h and T_c, and the ambient. The derivation follows.

4.6.1   *Derivation of irreversibility (also called Exergy destruction) for wall heat conduction*

Starting with the work generated by the Heat Engine working between T_h and T_o:

$$\dot{W}_t = \dot{Q}_h \times \left(1 - \frac{T_o}{T_h}\right).$$

Continuing with the Carnot Refrigerator expression for the work required:

$$\dot{W}_c = \frac{\dot{Q}_h}{\left(\frac{T_c}{(T_h - T_c)}\right)}.$$

The irreversibility is determined from:

$$\dot{I} = \dot{W}_t - \dot{W}_c.$$

Substituting the equations for W_t and W_c into Equation (4.15) and carrying out the necessary algebra derives the following equation:

$$\dot{I} = \dot{Q}_h \times T_o \times \left[\frac{1}{T_c} - \frac{1}{T_h}\right].$$

The first important observation is that the irreversibility, I, or exergy destruction, E_d, is always a positive value because T_c is always less than T_h and the heat energy transfer has been considered positive during the derivation. It is then observed that irreversibility is zero when $T_h = T_c$. But of course, there would be no heat exchange then! This later observation may be considered by some to be a thermodynamics equivalent of a "Catch-22" situation. That is, the only way to achieve maximum work or power generation is when irreversibility is zero, and this can only happen when there is no heat exchange. But then this would mean that there is nothing to work with.

The measure of how well real-world engineers have done in achieving what perfect engineers can accomplish with the same heat transfer and hot temperature heat source, is identified as the second law efficiency, η_{2nd}. The equations for this law are not as easily defined as the first law efficiency, but it is important to calculate it and understand its significance in the engineer's attempt to achieve the best possible solution to a heat transfer thermodynamics process. For the example given here, the Second Law's calculation is the simple ratio of the amount of work that real-world engineers can accomplish compared to (i.e., ratio) the maximum work or power that perfect engineers can achieve. In a more general sense, the calculation of the terms used in the numerator and denominator must wait till the discussion of energy at the end of this section. Exergy will be equivalent to the maximum available work that the utopic world of precise engineering can accomplish, while exergy destruction is the potential to do useful work that is irreversibly lost. The difference between these two can be used to determine the maximum work that can be produced in the real world of engineering and thus obtain the second of the two variables needed to define the second law efficiency. Examples of this will be given after exergy is better defined.

A common real-world example of how everyone contributes to the increase in entropy and thus to an increase in exergy destruction (or irreversibility) is the simple and daily ritual of taking a shower or bath. Certainly, the intent is not to make power or work during the morning (or evening) ritual, but nevertheless, heat was transferred across a temperature difference several times during the thermodynamics process of getting the water hot before releasing it into the atmosphere for its final cool down. Regardless of how the water is first heated to a comfortable bathing or showering temperature, the source of the heat energy is at a higher temperature than the temperature of the water. For example, if the temperature of the furnace wall is as high as 500°F and the water temperature is set by a thermostat to be 110°F, the bath or shower water will eventually released into the ambient and thus the temperature of the water cools to 60°F, giving up its heat to the environment. In a utopic world, these temperature differences would be used to power perfect Carnot Heat Engines, starting with the flame temperature of the burning oil or natural gas at $1,500^{+}$°F.

The modeling of a utopic thermodynamics model depicting the heating of water from a hot heat source is straightforward and is best calculated by using a spreadsheet, where the rows of the spreadsheet represent micro-heat engines whose cold and hot temperatures are easily discerned from a graphic profile of the hot and cold temperatures that exist along the heat exchanger length as portrayed in Figure 4.6. Once these high and low temperatures are known, the application of the Carnot Heat Engine (shown in Figure 4.3; reproduced below for the reader's convenience) can be imagined to be many micro-Carnot Heat Engines as shown in Figure 4.6. The micro-heat engines can easily determine the micropower generated by the heat engines and the micropower consumed by the heat pumps with respect to the dead state, also known as the ambient temperature. The sum of these micro-heat engines and heat pumps provides the total power generated from the heat transfer at the hot temperature and the total power consumed to bring the same heat transfer back to the cooler walls of the heat exchanger. The difference

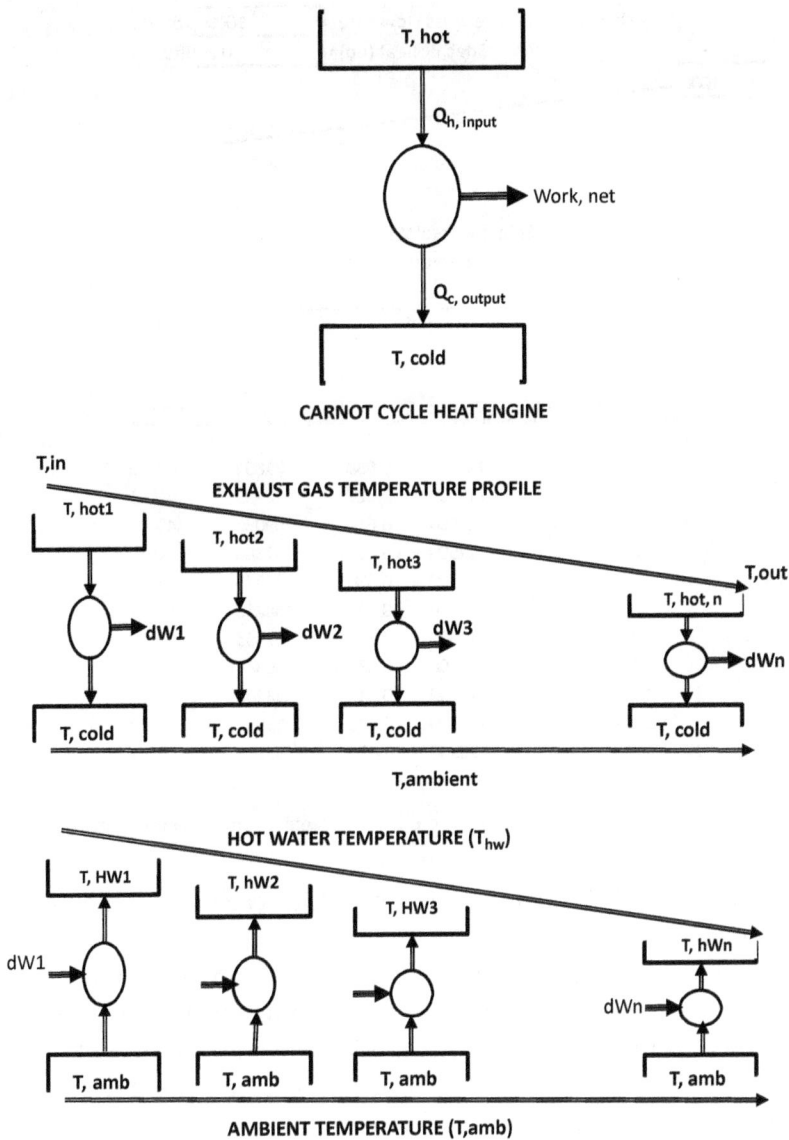

Figure 4.6. Illustrating the 2-step construction of how combustion gases can heat water using Carnot heat engines and refrigerators and heat pumps.

of these goals is the irreversibility (or exergy destruction) of entropy change of the Universe multiplied by the ambient temperature.

A sample calculation is shown in Figure 4.7 using a spreadsheet format to calculate only the Carnot Heat Engine micropower cycles shown in Figure 4.3. The corresponding calculation for determining the power required by the micro-heat pump cycles follow the same general protocol and is left

| | Exhaust Heat Source Mass Flow Rate= | | 10000 | Lbm/hr |
| | Specific heat (Cp)= | | 0.25 | Btu/Lbm/R |

1000

100

44.6%

| | Numerical Soln. for Total Wrev.sys. | 1,004,567 | Btu/hr |
| | Theoretical Wrev. | (1,004,269) | Btu/hr |

60

Tambient= To

DT, incre=   45

Section #	Texhaust	Tcold	Dq,hot	Carnot Eff.	Dw	Dq,cold
1	1000	60				
2	955	60	112500	0.638	71804	40696
3	910	60	112500	0.627	70489	42011
4	865	60	112500	0.614	69086	43414
5	820	60	112500	0.601	67586	44914
6	775	60	112500	0.586	65979	46521
7	730	60	112500	0.571	64253	48247
8	685	60	112500	0.555	62393	50107
9	640	60	112500	0.537	60384	52116
10	595	60	112500	0.517	58208	54292
11	550	60	112500	0.496	55841	56659
12	505	60	112500	0.473	53259	59241

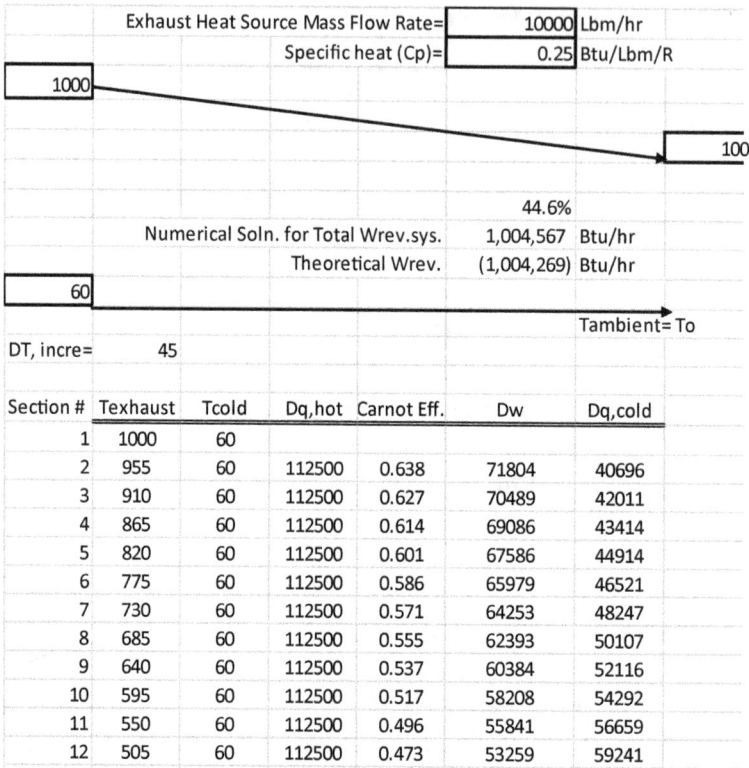

Figure 4.7.   Spreadsheet output for the micro-heat engine model shown in Figure 4.3.

as an exercise for the student. The protocol for calculating the micropower is as follows.

First, set up the columns as shown. These columns include the section number along the temperature change, the temperature of the hot exhaust gas and the cold temperature of that section, the amount of heat transferred in that section, the Carnot Cycle efficiency, the incremental work produced, and the heat rejected into the environment.

The amount of heat transferred from the hot heat source is found from the equation associated with a heat exchanger performance, or $Q = \dot{m} \times \Delta h$, where:

$$\Delta h = C_p \times (T_2 - T_1);$$

No phase change of the process fluid is encountered for this case.

The incremental amount of power produced is found by applying the Carnot Heat Engine cycle efficiency:

$$\eta = 1 - \frac{T_c}{T_h},$$

such that:

$$dW = dQ \times \left(1 - \frac{T_c}{T_h}\right).$$

The sum of all the incremental work determines the maximum amount of theoretically maximum power (or work) that could have been produced in the utopic world of perfect engineers. The fact that no work or power was produced from this heat source indicates that the sum of work or power is no longer available and that this work or power is irreversibly lost (or is destroyed exergy). In this instance the second law efficiency is zero because no work or power was produced that could be used in the real world to do useful things. Had there been an engine cycle that could have recovered the heat source at 1,000°F and produced a useful amount of power before discharging its waste heat to the water used in your bath, then the second law efficiency would be greater than zero, perhaps as large as 50%. This last figure is only an approximation by the author who uses a "rule of thumb" with regard to the Carnot Cycle efficiency and the performance of real heat engines. That "rule of thumb" simply states that the most practical heat engine that can be produced in the real world of imperfect engineers is about half of what could be produced in the utopic world of perfect engineers and engineering. One observes the range of Carnot Cycle efficiencies shown in Figure 4.7 to be from 64% to 47%. If one were to employ a reasonable "rule of thumb", engineers in the "real world" can only expect to achieve about half the performance that can be achieved in the utopic world of perfect engineering and engineers; in the real world, heat engine efficiencies may achieve efficiencies of 32% and 23.5%. Of course, in all applications of thermodynamics, it is best to calculate the efficiency of a real-world heat engine whenever possible.

A very close examination of Figure 4.8 reveals two calculations (circled in red) that are to be compared. The numerical solution is the result of the calculations performed by the numerical model. The second parameter, "Theoretical W_{rev}", is the theoretical calculation of the same power but based on a rather simple equation.

The equation is:

$$(e_{out} - e_{in} \equiv W_{rev.}) = h_{out} - h_{in} - T_o \times (s_{out} - s_{in}), \qquad (4.16)$$

where $(e_{out} - e_{in})$ refers to the exergy properties of the exhaust gas evaluated at the outlet/inlet pressure and temperature states, as given in this example's statement. Exergy is a thermodynamics property that is detailed later in this chapter. It is a measure of the potential work or power that can be accomplished by a fixed thermodynamics state of a system with

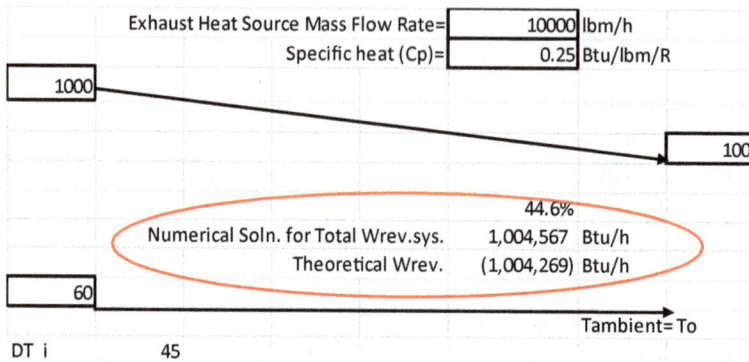

Figure 4.8.   Excerpt from Figure 4.7 showing the theoretical and numerical calculation of reversible power.

respect to the dead state or local environmental pressure and temperature. It is the thermodynamics property of a system that ultimately determines the irreversibility (or exergy destruction) inherent in any thermodynamics state of a system. It is mentioned here not only to prepare the reader for what is coming later in this chapter, but also because this property and the calculation shown in Equation (4.16) should be included in the numerical solution as a mathematical and thermodynamics check, and to impress upon the reader the equivalency of the exergy concept in thermodynamics and the more mechanical formulation of irreversibility (or exergy destruction) by using perfect Carnot Heat Engines and Carnot Refrigerators and heat pumps.

This previous example can be put into the values shown in Figure 4.7. The next example uses a condenser to exchange heat from condensing steam to coolant water, a very familiar engineering phenomenon in power plants in the real world of engineering applications. The results of the spreadsheet's programming are shown in Figure 4.9. Also included in Figure 4.9 is a calculation of the exergy for the condensing steam and the water coolant. The exergy equation is shown in the top left-hand corner of Figure 4.9. The entropy change for the condensing steam is observed to be negative because heat is leaving the steam as it condenses. However, the entropy of the water coolant that is receiving that same amount of heat is positive. The net change in the entropy after multiplying each of these entropies by their respective flowrates is observed to be a net positive, as it must be according to the Second Law of Thermodynamics.

Lastly, it is interesting to compare the numerical results of the spreadsheet calculation (shown as 2, 164 kWt in the spreadsheet with the theoretical

			Qcondensing =	3.49E+07	Btu/h
	Condensing Steam Heat Source Mass Flow Rate=			36000	lbm/h
	hsat.vap-hsat.liq =			970.7	Btu/lbm

212 F ————————————————————————————————→ 212 F

e=(h,out-h,in) - To x (s,out- s,in) ;   Exergy Change for Steam= | -2311.9 | kW

Entropy Change for Steam= | -1.445 | Btu/lbm-R

Coolant (Water) Mass Flowrate [lbm/h]= 1,747,347

Exergy Change for Water Coolant= | 186 | kW

80 F

60 F ————————— Entropy Change for Water= | 0.0378 | Btu/lbm-R

Numerical Soln. for Total Wrev.sys.= 7.385E+06 Btu/h;kW   2164

Dh, incre.= 48.5

DTcold, incre= 1.00   520 x {Mwater x Δs,water +Msteam x Δs,cond. Steam   7.253E+06 Btu/h;kW   2125   Error 1.8%

Section #	T,steam [R]	Tcold [R]	Dq,hot	Carnot Eff.	Dw	Dq,cold
1	672	520				
2	672	521	1747347	0.225	393933	1353414
3	672	522	1747347	0.224	391333	1356014
4	672	523	1747347	0.222	388733	1358615
5	672	524	1747347	0.221	386133	1361215
6	672	525	1747347	0.219	383532	1363815
7	672	526	1747347	0.218	380932	1366415
8	672	527	1747347	0.217	378332	1369016
9	672	528	1747347	0.215	375732	1371616
10	672	529	1747347	0.214	373131	1374216
11	672	530	1747347	0.212	370531	1376816
12	672	531	1747347	0.211	367931	1379416
13	672	532	1747347	0.209	365331	1382017
14	672	533	1747347	0.208	362731	1384617
15	672	534	1747347	0.206	360130	1387217
16	672	535	1747347	0.205	357530	1389817
17	672	536	1747347	0.203	354930	1392418
18	672	537	1747347	0.202	352330	1395018
19	672	538	1747347	0.200	349730	1397618
20	672	539	1747347	0.199	347129	1400218
21	672	540	1747347	0.197	344529	1402818

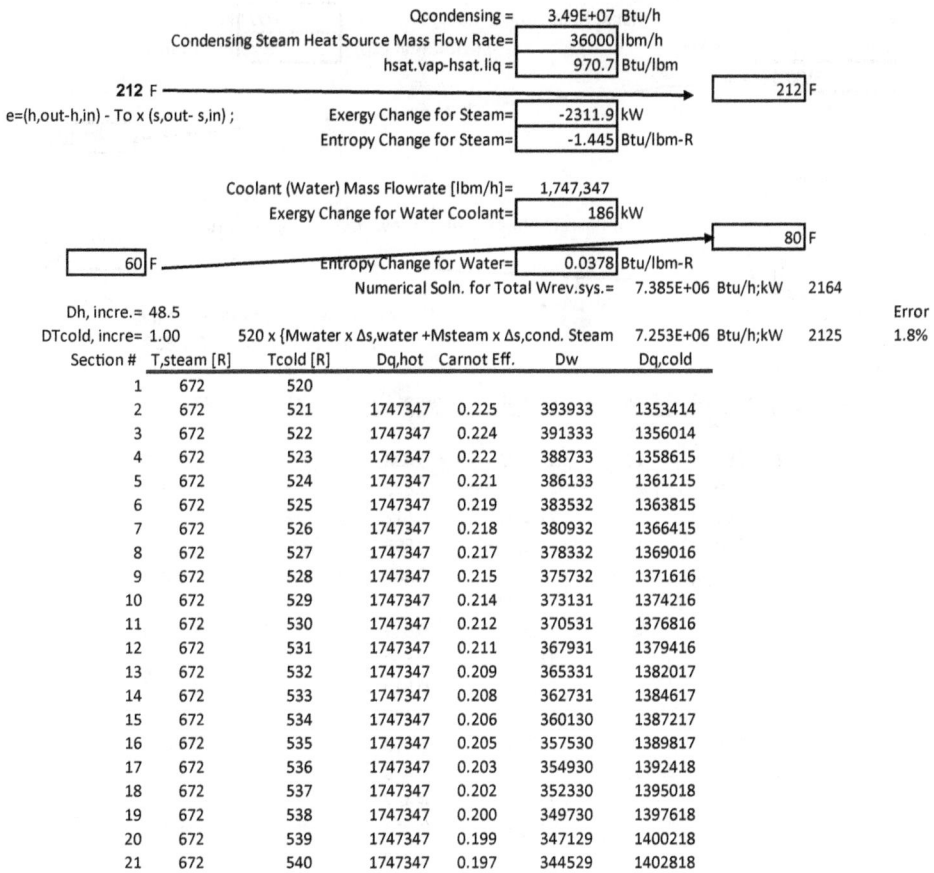

Figure 4.9.   Example of reversible work calculated for a condenser heat exchanger.

amount of power that could have been generated as a result of the condenser cooling, shown in the spreadsheet as 2,125 kWt or a 1.8% difference. The numerical calculation is based on adding the incremental power, dW shown in the left-column from the right. Each of these incremental power values is calculated using the Carnot equation applied to the "infinite" number of mini-Carnot Engine Cycles that could have been used between the source temperature of the condensing steam and the coolant temperature that appears in the left -most columns. The theoretical calculation is based on the product of the net entropy change for the condensing steam and the coolant water multiplied by the Dead State or ambient temperature, 520R. Thus, once again it is demonstrated that the very straightforward equations for exergy and exergy destruction and/or Irreversibility equal to the product of the Dead State Temperature, To and the entropy change

	Exhaust Heat Source Mass Flow Rate=	10000	lbm/h
	Specific heat (Cp)=	0.25	Btu/lbm/R

1000

100

43.4%

Numerical Soln. for Total Wrev.sys.   976,860   Btu/h

80

60

Water flow rate temperature profile

DThot, incre=	45			Water Flow rate=	112,500	
DTcold, incre=	1.00					
Section #	Texhaust	Tcold	Dq,hot	Carnot Eff.	Dw	Dq,cold
1	1000	60				
2	955	61	112500	0.638	71765	40735
3	910	62	112500	0.625	70368	42132
4	865	63	112500	0.612	68878	43622
5	820	64	112500	0.598	67284	45216
6	775	65	112500	0.583	65577	46923
7	730	66	112500	0.567	63742	48758
8	685	67	112500	0.549	61767	50733
9	640	68	112500	0.530	59633	52867
10	595	69	112500	0.510	57320	55180
11	550	70	112500	0.487	54806	57694
12	505	71	112500	0.463	52063	60437
13	460	72	112500	0.436	49058	63442
14	415	73	112500	0.407	45752	66748
15	370	74	112500	0.374	42097	70403
16	325	75	112500	0.338	38034	74466
17	280	76	112500	0.298	33492	79008
18	235	77	112500	0.252	28380	84120
19	190	78	112500	0.201	22584	89916
20	145	79	112500	0.142	15956	96544
21	100	80	112500	0.074	8305	104195

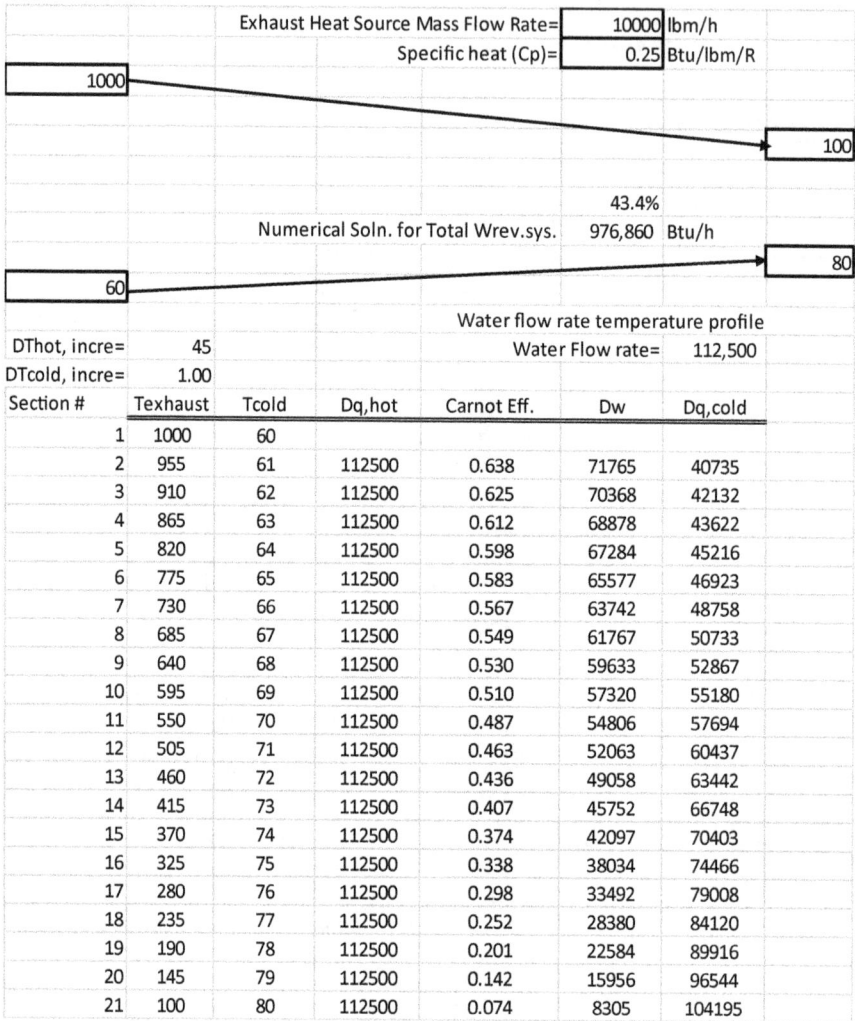

Figure 4.10.   Calculating reversible work for a typical generator.

of the universe gives the same result as what may be obtained by a proper graphical interpretation of how Carnot Heat engine and Carnot Refrigerators can be used to produce ideal, maximum power between temperature sources and temperature sinks.

A more general heat exchanger wherein both fluids change temperature can also be shown to be modeled by the utopian Carnot Cycle followed by the Carnot (perfect) heat pump as shown in Figure 4.10. The spreadsheet's programming procedure is very similar to the procedure used to calculate the Carnot maximum available work (or power) in Figures 4.7 and 4.8. In this

Exhaust Heat Source Mass Flow Rate= **10000** lbm/h
Specific heat (Cp)= **0.25** Btu/lbm/R

1000

100

43.7%

80 ← Numerical Soln. for Total Wrev.sys.   984,372 Btu/h

60

Water flow rate temperature profile

DThot, incre=	45			Water Flow rate=		112,500	
DTcold, incre=	-1.00						
Section #	Texhaust	Tcold	Dq,hot	Carnot Eff.	Dw	Dq,cold	
1	1000	80					
2	955	79	112500	0.625	70278	42222	
3	910	78	112500	0.613	68995	43505	
4	865	77	112500	0.601	67625	44875	
5	820	76	112500	0.588	66161	46339	
6	775	75	112500	0.574	64592	47908	
7	730	74	112500	0.559	62907	49593	
8	685	73	112500	0.543	61092	51408	
9	640	72	112500	0.526	59131	53369	
10	595	71	112500	0.507	57007	55493	
11	550	70	112500	0.486	54697	57803	
12	505	69	112500	0.464	52177	60323	
13	460	68	112500	0.439	49416	63084	
14	415	67	112500	0.412	46379	66121	
15	370	66	112500	0.382	43021	69479	

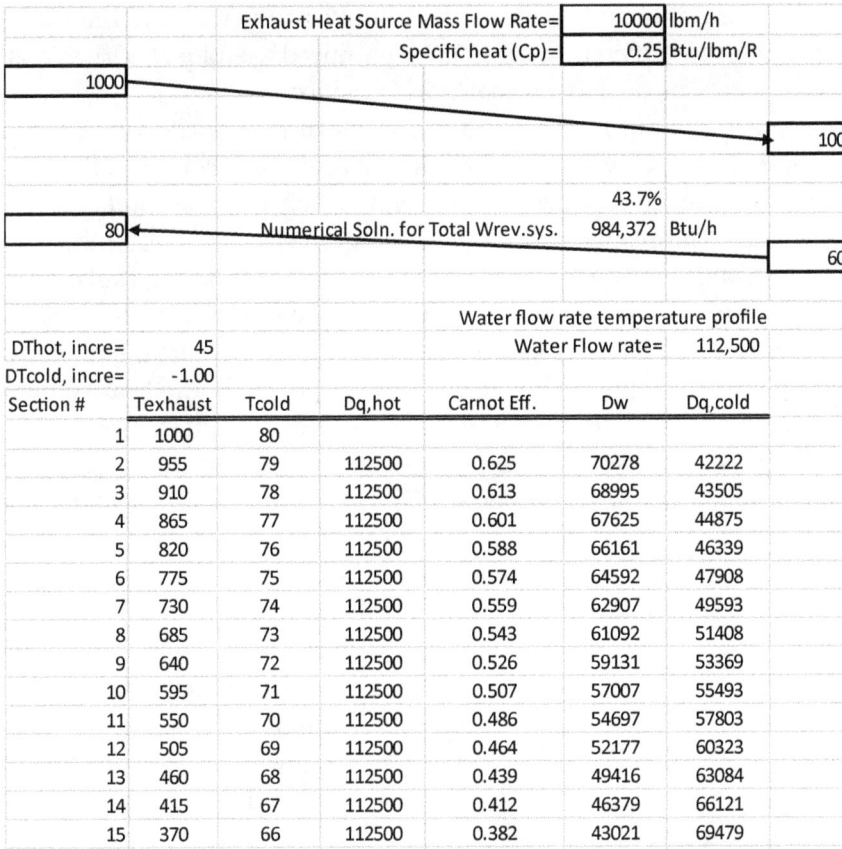

Figure 4.11.   Exergy destruction calculated for a parallel flow heat exchanger.

problem the temperatures of both the heat source and heat sink change. The result from the analysis indicates an irreversibility (or exergy destruction) of 975,663 Btu/h or 286 kW.

It is interesting to consider a slightly different heat transfer arrangement, as shown in Figure 4.11. This heat exchanger is conventionally called a counter-flow heat exchanger, compared to the previous heat transfer arrangement in Figure 4.10, which was labeled a parallel flow arrangement. The consequence of this arrangement is to have a larger exergy destruction due to the larger temperature differences between the inlet and outlets of the heat exchanger.

The modeling of these perfect engineering solutions using Carnot Heat Engines and heat pumps is left as an exercise for the reader and is strongly encouraged by the author. It provides the reader with a very strong graphical representation of what exergy destruction truly represents: work opportunity that is lost forever simply because the heat exchange that can only be done via a temperature difference, could have produced useful power that was not done completely or in part by real-world engineers.

In the last section, it will be shown how exergy destruction, E_d, can be determined from a very neat and elegant, closed form equation that can be readily applied to this application. The author left this equation till the end, so as to allow the reader to first gain an appreciation of the physical and graphical meaning of terms such as "exergy destruction", "availability" or "irreversibility", before determining real-world consequences of the Second Law of Thermodynamics. It is often useful to derive the maximum efficiency of a thermodynamic process or system that depends on the heat source and heat sink temperatures. To assist in this derivation it is often useful to model the heat transfer processes from the operating temperatures by graphically identifying how one or more perfect Carnot heat engines, heat pumps or refrigeration cycles can be configured to produce the desired net effect of power input or output and/or heat transfer input or output with respect to the system being studied.

Let us review several very familiar applications of thermodynamics — the determination of exergy destruction and the second law efficiency of the system — by applying a more graphical technique. It is hoped that this graphical approach will instill in the reader a more complete appreciation of the elegance and usefulness of the Second Law of Thermodynamics in the solution to problems.

Consider the wall introduced in Figure 4.5 and then repeated in Figure 4.12. The wall has a hot and cold surface and thus one can expect that heat energy will be transferred from the hot surface to the cold surface.

Let us then calculate an expression for the maximum work (or power) that could have been generated by the availability of the heat transfer, $\dot{Q} = \dot{Q}_{h1}$, that occurs across the hot and cold wall temperatures. The solution to this problem was outlined previously in this chapter. In that solution, a necessary constraint was to have $\dot{Q} = \dot{Q}_{h1} = \dot{Q}_{h2}$. But this is a necessary constraint only when the steady state (or thermal equilibrium) has been achieved. The steady state is defined as the time when the rate of change of temperature of any given part of the system is not changing with time. That is not to say that there is no temperature difference of any part of the system anywhere in that system. The steady state or thermal equilibrium occurs when there is no change in the temperature of a part of the system that is observed over time. Thus, in a utopic world, the steady state solution of the problem would be to complete the heat transfer across the wall without losing any heat transfer to the wall itself (i.e., the system is already at steady state).

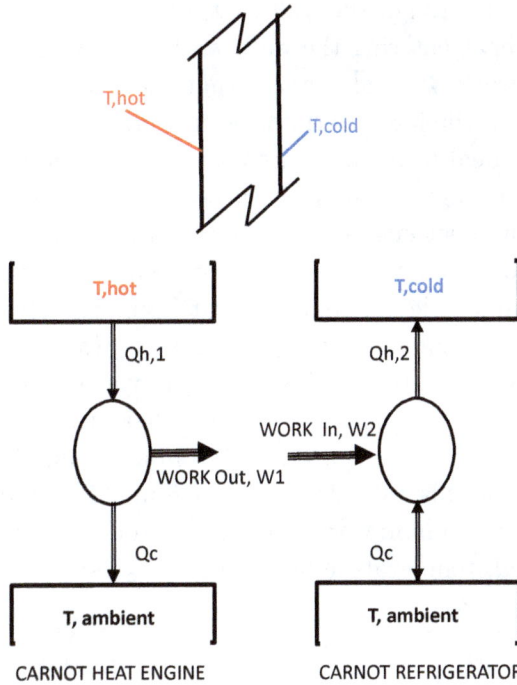

Figure 4.12. Carnot Cycle and heat pump representation of heat transfer through a wall.

Now let us derive a more general expression for the same problem, but with the heat transfer into the wall not being equal to the heat transfer out of it. This would occur when the wall is heated from a "cold start". That is, when the wall is at a uniform but cold temperature. For example, if the wall has the same thickness of an engine block and the engine has been left idle overnight, thus attaining a uniform temperature corresponding to the ambient temperature, T_o. After the engine has been started and the energy content of the fuel is used to heat the wall, there is less energy to be exchanged through the wall. During this transient period between the start of the engine and its steady state, there is a more general expression for irreversibility. The derivation follows.

4.6.2 *Derivation of wall heat conduction with system energy storage*

As before, in the utopic world of perfect engineering, the engineers can accomplish the transfer of heat energy by constructing two perfect Carnot machines: a heat engine and a refrigerator. It is noted from the diagram of the heat transferred through the wall that a Carnot Heat Engine is used to indicate that the heat entering the wall can be re-directed to a heat engine to produce ideal work. It is also noted that the heat that would leave the wall is modeled as if the ideal world of perfect engineers will need to supply this heat from an ideal heat pump. Essentially, the modeling of the wall by the two perfect heat engine systems is successful when the wall is eliminated and replaced by the heat engines. It is as though the perfect engineers were saying: "If you need to supply heat to the system at a given surface temperature to get an equal or less amount of heat from that system at a different temperature, then let us use that heat energy to produce idealized work or power. However, we will deliver this heat that you would have received at the temperature you required it to be." The heat that they deliver at the different (lower) temperature is produced from an ideal, Carnot heat pump.

The equations for this modeling are shown in this section.

The theoretical maximum work that can be produced by the heat transfer, Q_h, exiting the high temperature heat source, T_c, is:

$$Work\,out\,(W1) = \dot{Q}_h \times \left(1 - \frac{T_c}{T_h}\right).$$

Then, from the definition of a Carnot heat pump:

$$COP_{carnot\,min.} = T_{cold}/(T_{cold} - T_{ambient}).$$

But also, it is true that: $COP_{carnot\,min.} = Q_{h2}/(Work_{in})$.

Combining these two equations to get an expression for the work into the Carnot Refrigerator: $Work_{in\,W2} = \dot{Q}_{h2}/(T_{cold}/(T_{cold} - T_{ambient}))$.

Subtracting these two work terms and setting the difference to the irreversibility, $\dot{I}$:

$$\dot{I} = Work_{Out(W1)} - Work_{in\,W2}.$$

And completing the algebra to get a very useful and revealing equation:

$$\dot{I} = \dot{Q}_{h1} \times \left(1 - \frac{T_{ambient}}{T_h}\right) - \dot{Q}_{h2} \times \left(1 - \frac{T_{ambient}}{T_c}\right) + T_o \times \Delta s_{system}. \quad (4.17)$$

However, the analysis must now account for the change in entropy of the wall material. As it gets hotter with time, its entropy is changing: $\Delta s_{systen} = C \times \ln \frac{T_{t+1}}{T_t}$ referencing Equation (4.3) at constant pressure.

The same solution can be obtained by applying the Second Law directly:

$$\Delta S_{hot\,wall} = +\frac{\dot{Q}_h}{T_h},$$

$$\Delta S_{cold\,wall} = -\frac{\dot{Q}_h}{T_c},$$

$$\dot{I} = T_{ambient} \times (\Delta S_{hot\,wall} + \Delta S_{cold\,wall})$$

$$+ (e_{after\,time\,stemp,t+\Delta t} - e_{before\,time\,step,t}),$$

where it may be observed that the exergy change for the system must be used during the transient period before steady state is achieved.

Once again, the reader is reminded that this same equation will be (more) easily produced by an elegant equation that involves a term called exergy. The detailed description of exergy will be presented at the end of the chapter. Suffice to say, the property exergy must never be confused with the thermodynamics definition of energy. Exergy should be thought of as a measure of the potential for the state of a system to be able to produce theoretical work or power. Perhaps the greatest difference between exergy and energy is that exergy can be destroyed, as witnessed by the equation:

$$E_d \equiv T_o \times \Delta S_{\text{"universe"}} \equiv \dot{W}_{reversible,\,maximum,\,utopian}$$

$$-\dot{W}_{actual,\,real-world,\,dyno.measured}. \tag{4.18}$$

It is thus seen as the objective of the engineering analyst to recommend a process that reduces the destruction of exergy. Therein, lies its true value.

There are many practical applications in engineering thermodynamics for this equation of irreversibility (or exergy destruction). It can be used at any time where there is a heat transfer between a high and lower temperature difference with no energy lost. The wall example that started this example is a very common example of its application. But generally, any two boundaries that are at a temperature difference will experience heat transfer and consequently produce irreversibility, which decreases when the temperature difference is reduced to zero. But as the temperature difference approaches zero, so too does the heat transfer and thus its benefit is lost. There is a way of improving this heat transfer by increasing the surface area that promotes the heat transfer, according to the well-founded heat transfer equation:

$$\dot{Q}_{convection} = h_{coefficient} \times Area \times \Delta T.$$

However, any increase in surface area means occupying more floor space as well as being much more costly than a smaller heat exchanger. Usually, the optimization of selecting the correct surface area and the minimum temperature difference usually proceeds by minimizing the cost per $kW_{thermal}$ ($/kw, thermal).

A careful study of the irreversibility equation also reveals that any surface that has a temperature above the ambient will result in irreversibility. A re-look at the example shown previously in this chapter will indicate that the irreversibility of the temperature difference between the heat source and the ambient can be calculated using this equation with the same result, as was determined by the numerical analysis shown in Figure 4.9.

A similar utopic graphical construction can be produced to model an absorption chiller system using ideal Carnot Heat Engines and heat pumps. However, the reader must first know that an absorption chiller system is a very unique system that can provide chilled water by using only relatively low-grade heat to produce the chilled water. In most contemporary absorption chiller systems, the heat input is derived from condensing steam at temperatures of 250°F, corresponding to 15 psig steam. The absorption process is a very sophisticated system that mixes a refrigerant with a second catalysis-like fluid in order to maintain the mixture in a liquid state. As a liquid the pumping work necessary to increase the refrigerant mixture to the condenser pressure is much less than if the refrigerant was left as a simple vapor and a compressor is used to compress the vapor to the higher, condenser pressure. However, heat is needed to separate the refrigerant from the catalyst-like fluid. Thus, the absorption system replaces the compressor that is typically used in the refrigeration cycle. The absorption system will be described in more detail in Chapter 5. It is sufficient to indicate here that the chilled water was produced by the desired heat from a temperature that was lower than the local ambient and that the heat input came from a heat source that was at a temperature higher than the local ambient temperature.

The graphic representation of an absorption chiller system that uses only a Carnot Heat Engine and a Carnot Refrigerator is shown in Figure 4.13. As shown in Figure 4.13, only the work output from the Carnot Heat Engine is used to drive the Carnot Refrigerator.

The work out of the Carnot Heat Engine is determined from the now familiar Carnot Equation applied to the heat transfer into the ideal heat engine, $\dot{Q}_h$:

$$Work_{out} = \dot{Q}_h \times \left(1 - \frac{T_{ambient}}{T_{absorption}}\right)$$

In a similar manner, the work, $Work_{in}$, needed by the ideal thermodynamic refrigerator is determined from knowing how much energy is to be transferred

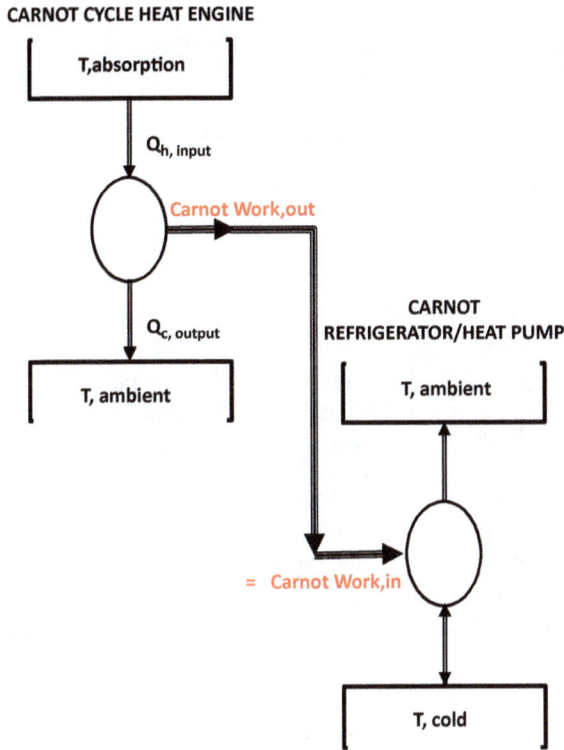

Figure 4.13. Graphical representation of an absorption chiller system using a Carnot Heat Engine and Refrigerator.

from the cold space, $\dot{Q}_{cold}$:

$$Work_{in} = \dot{Q}_{cold}/(T_{cold}/(T_{cold} - T_{ambient})).$$

Setting these two equations equal and solving for:

$\dfrac{\dot{Q}_c}{\dot{Q}_h}$, which is recognized as the COP_{ideal} for the system.

With the results:
$$COP_{ideal} = \frac{\dot{Q}_c}{\dot{Q}_h} \times \frac{1 - \dfrac{T_c}{T_{absorption}}}{1 - \dfrac{T_c}{T_{ambient}}}. \qquad (4.19)$$

Let us apply the utopic method of using only perfect Carnot heat engines, heat pumps and refrigeration to analyze the irreversibility of another very common engineered system: a furnace. In fact, a more contemporary application would be a solar heated panel that is used to heat water. If the reader prefers to think that a furnace implies that very high temperature levels are

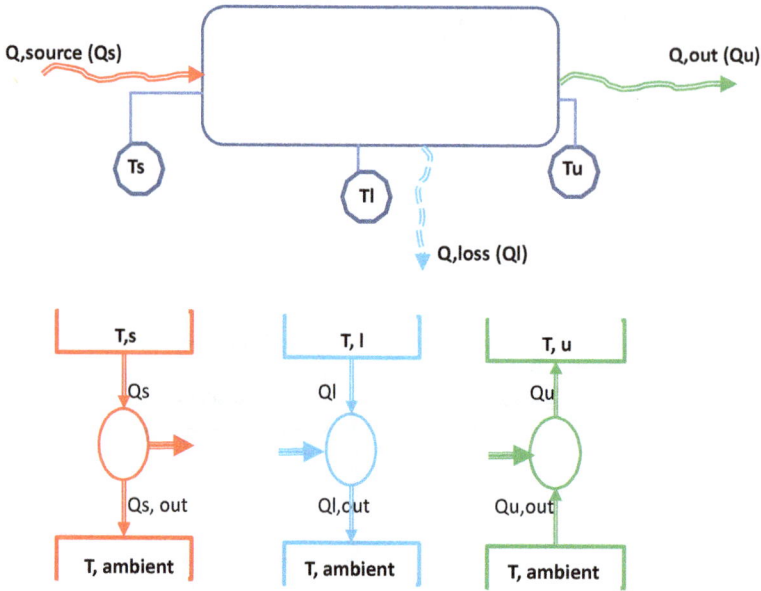

Figure 4.14. Model of a furnace with (a) a Carnot Heat Engine and (b) heat pumps.

produced, then he or she should know that a concentrating solar collector is used to heat molten salts to the very high temperature of 600–650°C. Thus, the reader is reminded that though many thermodynamics solutions may be applied to a number of seemingly different applications, these applications are very similar from the point of view of thermodynamics application.

A furnace or a solar heated panel is represented in Figure 4.14(a) and (b), where a heat transfer occurs at a given temperature, T,s. In a typical furnace some of the heat input is lost to the environment at a wall temperature, T,l, while the balance of the heat transfer proceeds to be used for its intended purpose at a temperature, T,u.

A heat engine–heat pump model showing how it would have been modeled in a utopic world is shown in Figure 4.15. The heat input is diverted to a heat engine and serves as the heat source for the perfect Carnot Heat Engine. The heat losses are then provided at the given surface temperatures shown. It is imperative to understand that for this model to work, the heat transfer must always be used as the heat input into the heat engines, and the heat loss (or removal from the system) is always modeled using a Carnot heat pump, the output of which is directed to the same heat sink that the original heat losses were to be directed to, as shown in Figure 4.15.

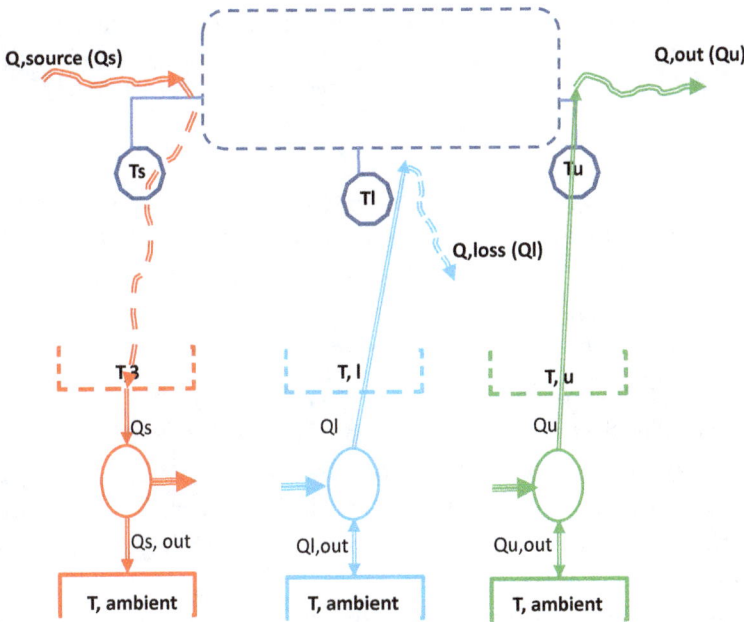

Figure 4.15.   Representation of a furnace using the Carnot cycle and heat pumps.

The single Carnot Heat Engine and the two Carnot Refrigerators can be expressed with the following equations:

$$Work_{out} = \dot{Q}_h \times \left(1 - \frac{T_{ambient}}{T_s}\right),$$

$$Work_{in} = \dot{Q}_l/(T_l/(T_l - T_{ambient})),$$

$$Work_{in} = \dot{Q}_u/(T_u/(T_u - T_{ambient})).$$

With these expressions for ideal work output and work inputs, an equation can be written that identifies how much of the ideal input and output work from the Carnot engines and the actual work are not balanced, thus leading to a term known as exergy destruction, E_d, previously known as irreversibility, I.

E_d (or irreversibility)= {Carnot Work$_{out}$ + (-Carnot Work$_{in}$)} – {Actual Work$_{Input}$ + Actual Work$_{Output}$}.

For this application, actual work input and output are zero.

This results in the equation for E_d, as follows:

$$E_d = \dot{Q}_h \times \left(1 - \frac{T_{ambient}}{T_s}\right) - \dot{Q}_l/(T_l/(T_l - T_{ambient}))$$

$$- \dot{Q}_u/(T_u/(T_u - T_{ambient})). \tag{4.20a}$$

And with some rearranging, a more easily remembered equation:

$$\mathbf{E_d} = \dot{Q}_h \times \left(1 - \frac{T_{ambient}}{T_s}\right) - (\dot{Q}_l \times (1 - T_{ambient}/T_l))$$

$$- (\dot{Q}_u \times (1 - T_{ambient}/T_u)). \tag{4.20b}$$

It is noted that the values for the heat transfer should be all positive as the subtraction signs account for the direction of the heat transfer.

The relationship of the heat input and output from the First Law must still be used. Thus, $\dot{Q}_h = \dot{Q}_l + \dot{Q}_u$.

This leads to an interesting comparison of the first $\left(\dot{Q}_h \times \left(1 - \frac{T_{ambient}}{T_s}\right)\right)$ and last terms $(\dot{Q}_u \times (1 - T_{ambient}/T_u))$ in Equation (4.20b). The first term can be recognized as the maximum amount of work that "could have been" produced by the heat transfer, $\dot{Q}_h$, and the work that is delivered by the furnace or solar heater with a heat transfer of $\dot{Q}_u$ from their respective thermal reservoirs and with $T_{ambient}$ used as the dead state.

If these two terms are compared by dividing the second by the first, and with the recognition that the ratio of the heat transfer out with the heat transfer into the system is the efficiency of the furnace or solar heater, $\left(\eta = \frac{\dot{Q}_u}{\dot{Q}_h}\right)$, then the second law efficiency of the furnace system can be defined as:

$$\eta_{2nd} = \frac{\dot{Q}_u}{\dot{Q}_h}\left(1 - \frac{T_{ambient}}{T_u}\right)/(1 - T_{ambient}/T_s). \tag{4.21}$$

4.7 Furnace/solar panel application

Let us apply the proceeding equations to two similar problems associated with the heating of a fluid from a heat source at a high temperature.

Consider the following steam boiler application. A boiler is designed to produce 1,000,000 Btu/h of heat transfer to boil water to a saturation vapor temperature of 500°F. The furnace is 90% efficient. That is, the actual heat input to the boiler from a combustion heat source is 1 MBtu/h/0.9 = 1.1 MBtu/h. The combustion source temperature is 1,500°F and is cooled to 600°F in order to produce the saturated vapor steam at 500°F.

Using Equation (4.21), the second law efficiency is calculated as follows:

$$\eta_{2nd} = \frac{\dot{Q}_u}{\dot{Q}_h}\left(1 - \frac{T_{ambient}}{T_u}\right)/(1 - T_{ambient}/T_s),$$

$$\eta_{2nd} = 0.1\left(1 - \frac{520}{1060}\right)/\left(1 - \frac{520}{1960}\right) = .069.$$

From the wording of the problem, it is cautioned that this is the second law efficiency for the furnace with respect to the ambient and loss of considerable source temperature (1,500°F to 600°F) that apparently results in a 90% loss of heat energy. With these parameters as inputs, the second law efficiency is 6.9%. This may seem very low and could cause some validity concerns during the last step — the "sanity" check — in the solution process. In fact, there may very well be an inconsistency in these given values. For example, for 90% of the heat source energy to have been "lost", questions need to be asked on how the 90% was referenced? In fact, the heat source temperature is given as 1,500°F, but its flow rate and specific heat are not. If the heat source is hot air (i.e., sourced from the products of combustion), then the amount of heat available would be:

$$\dot{Q}_{maximum} \equiv \dot{m} \times C_p \times (T_{in} - T_{out}) = \dot{m} \times C_p \times (1,500 - T_{ambient}),$$

then the 90% heat loss would mean that 90% of $\dot{Q}_{maximum}$ was permanently lost to the environment as useful heat — or in other words the temperature exiting the furnace would be based on the following equation:

$$\dot{m} \times C_p \times (1,500 - T_{ambient}) \times 0.9 = \dot{m} \times C_p \times (1,500 - T_{u,furnace\,out}).$$

Solving for $T_{u,\,furnace\,out}$ from this equation finds $T_{u,\,furnace\,out}$ = 204°F, and not 600°F.

For the 600°F stated in the problem, the heat loss through the furnace walls would only be 60% and the second law efficiency would then be 28%. The lesson learnt is to always check the assumptions and the data given in the problem statement for consistency.

Notice that the steam vapor temperature of 500°F was not used in the equation. To determine the second law efficiency of the heat source, now at 600°F, with respect to the saturated steam temperature, another set of Carnot heat cycles must be used to model this process arranged as shown in Figure 4.8. For that modeling, the heat source will need to be cooled to a lower temperature that must still be above the desired steam vapor temperature of 500°F.

Next, consider applying Equation (4.21) to a solar panel (see Figure 4.16), which is designed to produce hot water of 100°F from water of 80°F. The solar panel is heated by a solar energy heat flux (200 Btu/h/ft^2) that heats the panel surface to 140°F. The second law efficiency, by using Equation (4.21), is calculated to be 37% with only a 10% convection and radiation loss from the solar panel to the ambient. This value can be compared to the numerical solution presented earlier, which is also illustrated in Figure 4.8.

This section derives an expression for the second law efficiency and exergy for a furnace or solar panel. A similar but different derivation can also be

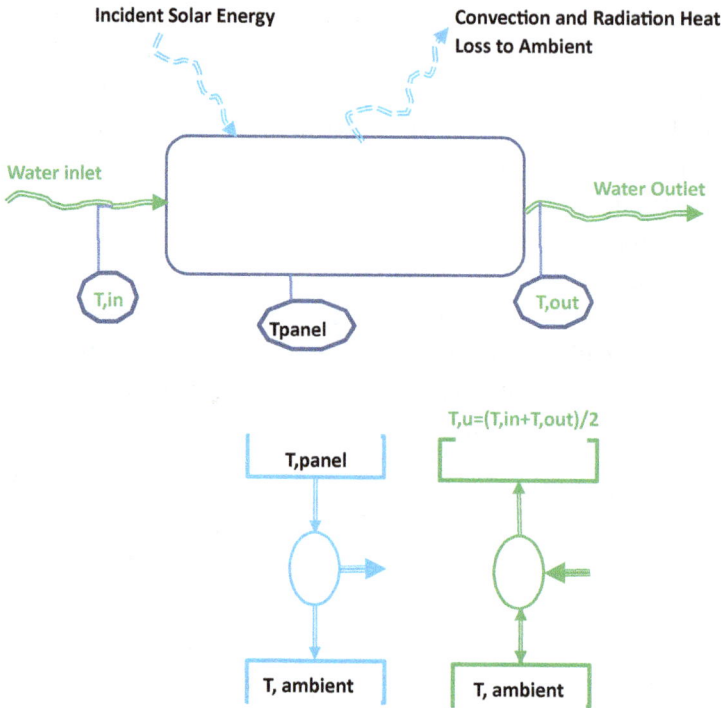

Figure 4.16.   A solar powered water-heating system using a solar panel as a heat exchanger for heating the water.

performed for a system that represents the mixing of the two heat streams. Figure 4.17 illustrates such an application.

The graphic models shown in Figure 4.17 illustrates the cooling of a hot fluid to T,mix using the Perfect, Carnot Heat Engines. Coupled with this is a graphic depiction using perfect Carnot refrigerators that heat the cold fluid to the same T,mix as using the heat removed from a cold temperature reservoir.

The ideal amount of available energy for the Carnot Heat Engine Cycles and heat pumps can be solved by using a numerical solution technique that was described in this chapter. The need to "sum-up" the incremental ideal work outputs from Figure 4.17(a) requires a calculus to be used to integrate the entire operating temperature range. It will be shown that the second law efficiency is determined from Equation (4.22).

$$\eta = \frac{\left| Q - C_p \times T_{ambient} \times \ln\left(\frac{T_{mix}}{T_{cold,in}}\right) \right|}{\left| -Q - C_p \times T_{ambient} \times \ln\left(\frac{T_{mix}}{T_{hot,in}}\right) \right|},\tag{4.22}$$

where $Q = (\dot{m} \times C_p)_{cold\ flow} \times (T_{mix} - T_{cold})$.

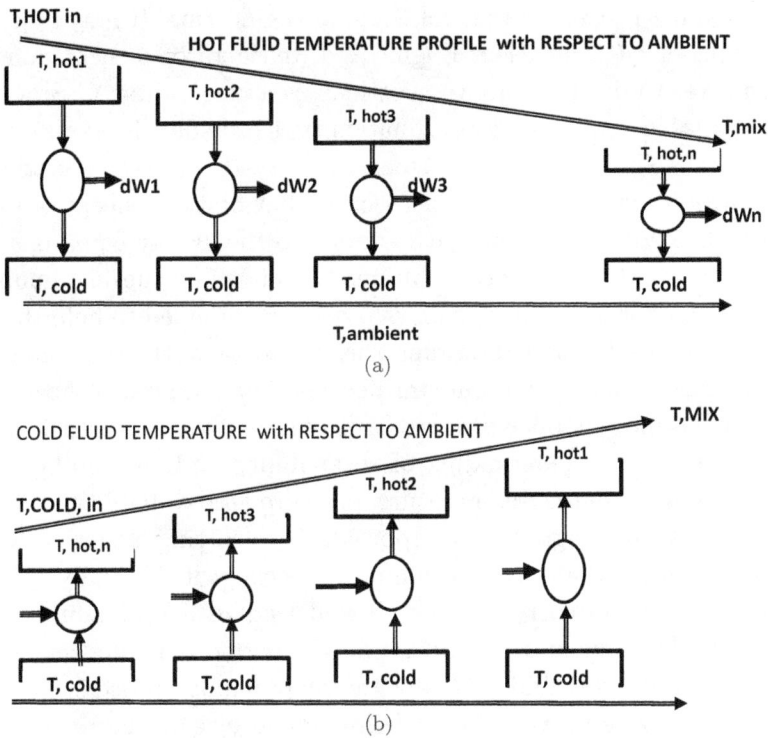

Figure 4.17. Graphic model of two streams that are mixed to an equilibrium temperature of T,MIX.

Although the same answer can be obtained through numerical integrating the Carnot Cycles shown in Figure 4.17, it is always helpful to produce a "closed" equation such as Equation (4.22), that could easily calculate the second law efficiency, as well as "see" how the various parameters affect the second law efficiency by using the equation in a parametric study.

Equation (4.22) can be derived based on the concept of reversibility given in the next section.

4.8 Reversible work and irreversibility

The enthusiastic reader may be asking: Would I always need to produce a combination of heat engines and heat pumps in a utopic world to determine the network $W_t - W_c$ and thus the exergy destruction, or is there a more simple and elegant equation that can provide the same answer? After all, combining a single Carnot Heat Engine and a single heat pump may not be sufficient to model a real-world energy exchange that constitutes a complex thermodynamics process in the real world of engineering. It may take several combinations of a Carnot Heat Engine and Refrigerator or heat pump.

The answer to both the above questions, is a resounding YES! More elegance is available in a closed form equation that provides the exergy destruction for both a Closed Mass and Open Flow System. These equations and their derivations are provided in this section, but first, a concept of the dead state must be explained because it factors into the exergy equations.

The following diagrammatic solutions to determine the maximum work that is available for a thermodynamics process is intended to help the reader understand the basis for determining the reversible work (maximum power output for heat engines or minimum power input required of heat pumps) for a given thermodynamics process. This diagrammatic approach will help fortify the reader's understanding of reversibility and the limits of power generation, given a source temperature, pressure and available heat transfer.

As useful as this diagrammatic approach is to solving a thermodynamics problem, it is somewhat time consuming, especially if the analyst does not diagram the problem correctly, as this will lead to a misleading solution. Fortunately, the First and Second Laws of Thermodynamics are available to determine the reversibility of any thermodynamics process. Combining these two basic laws provides several simple and elegant expressions of the reversibility that is available from a thermodynamics process when the start and the finishing points of that process are known. In fact, the elegant expression of the reversibility of the thermodynamics initial state can always be derived using what has been labeled the thermodynamics dead state at the end of the thermodynamics process. The dead state is the natural environmental pressure and temperature that is prevalent where the heat engine or refrigerator system is installed. In this real world, the ambient temperature and pressure may vary but it is generally accepted as 520°R and 14.7 psia. In any other place in the Universe, the dead state pressure and temperature will be different.

The dead state is defined as that environmental temperature and pressure to which all processes will conclude if left alone without additional effort (work or heat transfer) provided by the analysis/engineer. For example, if a container, at a pressure and temperature above atmospheric pressure and temperature, was allowed to come to equilibrium with the environment, it is reasonable to conclude that the contents of the container will achieve the environment pressure and temperature. It is not expected that the contents will equilibrate below atmospheric pressure or at a higher temperature than the environment if they are left without any environmental intervention of work or heat transfer.

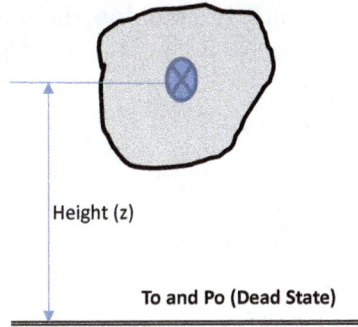

Figure 4.18.   Model of energy state in a CV system.

The dead state is important because any state of a system that is not at the dead state has the potential to do mechanical or thermal work if it is released to do so in the most efficient manner possible. This can be visualized by considering a CV (shown in Figure 4.18) that is elevated above the dead state. The reader should be careful and not take the word "elevated" too seriously. While the system may be considered to have pent-up potential energy or exergy, it is more than just a CV with potential energy. It has, in fact, not only the same "old" potential and kinetic energy, but also a pressure and temperature mechanical and thermal exergy (with respect to the dead state as the thermodynamics reference), and presumably the place where a datum for the potential energy due to height is considered, although this would not be necessary, as the reader will learn shortly.

Therefore, it can always be assumed that for any initial thermodynamics state, the dead state is available for which the magnitude of reversible work can be determined. Of course, this amount of reversible work must always be accepted with the realization that this reversible work is only available in a utopic world of perfect engineering and engineers. To determine what "likely power" is available from "real world" engineering endeavors, one needs to apply a reasonable correction factor. This correction factor can be identified as the second law efficiency for the process. More often than not, the second law efficiency (or the correction factor) is based on an engineer's "rule of thumb" experience with similar heat engines or heat pumps. A typical "rule of thumb" for the correction factor used with heat engines is 50%.

4.9   Derivation of reversibility equation from the first and second law

The following derivation of the availability or exergy relationships follows the derivation as proposed by William Black and James Hartley in their textbook *Thermodynamics*. This derivation is among the more straightforward derivations that will neatly combine the First and Second Laws to achieve a more universal expression for both a Closed Mass and an Open System in thermodynamics.

The First Law is expressed as:

$$\sum \dot{Q} + \overset{n, number\ of\ inlet\ flows}{\underset{in}{\sum}} \dot{M} \times (h + KE + PE)$$

$$= \sum \dot{W} + \overset{m, number\ of\ outlet\ flows}{\underset{out}{\sum}} \dot{M} \times (h + KE + PE) + \frac{\partial U}{\partial t})_{cv}$$

$$\dot{Q} = -\sum \dot{Q}_k,$$

where the subscript "k" indicates heat transfer relative to the surroundings. Next, also consider the total rate of entropy change for the system:

$$+ \frac{\sum \dot{Q}_k}{T_k} \geq - \left[\sum \dot{m}s_{exit} - \sum \dot{m}s_{inlet} + \frac{\partial S_{system}}{\partial t} \right].$$

Combining these two equations for the First and Second Law of Thermodynamics leads to:

$$\sum \dot{W} \leq - \left[\overset{m, number\ of\ exit\ flow\ streams}{\underset{exit}{\sum}} \dot{M} \times (h + KE + PE - T_o s) \right.$$

$$\left. - \overset{n, number\ of\ inlet\ flow\ streams}{\underset{inlet}{\sum}} \dot{M} \times (h + KE + PE - T_o s) \right]$$

$$- \frac{\partial (U - T_o S)}{\partial t} \bigg)_{system} - \sum \dot{Q}_k \times \left(1 - \frac{T_o}{T_k} \right).$$

When the equality is used, the power ($\dot{W}_{reversible}$) is the ideal or reversible power that the system can produce or consume.

The useful work that can be done by the system has more real-world significance because it can be measured by a dynamometer and provide some useful benefits. This useful work must also consider the work done by or on the system by the environment with a pressure P_o.

The expression for the useful work then becomes:

$$\sum \dot{W}_{useful} \leq - \left[\overset{m,\,number\,of\,exit\,flow\,streams}{\underset{exit}{\sum}} \dot{M} \times (h + KE + PE - T_o s) \right.$$

$$\left. - \overset{n,\,number\,of\,inlet\,flow\,streams}{\underset{inlet}{\sum}} \dot{M} \times (h + KE + PE - T_o s) \right]$$

$$- \frac{\partial (U + P_o V - T_o S)}{\partial t} \Bigg)_{system} - \sum \dot{Q}_k \times \left(1 - \frac{T_o}{T_k} \right).$$

The irreversibility $\dot{I} = \dot{W}_{reversible\,useful} - \dot{W}_{useful,\,dyno\,measureable}$.

It is necessary to observe that the useful, dyno (for dynamometer) measurable work that is used in the previous equation for irreversibility, I, already has the work done by or to the environment as expressed as $P_o \times \frac{\partial V}{\partial t}$.

It is also necessary to point out that this single equation actually has two distinct parts that can be used to determine the reversible work for a Closed Mass and an Open Flow System.

The open flow part of the equation shown here is defined as the exergy or availability of an Open Flow System.

$$e_{cm} = [(h + KE + PE - T_o s)].$$

The closed mass part of the equation shown here is defined as the exergy or availability of a Closed Mass System.

$$e_f = (U + P_o V - T_o S).$$

It is clear that the irreversibility, $\dot{I} \geq 0$, for the system would violate the First and/or Second Law of Thermodynamics. In contemporary texts and engineering discussions, irreversibility as expressed in Equation (4.23) is identified as the exergy destruction, E_d.

The term "exergy", E, is equivalent to the expression of "availability", A, of reversible work, which is used in a number of textbooks and technical papers. Irreversibility (or exergy destruction) can also be expressed as the product of the dead state temperature, T_o, and the entropy change of the Universe.

$$I = E_d = T_o \times \Delta S_{entropy\,change\,of\,the\,Universe}. \tag{4.23}$$

The entropy change of the Universe is the sum of the entropy change of the surroundings and the system, as discussed previously in this chapter.

The following equations for irreversibility (or exergy destruction) can also be derived from the combination of the First and Second Laws of Thermodynamics:

$$\dot{I} = T_o \times \left[\sum \dot{m}s_{exit} - \sum \dot{m}s_{inlet} + \frac{\partial S_{system}}{\partial t} + \frac{\sum \dot{Q}_k}{T_h} \right],$$

$$\dot{E}_d = \sum \dot{m}e_{inlet} - \sum \dot{m}e_{exit} - \frac{\partial E_{cv}}{\partial t}$$

$$+ \sum \dot{Q}_j \times \left(1 - \frac{T_o}{T_h} \right) - \left[\dot{W}_{cv} - P_o \times \frac{\partial V_{cv}}{\partial t} \right]$$

$$E_{destruction} = T_\infty \times \sigma_{universe}.$$

Exergy is in fact energy that can be destroyed. In a very real sense, exergy was invented in order to measure the degree of ineffectiveness in a process when compared to the most perfect process that could be done under the same system constraints by a perfect engineer. Recall that the amount of destruction is based on the comparison of the utopic world's ability to perform a task with perfect engines and refrigerators. Examples of this were given in the previous section. In this section we found an alternate method for finding the exergy of a system (and thus the exergy destruction) in the form of an elegant, closed-form equation for a Closed Mass System, as shown in Equation (4.24).

The derivation of the reversible work expression presented in the previous section provided the first indication of an energy term that expressed the measure of reversibility on a thermodynamics system. This energy term is labeled exergy, E, for both Closed Mass and Open Flow Systems, and is formally defined by the following equation with respect to the ambient temperature and pressure, also recognized as the dead state.

For a closed mass system, the equation is:

$$E_{closed\,mass} = (u - u_o) + P_o \times (v - v_o) - T_o \times (s - s_o). \qquad (4.24)$$

And for an Open Flow System:

$$E_{flow} = (h - h_o) - T_o \times (s - s_o).$$

Let us step back for a moment, now that a number of "second law" equations have been presented, to reflect on the elegance and the power of these equations. It can easily be argued that the equations for a perfect Carnot Heat Engine Cycle and Refrigerator and heat pump, and the equation that relates irreversibility to the entropy change of the Universe with the factor of the ambient temperature, are as powerful a result of thermodynamics science as was Einstein's famous equation, $E = MC^2$. These Exergy equations are also used in nuclear physics. They provide a deeper understanding of the significance of temperature and its influences on energy transformations in real-world engineering applications and thermodynamics.

Elegance is apparent in the simplicity of these equations. Their use is where the power of their application can be realized to great effect in deciding what engineering design should be chosen to perform a particular thermodynamics application with respect to cost and performance, as identified by either maximum power, efficiency or component size. All these three system parameters come into consideration to some extent in the decision process on how to proceed in the fabrication and implementation of a design for a thermodynamics system.

The example just described indicates that a heat transfer intended to occur during the implementation of a process must necessarily occur across a temperature difference. The mere fact that the temperature difference must be present to cause heat transfer (i.e., an energy exchange due to a temperature difference) also implies that there will always be an irreversible loss of exergy (energy). BUT this energy that is now recognized as work or power that could have been performed but was not done for some reason — either some real-world work or power was done or none of it was attempted or even desired — the association of exergy destruction (or irreversibility) or unavailable work (or power) is key to understanding how the Universe works and how entropy must be positive via the simple and elegant Equation (4.24). There can never be a negative temperature on the absolute scale, for then there could be a thermodynamic cycle that would have an inefficiency greater than 100% per the Carnot Cycle equation (*Carnot Eff.* $= 1 - \frac{T_{cold}}{T_{hot}}$) which has the consequence of more power being generated than the power input to the cycle; literally the definition of a perpetual motion machine. The existence of a negative absolute temperature would also allow the net entropy change of the "universe" could be greater than or less than zero; essentially removing the Second Law of Thermodynamics as one of the basic Laws of the Universe. Thus the cold glass of water could get colder without any other external effect from the environment otherthan heat transfer. Such an eventuality would never stop once initiated and lead to an ultimate catastrophe for the Universe. That the anthropic principle is in play is to indicate that we are here and therefore the Second Law must be a universal principle stance as the only derived proof of the Second Law of Thermodynamics.

To understand the concept of exergy destruction, consider the heat engine diagram shown in Figure 4.19. This familiar representation of a heat engine

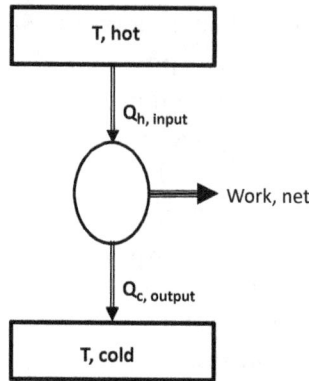

Figure 4.19. Carnot Cycle heat engine.

working between two temperature reservoirs is nothing new to the reader. Assume that the heat input is 1,000 Btu and the work output is 300 Btu. Thus, the engine efficiency is $300/1,000 = 30\%$. Using the equation for a Carnot Cycle, the Carnot Cycle efficiency is: $1-600/1,200 = 50\%$. The Second Law indicates that the best possible heat engine cannot exceed 50%. Given that the engine we are considering has an efficiency of 30%, we know that the Second Law has not been violated. But now consider that IF the heat engine had been built to have an efficiency of 50%, then $1,000 \times 50\% = 500$ Btu of work could have been achieved. The fact that in the real world of imperfect machines, where there is friction, drag, heat loss and everything else that make real world engines less efficient than ideal engines, then $500 - 300 = 200$ Btu of possible energy that could have been generated will never be generated. The destroyed availability of this 200 Btu of energy is lost forever. However, because energy cannot be lost forever, exergy destruction, which looks and almost sounds like energy and has the units of energy, is adopted for thermodynamics analysis because it can be used to measure the irreversibility of a process; exergy destruction refers to energy that has been lost. The comparison of the perfect engine's perfor-mance to its performance in the real world is very visual and experimental and — dare it be said — an intuitive representation of what exergy destruc-tion is as well as how the entropy change of the Universe must always be greater than or equal to zero. How does entropy come into the picture? The answer is straightforward: Equation (4.25) links entropy change and exergy destruction (or irreversibility).

$$Irreversibility \equiv E_d \equiv T_o \times \Delta S_{universe} \equiv W_{Ideal} - W_{actual,real-world}. \quad (4.25)$$

4.10 Closed Mass System availability, A, or Exergy, E_{cm}

An example of the application of the exergy equation for a Closed Mass System is given here. Consider the closed mass tank shown in Figure 4.20, which contains a gas. The temperature, pressure and volume of the tank at its initial state are known.

T_1 = Temperature of fluid in the tank

P_1 = Pressure of fluid in the tank

Ambient pressure and temperature (dead state) = P_o and T_o

The closed mass exergy equation for the Closed Mass System is given by the Equation (4.26).

$$E_{cm} = M \times [(u1 - u_o) + P_o \times (v1 - v_o) - T_o \times (s1 - s_o)], \quad (4.26a)$$

$$E_{cm} = [(U1 - U_o) + P_o \times (V1 - Vo) - T_o \times (S1 - S_o)]. \quad (4.26b)$$

Note that the specific volume and energy terms have been replaced with the absolute values of internal thermal energy, volume and entropy.

The entropy change can be substituted by the equations shown in Equation (4.26) where:

$$\Delta s = C_v \times \ln \frac{T_1}{T_o} + (R_u/MoleWt) \times \ln \frac{V_1}{V_o}.$$

The Perfect Gas Law can be arranged so that: $P_1 V_1/T_1 = P_2 V_2/T_2$. With the necessary algebra and collecting terms:

$$E = M \times C_v \times \left[T_1 - T_o \times \left(1 + \ln \frac{T_1}{T_o} \right) \right] + V_1 \times P_o$$

$$\times \left[1 - \frac{P_1 T_o}{P_o T_1} - \frac{P_1 T_o}{P_o T_1} \times \ln \frac{P_o T_1}{P_1 T_o} \right].$$

The following application is a real-world example of a Closed Mass System's availability analysis. The vehicle is powered by only compressed air energy stored in small tanks inside the vehicle. The question to be answered

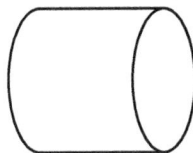

Figure 4.20.   Closed mass tank.

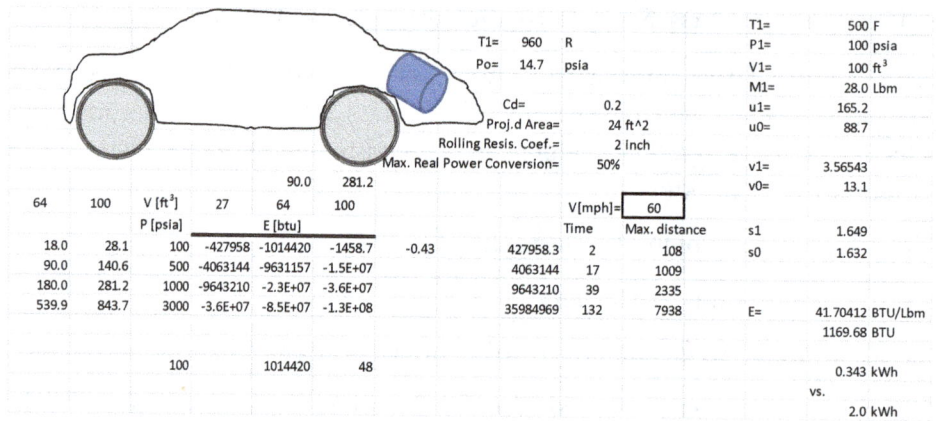

Figure 4.21. Example of a real-life availability analysis for vehicle powered by a closed mass system.

is how large the tanks need to be, such that the vehicle can be powered at "x" mph for "y" hours.

The solution for the exergy, E, shown in the spreadsheet's output is given in the three columns labeled E [Btu], and have been calculated for the storage pressures shown as 100, 500, 1,000 and 3,000 psi and for the tank volumes of 27, 64 and 100 ft³. This available energy must be reduced by a propulsion efficiency that is known from the efficiency of the electro-mechanical drive system that is turning the wheels. The reduced energy can then be equated to the power of the propulsion system multiplied by the time of use. The power is equal to the drag forces exerted on the vehicle multiplied by the speed of the vehicle.

4.11 Entropy analogy

Many students have a difficult time understanding the concept of irreversibility (or exergy destruction/loss of availability). The question that always comes up is: "Is exergy destruction something that is tangible or is it really 'forever lost'?"

Perhaps an analogy would be helpful towards explaining the concept.

Consider that the reader is in the market for a new pair of shoes. The newspapers on Sunday morning carry a shoe advertisement that announces an upcoming sale offering one free pair of shoes worth the value of the pair that is purchased. However, this deal is only good for those who buy the shoes on Monday. Tuesday will be too late as the store is going out of business and will not be open after Monday. Also, given that they are going out of business, they want all transactions in cash.

You also see in the advertisement a pair of shoes that you like. Not only that, this pair of shoes is in your size AND in your preferred colors. You make a mental note to withdraw some cash from the ATM near your workplace and you create reminders to help you remember.

Despite these preparations, while driving home from work on Monday evening you remember that you didn't visit the ATM after all — in other words, entropy has happened! It is now too late to find another ATM as the store closes at 6 pm, and there is only enough time left to get to the store with the cash in your wallet. The total cash that you can come up with is still a dollar short of the full price for this pair of shoes. Regardless of the shortfall, you proceed to the store and find that the owner is happy to sell you the pair of shoes for a dollar less than advertised, BUT he/she will not give you the extra pair for free. There are plenty of people in the store who can and will make use of the promotional offer to buy a pair of shoes in order to get a free pair of shoes. These individuals should certainly not be penalized from getting a free pair of shoes because they took the trouble to be more productive and alert to the seller's requirements. Just as it is to the benefit of the engineer/entrepreneur to also understand the full meaning of the First and Second Laws of Thermodynamics and with the complete knowledge of the natural "rules" of the Universe, proceed to achieve the maximum possible result.

Here's the question: Did you forfeit the cost of one pair of shoes, which is the free pair that you did not get, because you did not pay the full amount for the first pair? Or did you save a dollar because you managed to get the pair of shoes even though you didn't pay the full price for it? Or, did you save 20% of the full price you would have paid for the pair of shoes you purchased if not for the "going out of business" price? Whichever the case, did your bank account mysteriously increase in money in an amount equal to the amount of the money 'saved'? 'Saved' money that you never had to begin with? This is the same as implying that all the irreversible energy loss (or exergy destruction) due to the Second Law of Thermodynamics that came into existence since the Big Bang has been stored somewhere in the Universe, ready to be tapped.[2]

Perhaps the larger question would be: "Of what use is this information if I assume that I haven't lost anything?" After all, you did get a pair of shoes for 20% off, plus an additional dollar in savings courtesy of the owner. The answer to this question is perhaps the crucial point. There is considerable

---

[2]The author is working on a novella that has this as its plot interlaced with a primary theme of teaching thermodynamics and its relevance to the human condition.

value in knowing that there is a sale going to happen. More to the point, there is a maximum benefit to every experience, whether that experience is in the shoe store or the real world of thermodynamics. To end the story on an analogous note, for the shoe customer who could improve his planning to get the free pair of shoes the next time there is a similar shoe sale, so too could the thermodynamics analyst in preparing a more efficient cycle up to the point the Second Law of Thermodynamics indicates is the limit of maximum performance or opportunity.

Chapter 5

Heat Engines and Thermodynamics Cycles

Chapters 1 to 4 have all led up to this chapter. The previous chapters have set the foundation needed to enable the reader to use thermodynamics in innovative ways and to solve thermodynamics problems in a methodical way by using the basic tools of the First and Second Laws of Thermodynamics. By now, the reader should have a relatively comfortable understanding of the Closed Mass and Open Flow Analysis techniques used for solving thermodynamics problems. For each of these methodologies, there is a finite number of common processes and/or mechanical elements that constitute the entire mechanical engineering thermodynamics experience required in the real world. For a Closed Mass System Analysis, the process includes: constant temperature, constant pressure, constant volume, no heat transfer and almost no heat transfer processes. In an Open Flow System Analysis, the mechanical elements consist of turbines, compressors, pumps, valves, heat exchangers, nozzles and diffusers. A mixing tank (as suggested by the author) or reservoir may even be added to this list of elements; the latter being a device that allows the exchange of heat between one or more fluids while also providing a large inventory for storing the fluid. Given the proper application of the First and Second Laws of Thermodynamics with these mechanical elements or processes, the reader will soon be able to apply these useful thermodynamics processes to construct thermodynamics cycles or chemical processes.

The only missing tool in the reader's toolbox is "experience", which the worked examples in this textbook hope to provide. However, whatever the experience may be gained by reading this textbook, it is only the start of the reader's journey to understanding these thermodynamics devices. Other examples from the chapters will describe thermodynamics as a means to identify a single process or several processes that are coupled together. There should be a clear beginning and an end. More to the point, the analyst will be able to identify the beginning and fix the ending of a process based on the nature of the problem at hand.

This chapter will continue this experience by familiarizing the reader with a set of processes that close upon themselves and thus can be correctly identified as a cycle. Any three processes that are connected head-to-tail can construct a cycle. That is, the processes start and stop at the same point. The reader may even suggest that two processes constitute a cycle and he or she would be correct, with the understanding that neither of these two processes need to be ideal. However, the reader is reminded that Chapter 2 states there are only five basic processes or pathways — constant volume, constant temperature, constant pressure, the adiabatic (i.e., no heat transfer) process, and the more realistic polytropic (i.e., "almost" no heat transfer) process — that are used in the real world of engineering. In this case, it is very likely that at least three processes are needed to complete a cycle. The analyst need not be concerned about any additional difficulty of applying thermodynamics to a cyclic process, other than what is needed to analyze a single or double process. The thermodynamics principles remain the same. It is one of the main objectives of this textbook: to help the reader gain confidence in applying the basic tenets of thermodynamics to any problem that lands on an engineer's desk.

This chapter will also demonstrate the strength of the thermodynamics principles applied to several very common thermodynamics cycles, which provide the real world with power, heating and cooling.

However, there is an even more important and significant objective: to strengthen the reader's knowledge of basic and common cycles and to help him or her build upon such cycles and perhaps generate new and improved ones. The added benefit of increasing the reader's vocabulary in thermodynamics will also help him or her to understand technical papers. This growing experience will be further strengthened when all the Case Studies in this book are carefully reviewed. In these case studies, all this hard-won, newly digested knowledge on thermodynamics will be put to effective use in real-world engineering problems.

5.1   The Rankine Cycle

The Rankine Cycle is arguably the oldest power generation cycle that both nature and man have used to generate power. The Rankine Cycle as deployed by Nature is usually recognized as a hydropower turbine system used to generate electric power. The first human-engineered Rankine Cycle, even before it was so-named, was developed by engineers in the 18th century with the sole purpose of providing continuous power to pumps that were drawing water from coal mines. This Rankine Cycle used a boiler and what can be best described in today's terminology as a condenser (but is more correctly identified as a piston chamber) into which cold water was sprayed in order to quickly condense the steam vapor after the power stroke and thus return the piston to its minimum volume position.

Before describing the Rankine Cycle in detail, it is interesting to see how nature has deployed the basic processes found in it. It starts with a series of processes that begins with the evaporation of water from any body of water, followed by the movement of the evaporated water to a higher elevation of land, before it finally ends with the condensation of the evaporated water into rain, which fills basins, lakes and rivers. That water then flows to lower gravity gradients where it can be converted into mechanical kinetic energy or electrical power, if the rivers and lakes it feeds are dammed, and the water is controlled to flow through hydro turbines. In a similar manner, the action of the Moon–Earth system causes water to be pulled by gravity towards the Moon twice every day. This action results in a phenomenon called tides, which also fills geophysical basins that are either natural or man-made. During high tide, the water can be trapped using movable dikes or gates until the tide recedes. Then, with the trapped water higher than the water level on the opposite side of the dike, the water can be controlled to pass through hydro-turbines that produce electric power. The efficiency may be lower and the process cycle time much longer than a typical cycle in a contemporary Rankine Cycle, but the net result is to have nature produce mechanical energy from thermal (solar) energy input.

The classic Rankine Cycle is shown in Figure 5.1(a), where the block diagram of the basic mechanical elements and the process diagram for the cycle are shown. These block diagrams can be readily drawn using the tools available on Microsoft Excel, from its SHAPE toolbox located in the INSERT pull-down in the top tool bar menu. This feature is very useful for drawing the graphics of any thermodynamic cycle or process. The Rankine Cycle uses two processes that are both at constant temperatures, in order to have heat input at high temperatures and heat rejection at low temperatures. The utilization of constant temperatures where heat is either added or removed is in fact one of the basic tenets of thermodynamics, as expressed in the Second Law of Thermodynamics. That is, if heat can be transferred during a constant temperature process the entropy change of the heat exchange process has the greatest chance of being performed with entropy change approaching zero. As explained in Chapter 4, the Carnot, Sterling and Ericson Cycles are

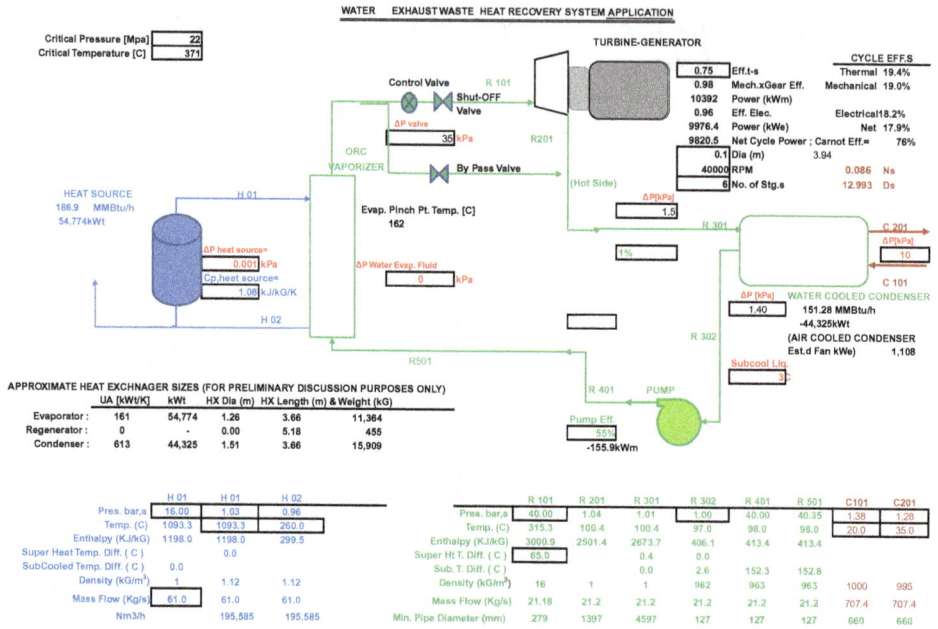

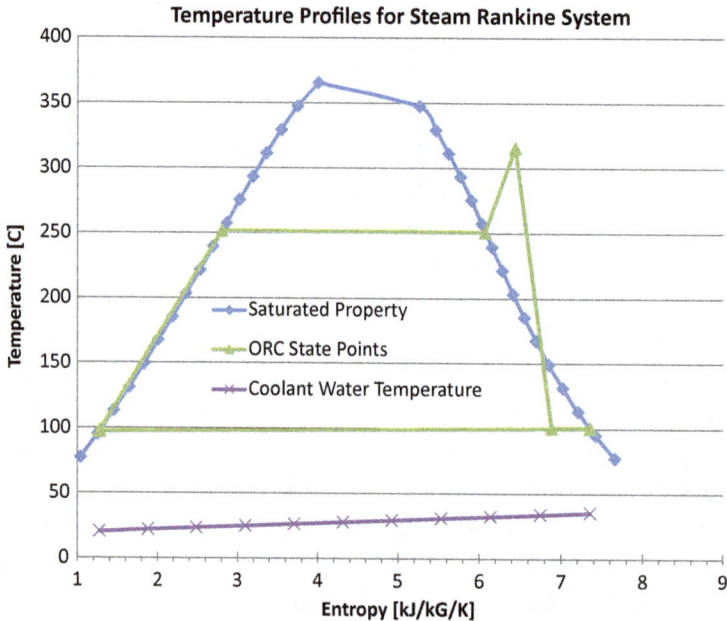

(b)

Figure 5.1.   (a) Rankine Cycle with regenerator, and (b) temperature profiles for the Rankine Cycle System, including waste heat and condensing coolant streams.

all equally capable of producing the highest efficiency when heat input and heat rejection occur only at a constant high temperature and a constant low temperature, respectively. The cycle efficiency is expressed elegantly in the famous Carnot Equation, an equation that has a prominent and well-deserved place in the foundations of the Second Law and concepts of availability and/or exergy, as was explained in Chapter 4.

$$\eta = 1 - \frac{T_c}{T_h}. \tag{5.1}$$

As noted in Chapter 2, it is the nature of any fluid to change phase from liquid to vapor or from vapor to liquid at a constant pressure and at constant temperature. This is depicted in Figure 5.1(b). It may be recalled that a heat exchanger maintains a process at constant pressure. Heat input to the Rankine Cycle occurs by placing an evaporator or boiler that enables the working fluid of the cycle, typically water, to absorb heat energy in a process path that is constant pressure and for most of the heat transfer, also at constant temperature. The liquid thus changes phase from liquid to at least a saturated vapor state and often times, to a superheated vapor state in order to increase the cycle efficiency. The heat rejection from the Rankine Cycle occurs by placing a condenser at the discharge of the turbine (which means the turbine operates at the lowest pressure in the cycle). The condenser rejects heat from the fluid which may still be in a vapor phase or possibly a two-phase, liquid-vapor phase, which is typical of a steam Rankine Cycle system. The fluid at the outlet of the condenser is entirely a liquid and in fact is usually subcooled by 10 to 20°F (5 to 10°C) in order to help improve the performance of the feed-pump which is installed immediately downstream from the condenser. Although the evaporation and condensation can occur at the same pressure, to generate power with a Rankine Cycle, the evaporator operates at a pressure that is higher than the pressure maintained in the condenser. In-between these two constant pressure and constant temperature processes, a turbine is placed on the right side of the T-s saturated dome while a pump is placed on the left side of the dome. It is noted in Chapter 3 that the turbine's inherent function is to take a high-pressure state point and reduce the pressure to a lower pressure state point, while converting the thermal energy into mechanical energy or mechanical power as represented by the rotating shaft shown with the turbine block diagram in Figure 5.1(a). Chapter 3 also explains the inherent function of a pump to pressurize a liquid state of a fluid. Thus, the pump (most often called a feed pump in the power generation industry because it "serves" to feed the evaporator with the liquid) is placed on the left side of the T-s diagram, where the fluid is known to always be in a liquid state. The complete cycle thus proceeds by having the feed pump pressurize the working fluid and force it through the boiler or evaporator, where heat energy from any number of heat sources is used to boil or evaporate the fluid into a saturated vapor state point or, most often, into a superheated vapor. The process continues by having the fluid pass through the turbine where the turbine extracts the thermal energy and produces rotary mechanical power. The fluid is then processed to the condenser where it is condensed and thus returned to a liquid state. The process has effectively converted the thermal energy into mechanical power. The source of the heat energy can vary, for example, from solar energy, to fossil fuels, nuclear energy or even friction heat caused by rubbing two sticks together. In fact, it is sufficient to heat paper to a temperature of 451 °F for it to spontaneously combust and burn until there are only ashes remaining.

A sample calculation that demonstrates the analysis involved with the basic Rankine Cycle is given in Example 5.1.

5.1.1   *Example 5.1*

Consider a heat source of exhaust gases from an engine. The mass flow rate of the exhaust is 10 lbm/s and the temperature of the waste heat is 950°C. This exhaust gas is cooled to 300°C as shown in Figure 5.2. Water is to be boiled at a pressure of 300 psia and superheated to a temperature of 700°F. The steam is expanded to a pressure of 14.7 psia using a turbine. The steam is then condensed and pumped to the evaporation pressure equal to the desired pressure at the inlet to the turbine.

The first step is to construct a Property Table in order to identify all the state point properties for the inlet and outlet conditions of each component, the most important property being enthalpy. The property table for the example cited above is shown in Figure 5.2(c). All the properties are determined by (1) knowing at least two properties for each state point, and then (2) looking up the remaining properties using the property table for water, in this example. The reader can follow the instructions given in Chapter 2.

This Property Table defines the cycle regardless of the power output requirements. However, in order to be able to define a specific power output or a specific heat input rate, and thus to be able to size the components, at least one heat input, power output or the heat rejection must be known in order to fix the flow rate and thus determine the power (or heat transfer rate) for all the components.

By using the simplified First Law expressions for each of the components that make up the Rankine Cycle, the heat transfer and the power

water   EXHAUST WASTE HEAT RECOVERY SYSTEM FROM HOGAT APPLICATION

(a)

| Critical Pressure [Mpa] | 22 |
| Critical Temperature [C] | 371 |

TURBINE-GENERATOR

0.82	Eff.t-s
0.98	Mech.xGear Eff.
1051	Power (kWm)
0.96	Eff. Elec.
1009.0	Power (kWe)
1000.0	Net Cycle Power
0.2	Dia (m)	7.87
3600	RPM
2	No. of Stg.s

CYCLE EFF.S
Thermal 17.7%
Mechanical 17.3%
Electrical 16.6%
Net 16.5%

0.001 Ns
94.887 Ds

Control Valve   R 101
Shut-OFF Valve
ΔP valve   35 kPa   R201

ORC VAPORIZER
By Pass Valve   (Hot Side)

HEAT SOURCE   H 01
20.7 MMBtu/hr
6,061 kWt

ΔP heat source= 0.001 kPa
Cp.heat source= 1.08 kJ/kG/K

H 02

Evap. Pinch Pt. Temp. [C]   202
ΔP ORC Evap. Fluid   45 kPa

ΔP[kPa]   1.5

REGENERATOR
(Cold Side)
ΔP[kPa]   1.5

R 301   C 201
Regen. Effectiveness   ΔP[kPa]
1%   10

ΔP [kPa]   WATER COOLED CONDENSER
1.40   17.06 MMBtu/hr
-4,998 kWt
(AIR COOLED CONDENSER
Est.d Fan kWe)   125

R 302   C 101
Subcool Liq.
3 C

R501   R 401   PUMP
Pump Eff.
55%
-9.0 kWm

APPROXIMATE HEAT EXCHNAGER SIZES (FOR PRELIMINARY DISCUSSION PURPOSES ONLY)

	UA [kWt/K]	kWt	HX Dia (m)	HX Length (m) & Weight (kG)	
Evaporator :	17	6,061	0.41	3.66	1,364
Regenerator :	0	-	0.00	5.18	455
Condenser :	69	4,998	0.51	3.66	1,818

(a)

TURBINE-GENERATOR

0.82	Eff.t-s
0.98	Mech.xGear Eff.
1051	Power (kWm)
0.96	Eff. Elec.
1009.0	Power (kWe)
1000.0	Net Cycle Power
0.1	Dia (m)
36000	RPM
1	No. of Stg.s

CYCLE EFF.S

Thermal 17.7%
Mechanical 17.3%

Electrical 16.6%
Net 16.5%

3.94

0.009 Ns
54.966 Ds

R201

ve

(Hot Side)

(b)

	H 01	H 01	H 02
Pres. bar.a	1.03	1.03	0.96
Temp. (C)	950.0	950.0	300.0
Enthalpy (KJ/kG)	1043.5	1043.5	342.6
Super Heat Temp. Diff. (C)	0.0		
SubCooled Temp. Diff. (C)	0.0		
Density (kG/m³)	1	1.12	1.12
Mass Flow (Kg/s)	8.7	8.7	8.7
Nm3/hr		27,747	27,747

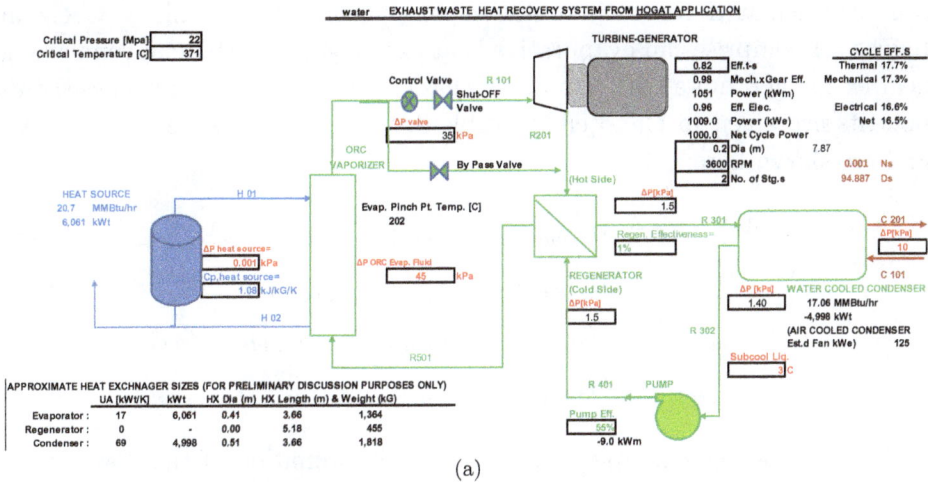

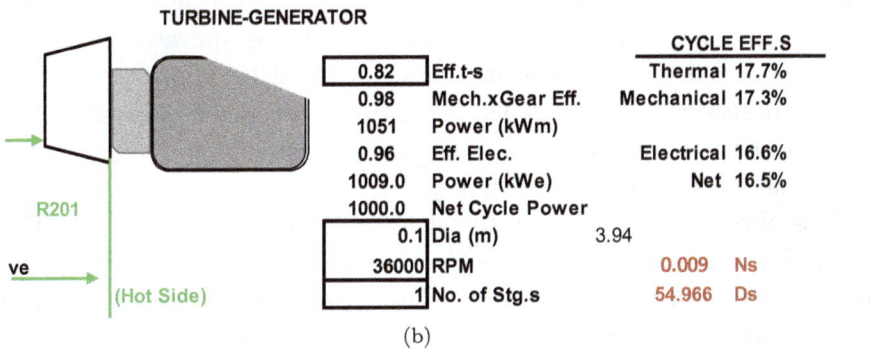

	R 101	R 201	R 301	R 302	R 401	R 501	C101	C201
Pres. bar.a	20.69	1.06	1.03	1.01	21.16	21.50	1.38	1.28
Temp. (C)	284.6	100.8	100.8	97.4	97.9	97.9	20.0	35.0
Enthalpy (KJ/kG)	2983.0	2529.4	2674.3	407.7	411.5	411.5		
Super Ht T Diff. (C)	70.0		0.4	0.0				
Sub. T. Diff. (C)			0.0	2.6	117.3	118.2		
Density (kG/m³)	9	1	1	962	982	962	1000	995
Mass Flow (Kg/s)	2.4	2.4	2.4	2.4	2.4	2.4	79.8	79.8
Min. Pipe Diameter (mm)	127	482	1524	50	50	50	228	228

	H 01	H 01	H 02
Pres. psia	14.94	14.94	13.94
Temp. (F)	1742.0	1742.0	572.0
Enthalpy (Btu/Lbm)	449.4	449.4	147.5
Super Heat Temp. Diff. (F)	0.0		
SubCooled Temp. Diff. (F)	0.0		
Density (Lbm/ft³)	0.07	0.070	0.070
Mass Flow (Lbm/s)	19.04	19.04	19.04
Cp (Btu/Lbm/F)	0.26		

	R 101	R 201	R 301	R 302	R 401	R 501	C101	C201
Pres. psia	300.0	15.3	14.9	14.7	306.8	312	20.01	18.54
Temp. (F)	544	213	213	207	208	208	68	95
Enthalpy (Btu/Lbm)	1284.6	1089.2	1151.7	175.6	177.2	177.2		
Super Ht.T Diff. (F)	126		1					
Sub. T Diff. (F)				4.7	211	213		
Density (Lbm/ft³)	0.53	0.04	0.04	59.95	59.98	59.98	62.3	62.1
Mass Flow (Lbm/s)	5.2	5.2	5.2	5.2	5.2	5.2	175.5	175.5
Min. Pipe Diameter (inch)	5	19	60	2	2	2	9	9
Volume Flow rate (ft³/s)	9.7	133.7	137.4	0.1	0.1	0.1		

(c)

Figure 5.2. (a) Computer model solution of Example 5.1, (b) details of the turbine-generator specification, and (c) Property Table for the Rankine Cycle of (a).

can be calculated for each of these components. The Rankine Cycle in Figure 5.1 comprises an evaporative heat exchanger, a turbine, a condensing heat exchanger and a feed pump. The basic equations for each of these components are given in Chapter 3's Table 3.8, where they were first derived, and are shown here:

$$\dot{W}_{turbine} = \dot{M}_{fluid} \times (h_{turbine\ in} - h_{turbine\ out}), \qquad (5.2a)$$

$$\dot{W}_{pump} = \dot{M}_{fluid} \times (h_{pump\ out} - h_{pump\ in}), \qquad (5.2b)$$

$$\dot{Q}_{condenser} = \dot{M}_{fluid} \times (h_{condenser\ in} - h_{condenser\ out}) \qquad (5.2c)$$

$$\dot{Q}_{evaporator} = \dot{M}_{fluid} \times (h_{boiler\ out} - h_{boiler\ in}). \qquad (5.2d)$$

To continue the example, consider the determination of the flow rate in order to have an output of 1,000 kWe. The equation that determines the flow rate is the energy equation across the turbine, as given in Equation (5.2a).

In a similar manner, if the heat rate into the evaporator is known, the flow rate of the working fluid can be determined by applying the first law energy conservation equation across the boiler, as shown in Equation (5.2d).

Once the flow rate is known, the heat rate or the power output for all the other components can be determined.

It is also usually necessary to calculate the cycle efficiency and then go one step further to calculate the Carnot Cycle efficiency for the given high and low temperatures in the cycle. Of course, the Carnot Cycle efficiency must always be higher than the actual cycle's efficiency, and any deviation from the Second Law of Thermodynamics must be checked for a simple calculation error (or an error in an assumption) made for one or more of the components that make up the cycle.

A very important contribution to the cycle analysis would be to include an energy and exergy balance. An energy balance ensures that the net amount of heat transfer that enters the control volume (CV) that is drawn by the analyst around all parts of the system is equal to the net power produced or absorbed through that CV boundary. A simple explanation of the exergy analysis is that the analyst can identify where there is a significant mismatch of the temperatures used in any heat transfer within that component in the cycle. A more detailed explanation is offered in Chapter 3. It is interesting to note here however, with the worked examples providing the numerical analysis, that an Exergy Balance can define the maximum amount of work that could be achieved with a given heat source and with the given temperature of that heat source. For example, an Energy and Exergy balance for the Steam Rankine Cycle shown in Figure 5.3 is given in Table 5.1.

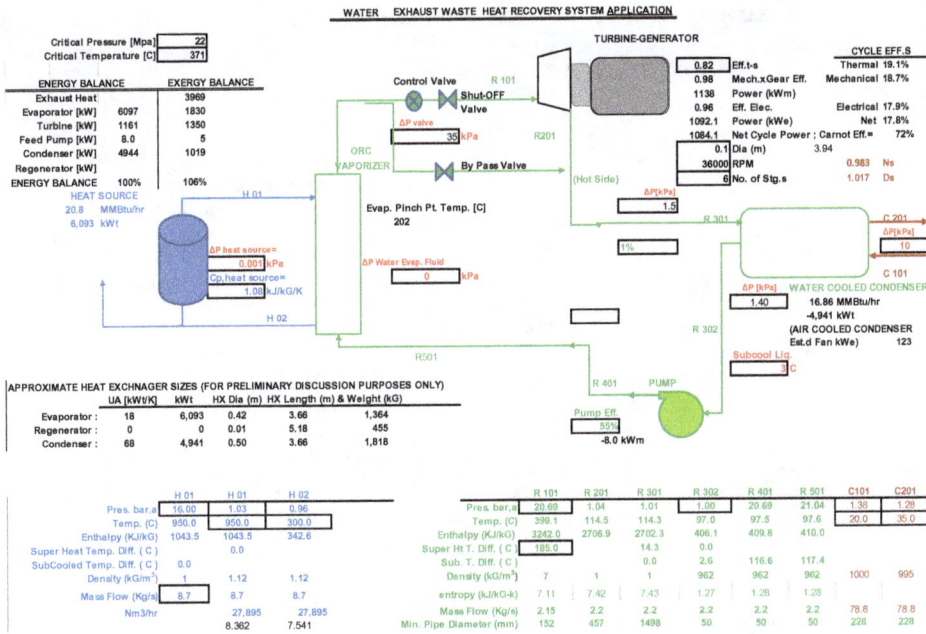

Figure 5.3. Steam Rankine Cycle System with Energy and Exergy Balance Shown.

Table 5.1. Summary of Energy and Exergy balance for the Steam Rankine Cycle in Figure 5.2.

	ENERGY BALANCE	EXERGY BALANCE
Exhaust Heat		4207
Evaporator [kW]	6097	1830
Turbine [kW]	1161	1350
Feed Pump [kW]	8.0	5
Condenser [kW]	4944	1019
Regenerator [kW]		
ENERGY BALANCE	100%	102%

The Brayton Cycle or gas turbine engine cycle shown in Figure 5.4 also includes an exergy and energy balance.

To better explain how exergy and energy balances are useful in optimizing the cycle with respect to energy transfers, it is worth-while to review Figures 5.5(a) and (b). Figures 5.5(a) and (b) illustrate the extreme heat source and heat sink temperatures but also the necessary temperature differences that must occur between components in a thermodynamic cycle in order to exchange energy via hat transfer. The net result of requiring these temperature differences causes exergy to be unavailable or destroyed as noted in Chapter 3.

STANDARD BRAYTON CYCLEOR GAS TURBINE ENGINE

FUEL FLOWRATE= 13837.38   6.665

COMBUSTOR

T2 (R)=	1187	Q= 2.9E+08 Btu/h.
T2 [F]=	727	

2520 =Tmax. (R)

Nc= 1.41   3   Generator Eff.=   0.95

EFF.= 0.84   **TURBINE GENERATOR:**

POWER=

50003 hp   POWER,net=

COMPRESSOR   36859 hp   27486 Kwe

127.8   1   32% EFF.

Ne= 1.41   **-OR- GAS TURBINE, JET ENGINE:**

EFF.= 0.88   VEL.out=   2489   (ft./s.)

Pinlet=	14.5 psia	POWER=	88802 hp	4	THRUST=	17150	lbf,static
Tinlet=	540 R		Poutlet=		IMPULSE=	77	lbf-s./lbm

PRES. RATIO=   12   Poutlet=   14.5 Psia

221.91 lbm/s.   Texhaust=   ↓ 1499 R

1039 F

FLUID:	AIR		P [psia]	T [F]	h [Btu/lbm]	Entropy
MOLE. WT.	28.966	1	14.5	80	129.1	1.642
Cp=	0.24 Btu/lbm/R	2	174	727	288.4	1.664
Cv=	0.17 Btu/lbm/R	3	174	2060	652.1	1.869
K=	1.41	4	14.5	1039	369.3	1.896

ENERGY BALANCE		EXERGY BALANCE: Perfect Gas			"Real" Air
Combustor Heat Transfer [kWt]=	85145	Combustion [kWt]=	58284		59432
Compressor Power [kWm]=	37288	Compressor Power [kWm]=	34452	2836	34452
Turbine Power [kWm]=	66221	Turbine Power [kWm]=	69646	3425	69646
Ambient Heat Rejection [kWt]=	56212	Cooling [kWt]=	22893		24144
Energy Balance=	100%	Exergy Balance=	100%		100%
Net Power [kW]=	28933	Net Exergy Power [kW]=	35194		35194

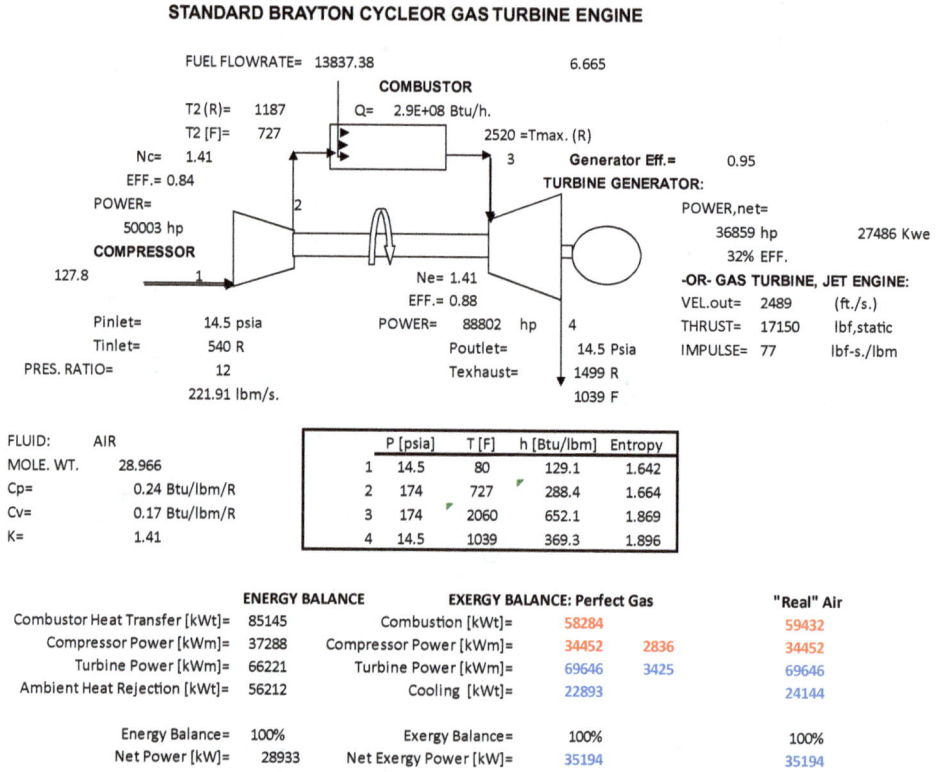

Figure 5.4.   Brayton (gas turbine engine) Cycle including Property Table and Exergy and Energy Balances.

Figures 5.5(a) and (b) are very similar except that there is a very important distinction between the two ideal cycles that further demonstrates the increase in entropy principle of the Second Law of Thermodynamics. The distinction between these two figures is the use of an additional power cycle in Figure 5.5(b) between temperature reservoirs T2,A and T2,B, which is absent in Figure 5.5(a). The addition of a power cycle that recovers heat energy between these two temperature reservoirs and produces additional net power is clearly a means of reducing the exergy destruction that otherwise occurs in the cycle shown in Figure 5.5(a).

The cycles shown in Figures 5.5(a) and (b) represent perfect reversible cycles; cycles that the utopian engineers often mentioned in Chapter 3 would use to produce the maximum power for the heat supply, Q_h, from the high temperature heat source, T_{hot}. Unfortunately the real-world cycle shown on the right in Figure 5.5(a) reveals a discontinuity between the temperature T2,A and T2,B (which represents a heat exchanger that is often necessary in heat engine cycles), for example, to exchange the heat between the heat

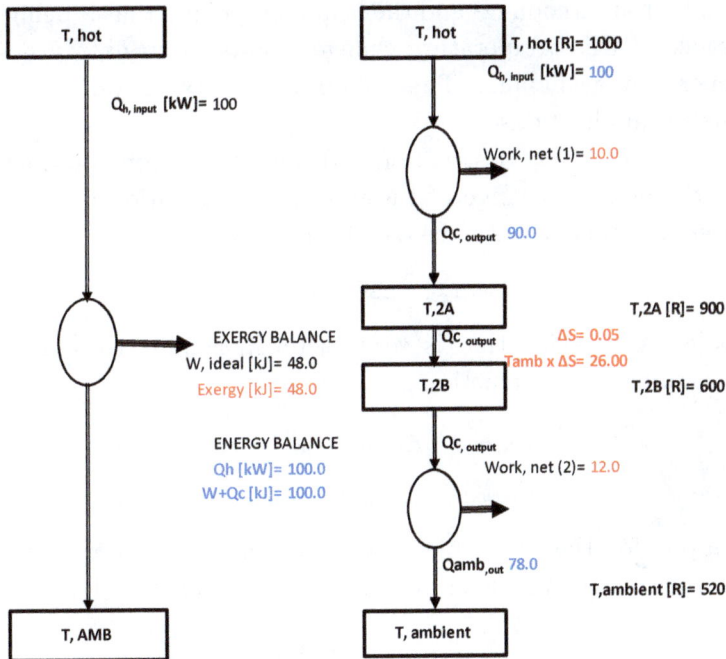

(a)

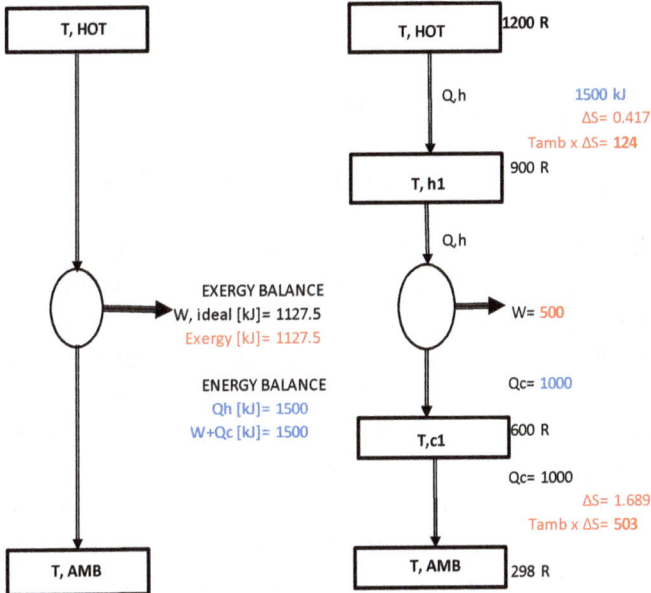

(b)

Figure 5.5. Illustrations of heat engine interaction between hot and cold temperatures within heat engine cycles. (a) Cycle with only a heat exchanger with high (T,2A) and low (T,2B) wall temperatures shown; (b) Cycle shown with a heat engine used between high and low temperature reservoirs.

source (a fluid that is cooling) and the heat sink (a fluid that is being heated). The presence of this temperature change results in an exergy destruction (Ed) or energy unavailability. That is, power that can never be made available for useful applications.

Recall that the exergy destruction and the total entropy production (also known as the entropy change of the universe, as mentioned in Chapter 3) for a process are related with the equation:

$$E_d = T_\infty \times \Delta S_{total\,production}$$

An equivalent expression that is very useful in helping to understand the significance of exergy destruction is:

$$E_d = W_{ideal,\,max,\,Utopian\,energy\,of\,power\,generation}$$

$$- W_{actual,\,real-world,\,energy\,or\,power\,produced}$$

The equations for the reversible power and the exergy destruction, when taken together, accounts for an exergy balance as follows.

$$Exergy_{destruction} = T_{ambient} \times \left(-\frac{Q_c}{T_{2a}} + \frac{Q_c}{T_{2a}} \right) \tag{5.3a}$$

$$Exhaust\,Gas\,Exergy_{availability} = \Delta h - T_{ambient} \times \Delta s \tag{5.3b}$$

Or the equivalent:

$$Ex.\,Exergy_{availability} = \dot{M}_{exhaust}$$

$$\times \left(C_p \times (T_{out} - T_{in}) - T_{ambient} \times C_p \times \ln \frac{T_{out}}{T_{in}} \right) \tag{5.3c}$$

An exergy balance for a cycle is a determination of how each of the components in the cycle would operate if the components were reversible; that is, operating without any entropy losses. As such, the results can be compared to how well each of the components are expected to operate with known efficiencies less than 100%; whether for the compressors, turbines, pumps, nozzles, diffusers or valves that may be used in the cycle. By their very nature of exchanging heat between two flow streams that must be at different temperatures, the heat exchangers used in any cycle are producers of entropy and thus are producers of exergy destruction. This is clearly demonstrated in Figures 5.5(a) and (b) wherein the exergy production is given by Equation (5.3aa). All other components have the exergy production determined by applying Equation (5.3ab) or equivalently, Equation (5.3ac).

It may help to understand the exergy balance by looking at Figures 5.5(a) and (b) as columns of potential energy if only as an analogy to the more familiar form of energy. Just as the energy potential to do work present in a mass raised in a gravity field with respect to a datum, exergy is larger in magnitude if it has a temperature (thermal energy), pressure (mechanical energy) and entropy above a datum. In the case of exergy, the datum is the dead state represented by the local ambient pressure and temperature. Each of the components in a thermodynamic cycle acts upon the thermal or mechanical energy and utilizes it to produce a desired result, typically power or heat transfer. Thus, after each component's action, the exergy is reduced. After all of the components have acted upon the system, the exergy is found to be reduced to the dead state. Comparing the actual performance of the components with the exergy change is the best way of understanding how each of the cycle components could have performed better, up to a maximum effect. This result is a better comparison of the how well the cycle is performing relative to a much simpler energy balance for the cycle. In fact, an energy balance can only indicate if all of the energy exchanges have been accounted for; it does not make known how much each component may be deficient in their individual performance.

In Figures 5.5(a) and (b), the exergy destruction is shown as being produced due to a heat transfer across a temperature difference — such as in the operation of a heat exchanger — or as the result of the inefficiency of any of the components used in the cycle to generate power. This inefficiency can be due to contact friction, drag losses or vortices in the flow streams, heat loss from components due to poor insulation and, in general, the increase in entropy anywhere within a component. The summing of these exergy destruction terms is observed in Figures 5.5(a–b) as equal to the total exergy that was available from an energy source at the temperature of that source, T,hot with respect to the ambient temperature, T,amb also known as the "Dead State Temperature". Thus, the presentation of an exergy balance in addition to the first law energy balance for all of the components in the cycle, can help the analyst to quickly access which of the components is not performing to its theoretical optimum. This realization can lead to an engineering improvement of the effected components.

5.2 Modified basic Rankine Cycle

While the basic Rankine Cycle has not changed, many modifications have improved upon its man-made version — for example, by increasing the cycle's efficiency and/or the power output for given component sizes. The reader should consider using one or more of the seven basic mechanical elements in implementing such improvements.

Critical Pressure [Mpa]	22
Critical Temperature [C]	371

WATER   EXHAUST WASTE HEAT RECOVERY SYSTEM APPLICATION

TURBINE-GENERATOR

ENERGY BALANCE		EXERGY BALANCE	
Exhaust Heat [kW]		14318	
Reheater Heat [kW]		1887	
Reheater [kW]	2877	2850	
Evaporator [kW]	21834	1703	
Turbines [kW]	4938	5347	
Feed Pump [kW]	83.2	53	
Condenser [kW]	19856	5731	
Regenerator [kW]			
ENERGY BALANCE	100%	97%	

HEAT SOURCE
74.5 MMBtu/hr
21,820 kWt

CYCLE EFF.S

0.75	Eff.t-s	Thermal 22.6%
0.98	Mech.xGear Eff.	Mechanical 22.2%
4839	Power (kWm)	
0.96	Eff. Elec.	Electrical 21.3%
4645.5	Power (kWe)	Net 20.9%
4582.3	Net Cycle Power ; Carnot Eff.= 72%
0.35	Dia (m)	13.78
8000	RPM	0.243 Ns 0.10
6	No. of Stg.s	4.100 Ds 9.511

ΔP vatve   0 kPa
Control Valve   R 101
SHUTOFF Valve

REHEATER
R 201
By Pass Valve   ΔP[kPa]
ORC VAPORIZER   0   R202

REHEAT T

Vap. Pinch Pt. Temp. [C]
199

ΔP heat source= 0 kPa
Cp.heat source= 1.121 kJ/kG/K

P Water Evap. Fluid   0 kPa

H 01
H 02

R301

C 201
ΔP[kPa]   0

WATER COOLED CONDENSER
ΔP [kPa]   0.00   56.84 MMBtu/hr
-17,240 kWt
(AIR COOLED CONDENSER
Est.d Fan kWe)   431

C 101

R 302
R501

Subcool Liq.   0 C

APPROXIMATE HEAT EXCHNAGER SIZES (FOR PRELIMINARY DISCUSSION PURPOSES ONLY)

	UA [kWt/K]	kWt	HX Dia (m)	HX Length (m) & Weight (kG)	
Evaporator :	69	21,820	0.82	3.66	5,000
Regenerator :	0	-	0.00	5.18	455
Condenser :	240	17,240	0.95	3.66	6,364

R 401   PUMP

Pump Eff.   55%
-83.2 kWm

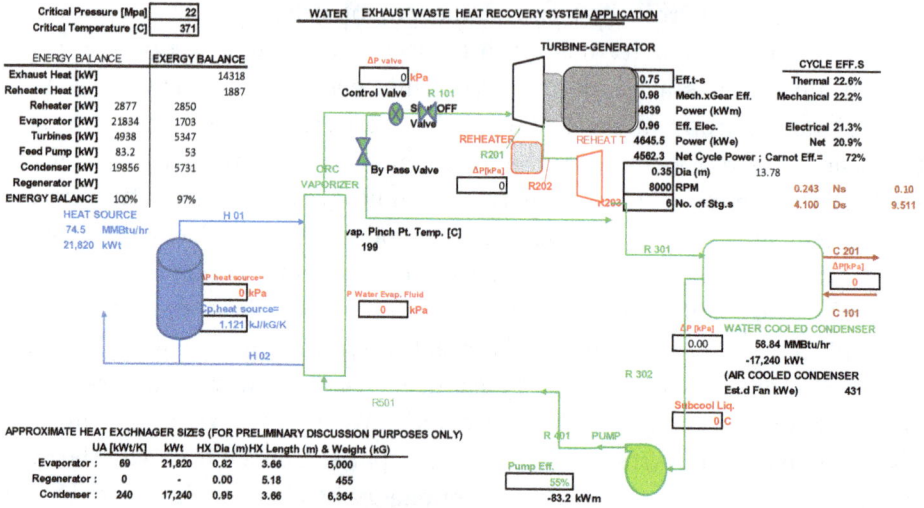

Figure 5.6.   Steam Rankine Cycle System using a reheater and reheat turbine.

One common improvement is the introduction of a second, third or fourth turbine and placing a heat exchanger in-between the turbines in the cycle. Thus, if two turbines are used, one heat exchanger is needed. If four turbines are added to the cycle, then three heat exchangers are added. A process modification to the Rankine Cycle using two turbines and one heat exchanger is shown in Figure 5.6. The net effect of adding one or more turbines followed by heat exchangers (also called reheat turbines and reheaters in the power industry) is to continue the heat input with as constant a temperature as possible. By maintaining a relatively constant temperature as heat is added, the heat input process starts to conform to the tenet of the Second Law of Thermodynamics. Theoretically, the more turbines and reheaters there are, the greater the increase in thermal efficiency. However, engineering in the real world must contend with the cost, space, complexity and reliability of adding too many turbines or heat exchangers. Thus, the economics of the cycle may be improved by adding one "set" (comprising one reheating turbine and one reheater) but not two or more despite the theoretically possible increase in cycle efficiency with each additional set. It can be shown that the cost effectiveness of any cycle is in proportion to the cost per kilowatt-hour ($/kWe) of the cycle, and is even more affected by the operation and maintenance (O&M) cost of keeping the system operational over any period of time. The simple payback and a relatively new benchmark called the levelized cost of electric power will be described in greater detail toward the end of this chapter, after more thermodynamics cycles have been reviewed.

The improvements to the cycle continue with the addition of another heat exchanger, called a recuperator or regenerative heat exchanger, shown as phantom lines in the process diagram in Figure 5.6. The regenerative heat exchanger is also used to increase the cycle's efficiency. It achieves this by recovering the sensible heat from the superheated fluid exiting the turbine before it enters the condenser. This heat can be used to pre-heat the flow rate entering the evaporator. By using this heat, less energy is needed from the fueled energy source while achieving the same power output. This results in an increase in the efficiency of the cycle. It is noted, however, that the flow rate of the fluid into the turbine is increased if the same amount of energy is input into the evaporator before the regenerator was added. Thus, the power output and the cycle efficiency are increased with a regenerator, compared to the non-regenerated cycle.

However, it is necessary to point out that not all Rankine Cycles can take advantage of using a recuperator. For example, if the working fluid is water, it is very likely that the exit of the turbine is in the two-phase region of the water. In this case, there is no superheat available for the recuperator to recover. For Rankine Cycles that use non-water working fluids such as common refrigerants, there is an opportunity for a recuperator to be used because the exit of their turbines would be located in the superheated region of the vapor. Figure 5.7 illustrates one such case using R245fa as the working fluid.

It is also necessary to point out that in some cases where the heat source is exhaust gases, the exit temperature of the hot exhaust gases will be forced to be increased when a recuperator heat exchanger is used. This occurs because the recuperator is intended to increase the temperature of the liquid working fluid that enters the evaporator (boiler). This increase in temperature may require the evaporator exit temperature to also be increased in order to avoid a significant increase in the size of the evaporator for the same heat transfer duty. The reason for this is easily understood by observing the temperature profiles of the hot waste heat and the evaporating working fluid, as shown in Figure 5.8. It is also noted that the heat source is a condensing fluid like steam, as evidenced by the constant temperature heat removal.

The heat exchanger design is based on four factors: the amount of heat transfer required, the overall effective temperature difference between the hot and cold fluids, the surface area of the heat exchanger and the heat transfer coefficient. The relationship of these parameters is elegantly expressed in the heat exchanger design equation given here:

$$\dot{Q}_{heat\,transfer} = U_{coefficient} \times A \times (T_{hot} - T_{cold}), \qquad (5.4a)$$

$$\frac{1}{UA} = \frac{1}{h_{inside} \times A_i} + R_{inside\,fouling} + \frac{\ln \frac{D_o}{D_i}}{2 \times \pi \times k_{thermalconductivity} \times L_{oa}}$$

$$+ R_{outside\,fouling} + \frac{1}{h_{outside} \times A_o}. \qquad (5.4b)$$

r245fa   CONDENSING STEAM WASTE HEAT RECOVERY SYSTEM

Critical Pressure [Mpa]	3.7
Critical Temperature [C]	154

TURBINE-GENERATOR

CYCLE EFF.S

0.8	Eff.t-s	Thermal 16.5%
0.98	Mech.xGear Eff.	Mechanical 16.2%
3857	Power (kWm)	
0.96	Eff. Elec.	Electrical 15.5%
3702.3	Power (kWe)	Net 14.2%
3394.3	Net Cycle Power	
0.2	Dia (m)	
12000	RPM	0.32 Ns
1	No. of Stg.s	3.397 Ds

ENERGY BALANCE / **EXERGY BALANCE**

	ENERGY BALANCE	EXERGY BALANCE
Exhaust Heat		8117
Evaporator [kW]	23846	1841
Turbine [kW]	3935	4967
Feed Pump [kW]	308.0	-175
Condenser [kW]	20052	709
Regenerator [kW]		547
ENERGY BALANCE	99.3%	97.2%

Control Valve

S- -OFF
R 101   Valve

ΔP valve   10 kPa   R201

ORC VAPORIZER

HEAT SOURCE

81.33	MMBtu/hr
23,831	kWt

H 01

Evap. Pinch Pt. Temp. [C]   55

ΔP heat source   2 kPa

ΔP ORC Evap. Fluid   20 kPa

H 02

(Hot Side)   ΔP[kPa] 1.5   R 301

Regen. Effectiveness= 65%

REGENERATOR (Cold Side) ΔP[kPa] 1.5   R302

ΔPcond. [kPa] 2

C 201   Cp (p/v) 10

C 101

CONDENSER
68.39 MMBtu/hr
-20,039 kWt
(AIR COOLED CONDENSER
Est.d Fan kWe)   689
Pinch Pt.=   7.2

R501

R 401   PUMP

Subcool Liq. 3.0 C

Pump Eff. 55%
-308.0 kWm

	UA [kWt/K]	kWt	HX Dia (m)	HX Length (m) & Weight (kG)	
Evaporator :	364	23,831	1.90	3.66	25,000
Regenerator :	143	5,412	1.01	5.18	10,000
Condenser :	1364	20,039	2.25	3.66	35,455

	H 01	H 01	H 02			R 101	R 201	R 301	R 302	R 401	R 501	C101	C201
Pres. bar,a	13.00	13.00	12.98	Pres. bar,a		26.00	2.56	2.52	2.50	26.22	26.30	1.38	1.28
Temp. (C)	261.6	261.6	190.5	Temp. (C)		185.7	126.0	69.6	37.2	39.2	80.6	20.0	30.0
Enthalpy (KJ/kG)	2893.1	2893.1	791.9	Enthalpy (KJ/kG)		564.6	520.7	462.8	248.5	251.8	309.7		
Super Heat Temp. Diff. (C)	70.0	70.0		Super Ht T. Diff. (C)		50.0		29.4	0.0				
SubCooled Temp. Diff. (C)			1.0	Sub. T. Diff. (C)				0.0	2.8	96.7	55.5		
Density (kG/m³)	1	1.12	1.12	Density (kG/m³)		116	11	13	1307	1310	1183	1000	997
Mass Flow (Kg/s)	11.11	11.11	11.11	Mass Flow (Kg/s)		93.6	93.6	93.6	93.6	93.6	93.6	479.74	479.74
Nm3/hr		35,622	35,622	Min. Pipe Diameter (mm)		228	711	2108	228	228	228	558	558
Entropy [kJ/kg-K]	6.838	6.838	2.239	Volume Flow Rate (m³/s)		0.810	8.778	7.434	0.072	0.071	0.079		
				Entropy [kJ/kg-K]		1.975	2.002	1.847	1.166	1.171	1.345		
Pres. psia	188.50	188.5	188.2	Pres. psia		377	37.1	36.5	36.3	380.2	381	20.01	18.54
Temp. (F)	502.9	502.9	374.9	Temp. (F)		366	259	157	99	103	177	68	86
Enthalpy (Btu/Lbm)	1272.7	1272.7	348.4	Enthalpy (Btu/Lbm)		243.1	224.2	199.3	107.0	108.4	133.4		
Super Heat Temp. Diff. (F)	126	126		Super Ht.T Diff. (F)		90		53					
SubCooled Temp. Diff. (F)	0	0	1.8	Sub. T Diff. (F)					5.0	174	100		
Density (Lbm/ft³)	0.07	0.070	0.070	Density (Lbm/ft³)		7.20	0.66	0.78	81.46	81.64	73.72	62.3	62.2
Mass Flow (Lbm/s)	24.44	24.44	24.44	Mass Flow (Lbm/s)		205.8	205.8	205.8	205.8	205.8	205.8	1055.4	1055.4
				Min. Pipe Diameter (inch)		9	28	83	9	9	9	22	22
Volume Flow Rate ft³/s)	59.21	59.21	0.45	Volume Flow Rate ft³/s)		28.58	309.74	262.33	2.53	2.52	2.79		

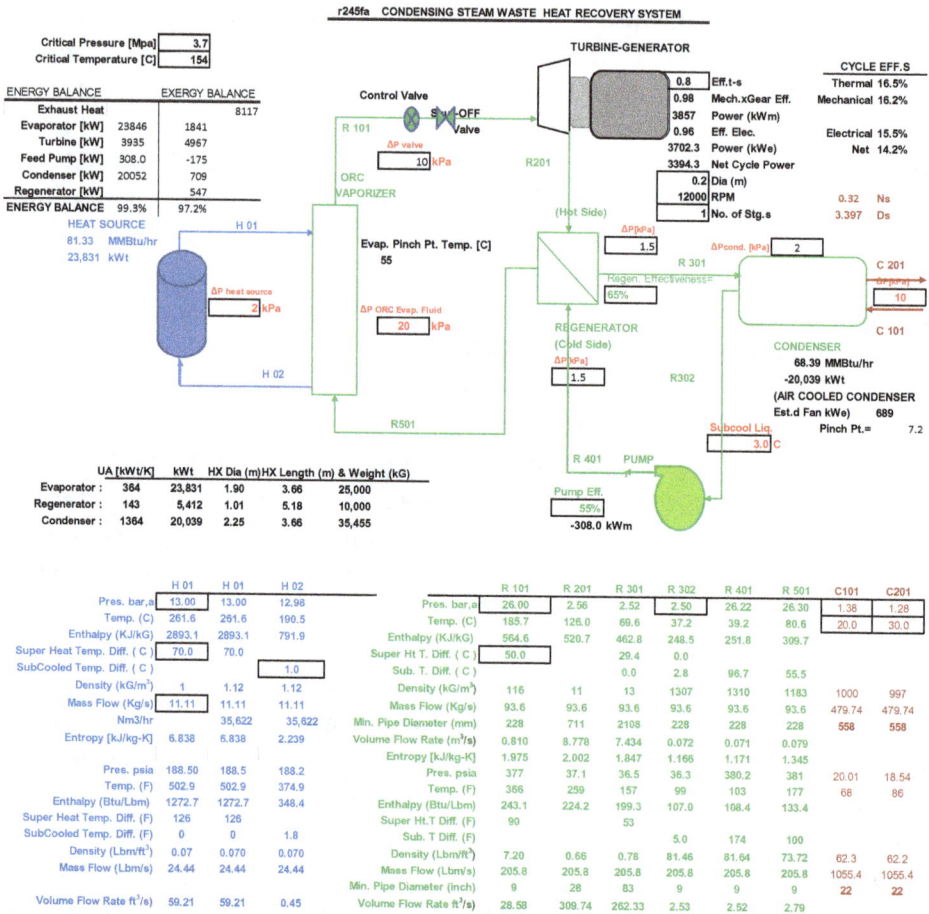

Figure 5.7. Organic Rankine Cycle System using R245fa and a regenerator heat exchanger to increase cycle efficiency.

The definition of a heat transfer coefficient is simply the ability of the heat transfer to be exchanged between two fluids, and is based primarily on the nature of the fluids, the velocity of the fluids and the geometry of the heat exchange.

As may be observed in Equation (5.4a), the heat transfer is increased or decreased by changes in any of these three parameters on the right side of the equation. The desire is to achieve the necessary heat transfer with a minimum surface area. To do this, the heat exchanger must have a large heat transfer coefficient or a large temperature difference. The heat transfer coefficient, as defined above, is usually limited for a given working fluid choice and the choice of the heat exchanger configuration. Thus, only the temperature difference between the hot and cold fluids will greatly affect

Temperature Profiles for ORC System with Regeneration to Increase Efficiency

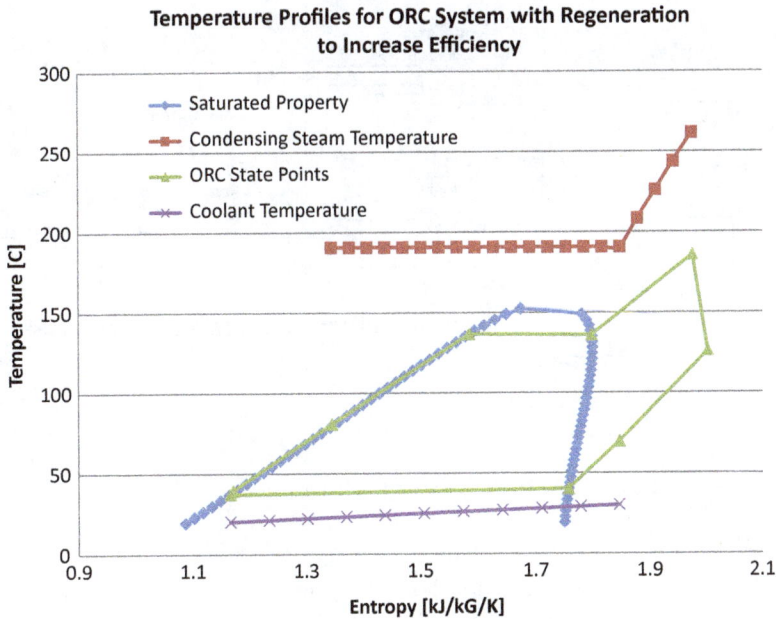

Figure 5.8. Temperature profiles for the waste heat source, ORC fluid and coolant.

the heat exchanger size and/or heat transfer. A recuperator with less heat transfer could cause a decrease in temperature, unless the surface area is increased. Unfortunately, it is also true that as temperature differences get smaller, there is more than a one-to-one increase in the surface area. That is, if the temperature difference decreases by 10% it can be shown that the surface area may increase by as much as 20 to 50%. This is can be demonstrated in any course in a heat exchanger design or in the careful study of heat transfer.

Whether one or more of the suggested methods of improving the Rankine Cycle efficiency is used, the method for calculating the specific power and heat transfer rate for any of the components is the same as the method given in the example.

5.3 Brayton Cycle

A Brayton Cycle is the second most useful thermodynamics cycle that generates mechanical and electrical power. This cycle is the basis for the gas turbine or jet engine that can be used for propulsion as well as for electric power generation. The basic Brayton Cycle consists of compression, expansion and two heat exchange processes, connected as shown in the line diagram

FUEL FLOWRATE=   40.76   6.665

COMBUSTOR

T2 (R)=   1191   Q=   8.6E+05 BTU/Hr.

2960 =Tmax. (R)

GENERATOR:

EFF.= 0.85

POWER=

     120 Hp

COMPRESSOR

POWER, net=

     100 Hp   75 Kwe

     30% EFF.

-OR- GAS TURBINE, JET ENGINE:

TURBINE

EFF.= 0.8

POWER=   220 Hp

VEL.out=   2874   (Ft./sec.)

THRUST=   50   LBf,static

IMPULSE=   89   LBf-sec./LBm

Pinlet=   14.7 PSIa

Tinlet=   560 R

PRES. RATIO=   10

     0.56 LBm/sec.

Poutlet=   14.7 PSIa

Texhaust=   1802 R

STANDARD BRAYTON CYCLE OR GAS TURBINE ENGINE

FLUID: AIR

MOLE. WT.   28.966

Cp=   0.24 BTU/LBm/R

Cv=   0.17 BTU/LBm/R

K=   1.41

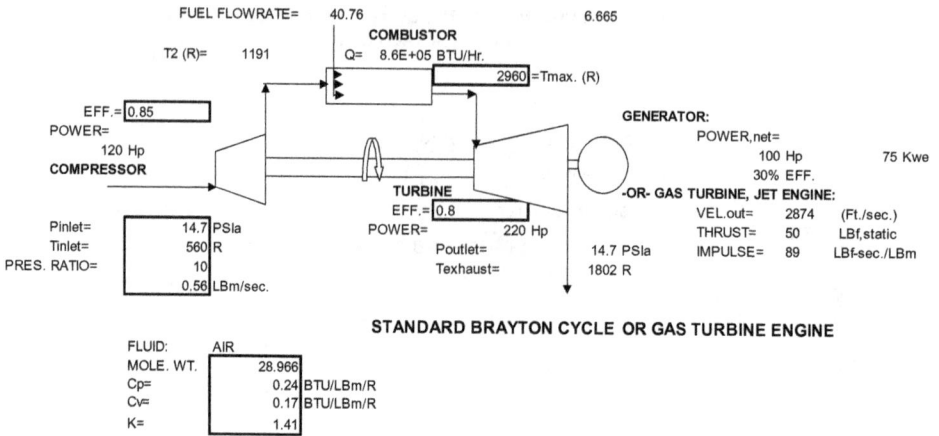

Figure 5.9.   Basic Brayton Cycle for power generation.

given in Figure 5.9. Figure 5.9 was also prepared using Microsoft Excel. The reason this cycle can produce net positive power is because the two constant pressure and heat exchange processes diverge as the temperature along the process curve increases. The higher the temperature, the more the divergence and therefore the larger the enthalpy difference between the inlet and outlet of the turbine expander. In the early years of the development and study of the net positive power Brayton Cycle, it was necessary to develop very efficient turbo machinery, compressors and turbines before the net power from the cycle could be net positive. The basic Brayton Cycle can achieve high cycle efficiencies if the inlet temperature to the turbine is increased to the limits that the materials of construction will allow before failing under thermal stress. The cycle efficiency also increases with pressure ratio, P_r, but its net power output will reach a maximum if the pressure ratio and the temperature into the turbine are selected according to the equation shown here:

$$P_r = \left(\frac{T_3}{T_1}\right)^{\frac{k}{(2\times(k-1))}}, \tag{5.5}$$

where T_3 is the turbine inlet temperature, °R or K, T_1 is the ambient temperature, °R or K, and k = specific heat ratio = $\frac{C_p}{C_v}$. For air, k = 1.41.

As an example, Figure 5.9 indicates a cycle efficiency of 30% with a pressure ratio equal to 10:1.

The following problem provides an example of the calculations involved with a power generation Brayton Cycle. As can be seen from the problem statement, it is necessary to provide several very basic parameters that define the performance of the cycle. The typical inputs include the pressure ratio, inlet pressure and temperature, mass flow rate and the maximum turbine inlet temperature. The inlet turbine temperature is usually restricted to the limits of the material properties used in the construction of the turbine. Typical maximum temperature values range from 2,500 to 2,700°F (i.e. 1400 to 1500°C). It is interesting to note that the combustion temperatures in a typical internal combustion engine may exceed these temperatures and still not suffer damage. This is due to the cyclic nature of the combustion process in the internal combustion engine, compared to the continuous combustion and hence relatively constant inlet turbine temperature during the power generation process. The cyclic combustion process provides an opportunity for the internal cylinder temperature to be reduced to a time average temperature that is considerably less than the maximum allowed temperature limit of the materials.

This is determined by first modeling the compression and expansion power by the simplified equations as given in Equations (5.6a) and (5.6b) with the compressor and turbine efficiencies shown.

$$\dot{W}_{turbine} = \dot{M}_{air} \times c_p \times T_{in} \times \left[1 - \left(\frac{1}{P_r} \right)^{(k-1)/k} \right] \times \eta_t, \quad (5.6a)$$

$$\dot{W}_{compressor} = \dot{M}_{air} \times c_p \times T_{in} \times \left[1 - P_r^{(k-1)/k} \right] / \eta_c. \quad (5.6b)$$

These equations are derived from the first law equation for a compressor and turbine: $W = M \times \Delta h$, but with an assumption that the working fluid, air, can be modeled as a perfect gas and hence the equation reduces to: $W = M \times C_p \times (T_{in} - T_{out})$. It has been shown in Chapter 2 that the inlet and outlet temperature ratios for an ideal, isentropic compression or expansion process is given by: $T_o/T_i = (P_o/P_i)^\wedge((k-1)/k)$, with the assumption that air can be modeled by the perfect gas equation:

$$P \times v = \frac{R_u \times T}{Mole_{weight}}.$$

Figure 5.10 uses Equation (5.7a) to determine that the maximum efficiency is 31% if the pressure ratio is increased to 17.4:1. Of course increasing the pressure ratio by 74% does not provide a proportionate increase in the cycle efficiency, while it most certainly increases the turbine and compressor costs.

$$\eta = \frac{\dot{W}_{net\,output}}{\dot{Q}_{heat\,input}}, \quad (5.7a)$$

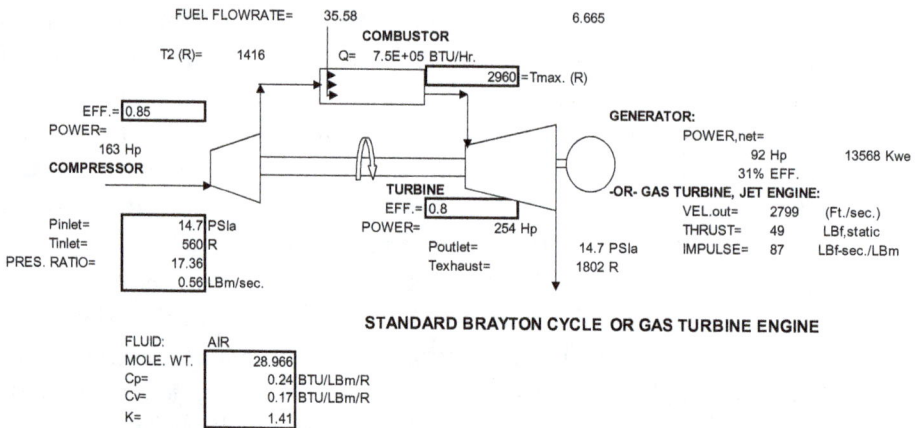

(a)

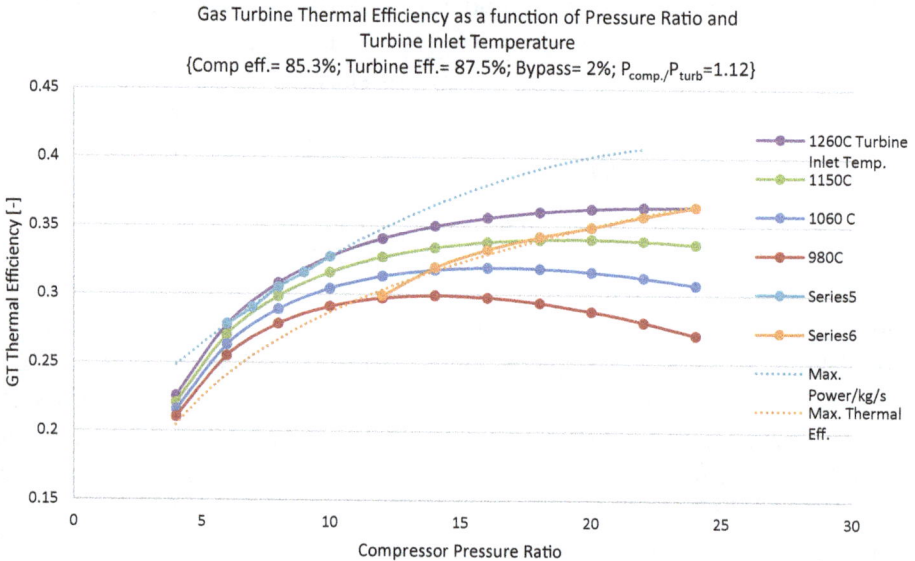

(b)

Figure 5.10.   (a) Standard Brayton Cycle (gas turbine or gas turbine jet engine) and (b) gas turbine efficiency as a function of the pressure ratio and turbine inlet temperature at fixed compressor and turbine efficiencies.

$$Maximum\ Brayton\ Cycle\ Efficiency = 1 - \frac{1}{P_r^{(k-1)/k}}, \qquad (5.7b)$$

$$\frac{T_{turbine\ inlet}}{T_{ambient}} = P_r^{(k-1)/2k}. \qquad (5.7c)$$

The above derivation involves algebra and the use of the perfect gas equation, along with the assumption that the compression and expansion processes are all proceeding with the specific heat ratio, k, for the working fluid, which is most often air. The Brayton Cycle is easily modeled using computer spreadsheet software like Excel or any other math-based software of the reader's choosing. This exercise is worth doing — if only for the interest to eliminate the need for deriving the relationship shown in Equation (5.6b), or for that matter, a similar optimization or minimization that commonly occurs in engineering and science concepts — and will be detailed in the following paragraph.

The exercise assumes that the model is constructed using Excel. The reader can use the output from the author's thermodynamics model of the Brayton Cycle to validate the reader's model. Once a thermodynamics model has been prepared and validated, there remains only the need to use the model in a parametric study to determine all optimizations and minimizations that are relevant for the design of the Brayton Cycle. The author will leave it to the reader to determine which of the two optimization analyses the reader is more comfortable and competent in completing — an algebraic derivation or a parametric study — that reveals an optimization. It is always interesting to remember that the latter methodology provides a ready-made set of graphs that can be used in a report.

It is also interesting how Equations (5.6a) to (5.6c) determine how the thermal efficiency of the cycle can be affected by the pressure ratio across the compressor and turbine, and the turbine inlet temperature. It is also possible to determine the maximum work per unit flow rate and the maximum efficiency as a function of these parameters. Figure 5.7(b) presents the result of this analysis. It may be observed that the maximum work per unit flow rate occurs at a different pressure ratio than the maximum thermal cycle efficiency. It is also necessary to note that in a real-world application of a gas turbine, the pressure ratio across the turbine is always less than the pressure ratio across the compressor. This is a consequence of actual pressure drops in the conduits and the combustor that connects the compressor and turbine when the inlet to the compressor and the outlet from the turbine is fixed by the ambient pressure.

Another useful and common diagram that is used to describe the operation of the Brayton Cycle is shown in Figure 5.11, which displays the temperature-entropy diagram for the cycle. It is useful because it graphically displays the temperature relationships between the working fluid (in this case, air) and the heat source (or heat sink). This is helpful because it immediately provides the analyst with a sense of the entropy increase and hence the second law efficiency of the cycle. That is, if the temperature profile between the heat source and the cycle process that is heated by the

Gas Turbine (Brayton Cycle) with Intercooled Compresors (88%) and Reheat Turbine (90%) using 1,500F Turbine Inlet

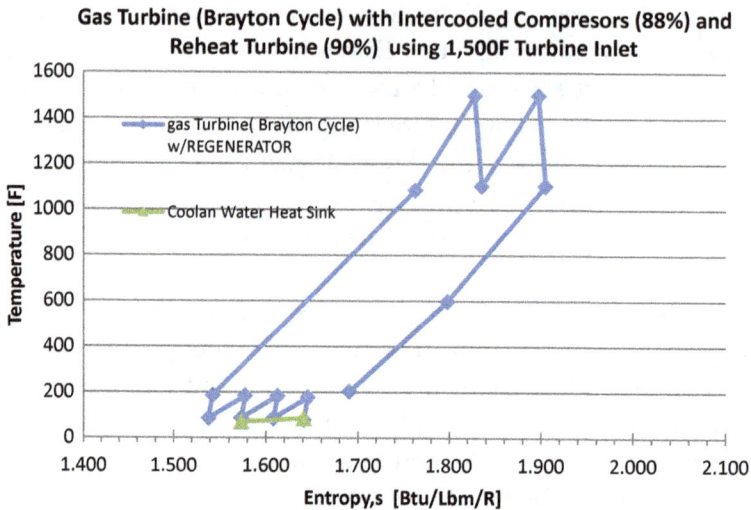

Figure 5.11.   Temperature–entropy diagram for the Brayton Cycle.

heat source is large, then there is a large entropy increase. A large entropy increase is equivalent to a loss in available energy that could have been converted into useful power. A more detailed description of entropy increase is given in the next chapter.

The SHAPE toolbox in Excel is useful for drawing the icons that represent the compressor, turbine and heat exchangers. The format used in Figure 5.9 to type in the cycle's parameters is the same as discussed in Chapter 2 when the cycle diagram for the reciprocating engine was detailed. The cells with black borders are the keystroke inputs to the model. All other cells may contain equations with dependent or independent variables.

The reader will notice that the Brayton Cycle uses a considerable amount of inputs.

The models in Figures 5.9 and 5.10 are known as the simple air standard model, which is so named because it does not use a regenerator (or reheat turbines) or intercooled, multi-stage compressors or waste heat recovery heat exchangers that could be used to increase the regenerative, energy utilization process. The air standard model assumes that the properties of air, such as the specific heats, are evaluated at room temperature. In fact, the use of the spreadsheet as the calculating tool will debunk this assumption, which somewhat limits the accuracy of the model. Thus, the specific heats and the actual properties of the air need not be limited to the properties with constant specific heats at any given pressure. It is easy to have the values of specific heat, C_p, input as a function of the temperature at each state point.

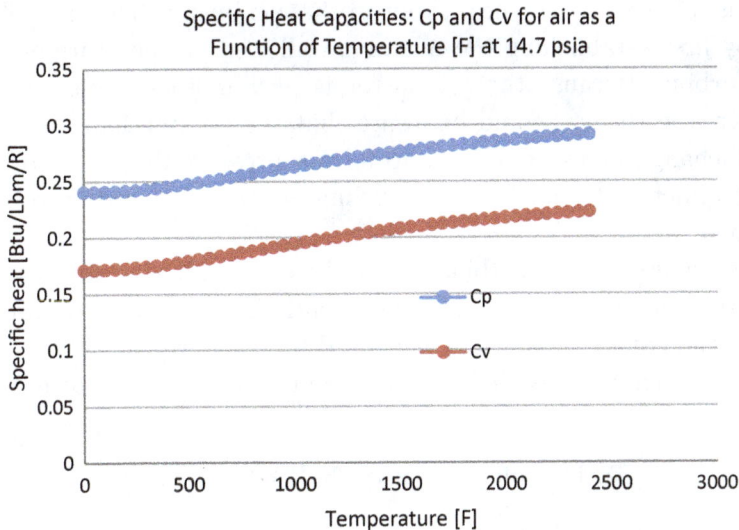

Figure 5.12. Specific heat for air as a function of temperature.

For example, Figure 5.12 provides the specific heat of air as a function of temperature. Similar functional relationships between specific heat and temperature are available for other gases. It should be noted that the operating pressure has negligible effect on the air for most gases. By using curve fits, or better still, imbedded properties from any number of sources used as an "add-in" to the Excel spreadsheet, it is possible to reduce the drudgery and the extended time it often takes to "look up" the thermodynamics properties in data tables. A recommended add-in for properties is the Reference Fluid Thermodynamic and Transport Properties (REFPROP) Database by the National Institute of Standards and Technology (NIST), USA.

The inlet pressure, temperature, pressure ratio and flow rate to the compressor are given and the first calculation is to determine the discharge temperature, enthalpy and entropy of the compressor. The discharge enthalpy and temperature are determined using the methodology detailed in Chapter 3. That is, the enthalpy of the discharge of the compressor, assuming isentropic compression is determined first. The equation shown here is used to calculate the actual discharge enthalpy, assuming the compressor efficiency as given:

$$\text{Compressor eff.} = (h_{2,s} - h_{,in})/(h_{2,out} - h_{1,in}).$$

With the enthalpy and pressure known at the discharge of the compressor, the compressor discharge temperature and actual temperature can be determined from property tables or through the add-In property routine.

This is followed by the heat input in the combustor. This is easily determined by first establishing the maximum allowed temperature at the inlet to the turbine. Because the combustor is just another name for the heat exchanger, the reader should be aware that the heat exchanger/combustor does not change pressure (except to have a pressure drop due to friction). Thus, the inlet state points of the turbine — enthalpy and entropy — are determined.

The discharge of the turbine is found to be similar to the discharge of the compressor. That is, the isentropic enthalpy drop is determined first, and then the equation shown here is used to calculate the actual discharge enthalpy and entropy as well as the specific volume and internal thermal energy.

$$\text{Turbine eff.} = (h_{2,\text{out}} - h_{1,\text{in}})/(h_{2,s} - h_{\text{in}}).$$

The result from these three process calculations completes the property table that is strongly suggested in Chapter 2 as the goal of any/most thermodynamics calculations.

It is now only necessary to calculate the compressor power, turbine power, heat input and cycle efficiency to complete the model. It is also good practice to calculate the Carnot efficiency in order to check the results. The reader will recall that the Carnot Cycle efficiency is calculated using the highest and lowest temperatures (in absolute °R or K temperature units) in the equation:

$$\eta = 1 - \frac{T_c}{T_h}.$$

Of course, if the reader does not intend to acquire the Excel add-in REF-PROP Property Tables, the turbine and compressor power, as well as the discharge temperatures from the turbine and the compressor, can still be determined. The turbine and compressor power are calculated using what are now familiar equations:

$$Power_{turbine} = \dot{M} \times C_p \times T_{inlet} \times \left(1 - \left(\frac{P_{out}}{P_{in}}\right)^{\frac{k}{(k-1)}}\right) \times \eta_t, \quad (5.8a)$$

$$Power_{compressor} = \dot{M} \times C_p \times T_{inlet} \times \left(1 - \left(\frac{P_{out}}{P_{in}}\right)^{\frac{k}{(k-1)}}\right) / \eta_c. \quad (5.8b)$$

But as always, WATCH the UNITS!

The discharge (exit) temperature from the compressor and the turbine can also be easily calculated by using parts of these two equations:

For the turbine, the discharge temperature is determined from:

$$T_{out} = T_{inlet} - T_{inlet} \times \left(1 - \left(\frac{P_{out}}{P_{in}} \right)^{\frac{k}{(k-1)}} \right) \times \eta_t. \tag{5.9a}$$

For the compressor, the discharge temperature is determined from:

$$T_{out} = T_{inlet} + T_{inlet} \times \left(1 - \left(\frac{P_{out}}{P_{in}} \right)^{\frac{k}{(k-1)}} \right) / \eta_c. \tag{5.9b}$$

5.3.1 *The gas turbine jet: Producing high velocity discharge and not net power*

As mentioned at the start of this section, the Brayton Cycle can provide shaft power for not only either electrical or mechanical power, but also propulsion via a high-speed exhaust gas jet stream. The velocity that can be produced can be determined by using the nozzle equation that is repeated below:

$$\Delta h_{nozzle} \times \eta_{nozzle} = (V_{out}^2 - V, in^2)/2/gc. \tag{5.10}$$

However, it is first necessary to determine the enthalpy, Δh_{nozzle}, and this must take into consideration the amount of power that the compressor requires to compress the gas before it is admitted to the combustor, then to the expansion turbine that drives the compressor and only then to the nozzle that will produce the high speed jet that can be used for propulsion.

The enthalpy, Δh_{nozzle}, given in the above equation is the difference between the enthalpy that is available for power production from the inlet turbine pressure to the ambient pressure. The easiest way to calculate this value is to first determine the compressor requirements, $\Delta h_{compressor}$, and then subtract this value from the enthalpy at the inlet of the drive turbine.

The equation for the enthalpy required for the compressor was previously given in Equation (5.6b).

$$\Delta h_{compressor} = C_p \times T_{inlet} \times \left(1 - \left(\frac{P_{out}}{P_{in}} \right)^{\frac{k}{(k-1)}} \right) / \eta_c.$$

The total enthalpy between the turbine inlet pressure and temperature expanded to the ambient is as follows:

$$\Delta h_{total\ rxpansion} = C_p \times T_{inlet} \times \left(1 - \left(\frac{P_{ambient}}{P_{turbine\ in}} \right)^{\frac{k}{(k-1)}} \right) \times \eta_t.$$

Note, this equation provides a positive value for enthalpy, but it is understood that enthalpy is negative as power is input into the system.

The enthalpy at the inlet to the nozzle is thus:

$$h_{nozzle\,inlet} = h_{turbine\,inlet} - \Delta h_{compressor} \qquad (5.11a)$$

or more directly :

$$\Delta h_{nozzle} = \Delta h_{total\,expansion} - \Delta h_{compressor}. \qquad (5.11b)$$

This equation assumes that the power of the compressor has been determined from the previous steps to determine the mechanical power output of the Brayton Cycle, as it would be used to generate mechanical or electrical power. The difference in the available enthalpy is then used in Equation (5.10). The isentropic enthalpy drop is then multiplied by the nozzle efficiency. The nozzle efficiency is found from the manufacturers of nozzles that are used in jet engine propulsion systems. Good engineering practice from a survey of nozzle designs indicates that nozzle efficiency varies from 95 to 98%.

The result of accelerating the velocity is to provide thrust from the jet engine and not rotary shaft power that is normally used to generate electric power with a generator, or to power the propeller in a ship's propulsion system or the drive wheels in a land-based vehicle. The thrust is determined by recognizing (from the subject of fluid dynamics) that the force is caused by a change in the momentum of the fluid stream. Similarly, the torque can be generated by the conservation of the fluid stream's angular momentum. Recalling that momentum is a vector quantity, the propulsion force is therefore equal to the change in the direction and magnitude of the momentum. The change in momentum is determined by the vector product of mass flow rate and the change in the velocity of the fluid stream. It is essential to know and apply the vector change in the velocity. That is, the velocity change must include the directional vector of the velocity magnitude. As an example, consider the following case study.

5.3.2   *Brayton Cycle with a recuperator (a.k.a. regenerator) and air intercooler*

The basic Brayton Cycle System detailed in the last section can have an increase in efficiency if a regenerative heat exchanger is used to recover the exhaust gas energy that is exiting the turbine. That temperature is typically very high in temperature, about 900–1,100°F, and there is quite a bit of exhaust gas mass flow rate because of the considerable amount of excess air burned in the combustor. The concept of excess air is one that will be explained in Chapter 6, but for now it is sufficient to state that the chemically perfect amount of air that can burn with a known amount of fuel will cause the products of combustion to be very hot, hotter than what most metals can tolerate. It is for that reason that in the examples given above for the Brayton Cycle, the turbine inlet temperature was limited to 2,100–2,500°F. In order to reduce the temperature of the combustion air, it is necessary to add more air to the combustor than is necessary to completely burn the fuel. This is analogous to adding cold water to a bathtub filled with water that is too hot for bathing in without getting scalded. A combustor, a pressurized vessel that is designed to safely and efficiently burn fuel in air in a continuous combustion process, has a pressure as high as 20–22 atmospheres for most current gas turbine (air-Brayton Cycle) engines. This can be easily confirmed by applying Equation (5.5), with turbine inlet temperatures exceeding 2,500°F and ambient temperatures of 60°F. Of course, Equation (5.5) is only applicable to basic or simple Brayton Cycle systems; that is, Brayton Cycles without regenerators or air intercoolers, which can improve the cycle efficiency and that thus require a lower pressure ratio. Equation 5.5 is also strictly true when the compressor and turbine efficiencies are close to 100%.

The regenerative heat exchanger is a heat exchanger that uses the same working fluid to heat the colder fluid on some other part of the thermodynamics cycle. Usually the heat from the regenerator is used to preheat the fluid before it is to be heated by the combustion of fuel or some other energy source. In the case of the Brayton Cycle, the hot fluid is the exhaust that is exiting the turbine, and the cold fluid is the fluid that is exiting the compressor before that fluid is used in the combustor to oxidize (burn or combust) with the fuel. The location of the regenerator (the green colored heat exchanger) may be observed in Figure 5.13(a) when it is included in the cycle along with a compressor intercooler heat exchanger. The amount of heating that can be done by the hot exhaust gas, shown in gray between the two stages of the compressor, is by no means unlimited. It certainly cannot heat the cold fluid to a temperature higher than that of the hottest exhaust gas in the system. In fact, the amount of heat transfer that can be achieved is much smaller than this amount and is determined by the concept of heat exchanger effectiveness.

The concept of heat exchanger effectiveness is best and usually taught in a course on heat transfer. However, a discussion of heat exchanger effectiveness is appropriate here as it is necessary to understand the limitations that must be observed when applying a regenerator.

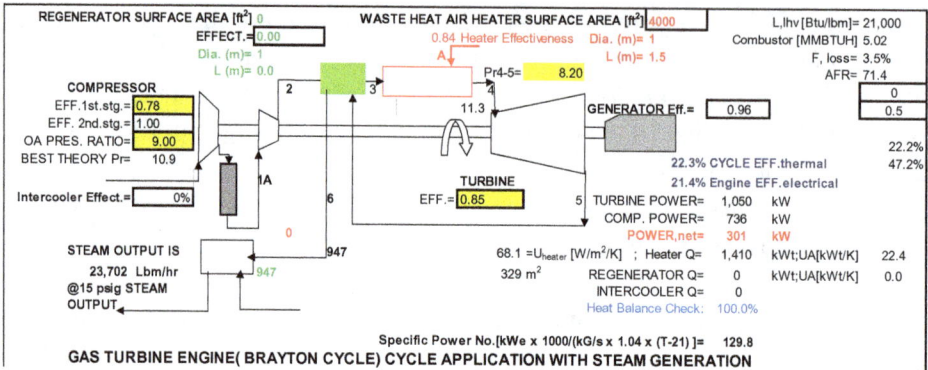

(a)

REGENERATOR SURFACE AREA [ft²] 0		WASTE HEAT AIR HEATER SURFACE AREA [ft²] 4000			L,lhv [Btu/lbm]= 21,000

GAS TURBINE ENGINE(BRAYTON CYCLE) CYCLE APPLICATION WITH STEAM GENERATION

Critical Pressure [psia]= 1070	73.8		Bar,a	FLUID: air				
Critical Temp. [F]= -221	-140.3		C	MOLE.WT.	28.96			
	1	1a	2	3	4	5	6	A
Pres. [psia]	14.7	132.3	132.30	131.90	120.6	14.7	14.7	14.9
Temp. [F]	80	671	671	671	1707	947	947	1907
Enthalpy [Btu/lbm]	129.1	274.1	274.1	274.1	552.0	345.2	345.2	608.3
Density [Lbm/ft³]	0.074	0.315	0.31	0.31	0.15	0.03	0.03	0.017
Flow rate [Lbm/s]	4.81	4.81	4.81	4.81	4.81	4.81	4.81	4.81
Entropy [Btu/Lbm/R]	1.6410	1.6709	1.6709	1.6711	1.8508	1.8781	1.8781	
Air Cp	0.2406	0.2534	0.2454		0.2683	0.2723	0.2616	0.28
Pres. bar,a	1.0	9.1	9.1	9.1	8.3	1.0	1.0	1
Temp. [K]	299.7	628.0	628.0	628.0	1203.4	781.4	781.4	1314.5
Enthalpy [KJ/kG]	300	637	637	637	1282	802	802	1413
Temp. [C]	26.7	355.0	355.0	355.0	930.4	508.4	508.4	1041.5
Density [kG/m³]	1.18	5.05	5.05	5.03	2.40	0.45	0.45	0.27
Mass Flow [Kg/s]	2.19	2.19	2.19	2.19	2.19	2.19	2.19	2.19
Volume Flow [Nm³/hr]	6516	6516	6516	6516	6516	6516	6516	6516
Pipe Size Estimate (mm)	278.3	134.5	134.5	134.7	194.9	449.5	449.5	579

(b)

Figure 5.13.   (a) Air-Brayton Cycle with a single regenerator and compressor intercooler heat exchanger, and (b) Property Table for the state points of Figure 5.13(a).

The effectiveness of a heat exchanger can be determined by one of two equations. These equations are:

$$\varepsilon_h = \frac{(T_{hot\,in} - T_{hot\,out})}{(T_{hot\,in} - T_{cold\,in})}, \tag{5.12a}$$

$$\varepsilon_c = \frac{(T_{cold\,out} - T_{cold\,in})}{(T_{hot\,in} - T_{cold\,in})}. \tag{5.12b}$$

The temperature designations are given in Figure 5.14 for reference. The choice of equation to use is dependent on the calculation of a term that is the product of the mass flow rate and specific heat for the flow stream, and this is done for both the hot and the cold fluid streams. The smaller of the two values of this product is used to identify which fluid effectiveness equation should be used.

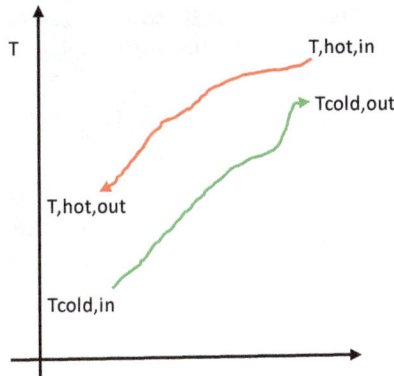

Figure 5.14.   Temperature profiles for heat exchangers.

An example is the most straightforward way of understanding this concept.

Consider the two fluid streams shown in Figure 5.14. The cold fluid stream has a mass flow rate and specific heat product of 10,000. The cold stream has a mass flow rate and specific heat product of only 8,000. Thus, the colder of the two streams is expected to achieve the largest amount of heat transfer and thus should be used to define the effectiveness of the heat exchanger by using Equation (5.11b), where the numerator product is the temperature difference experienced by that cold stream. If the hot stream were to have the lower mass flow rate multiplied by the specific heat product, $\dot{M}_{hot} \times C_{p,hot}$, then Equation (5.12a), with the temperature difference of the hot stream in the numerator, is used to define the effectiveness of the heat exchanger.

If the heat exchanger is a recuperator or regenerator, then the mass flow rates for the cold and hot fluids are the same. For such heat exchangers, the effectiveness of the heat exchangers is based on which fluid stream has the lowest specific heat. Usually the cooler fluid has the lower specific heat. Therefore, with a given effectiveness of the heat exchanger, the discharge temperature of the cold fluid out of the regenerator can be determined using Equation (5.11b).

With the discharge temperature and enthalpy difference across the cold fluid when it is being heated known, the same enthalpy difference is subtracted from the exhaust stream that exited the turbine and that was used to heat the cold stream. Thus, the enthalpy of the hot side of the regenerator is now known and therefore its temperature can be determined. The result from these calculations is to observe that the efficiency of the regenerative cycle has increased compared to the non-regenerative Brayton Cycle. However, that increase comes with the need to add a heat exchanger to the bill of

Heat Exchanger Effectiveness as a Function of N.T.U. =UA/(MxCp)$_{min}$.

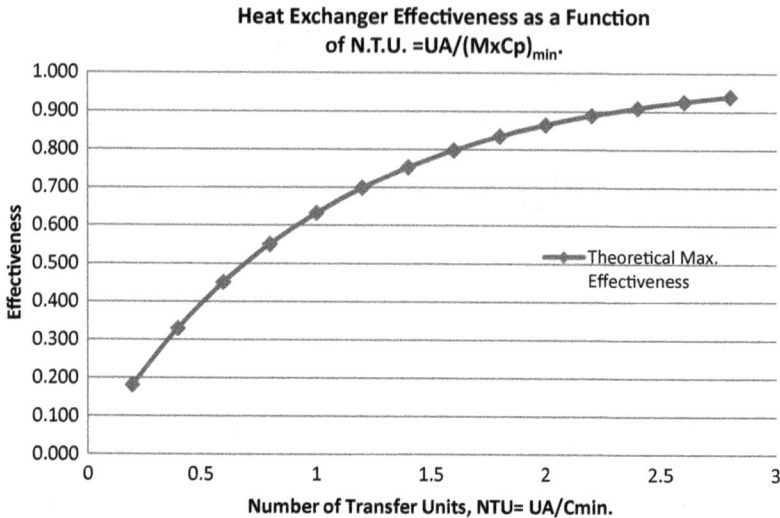

Figure 5.15.   Illustration of effectiveness as a function of NTU $= U \times A/(M \times C_p)_{min}$.

materials of the components in the cycle. An effectiveness of 100% can disproportionately add considerable size to the regenerator without returning an equivalent amount of cycle efficiency. That is, doubling the regenerator size may only increase the cycle efficiency by 5–10%, and this ratio of added size must be considered when choosing to increase heat exchanger effectiveness for cycle improvement. This effect is dramatically illustrated in Figure 5.15, which presents the effectiveness versus the Number of Transfer Units (NTU) parameter (i.e. the ratio of the heat exchanger size. This is expressed as UA, which is heat transfer coefficient multiplied by the heat exchanger area, divided by the product of mass flow rate and specific heat); the same term that was used to determine the choice between the Equations 5.12(a) and (b).

As may be observed from Figure 5.15, the effectiveness begins to asymptotically approach one. It is noted that the effectiveness chart shown in Figure 5.15 is for a pure counter flow heat exchanger. Such heat exchangers are very difficult to use because the configuration must always have the cold and hot fluids flowing throughout the system in opposite directions, with only a pipe wall separating the fluids. The only example of such a pure counter flow heat exchanger is a configuration that has a pipe within a pipe, with one fluid flowing in the inside pipe and another fluid flowing in the opposite direction in the annulus. All other heat exchanger designs would be a slight compromise of this perfect heat exchanger heat transfer. This compromise is usually handled numerically by applying a Heat Transfer Correction factor, Fc, to the product of U × A. In some heat exchanger configurations (not shown in this figure), the effectiveness may be limited to much less than one. Also important is the observation that the size, as given by the parameter NTU, increases many times faster than the effectiveness. Thus, the size of the heat exchanger may increase two- or three-fold, while the effectiveness increases only slightly, unless it is below the value that terminates the linear effectiveness versus NTU.

The equation for the curve shown in Figure 5.15 can be found in any Heat Transfer or Heat Exchanger reference textbook. The equation is:

$$\varepsilon = 1 - e^{-NTU},$$

where:
$$\text{NTU} = \frac{U \times A}{M \times C_p}.$$

The effect the regenerator effectiveness has on cycle efficiency can be very significant. For example, consider Figures 5.16(a), (b) and (c). The reader can, for the moment, ignore the air storage system which is shown in these figures as half circles. The use of this air storage system will be explained in the next section.

Figure 5.16(a) illustrates an air-Brayton Cycle with an output of almost 5,000 hp using a 75% effective regenerator and a turbine inlet temperature of 2,000°R. The cycle efficiency is shown to be 27%. Figure 5.16(b) illustrates the almost identical Brayton Cycle but with only the regenerator effectiveness reduced to 25%. With the reduced regenerator effectiveness, the cycle efficiency is reduced to 14%. It is noted that the power is the same but that is only because the heat input is increased to provide the same net power output from the cycle. In fact, this is done without additional or different parametric inputs to the cycle calculation. The net power output is simply the difference between the power output of the turbine and the power required by the compressor. The equations for the turbine and compressor are Equations 5.11(a) and (b).

$$\Delta h_{total\ expansion} = C_p \times T_{inlet} \times \left(1 - \left(\frac{P_{ambient}}{P_{turbine\ in}}\right)^{\frac{k}{(k-1)}}\right) \times \eta_t, \quad (5.11a)$$

$$\Delta h_{compressor} = C_p \times T_{inlet} \times \left(1 - \left(\frac{P_{out}}{P_{ambient}}\right)^{\frac{k}{(k-1)}}\right) / \eta_c. \quad (5.11b)$$

The net power is the difference between these two expressions multiplied by the flow rate through the cycle. In this example represented by

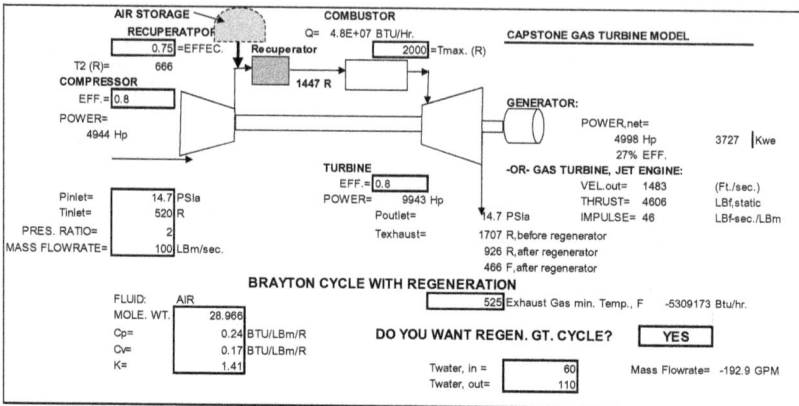

(a)

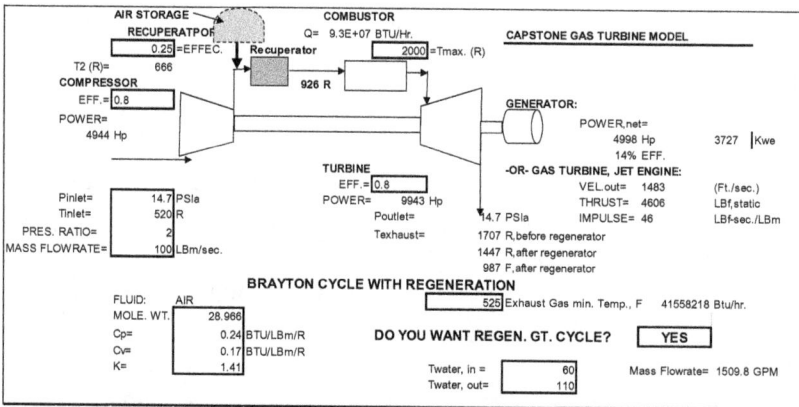

(b)

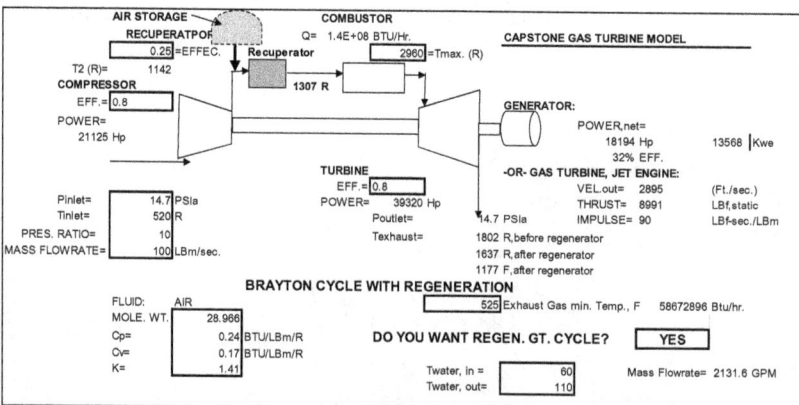

(c)

Figure 5.16. (a) to (b) Illustrating the effect that the recuperator/regenerator effectiveness has on the cycle efficiency of two Brayton Cycles with different heat inputs, and (c) the air-Brayton Cycle demonstrating high net output with improved cycle efficiency and higher turbine fluid inlet temperature.

Figures 5.16(a) and (b), none of the parameters in these two expressions have changed and thus the net power output is the same; only the heat input into the air heater or combustor differ. The reader is cautioned, however, not to think that if the heat input to the cycle is somehow limited due to a shortage of fossil fuel, limited renewable energy resources or just from the use of a small heat exchanger/combustor that the net power of the cycle would be the same. In the case where the heat input is constrained, then the turbine inlet temperature, shown in Figures 5.16(a) and (b) to be 2,000°R, cannot be achieved and the power output would be less due to the reduction in power from the turbine equation.

In fact, if the turbine fluid inlet temperature is increased to 2,960°R as shown in Figure 5.16(c), the net power from the cycle is shown to be considerably increased. This may be attributed to two effects on the cycle:

1. The increase in the turbine inlet temperature.
2. The increase in the turbine exhaust temperature, which then increases the regenerator outlet (or combustor inlet) temperature to improve the cycle efficiency.

5.4 Effect of compressor air intercooler on Brayton Cycle efficiency

The efficiency of the Brayton Cycle can also be increased if the compression process is split into many steps and each step is intercooled in-between each compression. Similarly, the turbine expansion process can be divided into many steps and a reheating process placed between each expansion step. The block diagrams of these advanced cycle processes are shown in Figure 5.16 with one compressor intercooler for the two compressors. The open air-Brayton Cycle, also commonly called the Gas Turbine Engine, can be operated as a closed cycle. In fact, the Brayton Cycle is intended to be a closed cycle and not opened to the atmosphere. Of course, if the cycle is closed then it can use any number of working fluids as might suit the application on hand. One such cycle is shown in Figure 5.17, with nitrogen as the working fluid. Of course, the difference between a nitrogen cycle and one that uses air is very small, considering that air is 79% nitrogen.

The closed Brayton Cycle shown in Figure 5.17 uses four compressors with three intercoolers plus a single regenerator. The switch from air to nitrogen is readily accomplished if the cycle is programmed on software platforms such as Mathcad, Excel, etc., where properties in a fluid can literally be changed with a keystroke. In the cycle shown in Figure 5.17, the efficiency is

N2 CLOSED BRAYTON CYCLE WITH REGENERATION & COMPRESSOR INTERCOOLERS

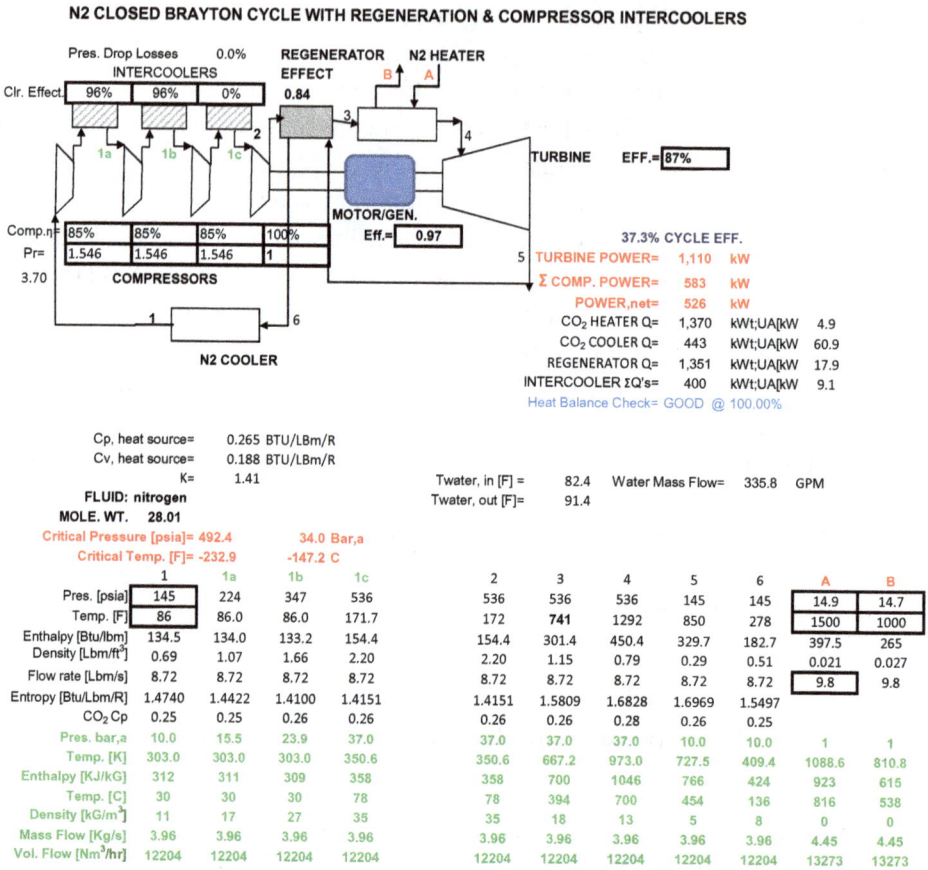

Pres. Drop Losses	0.0%	REGENERATOR EFFECT	N2 HEATER
INTERCOOLERS			

Clr. Effect: 96% | 96% | 0% | 0.84

TURBINE EFF.= 87%

MOTOR/GEN.

Comp.η= 85% | 85% | 85% | 100% — Eff.= 0.97

Pr= 1.546 | 1.546 | 1.546 | 1

3.70 — COMPRESSORS

N2 COOLER

37.3% CYCLE EFF.

TURBINE POWER= 1,110 kW
Σ COMP. POWER= 583 kW
POWER,net= 526 kW
CO₂ HEATER Q= 1,370 kWt;UA[kW 4.9
CO₂ COOLER Q= 443 kWt;UA[kW 60.9
REGENERATOR Q= 1,351 kWt;UA[kW 17.9
INTERCOOLER ΣQ's= 400 kWt;UA[kW 9.1
Heat Balance Check= GOOD @ 100.00%

Cp, heat source= 0.265 BTU/LBm/R
Cv, heat source= 0.188 BTU/LBm/R
K= 1.41

Twater, in [F] = 82.4　Water Mass Flow= 335.8 GPM
Twater, out [F]= 91.4

FLUID: nitrogen
MOLE. WT. 28.01
Critical Pressure [psia]= 492.4　　34.0 Bar,a
Critical Temp. [F]= -232.9　　-147.2 C

	1	1a	1b	1c	2	3	4	5	6	A	B
Pres. [psia]	145	224	347	536	536	536	536	145	145	14.9	14.7
Temp. [F]	86	86.0	86.0	171.7	172	741	1292	850	278	1500	1000
Enthalpy [Btu/lbm]	134.5	134.0	133.2	154.4	154.4	301.4	450.4	329.7	182.7	397.5	265
Density [Lbm/ft³]	0.69	1.07	1.66	2.20	2.20	1.15	0.79	0.29	0.51	0.021	0.027
Flow rate [Lbm/s]	8.72	8.72	8.72	8.72	8.72	8.72	8.72	8.72	8.72	9.8	9.8
Entropy [Btu/Lbm/R]	1.4740	1.4422	1.4100	1.4151	1.4151	1.5809	1.6828	1.6969	1.5497		
CO₂ Cp	0.25	0.25	0.26	0.26	0.26	0.26	0.28	0.26	0.25		
Pres. bar,a	10.0	15.5	23.9	37.0	37.0	37.0	37.0	10.0	10.0	1	1
Temp. [K]	303.0	303.0	303.0	350.6	350.6	667.2	973.0	727.5	409.4	1088.6	810.8
Enthalpy [KJ/kG]	312	311	309	358	358	700	1046	766	424	923	615
Temp. [C]	30	30	30	78	78	394	700	454	136	816	538
Density [kG/m³]	11	17	27	35	35	18	13	5	8	0	0
Mass Flow [Kg/s]	3.96	3.96	3.96	3.96	3.96	3.96	3.96	3.96	3.96	4.45	4.45
Vol. Flow [Nm³/hr]	12204	12204	12204	12204	12204	12204	12204	12204	12204	13273	13273

Figure 5.17.　Air-Brayton Cycle with multiple compressors and air intercooler heat exchangers.

37.3% with a pressure ratio of only 3.7:1, and not 10:1 as used in the cycles modeled in Figure 5.16. The turbine fluid inlet temperature is also considerably reduced to 1,750°R. Of course, the gain in overall cycle performance comes with a "price". The cycle improvement is "paid for" by the need to add several stages to the compression and expansion process, plus the need for the intercooler heat exchanger.

With or without intercooling, regeneration and/or reheating, the overall economics of the cycle is based not only on the need for maximum power for a given heat input, but also on the ratio of the final cost of the additional components with the power ($/kWe), which is one of several similar parameters used to determine the "simple payback" for an engine. The simple payback is the amount of time the engine must operate to accrue enough savings in displaced electric power or heating that it pays for the first cost of the engine. Its unit is expressed in years. Typically, investors or clients who use the power generation system will want a simple payback of less than 3 to 5 years.

It is interesting to note a resurgence in the study of the basic Brayton Cycle. The Department of Energy (DOE) of the United States is supporting the development of Brayton Cycles which can use supercritical CO_2 (sCO_2) as the working fluid. That is, the working fluid operates above the critical pressure and temperature of the fluid. Carbon dioxide has a critical pressure that is approximately 70 bar (or 1,050 psia) and a critical temperature of approximately 88°F. An example of a simple sCO_2 system is shown in Figure 5.18. The objective is to develop a heat engine cycle that can operate with an extremely high heat source temperature and thus operate at very high cycle efficiencies, typically above 50%. There is also the additional benefit that the sCO_2 cycles that develop power above 50 MWe can utilize turbines and compressors that are much smaller in size than a comparable air-Brayton Cycle.

The sCO_2 cycle is also being promoted in the industry due to its potential for very high efficiency if the heat input to the cycle is at a very high temperature. The temperatures often exceed 2,500°F and may be derived from either a nuclear reactor heat source or a concentrating solar energy application. At temperatures below 1,500°F, the cycle efficiency begins to be comparable to typical steam Rankine Cycle efficiencies, and the expense of a turbine, compressor, heater, cooler and regenerator may increase the simple payback for the system beyond what is financially viable.

The capability of the sCO_2 cycle to have high cycle efficiency is also due to the use of a recompression process in addition to the use of a regenerator. The recompression process includes an additional compressor and regenerator that, together, improve the heat recovery from within the cycle by strategically adding the recovered heat at the inlet to the sCO_2 heater. Figure 5.18 illustrates the recompression sCO_2 cycle. The simpler cycle eliminates the low temperature regenerator and the second compressor shown between process state points "F" and "Fh$_1$". The addition of the High-Pressure Compressor is an effective way of preheating the CO2 before the fluid enters the Heat Recovery HX. The "work of compression" for the compressor is returned to the cycle and thus improves the cycle efficiency while also decreasing the amount if SCO2 fluid passing through the turbine.

The sCO_2 cycle is being promoted as a high efficiency cycle for applications where the heat source temperature is high but does not greatly change during the heating of the sCO_2. This is often ignored, or at least not emphasized, in the promotion of the sCO_2 Brayton Cycle as being among the

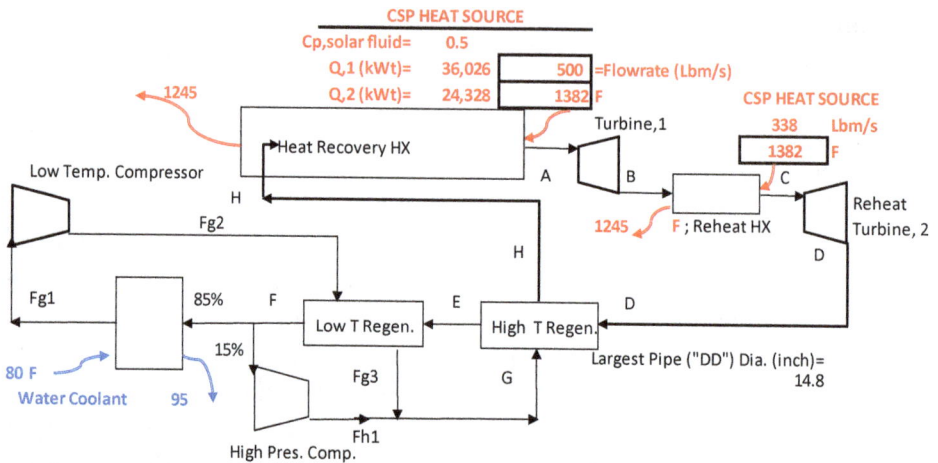

Figure 5.18. Illustration of a recompression, supercritical CO_2 Brayton Cycle with a reheater and second "reheat" turbine (see Case Study 5).

cycles with the highest cycle efficiencies. This requirement to have the heat source temperature not change much allows the regenerators to recover the maximum heat from the discharge of the turbine. The need for a maximum heat recovery to be returned to the sCO_2 before it enters the heater usually requires the recuperators to be extremely high in effectiveness, typically above 95%; consequently, the recuperator heat exchangers have to be very large in size in order to contain a maximum of heat transfer surface area.

5.5   The effect of water injection in gas turbine engines

Water injection into gas turbine engines is not uncommon. It is done to increase power output and also to reduce the parts per million (ppm) of NO_x emissions in the exhaust. The injection of water can be done at the inlet to the compressor. However, it is more common to inject water directly into the inlet of the combustor. In either location, the thermodynamics of the water injection process must be analyzed in order to affect the maximum desired results. This analysis is reviewed in Case Study 5 of this textbook. Figure 5.19 will be used with that analysis.

5.6   The ramjet engine

The ramjet belongs to a special class of engines that works principally on the Brayton Cycle process. The significant difference is that it has no moving parts. That is, the compressor and turbine are not necessary. The compression of the inlet ambient air stream is accomplished using a diffuser and

CN's AIR BRAYTON CYCLE SYSTEM WITH SINGLE REGENERATION and INTERCOOLED COMPRESSORS

				M,water inject [Lbm/s]=	DRIVE TURBINE			REHEAT TURBINE	
	0.0000	0.0000	0.0000	0.795	1546 ft/s	0.300	Dia (m)	0.350	1,503 ft/s
Wtr.Inject.	0.0%	0.0%	0.0%		40 Ns	30,000	RPM	25,000	56 Ns
[Lbm/s]=	0.000	0.000	0.000		2.43 Ds	1	No. of Stg.s	1	1.68 Ds
	Pres. Drop Losses		1.0%	REGENERATOR		0.63	V/Us	0.61	
	INTERCOOLER EFFECT.S		EFFECTIVENESS		#1		#2		
85%	85%	0%	0.98	COMBUSTOR #1	TURBINE		TURBINE		
					EFF.= 90%		90%		
					Pr= 2.76		2.73		
					Pr,oa= 7.54				

M,water inject [Lbm/s]= 0.795
POWER GEN. TURBINE

GEAR BOX EFF.= 0.97
GENERATOR EFF.= 96%

Comp.η=	88%	88%	88%	100.0%
Pr=	2.015	2.015	2.015	1.000

DRIVE TURBINE POWER Tsat= 167.6 COMBUSTOR #2 Tsat= 192.6

COMPRESSORS

fdibella

	Without Water Injection	With Water Injection
BRAYTON CYCLE EFF.	50.1%	45.9%

kWe=	270	Hd Coef. (ψ) 2.08
Nstages=	1	FlowCoef.(φ) 0.61
RPM=	30,000	Ns= 0.45
Diameter [m]=	0.160	Tip Spd.ft/s= 824
Pr=	2.015	

DRIVE TURBINE POWER	920	kW
POWER GEN. TURBINE	902	kW
AIR COMPRESSOR	822	kW
NET POWER	1,000	kW
Water recovery; COMB. #1 Q=	941	kW; UA[kWt/K
Air COOLER Q=	339	kW; UA[kWt/K
REGENERATOR Q=	1,822	kW; UA[kWt/K
INTERCOOLER ΣQ=	521	kW; UA[kWt/K
COMB. #2 Q=	919	kW;
THERMAL STORAGE Q=	339	kW;

		CARNOT Cycle Eff.=	
		2nd Law Eff.=	
		NET POWER [kW]=	1,182 / 1199
	3.1		
	10.9		
	58.5		
	4.5		
			1104

0.296	Cp, heat source=	0.316 BTU/LBm/R
0.226	Cv, heat source=	0.246 BTU/LBm/R
	K=	1.28

Water Injection at 70 F
Heat Balance Check= 100%

FLUID: air
MOLE. WT. 28.96
Critical Pressure [psia]= 548 — 37.8 Bar,a
Critical Temp. [F]= -221 — -140.6 C

Twater, in [F] = 70   Water Mass Flow= 391 GPM
Twater, out [F]= 85

WATER 9

	1	1a	1b	1c	2	3	4	5	6	7	8	9
Pres. [psia]	14.70	29	59	117	117	116	114	41	41	14.9	14.72	116
Temp. [F]	80.6	91.6	93.7	230.7	231	1087	1500	1097	1500	1105	248	933
Enthalpy [Btu/lbm]	129.3	131.8	132.1	165.1	165.1	382.2	494.4	384.7	494.2	386.7	169.6	1498.4
Density [Lbm/ft³]	0.07	0.14	0.29	0.46	0.46	0.20	0.16	0.07	0.06	0.03	0.06	0.140
Flow rate [Lbm/s]	7.95	7.95	7.95	7.95	7.95	7.95	7.95	7.95	7.95	7.95	7.95	-0.138
Entropy [Btu/Lbm/R]	1.6413	1.5986	1.5518	1.5576	1.5576	1.7615	1.8264	1.8344	1.8976	1.9053	1.7062	
Air Cp	0.24	0.24	0.24	0.24	0.24	0.27	0.28	0.27	0.28	0.27	0.24	
Pres. bar,a	1.0	2.0	4.0	8.1	8.1	8.0	7.9	2.8	2.8	1.0	1.0	8.0
Temp. [K]	300.0	306.1	307.3	383.4	383.4	859.2	1088.6	864.6	1088.6	868.9	392.9	773.8
Enthalpy [KJ/kG]	300	306	307	383	383	888	1148	893	1148	898	394	3480
Temp. [C]	27	33	34	110	110	586	816	592	816	596	120	501
Density [kG/m³]	1.18	2.30	4.58	7.31	7.31	3.23	2.52	1.14	0.90	0.41	0.90	2.25
Mass Flow [Kg/s]	3.62	3.62	3.62	3.62	3.62	3.62	3.62	3.62	3.62	3.62	3.62	-0.06
Vol. Flow [Nm³/hr]	10774	10774	10774	10774	10774	10774	10774	10774	10774	10774	10774	-0.226

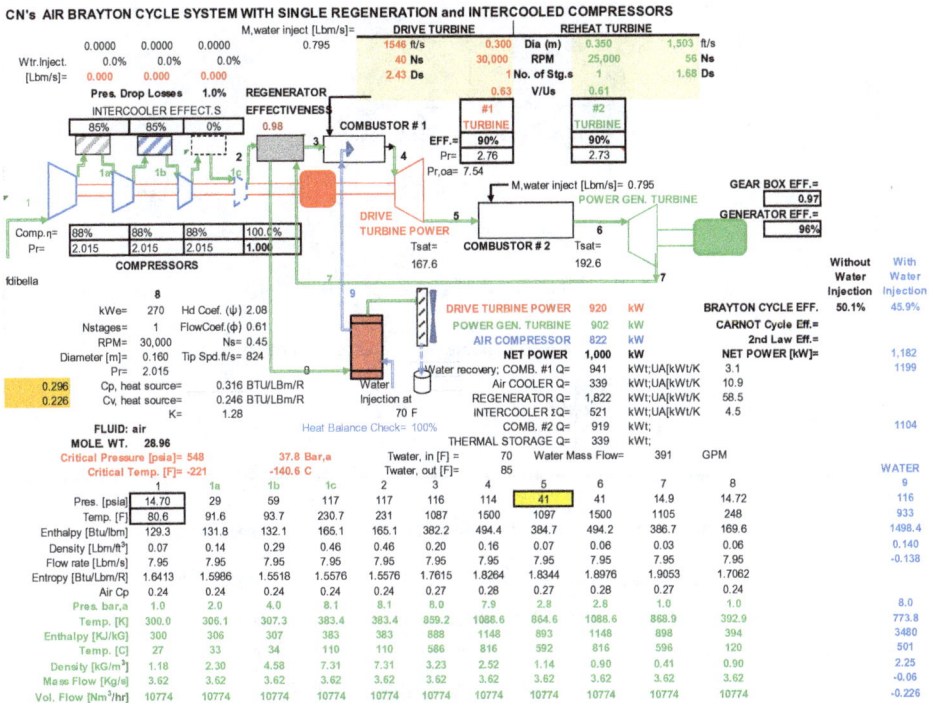

Figure 5.19. A gas turbine engine cycle shown with water injected into the combustor, but only after being converted to vapor using the heat from the turbine exhaust.

the high velocity of the inlet air stream. The turbine expansion part of the Brayton Cycle is not needed because the objective of the ramjet is to provide propulsion thrust. Without a compressor in the cycle to be powered, the mechanical turbine is replaced with an expansion nozzle that can produce a high velocity gas. The absence of the compressor and turbine enables the ramjet to be very light with respect to the thrust that can be generated. The diffuser and nozzle were two of the seven mechanical devices described in detail in Chapter 3. The ramjet engine uses the very distinctive ability of the diffuser to transform the kinetic energy of the inlet air velocity into a slightly higher pressure at the discharge of the ramjet which is immediately followed by the combustion system for this heat engine. For this to happen, however, the ramjet must be in motion, and therein lies the only drawback for the ramjet engine: it must already be in motion for the engine to develop a significant amount of thrust. Thus, the ramjet engine cannot be the prime propulsion engine, and must have an air inlet stream velocity that is high enough for the diffuser to be able to convert the stream's kinetic energy into potential energy in the form of higher pressure. The air speeds typically needed are 500 mph (733 ft/s) or higher. The ramjet engine is therefore

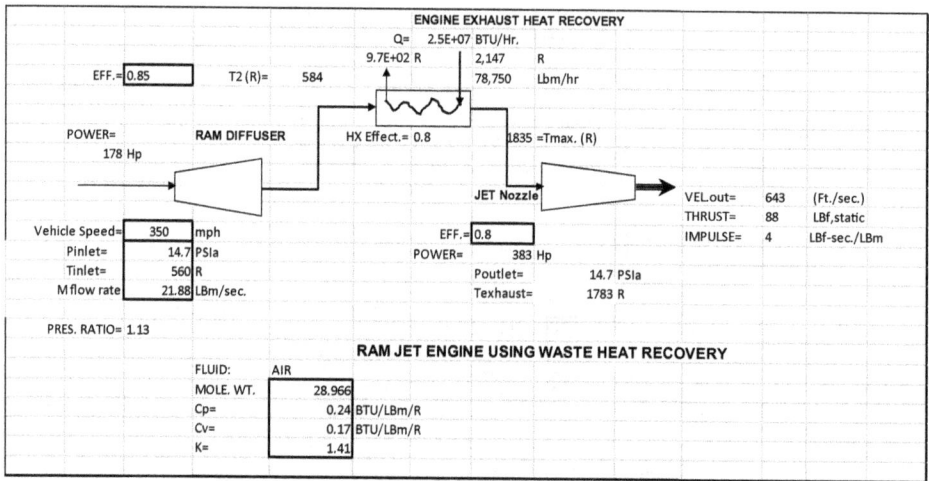

Figure 5.20.   Spreadsheet display of the solution to Example 5.3.

typically deployed in high-speed aircraft, usually military aircraft, that can provide a boost of thrust when it is most needed. The fuel for the heat input comes from the fuel supply on the aircraft. Unfortunately, the fuel efficiency of the ramjet engine is not very good unless the speed is extremely high and/or the combustion temperature of the compressed air stream can be increased to above the normal temperature constraint that is imposed due to the material limitations of jet engines that use a turbine to expand their working fluid before thrust is produced in their nozzles.

Consider the following example of the calculations involved in a ramjet engine.

5.6.1   *Example 5.2: Calculations involved in a ramjet engine*

Determine the thrust that can be produced by a ramjet engine attached to an aircraft that is flying at a speed of 350 mph. The ramjet heat input is used to increase the nozzle inlet temperature to 1,835°R. The local ambient pressure is 14.7 psia at a temperature of 560°R.

The solution to this example problem is illustrated in Figure 5.20 as it appears on an Excel spreadsheet workbook.

The basic equation for the diffuser and nozzle is shown in Equation 5.13(c) and (b).

$$\Delta h_{\text{diffuser}}/\eta_{\text{diffuser}} = (V_{\text{out}_2} - V_{\text{in}_2})/2/gc, \qquad (5.13a)$$

where: $$\Delta h_{diffuser} = C_p \times T_{inlet} \times \left(1 - \left(\frac{P_{out}}{P_{in}}\right)^{\frac{k}{(k-1)}}\right)$$

and inputs are $V_{,in} = 350\,\text{mph} = 513\,\text{ft/s}$, with $V_{,out} \sim 50\,\text{ft/s}$ and the objective is to solve for P_{out}.

$$\Delta h_{\text{nozzle}} \times \eta_{\text{nozzle}} = (V_{out2}^2 - V_{in2}^2)/2/gc, \qquad (5.13b)$$

where : $\qquad \Delta h_{nozzle} = C_p \times T_{inlet} \times \left(1 - \left(\dfrac{P_{out}}{P_{in}}\right)^{\frac{k}{(k-1)}}\right)$

and inputs are: $\Delta h_{\text{nozzle}} \times \eta_{\text{nozzle}}$ and the objective is to solve for V_{out}.

There is an interesting alternative to a ramjet engine that can be just as efficient by reducing or even eliminating the need for fuel input. This engine cycle is known as the Meredith Effect. The engine using the Meredith Effect is more efficient because the heat input into the ramjet is derived from the waste heat energy of another engine, perhaps another prime mover whose heat would otherwise be wasted. Case Study 6 will provide a more detailed analysis of one such application of a ramjet engine with waste heat recovery.

It is also interesting to note that the Brayton Cycle operating as a gas turbine engine with a mechanical compressor and turbine can also be used as a heat recovery engine. Figure 5.21 is an example of a gas turbine engine that has the combustor component replaced with a more conventional heat exchanger. The heat input is the waste heat stream from process "A" to "B". The replacement of the combustor with a heat exchanger imposes a constraint that limits the efficiency of the system. The constraint is known as the pinch point temperature difference between hot exhaust gas state point "B" and air (the working fluid in the cycle) at state point 3. This pinch point or minimum approach temperature is usually not allowed to be smaller than 75 to 100 °F. With temperature differences smaller than 75 °F, the heat exchanger's effectiveness will need to be over 85%, thus requiring a larger heat exchanger, as presented previously in this chapter.

5.7 Lenoir engine

The Lenoir Cycle is shown in Figure 5.22, including its process diagram and Property Table. The three processes include a constant pressure followed by a spike in the pressure and expansion of the high pressure and high temperature back to state point 3. The Lenoir Cycle is an open cycle engine similar to the gas turbine and ramjet engine described in the previous section. That is, the cycle uses air from the ambient, which is re-injected into the ambient after the heating and the power extraction processes have been completed; the cycle uses the same mass flow rate, certainly, but at a much hotter temperature due to the combustion process that has occurred. The Lenoir

FURNACE UNITS 3, 4 and 5

WASTE HEAT AIR HEATER SURFACE AREA 2500 ft²

L,lhv [Btu/lbm]= 21,000

MMBTUH= 8.16

REGENERATOR

B   A

SIZE: ~ 1m x 1m x 2 m

0.29   EFFECT.= 0.30

853 DT,pinch   AFR= 42.8   F, loss= 4.0%

0.91 Exh. HX Effectiveness

COMPRESSOR
EFF.1st.stg.= 0.80
EFF. 2nd.stg.= 0.80
OA PRES. RATIO= 4.00

2   3   Dp= 5   4

GENERATOR:

15.5% CYCLE EFF.

1   1A

Intercooler Effect.= 70%   B   591

TURBINE
EFF.= 0.82

5   TURBINE POWER=   1,299   kW
COMP. POWER=   905   kW
POWER,net=   393.9   kW

STEAM OUTPUT IS
23,383 Lbm/hr
@15 psig STEAM
OUTPUT

685
653.8   6

68.1 =U_heater [W/m²/K] ; Heater Q=   2,534   kWt;UA[kWt/K]   16.4
241   COOLER Q=   1,824   kWt;UA[kWt/K]   131.9
REGENERATOR Q=   513   kWt;UA[kWt/K]   1.3
INTERCOOLER Q=   308
Heat Balance Check: 99.7%

Specific Power No.[kWe x 1000/(kG/s x 1.04 x (T-21)]=   129.2

AIR BRAYTON CYCLE FOR NWG INCINERATOR APPLICATION WITH STEAM GENERATION

Twater, in [F] =   75   Mass Flowrate=   415,003 GPM
Twater, out [F]=   90

Msqrt(T)/p=   0.441

					FLUID: air				
Critical Pressure [psia]= 1070	73.8		Bar,a	MOLE.WT.	28.96					
Critical Temp. [F]= -221	-140.3		C							
	1	1a	2	3	4	5	6	A	B	
Pres. [psia]	14.7	29.4	58.80	58.62	53.6	14.7	14.7	14.9	14.7	
Temp. [F]	80	121	278	449	1247	848	685	2100	591.0	
Enthalpy [Btu/lbm]	129.1	138.8	177.0	218.7	425.0	319.3	277.6	663.6	254.0	
Density [Lbm/ft³]	0.074	0.137	0.23	0.17	0.08	0.03	0.016	0.016	0.038	
Flow rate [Lbm/s]	11.65	11.65	11.65	11.65	11.65	11.65	11.65	5.72	5.72	21,838
Entropy [Btu/Lbm/R]	1.6410	1.6108	1.6212	1.6724	1.8406	1.8590	1.8249			
Air Cp	0.2406	0.2412	0.2408		0.2586	0.2648	0.2559	0.29	0.25	
Pres. bar,a	1.0	2.0	4.1	4.0	3.7	1.0	1.0	1	1	
Temp. [K]	299.7	322.3	381.5	504.8	948.0	726.2	635.6	1421.9	583.6	
Enthalpy [KJ/kG]	300	322	382	508	987	741	645	1541	590	
Temp. [C]	26.7	49.3	108.5	231.8	675.0	453.2	362.6	1148.9	310.6	
Density [kG/m³]	1.18	2.19	3.70	2.79	1.36	0.49	0.56	0.25	0.61	
Mass Flow [Kg/s]	5.29	5.29	5.29	5.29	5.29	5.29	5.29	2.60	2.60	
Volume Flow [Nm³/hr]	15775	15775	15775	15775	15775	15775	15775	7742	7742	

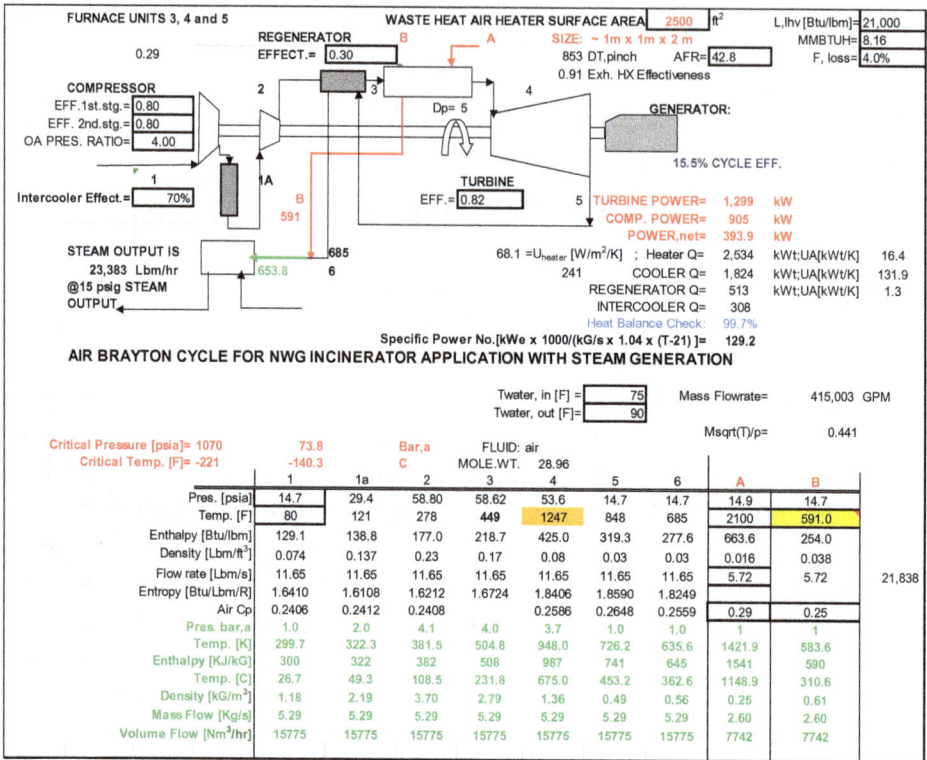

Figure 5.21.   An open air-Brayton Cycle used for waste heat recovery in place of the combustor.

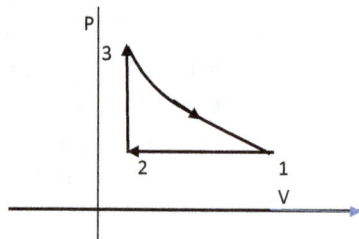

Figure 5.22.   Lenoir engine cycle.

Cycle engine is a special type of engine that can produce thrust or rotary shaft power using a combustion process that is actually produced by a continuous, timed, pulsed ignition of the fuel. It is assumed that the combustion is almost instantaneous and causes a pressure spike. The constant pressure is usually assumed to occur at atmospheric pressure as it fills the engine chamber before the air is used as the oxidant for the fuel. As in the case of the ramjet engine, the Lenoir Cycle would be simpler than the Brayton Cycle gas turbine engine. Its efficiency is also relatively low.

5.7.1 *Example 5.3: Analyzing the Lenoir Cycle*

Example 5.3 illustrates how the Lenoir Cycle is analyzed. For this study, the cycle is assumed to use using a positive displacement engine, much like the Otto or Diesel Cycles modeled in Chapter 2, and is analyzed using a Closed Mass Analysis.

The analysis process is the same as that used for Otto and Diesel cycles (see Chapter 2). The first step is to construct a Property Table and then a Process Table, as shown in Figure 5.23. Each of the three processes shown in Figure 5.23 are among the five basic, closed mass processes discussed in Chapter 2. The pressure, P, specific volume, v, and the temperature, T, can be calculated using the Perfect Gas Law if two of the three properties are known. The adiabatic (or almost adiabatic) process follows the (hopefully now) familiar relationship:

$$P_1 \times v_1^k = P_2 \times v_2^k.$$

The results of these calculations for the constant pressure, constant volume and adiabatic (or polytropic) process are shown in Figure 5.23.

The reader is reminded that each of these cycles could also be assumed to be constructed of several open flow components such as a heat exchanger (process path 1–2), an instantaneous combustion or detonation (process 2–3) and a turbine expansion (process 3–1). Figure 5.24 illustrates these processes. To be more specific, the heat exchanger process path 1 to 2 (1–2) can be represented by an open cycle that starts at state point 1 before proceeding to state point 3, and then concluding with state point 1 at ambient pressure.

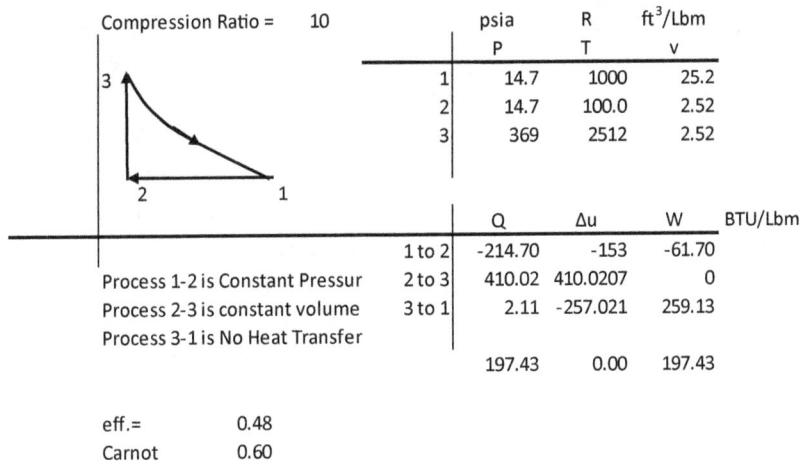

Compression Ratio = 10

		psia	R	ft³/Lbm
		P	T	v
	1	14.7	1000	25.2
	2	14.7	100.0	2.52
	3	369	2512	2.52

		Q	Δu	W	BTU/Lbm
	1 to 2	-214.70	-153	-61.70	
Process 1-2 is Constant Pressur	2 to 3	410.02	410.0207	0	
Process 2-3 is constant volume	3 to 1	2.11	-257.021	259.13	
Process 3-1 is No Heat Transfer					
		197.43	0.00	197.43	

eff.=	0.48
Carnot	0.60

Figure 5.23. Property and Process Tables for the Lenoir Cycle in Example 5.3.

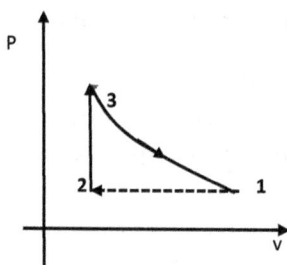

Figure 5.24.   Open Cycle Lenoir engine equivalent to a pulse detonation engine.

The open cycle engine that follows this process path is called a "detonation" engine. It is a very simple engine with no moving parts that works at very high speeds with respect to the air that it uses for combustion. The intake air is compressed by using a diffuser at the inlet to the engine that can raise the pressure slightly above atmosphere. The pressure spike to state point 3 is created by the detonation of the fuel. The open flow cycle that processes the analysis for each process is based on the enthalpy difference between the inlet and exit state points.

5.8   Cogeneration systems

Let us now extend the analysis' flexibility by combining any one of these basic thermodynamics power generation cycles with each other. This is now a common practice in energy conservation and engine efficiency improvement. The goal in combining one or more cycles is to utilize the waste heat stream from the lead engine to power the secondary engine. The lead engine is often called the prime mover while the secondary engine — if that engine is used to recover the waste heat energy that has been generated by the primary engine — is often called the bottoming cycle engine.

A common example of this arrangement is a reciprocating engine, as detailed in Chapter 3, or a gas turbine engine serving as the prime mover, and a Rankine Cycle system serving as the bottoming cycle to recover the waste heat from the exhaust stream of the engine and possibly the water jacket cooler's waste heat[1] to produce additional mechanical or electrical power. The integration of these two systems does not require any new thermodynamics concepts to be mastered — just a confident application of the

---

[1]The author's Total Recovery Energy Cycle (TREC) system can simultaneously recover high-grade energy from the hot exhaust waste gas streams, as well as the low-grade energy that is typically available from the water-cooled jacket of a reciprocating engine. The TREC will be reviewed in Case Study 13.

Table 5.2.   Specifications for the reciprocating engine cooling system.

COOLANT JACKET		
WATER MAX. INLET TEMPERATURE [C]	95	
NOMINAL [C]	90	
MIN. [C]	83	
WATER MAX. OUTLET TEMPERATURE [C]	99	
WATER WATER FLOWRATE [GPM]	771	@1000 RPM
	694	@900 RPM

thermodynamics that has been reviewed and learned. Indeed, this is the real message the author intends to convey to the reader: At this point the reader should be able to utilize both Closed Mass and Open Flow Systems and thus construct, perhaps for the very first time, a unique cycle or system whose combined operation is assured based on the strengths of the individual primary cycles.

To further emphasize this point, the following example indicates an Organic Rankine Cycle (ORC) system that recovers the exhaust gas waste heat energy and the water jacket waste heat energy from a CAT 3616 engine. The Rankine Cycle engine uses an organic working fluid and not water. This working fluid is a hydro-carbon-fluorine-chlorine fluid made available from any number of refrigerants. In this worked example the working fluid is R245fa, but it could be as easily chosen to be R134a or even ammonia. The choice is dependent on the operating temperature of the waste heat and the need to not exceed the thermal breakdown temperature of the organic fluid.

The cogeneration system (i.e., the combination of the 3616 engine and an ORC system) analysis is first completed for the recovery of heat from the engine jacket's rejected heat. The engine jacket's water's data is shown in Table 5.2. The analysis indicated a potential of 110 kW of recoverable power using an ORC System. The thermodynamics analysis indicates a similar power recovery of approximately 110 kWe, as shown in Figures 5.25(a–c). The turbine would be a single stage, 8-inch diameter axial turbine operating at 10,000 rpm.

Of additional interest is the part load performance of the ORC turbine. Figure 5.26 identifies the approximate part load performance expected of a typical ORC used with an engine exhaust heat source.

A similar analysis was completed by Concepts NREC, USA on the exhaust gas waste heat from the CAT 3616 engine, as shown in Table 5.3. The results indicate a power recovery potential of 290 kWe, as shown in Figures 5.26(a–d).

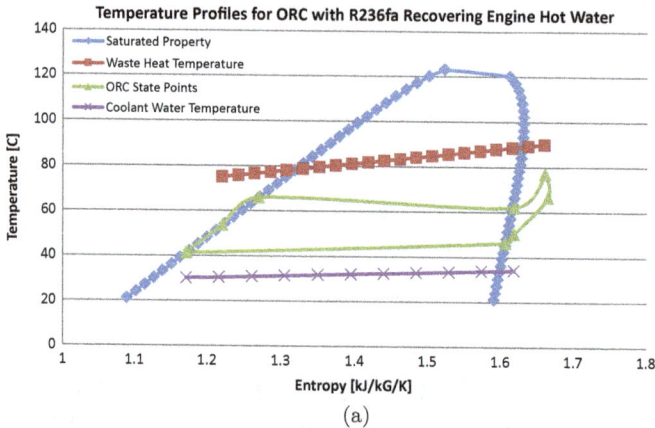

(a)

r236fa   EXHAUST WASTE HEAT RECOVERY SYSTEM FROM CEMENT PRODUCTION INDUSTRY

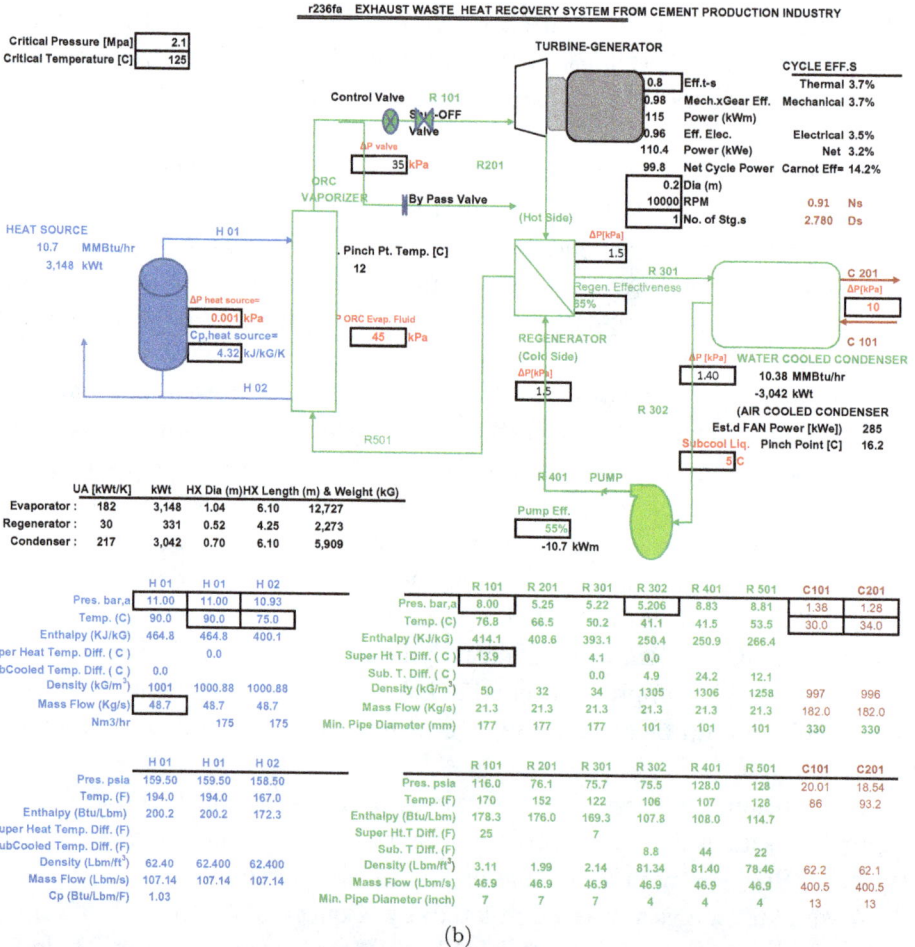

UA [kWt/K]	kWt	HX Dia (m)	HX Length (m) & Weight (kG)	
Evaporator :	182	3,148	1.04	6.10	12,727
Regenerator :	30	331	0.52	4.25	2,273
Condenser :	217	3,042	0.70	6.10	5,909

	H 01	H 01	H 02
Pres. bar,a	11.00	11.00	10.93
Temp. (C)	90.0	90.0	75.0
Enthalpy (KJ/kG)	464.8	464.8	400.1
Super Heat Temp. Diff. (C)		0.0	
SubCooled Temp. Diff. (C)	0.0		
Density (kG/m³)	1001	1000.88	1000.88
Mass Flow (Kg/s)	48.7	48.7	48.7
Nm3/hr		175	175

	R 101	R 201	R 301	R 302	R 401	R 501	C101	C201
Pres. bar,a	8.00	5.25	5.22	5.206	8.83	8.81	1.38	1.28
Temp. (C)	76.8	66.5	50.2	41.1	41.5	53.5	30.0	34.0
Enthalpy (KJ/kG)	414.1	408.6	393.1	250.4	250.9	266.4		
Super Ht T. Diff. (C)	13.9		4.1	0.0				
Sub. T. Diff. (C)			0.0	4.9	24.2	12.1		
Density (kG/m³)	50	32	34	1305	1306	1258	997	996
Mass Flow (Kg/s)	21.3	21.3	21.3	21.3	21.3	21.3	182.0	182.0
Min. Pipe Diameter (mm)	177	177	177	101	101	101	330	330

	H 01	H 01	H 02
Pres. psia	159.50	159.50	158.50
Temp. (F)	194.0	194.0	167.0
Enthalpy (Btu/Lbm)	200.2	200.2	172.3
Super Heat Temp. Diff. (F)			
SubCooled Temp. Diff. (F)			
Density (Lbm/ft³)	62.40	62.400	62.400
Mass Flow (Lbm/s)	107.14	107.14	107.14
Cp (Btu/Lbm/F)	1.03		

	R 101	R 201	R 301	R 302	R 401	R 501	C101	C201
Pres. psia	116.0	76.1	75.7	75.5	128.0	128.0	20.01	18.54
Temp. (F)	170	152	122	106	107	128	86	93.2
Enthalpy (Btu/Lbm)	178.3	176.0	169.3	107.8	108.0	114.7		
Super Ht.T Diff. (F)	25		7					
Sub. T Diff. (F)			8.8	44	22			
Density (Lbm/ft³)	3.11	1.99	2.14	81.34	81.40	78.46	62.2	62.1
Mass Flow (Lbm/s)	46.9	46.9	46.9	46.9	46.9	46.9	400.5	400.5
Min. Pipe Diameter (inch)	7	7	7	4	4	4	13	13

(b)

Figure 5.25.   (a) T-s diagram of the ORC system and its heat source and heat sink, (b) cycle state points for engine jacket heat recovery, and (c) details of the ORC turbine speed and diameter.

TURBINE-GENERATOR

R 101
Shut-OFF
Valve

|kPa   R201

By Pass Valve

(Hot Side)

mp. [C]

|kPa

0.8	Eff.t-s
0.98	Mech.xGear Eff.
116	Power (kWm)
0.96	Eff. Elec.
111.1	Power (kWe)
101.4	Net Cycle Power
0.2	Dia (m)
10000	RPM
1	No. of Stg.s

ΔP[kPa]
1.5

R 301

REGENERATOR
(Cold Side)
ΔP[kPa]
1.5

R 302

CYCLE EFF.S
Thermal 3.7%
Mechanical 3.7%

Electrical 3.5%
Net 3.2%

0.91   Ns
2.774 Ds

C 201
ΔP[kPa]
10

ΔP [kPa]   WATER COOLED CONDENSER
1.40   10.38 MMBtu/hr
            -3,042 kWt
            (AIR COOLED CONDENSER

C 101

Regen. Effectiveness=
85%

(c)

Figure 5.25. (*Continued*)

Table 5.3. CAT 3616 engine specifications.

Exhaust Gas		
Stack Temp	384	C
Gas Flow @ stack temp, 101.3 kPa	999.3	m3/min
Mass Flow	32228	kg/hr

5.9 Refrigeration cycles: Using mechanical power to heat or cool

The previous cycles were designed to produce mechanical power by providing heat energy input at the correct process within the cycle. The mechanical power is delivered to the demand by rotating a shaft, which can be used to generate electric power or drive a propeller for propulsion, or it can be linked to produce reciprocating motions. As described in detail in Chapter 3, plotting the cyclic processes on a P-v or T-v diagram was shown to produce a net power output if the sequence of processes proceeds in a clockwise direction. The area circumscribed in a P-v diagram is the net positive work that is produced by the thermodynamic cycle. It is also interesting to note that the same process path sequenced in the opposite, counter-clockwise direction will consume power, but in doing so, transfer heat energy from a cold temperature reservoir to a higher temperature reservoir. Put simply, if any of these cycles were to literally operate in reverse, the same thermodynamics cycle would produce a cooling effect. That is, by providing a mechanical input, the heat exchangers could absorb heat from a cold system and deliver that energy, plus the work input, to a higher thermal reservoir. The cold temperature reservoir can be anything from a kitchen refrigeration

Temperature Profiles for ORC with R236fa Recovering Engine Exhaust Heat from CAT 3616 Engine

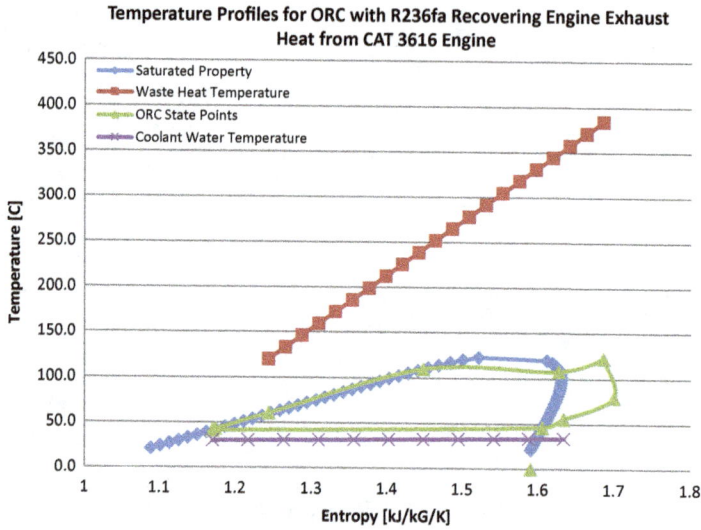

(a)

r236fa   EXHAUST WASTE HEAT RECOVERY SYSTEM FROM CAT 3616 ENGINE

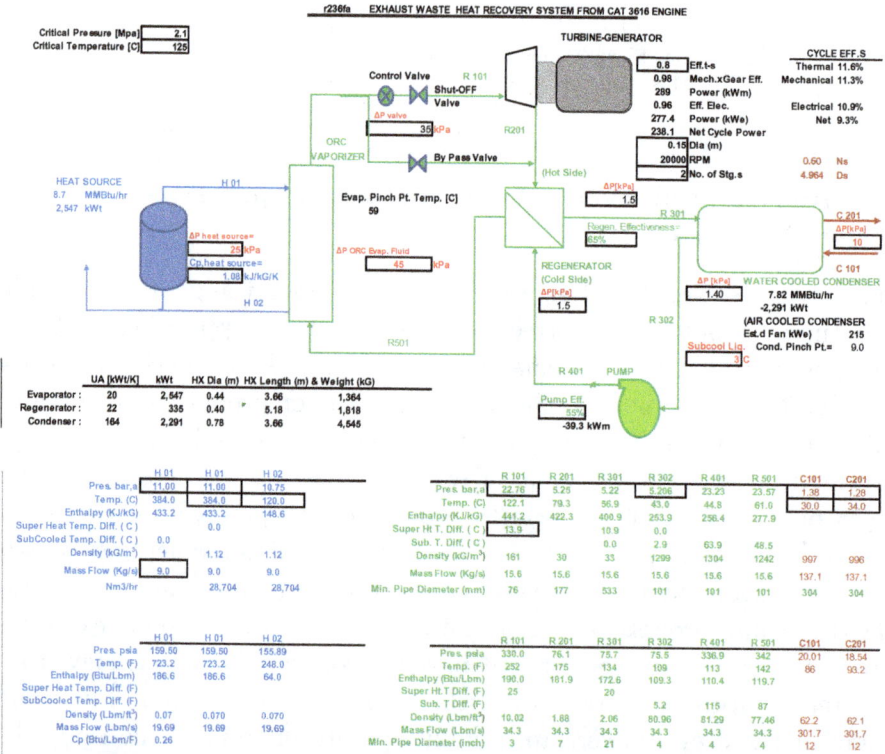

(b)

Figure 5.26.   (a) T-s diagram of exhaust gas waste heat recovery, (b) state points for the engine exhaust waste heat recovery, (c) details of the ORC turbine required for the CAT 3616 engine's exhaust gas waste heat recovery, and (d) approximate part load cycle performance of an ORC used with an exhaust gas heat source.

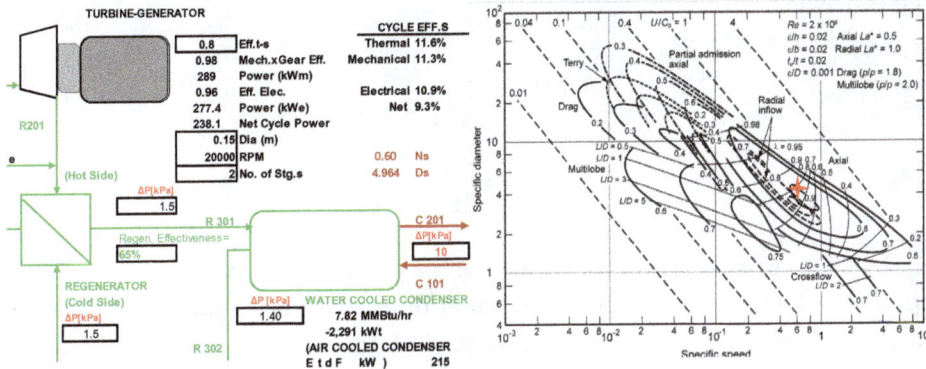

TURBINE-GENERATOR

0.8	Eff.t-s
0.98	Mech.xGear Eff.
289	Power (kWm)
0.96	Eff. Elec.
277.4	Power (kWe)
238.1	Net Cycle Power
0.15	Dia (m)
20000	RPM
2	No. of Stg.s

CYCLE EFF.S
Thermal 11.6%
Mechanical 11.3%

Electrical 10.9%
Net 9.3%

0.60 Ns
4.964 Ds

R201

(Hot Side)
ΔP[kPa]
1.5

R 301
Regen. Effectiveness=
65%

REGENERATOR
(Cold Side)
ΔP[kPa]
1.5

R 302

C 201
ΔP[kPa]
10

C 101

ΔP [kPa]   WATER COOLED CONDENSER
1.40   7.82 MMBtu/hr
               -2,291 kWt
              (AIR COOLED CONDENSER
              E t d F   kW)   215

(c)

**Part Load Turbine Inlet Pressure and
Part Load Engine Performance**

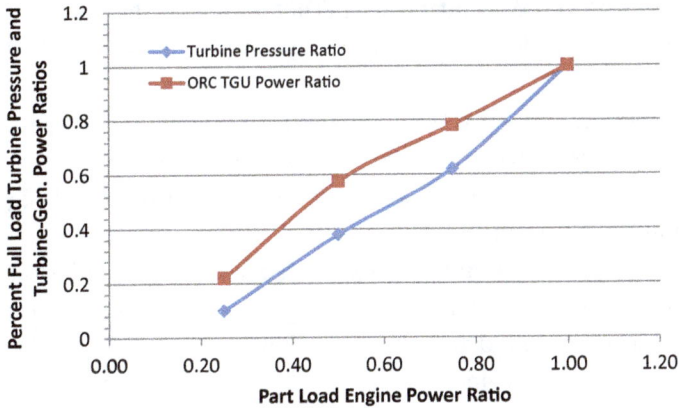

Percent Full Load Turbine Pressure and Turbine-Gen. Power Ratios

- Turbine Pressure Ratio
- ORC TGU Power Ratio

Part Load Engine Power Ratio

(d)

Figure 5.26. (*Continued*)

appliance to a cryogenic thermos. The kitchen refrigerator can keep food products fresh longer by maintaining a temperature of 45°F. The cryogenic thermos at temperatures near absolute 0 K can be used to keep electrical wires super cooled to reduce electrical resistance or to produce liquid air, nitrogen or helium. Similarly, the heat rejected to the atmosphere in these systems could either be pumped into the atmosphere outside a house, or the interior of a winter home that needs to be heated.

Chapter 3 also identified the equation that could determine the maximum Coefficient of Performance (COP) for any cycle. The equation is reproduced in Equations 5.14(a) and (b). Just as in the case of a heat engine that produces power, the efficiency of a cooling system depends on the nature of the processes.

$$COP_r = \frac{T_c}{(T_h - T_c)},$$ (5.14a)

$$COP_r = \frac{\dot{Q}_c}{\dot{W}_c}.$$ (5.14b)

There is certainly something in common between the cycles, whether they operate as heat engines to produce power or as thermodynamics refrigerators to produce cold. Their commonality is that the seven "magic" mechanical engineering devices — compressors and/or turbines, valves, heat exchangers, nozzles and/or diffusers — make up the components of the cycles, and these components must operate according to the First and Second Laws of Thermodynamics. The same equations, whether they are the complete equations or the simplified equations based on the assumptions as stated in Chapter 4, still apply.

Thus, the good news for the engineering student is that there is almost nothing new to be learned. However, let us review some very important cycles and processes that must be understood in order to design the most efficient cooling or heating system.

First let us review several important "thermodynamics dictionary" terms. For example, the cooling system is called a thermodynamics refrigerator. If the same cycle is intended to utilize the heat rejection from the condenser, then it is more proper to identify the cycle as a heat pump. A cooling cycle that produces "cold" will have its effectiveness (COP) identified as COP,$_r$, where "r" stands for "refrigeration". The same cycle that is intended to produce heat will have its effectiveness identified as: COP$_h$, where "h" stands for "heat pump". The same physical system could be used as both as a refrigerator for cooling or as a heat pump for heating, BUT usually not at the same time. A block diagram of the heat pump is shown in Figure 5.27. The reader will recall the very simple diagram of a refrigerator or heat pump that was expressed by only the high and low operating temperatures and the clear indication that work or power was required to be input into the system. This was made clear by the Clausius Statement of the Second Law of Thermodynamics, as detailed in Chapter 4.

Common names for thermodynamics refrigerators and heat pumps include "water chillers" or "air conditioners". In particular, a water chiller indicates that the system is used to cool water, which in turn is used to cool air that is emitted into the cold space for proper conditioning and tempering. In fact, coolant water streams shown entering the condenser and

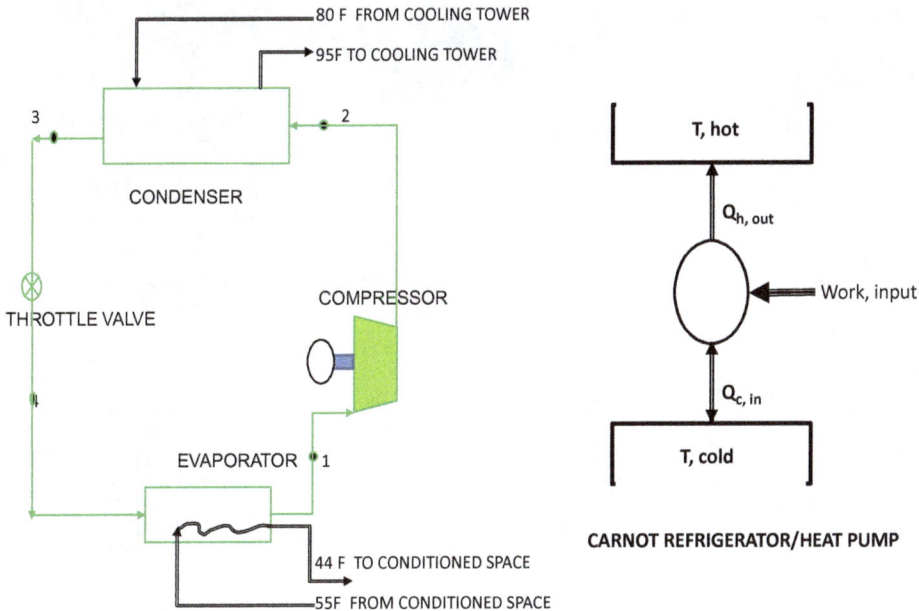

Figure 5.27. The basic (simple) vapor compression system with a classic thermodynamics model of the refrigerator.

the evaporator in Figure 5.27 have standard values of temperature that are typical of water chiller applications throughout the world and are well documented in the ASHRAE (American Society of Refrigeration and Air Conditioning) Standards Handbook.

The refrigeration cycle shown in Figure 5.28 provides a summary of the state points at each component. It may be apparent to the reader that the cycle appears very similar to the Rankine Cycle in its use of a high and a low constant temperature. In fact, the similarly is so strong that the refrigeration cycle is sometimes called the reverse Rankine Cycle rather than a vapor compression cycle, which is its more proper thermodynamics name. These constant temperatures are produced by the working fluid either condensing at a high temperature or evaporating at a lower temperature. The heat source and heat sink used in Figure 5.27 is reproduced in Figure 5.28 to describe the relative temperatures of these two constant fluid temperatures with respect to the source of the heat energy and the heat sink's temperature. The source of the heat is the system that is to be kept cold. Thus, for energy to leave that system it must be at a higher temperature than the constant cold working fluid temperature. As observed in Figure 5.28, the temperature difference between the heat source and the cold working fluid will essentially and almost unilaterally define the temperature of the working fluid to be

CONDENSER TEMPERATURE PROFILE [C]

46
36.0
30.9   21
18.3 [gpm/rTn] 7.10

EVAPORATOR TEMPERATURE PROFILE [C]

[gpm/rTn] 2.66   12.0
7.0
7.0
6.0

CONDENSER

Throttle Valve

COMPRESSOR SPECIFICATION SUMMARY

400 r134a	Comp. Inlet
P,in [bar,a]=	3.62
T,in [C]=	7.0
M [kg/s]=	8.81
Comp. Eff.=	0.77

Comp. Eff. Stg 1= 0.77
Mech. Eff.= 0.98
Elec. Motor Eff= 0.96

EVAPORATOR

	Comp. Outlet
P,out [bar,a]=	132.28
T,out [C]=	45.6
M [kg/s]=	8.81
Pr=	2.52

	1	2	3	4
P [bar,a]	3.62	9.12	9.12	3.62
T [C]	7.0	45.57	30.94	5.99
h [kJ/kG]	403.2	427.8	242.9	242.9
s [kJ/kG/K]	1.728	1.745	1.147	1.028
density [kG/m³]	17.63	42.08	1185.2	1275.22
Sat. Temp. [C]	6.1	36.17	36.16	6.15
Quality (x)				0.18
Flow [kG/s]	8.81	8.81	8.81	8.81
m³/s	0.50	0.21	0.01	0.01
Nm³/s	0.01	0.01	0.01	0.01

COP)r= 6.00
Motor Power [kWe]= 234.3

Qevap. [kWt]= 1406.5
Qcond. [kWt]= -1626.9
Qloss [kWt]= -13.87

UA,evap[kWt/K]= 495
UA,cond[kWt/K]= 1,021

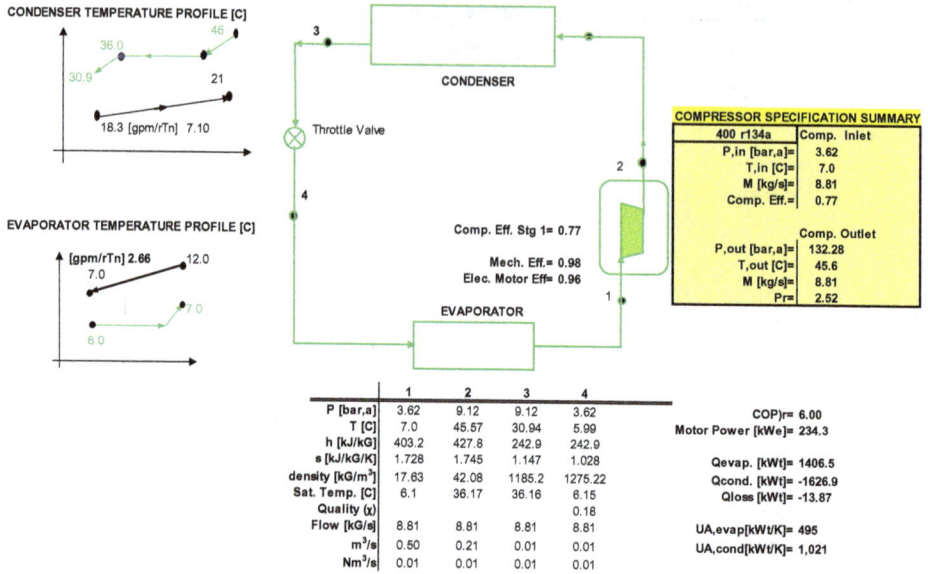

Figure 5.28.   Vapor compression (refrigeration) cycle with Property Table.

evaporated. Typical temperature approaches (or pinch points) are 2–3°F for industrial chiller systems and 10–20°F for room air conditioning (i.e. cooling) systems. Thus, for example, if the cold space to keep food produce in the kitchen appliance called a refrigerator cold is set to maintain a temperature of 45°F, then the evaporating working fluid temperature must be colder by 2–20°F. Once this temperature is known, then the operating pressure for this fluid evaporator is directly known, as the temperature and the pressure in a two-phase state must come in pairs.

The high temperature state point is similarly obtained. The heat that must be rejected by the thermodynamics refrigerator must be at a higher temperature than the local ambient for that heat to be rejected from the system. Thus, if the ambient temperature is 100°F, then the condensing temperature must be 10–20°F higher than the temperature of the air that exits the condenser. It is the exit temperature of the coolant flow stream that must be known because it will still be hotter than the ambient after the rejected heat from the cycle is absorbed by the coolant stream. Of course, the coolant can be the ambient air, but it can also be a stream of water that is in turn cooled in the air-to-water heat exchanger. Such air-to-water heat exchangers can be a wet cooling tower or a dry cooling tower. The wet cooling tower allows the water from the condenser heat exchanger stream to be intimately in contact with the ambient air, thus being allowed to cool to the ambient temperature. The dry cooling tower is a heat exchanger that does not allow intimate contact between the coolant and the cooling air, but rather, heat exchange is allowed only through a tube wall, hence its name. It is interesting to know, in the case of a wet cooling tower, that the coolant can be cooled to below the local ambient temperature. However, its cooling limit is the wet bulb temperature. The wet bulb temperature will be discussed in the next chapter when the concept of psychometrics or the thermodynamics of non-reacting mixtures is reviewed. If the ambient is 100°F and is heated to 130°F by absorbing heat from the condenser, then the condensing temperature of the working fluid is $130 + \Delta T$, where the ΔT can be 2–20°F, depending on the service requirements of the refrigerator.

The similarity between the Rankine Cycle and the vapor recompression or refrigeration cycle continues with the need for an evaporator and a condenser as the heat exchangers of choice and function. However, a compressor is used between the evaporator and the condenser instead of a turbine (which is used in the Rankine Cycle). The compressor's purpose is to boost the pressure from the discharge of the evaporator to the inlet of the condenser. This is done for one purpose: to increase the temperature of the working fluid until it is above the temperature of the ambient and the pinch point of approach temperature difference, as explained above.

With the cycle almost complete, there is a need to now determine the flow rate into the evaporator. The discharge of the condenser is passed through an expansion valve that controls the pressure difference that the fluid must experience in order to reduce the fluid pressure to the evaporator pressure with little or no energy loss. That is, like any valve that is functioning as planned, the enthalpy difference across it is zero or as close to zero as is practical.

With the fluid returning into the evaporator usually at a quality[2] of approximately 20–30%, the state point at the exit of the throttle valve will be inside the dome, between the saturated liquid and saturated vapor state points along the constant temperature process from the all liquid to the all vapor state points. The quality, χ, can be graphically represented as the "length" of a straight line along the constant temperature process between the state point inside the dome to the saturated liquid ratio with the complete straight line between the all vapor state point to the all liquid state point. The discharge of the throttle valve is at a lower temperature than the

---

[2] This percentage is the ratio of the mass of vapor in the two-phase mixture to the total liquid plus vapor water. This ratio is best pictured in the "Mind's Eye" by looking at a Temperature-Entropy diagram of the refrigerant such as that given in Figure 5.32.

temperature of the space or object that is to be kept cold. The space or the object that is to be kept cold will be the source of the heat which, though perhaps relatively cold to the human touch, is nevertheless hotter than the evaporating temperature of the refrigerant and thus enabling heat energy to "flow" from the space or object to the evaporating refrigerant. This results in the vaporization of the fluid and thus the absorption of the heat from the space that needs to be kept cool.

It is clear from the vapor compression's description that the evaporator and condenser temperatures of the refrigeration cycle are defined by the temperatures of the cold space and the environment. Once this is done, the enthalpy at each of the state points can be determined using the definitions of the four major components that constitute the cycle. These definitions and functions are defined in Chapter 3. For example, the inlet of the compressor is known because its pressure is the saturation pressure at the evaporator temperature and the evaporator temperature is the cold space temperature of -2 to $5\,°F$. The condenser pressure and thus the exit pressure of the compressor are equal to the saturation pressure at the condensing temperature, and the condensing temperature is -2 to $5\,°F$ above the ambient temperature. The compressor's discharge temperature is determined by knowing its efficiency and by using Equation (5.14) which was first presented in Chapter 3. The enthalpy at the exit of the valve is determined by knowing that it is equal to the enthalpy at the inlet to the valve. With all the enthalpies known, the refrigerant flow rate is calculated knowing the chiller capacity (typically expressed in refrigerant tons, $\mathrm{RT} = 12{,}000\,\mathrm{Btu/h}$) and the enthalpy drop across the evaporator by using Equation (5.15).

$$\eta = \frac{(h_{2s} - h_1)}{(h_2 - h_1)}, \tag{5.15}$$

$$\dot{Q}_c = \dot{M}_{evaporator} \times (h_1 - h_4). \tag{5.16}$$

It must be noted that the enthalpy at state point 4 in Figure 5.28 corresponds to a quality that is typically between 10 and 30%.

The heat balance across each of the components is easily demonstrated to be the product of the mass flow rate across the component multiplied by the enthalpy difference. Once again, it is appropriate to note that there is nothing very new in this analysis with respect to the application of the First and Second Laws for each of these four major components.

How is the performance of the refrigerator measured? The reader recalls that a heat engine cycle that produces power is measured by the ratio of the amount of power that can be generated to the amount of fuel or energy input in the form of heat. The cycle efficiency is always less than 100%. In fact, the Carnot Cycle efficiency is the maximum limit to what the engine cycle can achieve between two thermal reservoirs — a cold and a hot reservoir — by recalling the equation: cycle efficiency $= 1 - T_{cold}/T_{hot}$.

In the case of a refrigeration cycle, the efficiency of the cycle is called the Coefficient of Performance (COP). In fact, two COPs are assigned to a refrigerator. The first is the parameter that measures the efficiency of the cooling that can be the achieved with an input of mechanical energy or power to the refrigeration system.

For a perfect refrigeration cycle, the Carnot refrigeration cycle efficiency is defined by these equations:

For a cooling system:

$$COP_r = \frac{T_c}{(T_h - T_c)}.$$

For a heating system:

$$COP_h = \frac{T_h}{(T_h - T_c)}.$$

These equations can be remembered more easily if several diagrams are used to depict the refrigeration cycle. Figure 5.29 illustrates the refrigeration cycle that works between two heat reservoirs. It can be easily seen that if the heat recovered from the cold space is the desired quality of the refrigeration cycle,

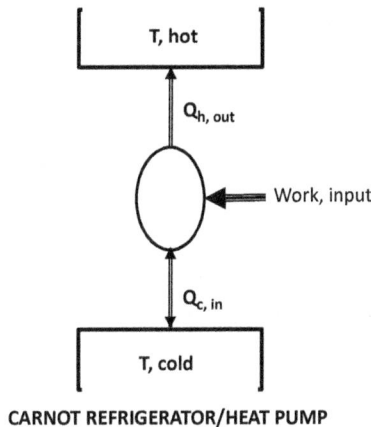

CARNOT REFRIGERATOR/HEAT PUMP

Figure 5.29. Simple thermodynamics representation of a refrigerator: a device that uses mechanical power enables the cooling or the heating of a space at temperatures, *T,cold* and *T,hot*.

then the equation for the COP is shown in Equation (5.17a).

$$COP_r = \frac{T_c}{(T_h - T_C)}. \tag{5.17a}$$

The equation can be determined from the ratio of heat recovered (cooling) to the work or power input as given in Equation (5.17b):

$$COP_r = \frac{Q_c}{Work_{in}}. \tag{5.17b}$$

Similarly, the refrigeration cycle, also known as a heat pump, that is used for providing heat into a space to be heated is given by the ratio of that heat transfer to the amount of work or power into the cycle, as shown in Equation (5.18a).

$$COP_h = \frac{T_h}{(T_h - T_C)}, \tag{5.18a}$$

$$COP_h = \frac{Q_h}{Work_{in}}. \tag{5.18b}$$

The alternative parameter defines the COP for the refrigeration system that is used as a heater. The heat energy that is released from the refrigeration cycle is the benefit that the user is interested in and therefore the COP_h, with the subscript "h" representing "heating", is determined by Equation (5.18b).

It is easily determined from these equations that: $COP_h = COP_r + 1$. It is also easily demonstrated by Equations (5.17a) or (5.18a) that the COP can be greater than one. This may seem like a violation of the First Law, but it is not. That it does not violate the First and Second Laws can be observed in Figure 5.29, in the representation of a cycle operating between two temperature reservoirs. The reader may observe more easily from this figure that a CV around the cycle represented by the closed circle clearly has heat transfer and work (a turbine shaft) that cross the boundaries of the CV. As discussed in Chapter 3, the sum of the heat transfer (take care to watch the signs of the heat transfer direction) must equal the internal energy change of the cycle, plus the sum of the works (or powers) performed. To be clear, the First Law identifies the conservation of energy and not power or the rate at which the energy is being transferred. The conservation of energy may be stretched to the conservation of power ONLY when the heat transfer and work, and internal energies changes are occurring at the same time. Clearly this is evident by taking the energy equation and dividing each term by a time increment. However, if the delivery of heat, power or internal energy change is not done at the same time, or not done within the same time interval, then only the energy is conserved and the instantaneous power may not conform to the First Law. This is commonly observed in any energy storage system, wherein the amount of energy stored over a time period, ΔT_{store}, may be extracted in a longer or shorter time interval, $\Delta T_{generate}$.

Continuing with the diagram in Figure 5.29 and the equation of the First Law for that CV, one can see that the sum of the heat transfer must be equal to the sum of the works performed, because the internal energy of the cycle is assumed to be equal to zero. It must be equal to zero because the "n" processes that make up the cycle start and stop at each state point. Thus, the internal energy change must be zero. With this understood, then it follows from the First Law of Thermodynamics that the output heat transfer is the algebraic sum of the heat input from the cold space and the power input. Thus, when the refrigeration cycle is used as a heat pump it is really heating the required space (usually the house or building) with the power into the cycle coming from the compressor, plus the heat extracted from the cold space, usually the very cold ambient.

Before we continue on to review a significant and very common advancement to a vapor compression refrigeration system, it is interesting how the "real world" effects of friction and motor inefficiencies may affect the overall COP performance of the refrigeration cycle. Figure 5.30 is almost identical to Figure 5.28, except that a slipstream of working fluid is removed from the condensate exiting the condenser to cool the electric motor that is powering the refrigerant compressor. The working fluid absorbs the heat that must be rejected from the electric motor. In this example, the motor efficiency is still very high at 95%, but the 5% of motor power that must be rejected into the environment is conveniently done (it may even be fairer to say that it is expected to be done) by the same refrigerant that is used in the cycle. The heat is thus rejected in the condenser and ultimately enters the environment by means of the condenser's water coolant stream. The temperature profile for the condenser coolant stream is shown in the plot of the temperature profiles in the top left-hand corner of Figure 5.30. The motor cooling stream (4*) that enters the motor is typically separated into two streams that ultimately exit the motor. One of these streams is a liquid condensate (shown as 7*), while the other is a slightly superheated vapor shown returning to the refrigerant cycle immediately after state point 1 in Figure 5.30. By adding the motor heat energy to the refrigeration cycle, the compressor must compress a higher fluid flow rate as may be observed by comparing the flow rates at state points 1 and 2. This results in higher compressor power and thus a lower COP. For example, the COP as calculated in this example is 5.87 compared to the 6.00 calculated in Figure 5.29. This is a rather small

CONDENSER TEMPERATURE PROFILE [C]

46
36.0
31.0   35
29.4 [gpm/rTn]   3.01

CONDENSER

V1

Comp. Eff. Stg 1 0.77
Comp.1 shaft kW 223.3
Mech. Eff. 0.98

EVAPORATOR TEMPERATURE PROFILE [C]

[gpm/rTn] 2.66   12.0
7.0
7.0
5.99

Elec. Motor Jacket Heat Loss = 2.5%
Elec. Motor Windings Heat Loss = 2.5%
Motor Winding Fluid Δ[F]= 25

EVAPORATOR

Motor Winding
Vapor Coolant

Motor Jacket Coolant
Condensate Return

	1	2	3	4	5	6	7	7*	
P [bar,a]	3.61	9.14	9.14	9.12	9.04	3.62	3.62	3.62	COP)r= 5.87
T [C]	7.0	45.80	45.80	31.00	31.00	5.99	5.99	6.15	Motor Power [kWe]= 239.7
h [kJ/kG]	403.3	428.1	428.1	242.9	242.9	242.9	242.6	242.4	kWe/RT= 0.599
s [kJ/kG/K]	1.729	1.745	1.745	1.147	1.147	1.029	1.153	1.152	Qevap. [kWt]= 1406.5
density [kG/m³]	17.56	42.09	42.09	1184.96	1184.91	92.73	93.64	94.06	Qcond. [kWt]= -1639.2
Sat. Temp. [C]	6.0	36.22	36.22	36.16	35.85	6.15	6.1	6.15	Heat Balance Chk.: 99.6%
Quality (x)						18.0%	17.7%	0	
Flow [kG/s]	8.77	8.87	8.87	8.44	8.44	8.44	8.77	0.33	UA,evap[kWt/K]= 397
m³/s	0.50	0.21	0.21	0.01	0.01	0.09	0.09	0.00	UA,cond[kWt/K]= 405
Nm³/s	0.0072	0.0073	0.0073	0.0070	0.0070	0.0070	0.0072	0.0003	

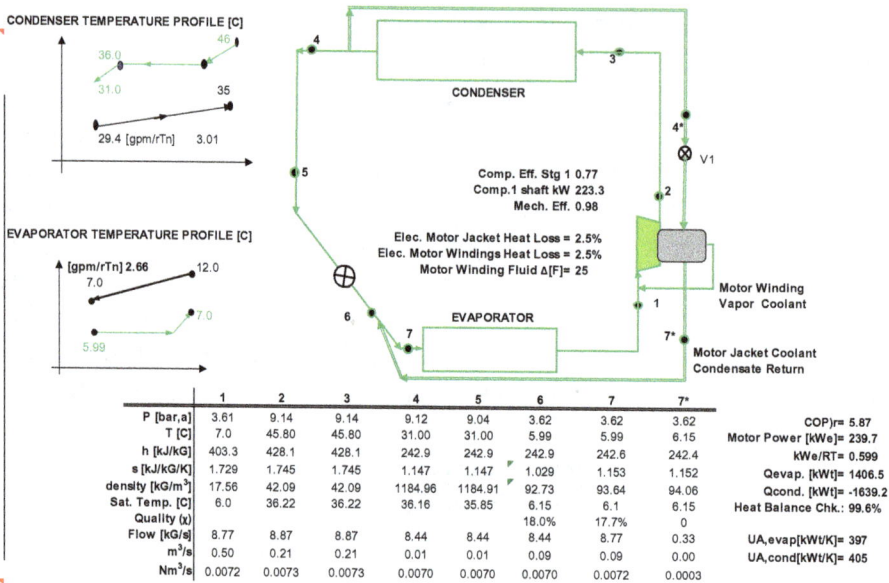

Figure 5.30. Refrigeration cycle that is similar to Figure 5.28, but with cooling of the electric motor using a refrigerant as the working fluid.

2.2% drop, but the user still has to pay for this slight increase in power when he or she pays for his increased electric bill. Even this small efficiency drop is enough to have engineers looking to improve the problem cycle's efficiency. One such improvement is the use of a cascade or economizer, as will be presented in the next section. Another improvement is the choice of refrigerant working fluid that can improve the cycle's COP, while also being safe for the environment. Case Study 8 provides a comparison of the current common refrigerant, R134a, with a new refrigerant, R1234ze, that is being touted as being the replacement fluid for chillers (both old and new) still in service after 2025.

5.10   Vapor compression cycle refrigeration using a cascade or economizer

Figure 5.31 illustrates a very common advancement for a vapor recompression cycle that results in a dual-pressure compressor to improve the COP of the refrigeration cycle. A careful review of Figure 5.31 will reveal that an additional component called a flash tank and a 2-stage compressor have been added to the cycle.

What exactly is a flash tank? While the author has not listed this component as being one of the seven mechanical devices described in Chapter 3,

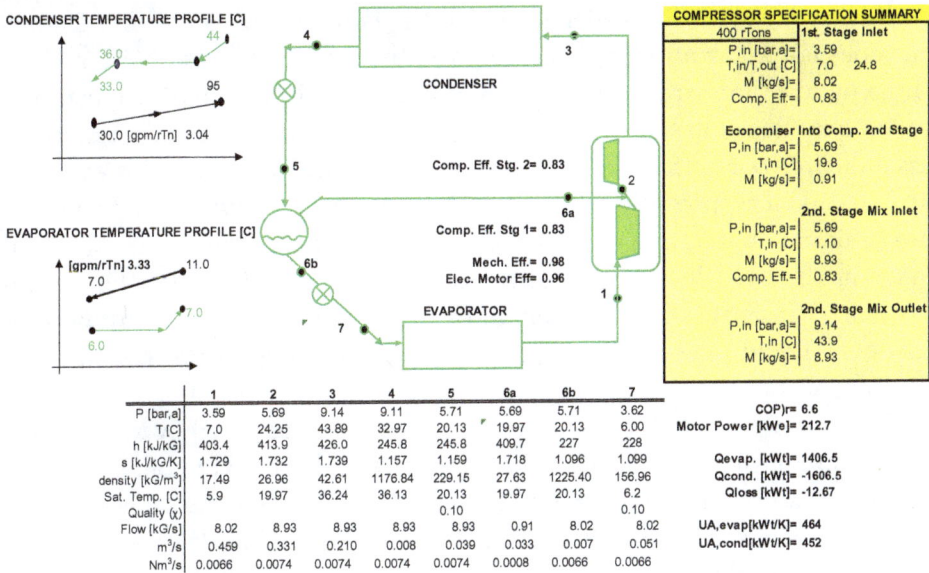

Figure 5.31. A vapor compression refrigeration cycle using an economizer or cascade system to improve its COP.

it would fall under the category of a "tank", which had been suggested as being an eighth mechanical device that collectively entails the necessary mechanical devices that make up any physical system. The flash tank always comes coupled with a throttle valve as shown in Figure 5.31. Usually, the valve is connected directly to the inlet port of the flash tank. The valve is one of the mechanical devices that have been studied. It may be recalled that this device will throttle or reduce the pressure of the fluid that enters it and delivers that fluid to the discharge piping. In this case the discharge piping is connected to the top of the flash tank, which has no moving parts and has only one job — to take in a two-phase fluid (liquid and vapor) and separate that fluid into two streams: one liquid stream and one vapor stream. The liquid drops to the bottom of the tank (because of gravity) and the vapor leaves the top of the tank to be used outside the vessel.

The flash tank in the vapor compression refrigeration application does exactly that. It takes the two-phased refrigerant at the now lower pressure and separates the fluid into a vapor and a liquid stream. The vapor stream is injected into the second stage of the compressor. The liquid stream is connected to a second throttle valve, where the pressure of the liquid is once again reduced one final time to the (low) evaporator pressure. BUT once again, the fluid stream coming out of the valve is two-phased.

But how would the reader know that it is in two phases? Very simply, one does not know until he or she looks at the state of the fluid with a given enthalpy and pressure. In both cases, the throttled fluid coming out of the valves has the same enthalpy as the liquid stream that came into the valve, assuming of course that the valve is well-insulated. Then with this enthalpy and the reduced pressure known (as the knowledge of two fluid properties are sufficient to determine all of the others) it is easily found that the fluid lands in the two-phase region of the dome; that is, it is both a liquid and a vapor. The flash tank then separates the two phases and the evaporator takes the liquid phase. It is helpful to represent the flash process on a T-s diagram, as shown in Figure 5.32.

The enthalpy at state points 1 and 2 are equal, as required by the function and nature of the throttle valve that separates the two state points. A check of the enthalpy at state point 1 will quickly indicate that the same enthalpy at state point 2 will fall "inside" the dome, that is, inside the two-phase regime. The quality of state point 2 can be calculated using the enthalpy at state point 2 and the saturated liquid and vapor enthalpies at the (lower) discharge pressure at state point 2.

The heat from the space that is to be cooled is used to evaporate only the liquid portion of the two-phase fluid at state point 2. It is important to completely understand this last sentence. The evaporator uses the liquid constituent to absorb the heat that is coming from the space that was supposed to be kept cool. It is ONLY the liquid that has a low energy (internal or enthalpy) that can be increased by absorbing (at constant pressure and

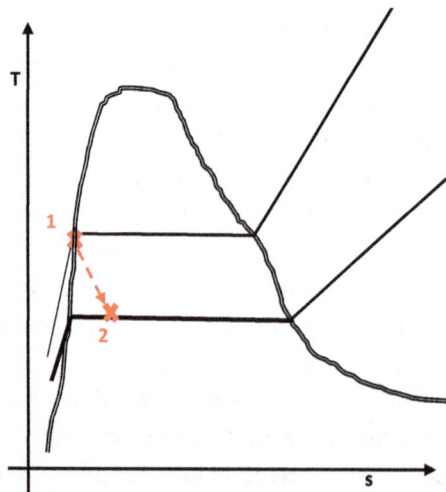

Figure 5.32.   Illustrating the process performed by the flash tank in Figure 5.31.

constant temperature) the heat that is to be removed from the cold space. The vapor constituent is ALREADY a vapor and is simply delivered to the inlet to the compressor just as it arrived into the evaporator. Thus, the heat absorbed by the evaporator may be expressed by the following equation:

$$\dot{Q}_{evap} = \dot{M} \times (1 - quality, x_2) \times \Delta h_{fg}, \tag{5.19}$$

which can be shown to be equivalent to the expression for the same heat transfer absorbed in the evaporator given by the equation:

$$\dot{Q}_{evap} = \dot{M} \times (h_{sat.v} - h_2).$$

The temperature difference between the cold space and the evaporating fluid causes a heat transfer to the liquid, allowing it to be evaporated. Certainly, the cold space is at a temperature similar to our body temperature and therefore does not "feel" cold. Nevertheless, the temperature of the evaporating fluid is much colder, relatively speaking, and thus enables an energy exchange (due to a temperature difference) in a process known as heat transfer.

One major benefit of the economizer cycle is the reduction in the net flow rate through the evaporator. This leads to an increase in the COP for the cycle. Typical efficiency improvements are approximately 8–12%. However, this improvement comes with the need for a 2-stage compressor, which clearly must be at an additional cost compared to a single stage compressor. However, the extra cost usually pays for itself by reducing the cost of the electric power that is powering the compressor over the lifetime of the system.

The reader is encouraged to check the enthalpy and entropy for each state point in Figure 5.32 and then determine the power required by the compressor and the heat transferred in the evaporator and condenser. With these last two parameters, the COP_r and COP_h for the system can be easily determined. Lastly, the reader should confirm the flow rate required to provide the cooling identified in this cycle as 400 RT (or 4,800,000 Btu/h).

5.10.1 *Example 5.4: A Heating, Ventilation and air Conditioning (HVAC) system*

A building's Heating, Ventilation and Air Conditioning (HVAC) system uses a chiller refrigeration system, as shown in Figure 5.33. Assume that the compressor is 85% efficient, the refrigerant is R134a, and that the cooling capacity of the chiller is 400 Refrigerant Tons (RT):

1. CHOOSE a **REASONABLE** condenser and evaporator pressure and temperature based on the information that is given for the evaporator

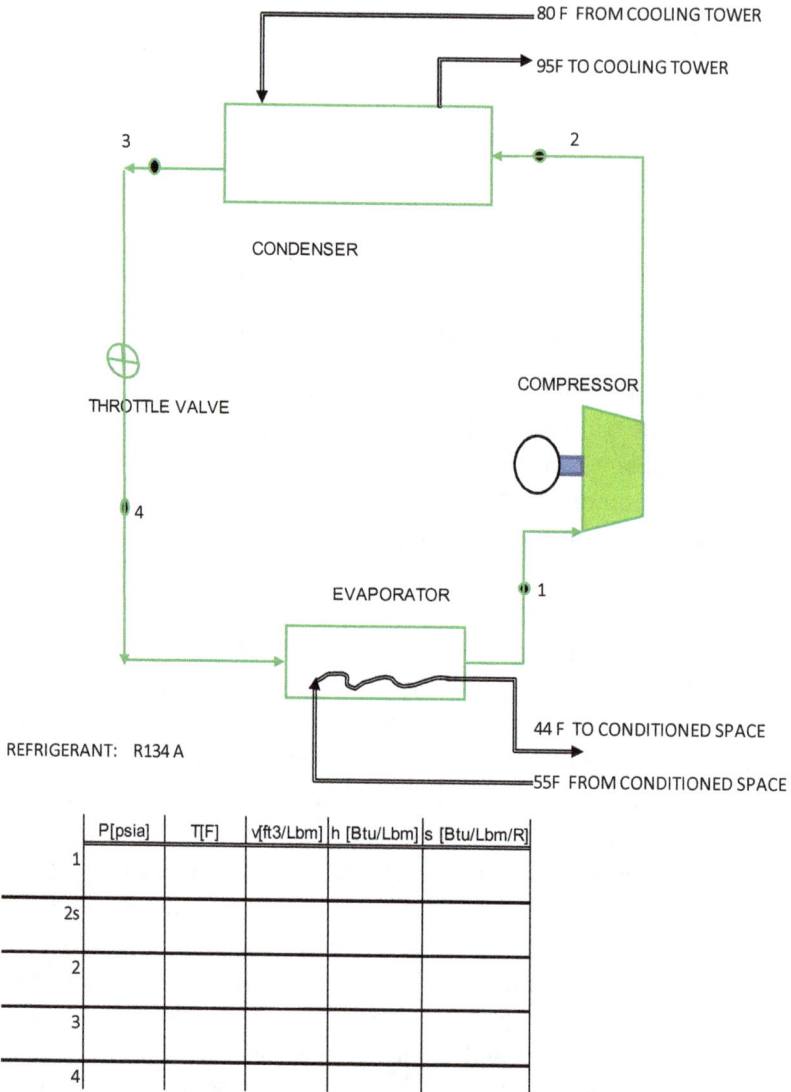

Figure 5.33.   Cycle diagram for use with Example 5.5.

and condenser coolant temperatures shown. Note: There is NO single solution. The coolant fluid is water. Insert the temperature and pressures into the appropriate places in the Property Table (see Figure 5.33).

2. Assume that the superheat temperature into the compressor is 10 °F. Determine state points 2s and 2 and fill in the Property Table.

3. What is the refrigerant flow rate required to absorb the cooling in the evaporator?

4. What is the power required for the compressor (kW or hp)?
5. What is the actual COP_h heat pump and the Carnot Refrigeration's COP_r for the cycle?
6. If the throttle valve is replaced with a perfect turbine (i.e. turbine with an efficiency of 100%), how much power could be generated?

5.11 Absorption chiller system

A new subsystem that is placed beside the economizer system illustrates another very common refrigeration system called an absorption chiller system. This system, shown in Figure 5.34, is actually very similar to the standard refrigeration cycle, except for the obvious differences as noted by the presence of the absorber and (vapor) generator.

The absorber and the generator are part of a complex chemical system that mixes two chemicals in the absorber. However, these two chemicals are later separated by heat in the generator, which liberates one of the chemicals in the form of a vapor; it evaporates the refrigerant into a vapor so that it may proceed to the condenser in the cycle. The two chemicals are typically either lithium-bromide and water (as the refrigerant), or water and ammonia (as the refrigerant). These chemical pairs are easily mixed to form a liquid at the prescribed temperatures and then easily separated by heat at the same temperature.

What is the purpose of the absorber? It is very apparent from a study of the cycle examples shown in Figures 5.30 and 5.31 that the compressor power is very significant in contributing to a high or low value of the COP. If the compressor power could be reduced, the COP would improve. As studied

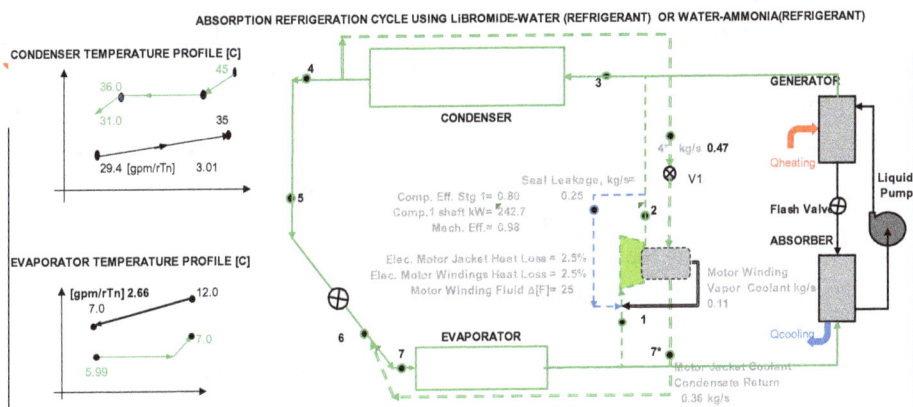

Figure 5.34. Absorption refrigeration cycle with (vapor) generator and refrigerant absorber replacing the compressor subsystem.

in Chapter 3, the compressor's power is dependent on the pressure ratio and the inlet temperature for any given vapor. The key word is "vapor", because if the fluid were a liquid, its compressing power would be better identified as the pumping power for pressurizing the liquid, which would be very small.

The absorber receives the vaporized refrigerant fluid from the evaporator, as typically done in a refrigeration system, and mixes this with the second chemical to produce a liquid mixture. This liquid mixture will be much easier to pressurize to the higher pressure required by the condenser. The pressurization requires much less power compared to the heat recovered in the evaporator or released in the condenser. However, before the condenser gets the same refrigerant fluid that it expects as a vapor, the refrigerant chemical that was mixed in the absorber to form a liquid must be separated from the liquid mixture and returned to its vapor state. This requires heat energy, which is a significant part of the heat that is recovered in the evaporator and thus must be accounted for in the COP calculation. The equation for the COP for the absorption chiller changes to:

$$COP_r = \frac{\dot{Q}_{cool}}{\dot{Q}_{input}}.$$

The result is a COP that is approximately 1, ranging from 0.7 to as high as 1.3 for some of the more contemporary systems that deploy more than one generator, each driven at different temperatures.

The net benefit in the market for an absorption chiller is therefore not the COP improvement, for it is clearly much less than the typical refrigeration system. The benefit is the ability to produce chilled water (or cooling in general) with the use of relatively low-grade heat. Low-grade heat is informally identified as heat energy at 250°F, or less, and a low-grade energy is typically available from a user's otherwise idle packaged boiler. The ability to use the low-grade heat from the packaged boiler is key here, because in many cities where cooling in the summer and heating in the winter is almost on a 50/50 basis, the very active boilers during winter are virtually idle during summer. Rather than have these boiler systems sit idle, which does not bode well for their long-term maintenance because of oxygen corrosion due to inactivity, they can be utilized in the summer to produce chilled water at very reasonable and much lower fuel prices.

The lower COP is rather troublesome in the era of fuel efficiency, particularly in the saving or reducing of fossil fuels. A theoretical analysis of what can be expected from an absorption chiller system does little to encourage its deployment. As indicated in Chapter 4, the maximum COP for an absorption chiller system can be rather straightforwardly determined by applying

CARNOT CYCLE HEAT ENGINE

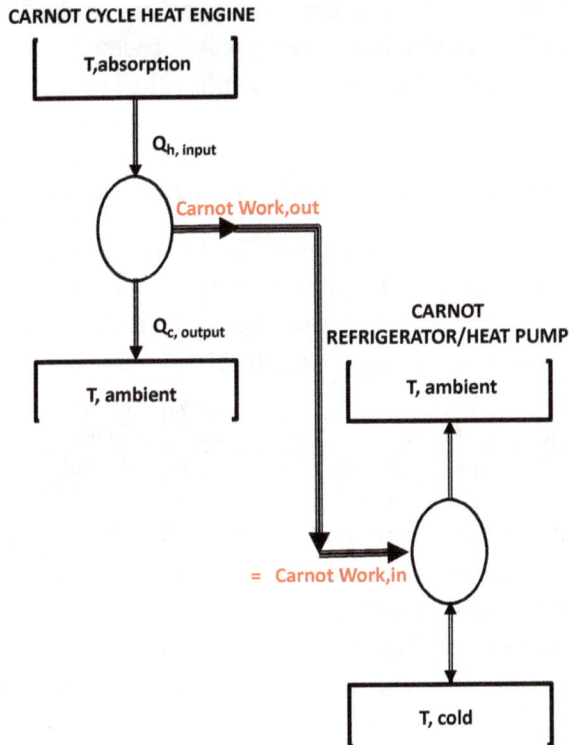

Figure 5.35. (from Figure 4.20). Graphical representation of an absorption chiller system using a Carnot Heat Engine and Refrigerator.

the Carnot Engine and Carnot Refrigerator equations together. Figure 5.35, reproduced from Figure 4.20 , helps the reader visualize how to formulate the derivation. An absorption chiller system can be modeled as a Carnot engine working between the hot (generator) low-grade heat energy that is input to the cycle, that could have been used to produce a maximum amount of power from an ideal Carnot engine. That maximum power could then have been used to power an ideal Carnot Refrigerator to cool the working space with a heat transfer of Q_c. By determining the ratio of $Q_{,cooling}$ to $Q_{,low\ grade}$ heat input, the equation shown here is the result:

$$COP_{ideal} = \frac{\dot{Q}_{heat\ in}}{\dot{Q}_{cooling\ capacity}} \times \frac{1 - \frac{T_c}{T_{absorption}}}{1 - \frac{T_c}{T_{ambient}}}.$$

To improve the theoretical actual COP of the absorption chiller, several progressively higher temperature heat sources can be used. However, given the limit to the overall pressure ratio that is needed in a typical refrigeration cycle, there are seldom more than three temperature levels. There are usually only 1 to 2 levels for practical and cost points of view.

The following example provides a sample calculation for an absorption chiller system.

5.11.1   *Example 5.5: An absorption chiller system*

The heat required for the absorption chiller is provided at a temperature of 250°F. The chiller must cool a space to 40°F. What is the heat input that must be provided, assuming that the Carnot COP is 2 for this absorption chiller system to achieve a cooling of 200 RT?

$$2.0 = \frac{\dot{Q}_{heat\,in}}{200 \times 12,000} \times \frac{1 - \frac{T_c=(460+40)}{T_{absorption}=(460+250)}}{1 - \frac{T_c=(460+40)}{T_{ambient}=(460+60))}}.$$

Note the magnitudes of the temperatures for T_c, $T_{absorption}$, and $T_{ambient}$ from the problem statement. With these values, the solution for $\dot{Q}_{heat\,in}$ using the equation above is straightforward.

With the magnitude of $\dot{Q}_{heat\,in}$ now known, it is also interesting to determine the actual COP for the absorption cycle by calculating the ratio: $\frac{\dot{Q}_{heat\,in}}{200 \times 12,000}$

Lastly, it is interesting to increase the value of $T_{absorption}$ and/or increasing the value of T_c to increase the Carnot Efficiency of the absorption refrigeration cycle or to decrease the amount of the required heat input, $\dot{Q}_{heat\,in}$

5.12   Reversed Brayton Cycle or air refrigeration cycle

The last refrigeration cycle is derived from the Brayton Cycle. Two refrigeration air-Brayton (reverse) Cycle Systems are shown in Figures 5.36 and 5.37. There are several versions to these two cycles that accomplish the same objective: the usage of air as a working fluid refrigerant to cool a space. The process proceeds by having the compressor draw down on the fluid stream exiting the turbine. In fact, the turbine discharge is drawn below atmosphere pressure by the suction from the compressor. This causes the turbine to operate between an inlet pressure of atmosphere pressure and a sub-atmospheric pressure. This may cause some losses if the system is not sealed against vacuum leaks. The discharge of the turbine is also at a very low temperature, as may be seen in Figure 5.37. This cooled air is inserted into the space that is to be cooled. The cooling is performed in a heat exchanger between the compressor and turbine, as shown in Figures 5.36 and 5.37, to separate the air from the cold space from the fresh air from the ambient air.

6.0

Eff.t= 0.55
Pr,t= 5

Eff.c= 0.65
Pr,c= 5.18

Motor Power [kW]= 1.5

COMPRESSOR SPEC.S
Speed [rpm] 45,000
Diameter [in.] 5.90
Ns 0.280

HX Effect.= 0.85

TURBINE SPEC.S		
Flow Rate	4.42E+01 lbm/hr	0.006 kg/s
	P,in 14 psia	0.10 Mpa
	T,in 50 F	283.0 K
	P,out 3 psia	0.020 MPa
	P,stage,out 3 psia	0.020 MPa
	55% Turbine Eff.	
Δh,issen.=	104.7 KJ/Kg	
kW	0.32	
	Ns= 0.52	
	Ds= 19.99	

Qcool= 0.01 rTon
COP)r= 0.03
COP)ideal= 10.60

0.075 Dia[m]&[in.] 2.95   580 ft/s
45000 RPM   0.055 Ns
1 No. of Stg.s   19.97 Ds

	P [psia]	T [F]	h [Btu/Lbm]	s [btu/Lbm/R]	Flow rate [Lbm/s]	[ft3/s]
1	14.70	86.0	130.62	1.64	0.0123	0.16893442
2	14.55	34.5	118.24	1.62	0.0123	0.1545022
3	14.41	50.0	121.96	1.63	0.0123	0.16096279
4	2.88	25.5	97.16	1.73	0.0123	0.76630366
5	2.85	64.1	109.55	1.75	0.0123	0.83572211
6	14.77	515.6	250.04	1.78	0.0123	0.3006143

	P [kPa]	T [C]	h [kJ/kG]	s [kJ/kG/K]	Flow Rate kg/s	m3/s
1	101.4	30	303.3	6.88	0.0055835	4.79E-03
2	100.4	1.4	274.6	6.78	0.0055835	4.38E-03
3	99.4	10	283.2	6.81	0.0055835	4.56E-03
4	19.9	-3.6	225.6	7.23	0.0055835	2.17E-02
5	19.7	18	254.4	7.31	0.0055835	2.37E-02
6	101.9	269	580.6	7.47	0.0055835	8.53E-03

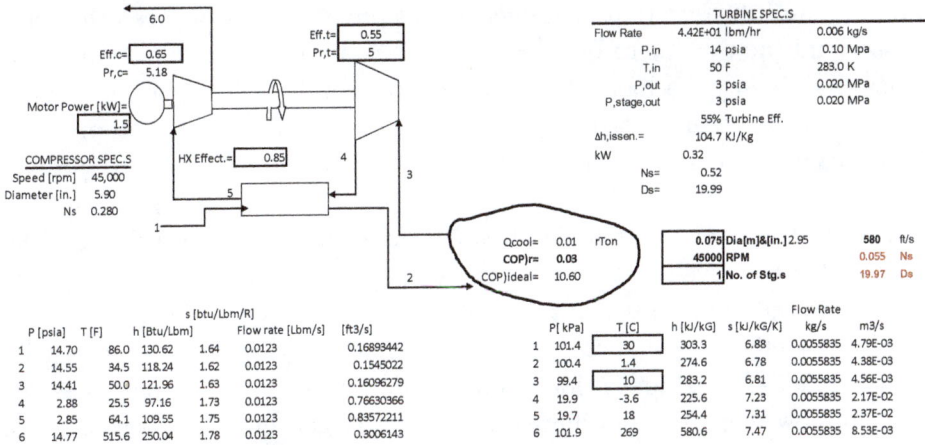

Figure 5.36. An air refrigeration system using a Reverse Brayton Cycle.

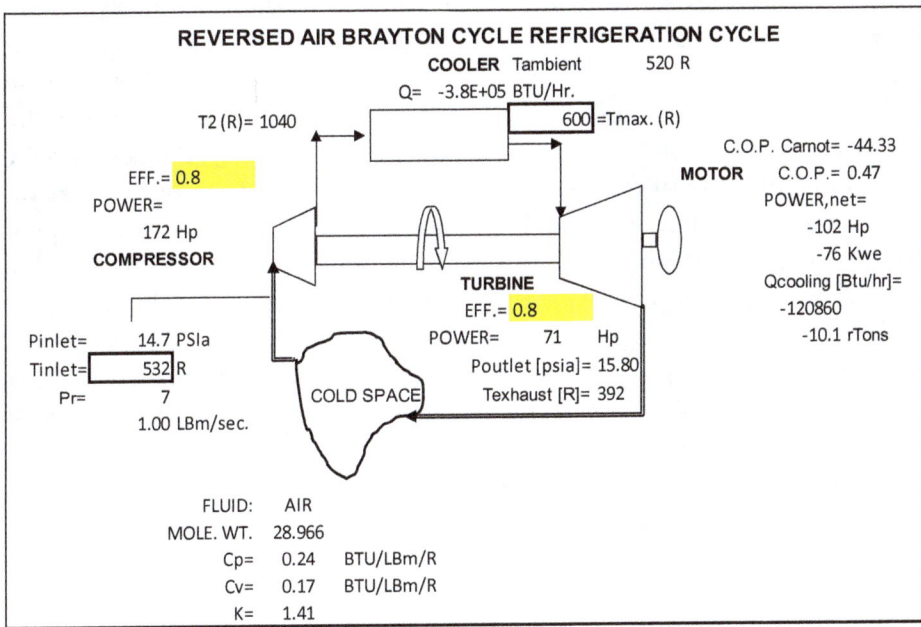

REVERSED AIR BRAYTON CYCLE REFRIGERATION CYCLE

COOLER Tambient   520 R
Q= -3.8E+05 BTU/Hr.

T2 (R)= 1040   600 =Tmax. (R)

C.O.P. Carnot= -44.33

EFF.= 0.8
POWER=
172 Hp
COMPRESSOR

MOTOR   C.O.P.= 0.47
POWER,net=
-102 Hp
-76 Kwe
Qcooling [Btu/hr]=
-120860
-10.1 rTons

TURBINE
EFF.= 0.8
POWER= 71 Hp
Poutlet [psia]= 15.80
Texhaust [R]= 392

Pinlet= 14.7 PSla
Tinlet= 532 R
Pr= 7
1.00 LBm/sec.

COLD SPACE

FLUID: AIR
MOLE. WT. 28.966
Cp= 0.24 BTU/LBm/R
Cv= 0.17 BTU/LBm/R
K= 1.41

Figure 5.37. An alternative air-Brayton Cycle refrigeration system.

In Figure 5.36, the air into the room is 100% fresh. There is an option to mix the room air with the fresh air, assuming appropriate filtration, etc.

The refrigeration process is achieved by having the Brayton Cycle processes occur in reverse for the energy from the cold space to be absorbed by the cycle, and then for that heat energy plus the net power between the compressor and the turbine to be released into the atmosphere. The compressor is powered by any mechanical or electrical means. Clearly, however, the net power must be an input or negative into the system in order to comply with the Clausius Statement of the Second Law of Thermodynamics, as described in Chapter 4.

It may be observed from Figures 5.36 and 5.37 that the COP for the reverse Brayton Cycle is also very low. In Figure 5.36, this is especially due to the very low temperatures that are attempted for the cold space. Using the Carnot efficiency equation for a refrigeration system, it can be easily shown that the COP for an ideal refrigeration cycle approaches zero as the high and low temperature difference becomes larger, even as the cold space temperature is extremely cold.

It is essential that the reader appreciate the function of the compressor and its role in achieving and maintaining this flow rate. It is also important to note that the reverse Brayton Cycle depends on the performance of the compressor, perhaps even more so than the typical reverse Rankine-type refrigeration system. The compressor can either be a positive displacement or a centrifugal compressor. For either design, the compressors volume flow rate (measured in ACFM or Nm^3/min) that is prescribed by the design point is for the compressor at an operating speed. If the compressor operates at the design speed, it will draw down the suction pressure to its design point pressure because it needs to draw into the compressor the volume flow rate that it is designed to compress. Once the suction pressure is maintained below atmospheric pressure, the turbine discharge is maintained at that reduced pressure and thus a pressure ratio across the turbine is established, which is sufficient to enable the turbine to generate power. As indicated above, the turbine power is not equal to or greater than the power required by the compressor, though power from the turbine helps reduce the net power required by the cycle. Thus, the COP for the cycle is improved.

Figure 5.38 illustrates an additional air refrigeration cycle that is slightly different from those in Figures 5.36 and 5.37. However, this cycle still operates on a reverse Brayton Cycle. The reader will notice the different placement of the heat exchanger and the different way the room's air stream is recovered. The room's air stream can either be compressed first, such as in Figures 5.37 and 5.38, or have the room ambient air enter the turbine first and then expanded to a slight vacuum before it is compressed, as shown in Figure 5.36. The reader is strongly encouraged to model these cycles, if only to experience firsthand that the rearrangement of the components neither increases the complexity nor the difficulty of determining a cycle's closed

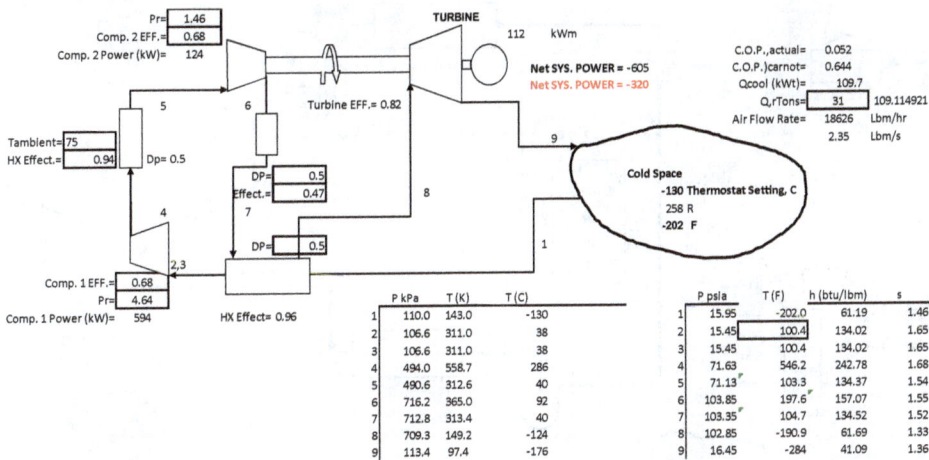

Figure 5.38. An alternative air-Brayton (Reverse) Cycle refrigeration system.

cycle analysis, cooling capacity or COP. It is also important to see how the state point temperature and pressures before and after each component change while still producing the desired effect of providing cooling to the room.

The last cycle that is used to produce refrigeration (or cool air) for conditioning a living space is shown in Figure 5.39. This cycle combines a solar powered Rankine Cycle that uses a reheat turbine and a vapor compression refrigeration cycle to produce cool air for a living space. The working fluid in both cycles is water, the cleanest and most environmentally safe fluid. Water vapor is generated by heat in the living space that is to be cooled. The water vapor at the inlet to the compressor is kept at a pressure of approximately 1 psia to keep the living space at a temperature of 72°F. The evaporating pressure is also very low because it is the saturation pressure that corresponds to the very low operating temperature. This cycle is reviewed in detail in Case Study 7.

5.13 Refrigerant fluids

Heat engine cycles that are operated in reverse can cool a space instead of produce power. Of course, the mechanical devices that may have been used in the heat engine cycle must be changed to the mechanical devices that operate in reverse. For example, a turbine in a heat engine is substituted with a compressor; a pump in place of a throttle valve; and a condensing heat exchanger instead of an evaporator.

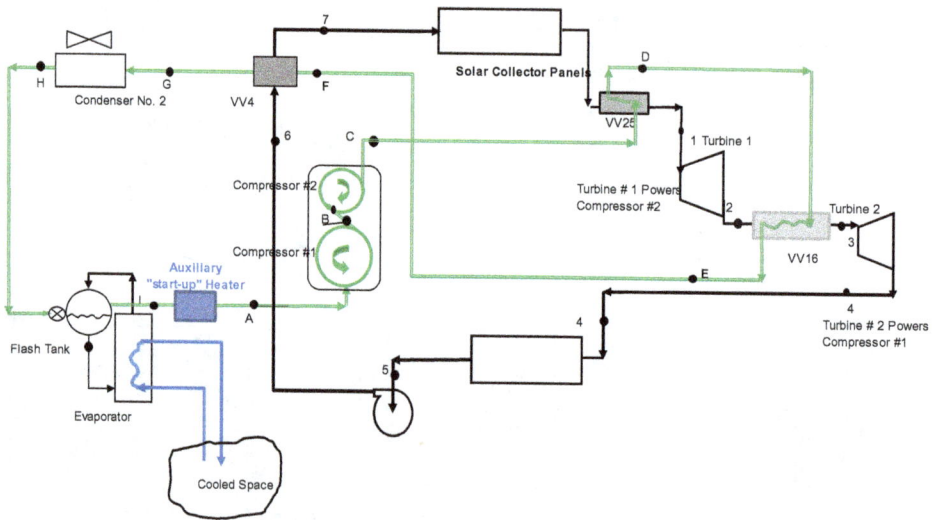

Figure 5.39.   Solar powered refrigeration cycle using water as the refrigerant powered by solar energy (see Case Study 7).

Table 5.4.   Summary of Common (Old and New) Refrigerants with ODP and GWP shown.

Refrigerant	Category	ODP (R11-1)	Atmospheric lief [years]	GWP $CO_2 = 1$ (100 years)
R11	CFC	1	45	4700
R12	CFC	1	100	10700
R113	CFC	0.8	85	6000
R22	HCFC	0.05	12	1800
R123	HCFC	0.02	1.3	76
R32	HFC	0	4.9	670
R125	HFC	0	29	3450
R134a	HFC	0	14	1400
R290	HC	0	3	3
R600	HC	0	–	4
R717	NH_5	0	1	0
R744	CO_2	0	100]	1
R1234yf	HFO	0	11 days	4
R1234ze	HFO	0	–	6

However, the working fluid for the cycle must be given serious attention if it is to operate with the highest COP. In the last several cycles, the refrigerants have varied from water to ammonia to hydrofluorocarbon chemicals. These working fluids are conveniently identified with a refrigerant number provided by the manufacturer. Table 5.4 identifies the most common past and present refrigerants. Table 5.4 also identifies several important characteristics that compares how that refrigerant may affect the atmosphere should they be released into the ambient.

Refrigerants have been the subject of significant controversy since the early 1970s, when research findings confirmed that some refrigerants aggravated the atmosphere by thinning Earth's ozone layer. Such damage causes the atmosphere to allow more solar energy of higher frequencies to pass through the atmosphere and reach Earth's land and sea surfaces, setting off a cascading series of negative effects. To begin to mitigate the use and the eventual release of the refrigerant into the atmosphere, a gathering of the major industrialized countries in Montreal (1987) and then again in Kyoto Japan (1997) with the objective of reaching a consensus on how to reduce the damaging consequence of the continuous release of refrigerants into the atmosphere. Each meeting concluded with at least Memorandums that listed suggestions on how to limit the production of certain refrigerants and their replacement with alternative refrigerants. These results were to be referred to as Montreal and Kyoto Protocols. A summary of these to meetings is shown in Figures 5.40 A and B.

Working fluids have been given two designated quantifiers to determine the relative extent of damage that they wreck on the atmosphere. The first is labeled the Ozone Depletion Potential (ODP). The ODP is a relative measurement of the chemical's ability to destroy the stratosphere's ozone layer compared to the effect of R11 on the stratosphere which is thus given a "baseline" ODP of 1. The second parameter is the Global Warming Potential (GWP). The GWP measures the longtime effect of how the chemistry of this "greenhouse" gas may affect the warming of the earth due to the continuous release of the chemical. With the Paris COP1 Protocol, several refrigerants such as R22 and R11 have been prohibited from being manufactured. Existing stockpiles of these fluids can only be depleted and not replenished. Several refrigerants have taken their place, such as R245fa and R134a. Unfortunately, the overall cycle system efficiency has dropped by 5–8%. Replacement fluids that are more environmentally friendly continue to be developed, though it is interesting to note that these fluids tend to reduce the efficiency of the cycle until comparable cooling conditions. However, this is a very general statement and is evident only through commercial observations and a theory that cooling system manufacturers will research and, through competitiveness, determine the best fluid to use in a cycle. Usually, this is simply based on which refrigerant fluid enables the highest efficiency (and thus requires the least amount of power) of the cycle. Market competitiveness thus determines the choice of fluid, but this choice may be based more on efficiency than on environmental effects.

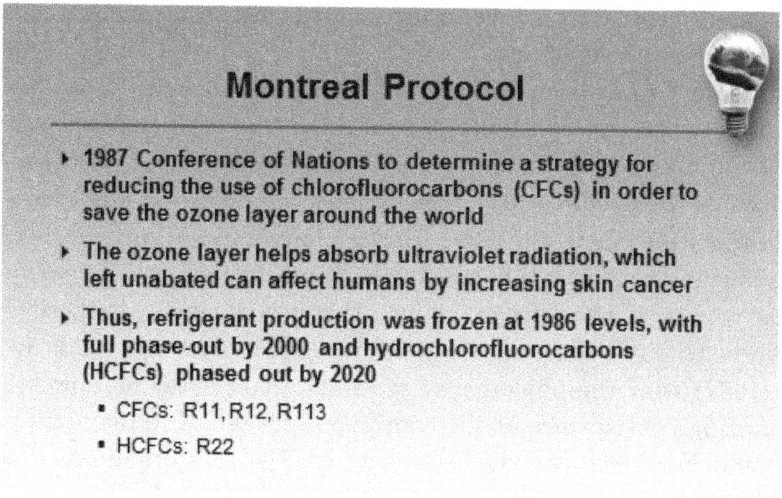

(a)

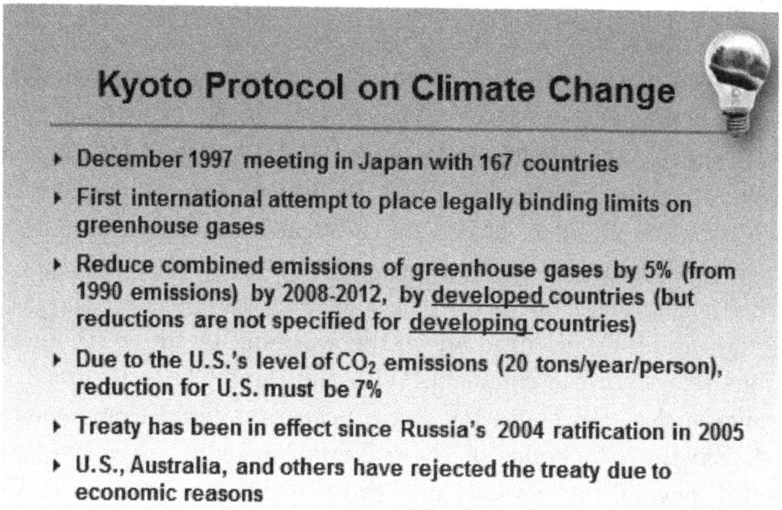

(b)

Figure 5.40.   A summary of the (a) Montreal and (b) Kyoto Meetings' guidelines on how to control the use of refrigerants throughout the world in order to reduce effects on the atmosphere.

Among the most recent refrigerants proposed for use in the marketplace is R1234ze. This is intended to be a replacement for R134a. A study of this fluid in use in the Rankine Cycle will indicate that the inlet volume flow rate into the compressor will be higher than the equivalent loading for R134a. Thus, the compressor impeller may be larger than necessary. However, the benefits in using a fluid with a lower impact on the environment are worth the extra cost and physical size of a larger compressor.

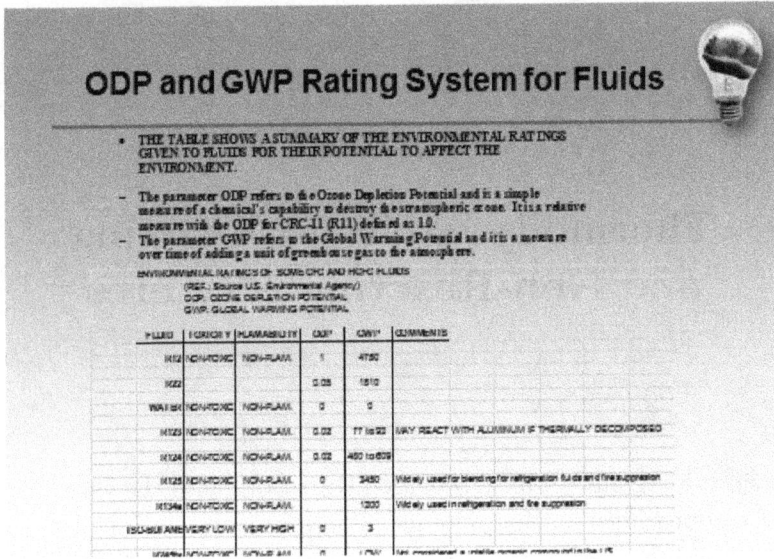

Figure 5.41. The Ozone Depletion Potential (ODP) and the Global Warming Potential (GWP) for several common refrigerants providing a relative measure of potential harm to the environment from these refrigerants if released into the atmosphere.

Chapter 6

Thermodynamics of Reacting
and Non-Reacting Mixtures

The previous five chapters have provided the reader with the very basic principles of engineering thermodynamics. These universal principles, together with the characteristics of the seven major mechanical engineering devices, provide a solid foundation to build useful systems and convert heat energy into mechanical energy. The conversion of different forms of energy requires a study of the thermodynamics of reacting mixtures, such as combustion processes and the thermodynamics of non-reacting mixtures, the most common example of which is the very air that we breath; a non-reacting mixture of water and air. The thermodynamics of these mixtures is useful in solving many routine problems that may be found in many engineering applications. Such applications range from air conditioning to the combustion of fuels. Air conditioning applications require the thermodynamics of a non-reacting mixture of dry air and low-pressure water vapor. The combustion of fuels in internal or external combustion engines that provides the energy to heat engines requires the thermodynamics of a reacting mixture of oxygen with a carbon–hydrogen-based fuel. The energy released during the reacting mixture provides the energy that can be transformed into mechanical, rotary shaft power.

The thermodynamics of non-reacting mixtures is the first of these two thermodynamics topics to be presented. This is considered the proper sequence of instruction because the information gained from a study of the thermodynamics of non-reacting mixtures will be used in the study of the thermodynamics reacting mixture analysis.

6.1 Dalton's law of partial pressure and Amagat's law of partial volume for non-reacting mixtures

Understanding the properties of non-reacting mixtures begins with understanding Dalton's law of partial pressures for gases and Amagat's law of partial volume. These are two similar and very useful laws that can be used to identify the proportions of a mass (or mole) of two or more constituent gases in a non-reacting mixture. Consider the mixture of two different gases that do not chemically react with each other: oxygen and nitrogen. These two are very familiar constituents of the air we breathe. Oxygen is used as an oxidizer while nitrogen is a noble gas that hardly reacts in a chemical reaction, except perhaps to function as a catalyst. Oxygen occupies approximately 21% of the space occupied by the total air mixture. The properties of air as a single substance are well known and can be determined from thermodynamics property charts or computer models such as REFPROP. Such properties are then used on computer platforms such as Excel, Math Cad, Basic and MATLAB, for example. The same may be said for the thermodynamics properties of nitrogen, which makes up the remaining portion of the non-reacting mixture. But at what pressure and temperature are these properties measured by if the mixture is held at a temperature, T, and total pressure, P, as measured by a pressure gauge on the container vessel?

Dalton's law of partial pressure must be used to determine the individual partial pressure for the constituent at the equilibrium temperature that is given for the two-gas system. Dalton's law simply states that the partial pressure of a constituent gas in a mixture is determined by assuming that all the constituents at the total volume of the container and at the mixture are at equilibrium temperature. Equation (6.1a) defines this statement.

$$p_n \times V = \frac{M_n \times R_u \times T}{Mole}.$$

$$(6.1a)$$

Amagat's law of partial volume is very similar and states that the partial volume of a constituent gas is found by using the Perfect Gas Law, the total equilibrium pressure and temperature of the system, and the individual mass and molecular weights of the constituent. Equation (6.1b) defines this statement.

$$P \times v_n = \frac{M_n \times R_u \times T}{Mole.wt.}.$$

$$(6.1b)$$

It is also true that the sum of the partial pressures is equal to the total equilibrium pressure of the system. Similarly, the total volume of the system is the sum of the individual partial volumes. Dalton's law of partial pressure and Amagat's law of partial volume are defined in Equations (6.2a) and (6.2b), respectively.

$$P = \sum_{1}^{n} p_n, \tag{6.2a}$$

$$V = \sum_{1}^{n} v_n, \tag{6.2b}$$

Clearly, a non-reacting mixture has the constituents at the same temperature, as per the Zero Law of Thermodynamics, which states that if Body A is in contact with Body B, then they are in thermal equilibrium — an equilibrium achieved over a period of time that is long enough for it to be established. Thermal equilibrium in this context means that the temperature of the two bodies are equal; i.e.

$$T_A = T_B.$$

The partial pressure of the oxygen may be found using the Perfect Gas Law or a software program that has tables of pressure as functions of specific volume and temperature. The state postulate, first introduced in Chapter 2, requires only two thermodynamic properties to be known in order to find all of the others. The state postulate applies for non-reacting mixtures as well and therefore if either the partial pressure or the partial volume is known with the mixture temperature, then the unknown property (partial pressure or partial volume) can be determined from property tables in electronic or hard copy formats. Of course, the use of the word "partial" will not be found in the tables as it is applied only when the constituent gas is part of a non-reacting mixture calculation. However, it is easy enough to determine the unknown property (p_n or v_n) by simply using the Perfect Gas Equation, as shown in Equation (6.3a), using the procedure and calculations shown in the following example. The specific volumes of the constituent gases are determined by knowing the mass of the constituent and the total volume of the container, thus $v_n = V/M_n$.

$$P \times v_n = \frac{M_n \times R_u \times T}{Mole}, \text{ or } P \times v_n = N_{moles} \times R_u \times T. \tag{6.3a}$$

$$\frac{P \times v_{O2}}{P \times v_{N2}} = \frac{\frac{M_{O2} \times R_u \times T}{Mole.O2}}{\frac{M_{N2} \times R_u \times T}{Mole.N2}},$$

$$\frac{v_{O2}}{v_{N2}} = \frac{\frac{M_{O2}}{Mole.O2}}{\frac{M_{N2}}{Mole.N2}},$$

$$\frac{v_{O2}}{v_{N2}} = \frac{\frac{M_{O2}}{32}}{\frac{M_{N2}}{28}},$$

$$\frac{v_{O2}}{v_{N2}} = 0.875 \times M_{O2}/M_{N2}.$$

Also:

$$v_{Total} = v_{N2} + v_{O2}.$$

But for air it is known that:

$$\frac{v_{O2}}{v_{Total}} = 0.21 \; and \; \frac{v_{N2}}{v_{Total}} = 0.79.$$

Therefore:

$$\frac{v_{O2}}{v_{N2}} = \frac{0.21}{.79} = 0.265.$$

Combining the equations, we find:

$$0.265 = 0.875 \times \frac{M_{O2}}{M_{N2}} \; and \; therefore: \; \frac{M_{O2}}{M_{N2}} = \frac{0.265}{.875} = 0.303.$$

Similarly, by using partial pressures:

$$p_n \times V = N_{moles} \times R_u \times T,$$

$$\frac{p_{O2} \times V}{p_{N2} \times V} = \frac{N_{molesO2} \times R_u \times T}{N_{molesN2} \times R_u \times T}.$$

But then:

$$\frac{p_{O2}}{p_{N2}} = \frac{N_{moles\,O2}}{N_{moles\,N2}} \; and \; therefore: \; \frac{p_{O2}}{p_{N2}} = \frac{M_{O2}}{M_{N2}} \times \frac{Mole.Wt.\,N2}{Mole.Wt.\,O2}.$$

And from the above values:

$$\frac{M_{O2}}{M_{N2}} = 0.303. \tag{6.3b}$$

It is also true that:

$$M_{total} = M_{O2} + M_{N2}.$$

Then also:

$$\frac{M_{total}}{M_{N2}} = \frac{M_{O2}}{M_{N2}} + 1 \; and \; therefore: \; \frac{M_{total}}{M_{N2}} = 1.303,$$

$$or: \; \frac{M_{N2}}{M_{total}} = 0.7675 \; \& \; \frac{M_{O2}}{M_{total}} = 0.2325,$$

$$\frac{M_{N2}}{M_{total}} = 0.7675 \; \& \; \frac{M_{O2}}{M_{total}} = 0.2325. \tag{6.4}$$

$$\frac{Mole.Wt\,N2}{Mole.Wt\,O2} = \frac{28}{32}.$$

Therefore:

$$\frac{p_{O2}}{p_{N2}} = 0.303 \times \frac{28}{32} = 0.265.$$

$$\frac{P_{atm.}}{P_{N2}} = \frac{p_{o2}}{p_{N2}} + 1, \text{ and therefore: } \frac{P_{atm.}}{P_{N2}} = 0.265 + 1.$$

But:

$$P_{atm.} = p_{o2} + p_{N2},$$

$$p_{N2} = \frac{14.7}{1.265} = 11.62. \tag{6.5a}$$

The calculation of the oxygen partial pressure is therefore:

$$p_{O1} = 14.7 - 11.62 = 3.08. \tag{6.5b}$$

Why is this important? It is important because with the partial pressure and temperature now known, all of the thermodynamics properties of enthalpy, entropy and internal energy can be found for the mixture by using these separate constituent values.

As an example, let us continue with the example of the oxygen and nitrogen. It was determined that:

$$p_{N2} = 11.62 \text{ psia and } p_{O1} = 3.08 \text{ psia}.$$

It can be found from the look up tables of software programs such as REFPROP, that the enthalpy value of oxygen is $126.79 \text{ Btu/lbm}, O_2$ and that of nitrogen is $131.33 \text{ Btu/lbm}, N_2$, at a temperature of 70°F. The total enthalpy of the mixture can be determined by adding the total enthalpy (units of Btu) and then dividing this number by the total mass of both constituents. This is made even easier by using the mass fractions of each constituent in the mixture as shown in Equation (6.4).

$$h_{mixture} \equiv 126.79 \times 0.2325 + 131.33 \times .7675 = 130.27 \text{Btu/Lbm}_{mixture}.$$

The properties of specific heat with respect to pressure, C_p, and specific heat with respect to volume, C_v, can be found in the same way, given the value for each constituent separately. All of the properties can be found in a similar manner.

As stated and demonstrated numerous times in the previous chapters, it is important to determine the enthalpy and entropy values of the system in order to complete the thermodynamics analysis. For example, as indicated in Chapter 3, in an Open Flow Analysis case, the power or heat transfer

Table 6.1. Calculation of oxygen, nitrogen and water content in moist air with 0.006 lbm$_{water}$/lbm$_{air}$ ratio.

Ratio of Water Content in Air (by Mass)=	0.006	
Ratio of Oxygen to Nitrogen (by Mass)=	0.303	
Oxygen Content [Lbm]=	1	0.231 = Oxygen O$_2$ Mass/Total Mix. Mass
Nitrogen Content [Lbm]=	3.300	0.763 = Nitrogen N$_2$ Mass/Total Mix. Mass
Water Content [Lbm]=	0.0258	0.006 = Water Mas/Total Mix. Mass
	4.326	1.000

from the start to the end of a process is determined by knowing the enthalpy difference between the start and end of the process. In a Close Mass Analysis, the internal thermal energy of the mixture is important. All of the work, heat transfer and internal energy changes must still conform to the First Law of Thermodynamics. The thermodynamics analysis then becomes identical to the procedure that was used in Chapters 3 and 4 in all instances.

Another useful example of the use of Dalton's law applied to a very common non-reacting mixture is when water vapor and air are mixed together. The previous example of the oxygen and nitrogen mixture being identified as the air we breathe, should in fact have included some measure of water vapor. The air that we breathe contains humidity that can be either a torment or life-saving, depending on whether the air is too dry or wet with water content.

The next example will help clarify this point. Consider the same amount of air as before, where 21% is oxygen and 79% is nitrogen, respectively. However, the mixture should also have considered the presence of water vapor. Therefore, the actual proportion of oxygen and nitrogen in the air would be slightly less than the 0.2625 for oxygen and 0.7375 for nitrogen previously calculated (see Equation (6.4)). Assume that the ratio of water vapor in the air is 0.006 lbm$_{water}$/lbm$_{air}$. This ratio is given the name "specific humidity" and is not to be confused with the more common parameter "relative humidity".

Proceeding as before, the partial pressures of the water vapor for the three constituents are shown in Equation (6.6a). This allows the enthalpies and entropies for each of the constituents to be determined as shown in the procedure above. The result is shown in Equation (6.6b).

$$h_{mixture} \equiv 126.79 \times 0.231 + 131.33 \times .763 + 1082.79$$

$$\times .006 = 136.0 \, \text{Btu/lbm}_{mixture}, \tag{6.6a}$$

$$s_{mixture} \equiv 1.74387 \times 0.231 + 1.6472 \times .763 + 2.1539$$

$$\times .006 = 1.673 \, \text{Btu/lbm}_{mixture}. \tag{6.6b}$$

The total enthalpy of the mixture is thus 136 Btu/lbm, a difference of 5%. Any thermodynamics process will usually require the entropy and enthalpy to be known and readily handled by Equations (6.6a) and (6.6b).

However, care must be taken that the constituent masses of oxygen, nitrogen and water vapor do not change. It is unlikely that the constant masses of the oxygen and nitrogen will change for any given size of mixture volume. However, if the temperature and pressure relationships for the water vapor are such that the water vapor starts to condense (this would be called rain or fog in the real world), then the amount of water vapor that remains in the mixture must be determined and the partial pressures, etc., must all be recalculated before the thermodynamics process continues.

6.2   Converting mole fractions to mass fractions

It will be useful to consider the method for converting a mixture that is given in moles (or mole fraction) to masses (or mass fraction) or partial molar volume (or volume fraction). First, what are moles, mass fraction or volume fraction?

Any mixture of two or more non-reacting constituents can be identified in terms of volume, mass or mole fractions. That is, if $1 \, \text{ft}^3$ (the total volume) of a mixture is known to consist of 21% of substance A and 79% of substance B, then the partial volume or volume fraction of substances A and B are 21% and 79%, respectively.

The conversion from volume fractions can be very straightforward, as shown in the spreadsheet in Table 6.2. The first two columns on the left of the table list the molecular weight and density of each constituent gas. The volume fraction of each gas is the given input for the calculation. With the volume given for each gas, the mass for each of the volumes can be determined by using the density for each gas in the equation:

$$M_n = \frac{volume\ fraction_n}{density_n}.$$

With the mass fraction of each constituent now known, the number of moles of each gas can be determined using the equation:

$$\# of\ Moles_n = \frac{Mass\ fraction_n}{Mole\ Weight_n}.$$

Table 6.2.   Example of a conversion from volume fractions to mass and mole fractions.

P,total	14.7	psia					
T,total	60	F					
Mol. Wt.	Density	vol. fraction		mass	mass frac.	moles	mole fraction
1	32	0.084	oxygen	0.200	0.016882	0.380972	0.000528	0.200
2	28	0.074	nitrogen	0.300	0.022158	0.500039	0.000791	0.300
3	4	0.011	helium	0.500	0.005273	0.118988	0.001318	0.500
			SUM	1.000	0.04431	1.00000	0.00264	

Lastly, the mole fraction can be determined by dividing the number of moles of each gas constituent by the total sum of the moles.

$$Mole\ fraction_n = \frac{Mole_n}{\sum Mole_n}.$$

It is interesting to note that the mole fraction is equal to the volume fraction of each constituent. This fact is useful when working with chemical formulae because the chemical formula clearly shows how many moles of each constituent is being chemically changed from the reactants side of the equation to the products of the chemical reaction. Thus, a ratio of the moles of each constituent in the products of reaction to the sum of the moles of all of the products is also equal to the volume ratio and therefore also the partial volume of each constituent. The utility of this fact will become evident in the last section of this chapter when the chemistry and thermodynamics of reacting mixtures are reviewed.

6.3   Psychrometric analysis

One of the most important analyses to be performed in thermodynamics is the thermodynamics analysis needed for air-conditioning applications. In order to perform this analysis on the air that is to be heated, cooled, humidified or dehumidified, it is necessary to first determine the thermodynamics properties of humid air, which is a non-reacting mixture of air and superheated water vapor, where the latter is in a thermodynamics state of ambient temperature but at a relatively low partial pressure. The pressure is low enough to observe that the water vapor is in fact superheated at just room temperature. In order to determine the enthalpy, entropy and density of the humid air, it is necessary to utilize the non-reacting mixture analysis that was presented in the first part of this chapter.

To proceed, consider that you, the reader, have a volume of air, V, at a temperature of 70°F and pressure of 1 atm (or 14.7 psia). This volume of air has a very small amount of water vapor, though that must be considered as it affects the size(s) of the heat exchangers, fans, humidifiers and dehumidifiers needed to change the temperature of the air mixture. For humidifiers that need to be designed to heat, cool or moisturize the air, we will start with Dalton's law of partial pressures. That is, consider that the air's partial pressure is represented by Equation 6.7(a).

$$p_{air} \times V = \frac{M_{air} \times R_u \times T}{Mole.Wt_{air}}. \tag{6.7a}$$

Similarly, the partial pressure of water is determined from Equation (6.7b).

$$p_{water} \times V = \frac{M_{water} \times R_u \times T}{Mole.Wt_{water}}. \tag{6.7b}$$

The first step is to divide these two equations and then substitute the equation that relates the partial pressure of air and water with the total pressure of one atmosphere.

$$\frac{p_{air}}{p_{water}} = \frac{M_{air}}{M_{water}} \times \frac{Mole.Wt._{water}}{Mole.Wt._{air}}.$$

These results are shown in Equation (6.8), solved for the ratio of the mass of water to the mass of air in the volume of air.

$$\frac{M_{water}}{M_{air}} = \frac{p_{water}}{p_{air}} \times \frac{Mole.Wt._{water}}{Mole.Wt._{air}}, \tag{6.8}$$

$$\frac{Mole.Wt._{water}}{Mole.Wt._{air}} = \frac{18}{28.966} = 0.622.$$

The result is an equation of water to air mass ratio:

$$\frac{M_{water}}{M_{air}} = 0.622 \times \frac{p_{water}}{p_{air}}.$$

It is also useful to consider the fact that:

$$p_{atm} = p_{water} + p_{air}.$$

But with atmospheric pressure equal to 14.7 psia, the relationship for the partial pressure of air can be used as:

$$p_{air} = 14.7 - p_{water}.$$

Combining equations:

$$\frac{M_{water}}{M_{air}} = 0.622 \times \frac{p_{water}}{14.7 - p_{water}}. \tag{6.9a}$$

The ratio of the mass of water to the mass of air is given by w and is labeled as the specific humidity.

$$w = \frac{M_{water}}{M_{air}}. \tag{6.9b}$$

The specific humidity or mass ratio is the amount of water that can be statically held in suspension by 1 lbm mass of air. It should not be confused with relative humidity, which is a term more often used in weather forecasts.

The relative humidity (φ) is the ratio of the partial pressure of the water vapor in the air volume to the saturated pressure of air at the temperature of the air volume. The following example describes this ratio. Consider a mixture of air and water (or moist air) with a temperature of 80°F and relative humidity of 60%. By the definition of relative humidity, the 60% (or 0.60) is equal to the partial pressure of the water in the air to the saturation pressure of water at the same temperature. This may be represented by the equation:

$$Relative\,Humidity\,(R.H.),\ \varphi = \frac{p_{water}}{p_{sat.at\,T}} \tag{6.9c}$$

Using the property tables for water, the saturated water pressure at 80°F is found to be 0.5075 psia.

Using Equation (6.9), the partial pressure of water vapor at the temperature given is determined as shown:

$$\varphi = \frac{p_{water}}{(p_{sat.\,at\,T} = .5075)} = 0.6, \tag{6.10}$$

and thus: $p_{water} = 0.6 \times 0.5075 = .305$.

With the partial pressure of water now known, the specific humidity ratio can be determined from Equation (6.10) and is shown here:

$$\frac{M_{water}}{M_{air}} = 0.622 \times \frac{p_{water} = 0.305}{14.7 - (p_{water} = 0.305)},$$

$$\frac{M_{water}}{M_{air}} = 0.01316.$$

Given that the partial pressure of water is 0.305 psia, the wet bulb temperature is determined to be 65°F because this is the temperature of the saturated water vapor at 0.305 psia.

With the specific humidity (or the ratio of water to 1 lbm air) in the volume of air, the enthalpy can be determined. Of course, if the relative humidity is not known, this sequence of calculations cannot be made, and some other property would need to be known in order for the complete set of humid air properties to be known.

For the example given, 0.01316 lbs of water can be suspended in 1 lbm of air. The enthalpy for this mixture can be found on a per mass of dry air basis by looking up the enthalpy for each of the constituents in their respective property charts. For example, the enthalpy for the water vapor at a partial pressure of 0.305 and a temperature of 80°F is 33.63 Btu/lbm. The enthalpy for the air can be found as 0.24 Btu/lbm°F. However, the combined enthalpy for the mixture per unit lbm of air is determined by applying the procedure mentioned earlier in the chapter and shown in Equation (6.11).

$$h \left[\frac{Btu}{Lbm}, dry\, air \right] = h_{dry\, air} + \omega \times h_{water}. \tag{6.11}$$

$$h_{dry\, air} \sim f(p_{air}, T_{ambient}); p_{air} = 14.7 - .305 = 14.395; T_{ambient} = 80°F,$$

$$h_{water} \sim f(p_{water}, T_{ambient}); p_{water} = 0.305; T_{ambient} = 80°F.$$

The choice of using the thermodynamics properties on a per pound of dry air basis is arbitrary, but one that has been universally adopted by the heating and ventilation industry. The convention must be used in order to properly size the heat exchangers and other equipment that are used in air conditioning applications to industry standards.

The calculation of the enthalpy of the non-reacting mixture is relatively straightforward and requires the use of the property tables for water. With the ubiquitous presence of laptops and desktop computers this chore is no longer a burden. However, from the HVAC industry, a curve fit has been universally accepted that provides a very accurate $(+/-2\%)$ value for the enthalpy whose quantity is given in Equation (6.12).

$$h \left[\frac{Btu}{Lbm}, dry\, air \right] = 0.24 \times T_{DB} + \omega \times (1061 + 0.444 \times T_{DB}). \tag{6.12}$$

The terms in the equation should now be apparent. For example, the 0.24 is clearly the specific heat for liquid water with units of Btu/lbm/°R (or kJ/kg/K) and the second term is actually a curve fit for the latent heat of evaporation of water, with the superheat expression for water vapor being $0.444 \times T$. The T in the equation is the mixture temperature that is identified as the dry bulb temperature. The name "dry bulb" will be explained in more detail in the following section. For now, it is sufficient to know that it is identical to the temperature that can be measured by either a simple bulb thermometer or a sophisticated thermocouple instrument. With the enthalpy of the mixture of water and air now known, it can be evaluated for any process that the air mixture is subjected to during the conditioning of the air.

The thermodynamics of moist air (or psychrometric) also involves several other temperature definitions, in addition to the parameters that we have discussed so far, including enthalpy, entropy, specific volume, and several more temperatures that are uniquely defined for moist air calculations. In addition to the dry bulb temperature, there are also the wet bulb (T_{wb})

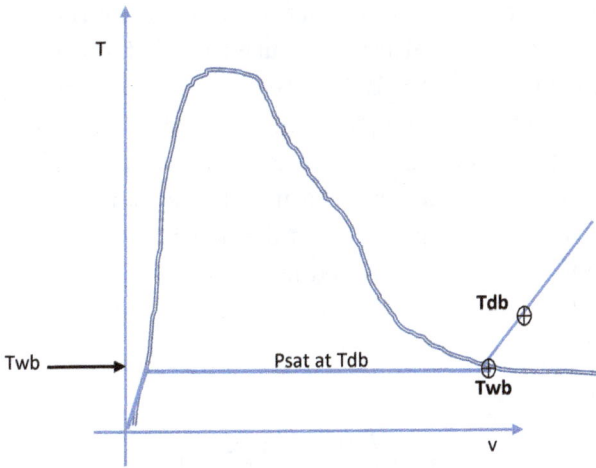

Figure 6.1.   T-v diagram of water (not to scale) illustrating the wet bulb and dry bulb temperatures.

and the dew point temperatures (T_{dp}). "Dry bulb temperature" is simply another name for the temperature measurement that we have been using throughout the textbook.

Figure 6.1 provides a graphical description of the dew point temperature. It also displays the familiar T-v diagram with a constant pressure curve shown. The partial pressure of water is shown with the dry bulb temperature (T_{db}) obviously in the superheat region. This assumes that the relative humidity, as defined by Equation (6.10), is shown to be less than one. Thus, the partial pressure of water is less than the saturation pressure of water at the dry bulb temperature.

Assume that the constant pressure curve has a value equal to the partial pressure of the water determined in the previous example, or 0.305 psia. Starting at the temperature of 80°F (as used in the previous example), "ride down" the curve until the pressure curve intersects the saturated temperature on the dome. The temperature corresponding to this saturation pressure is the dew point temperature, which is the temperature where water starts to condense. In weather reports this condensation would be simply called "rain" or "fog", a somewhat quasi-equilibrium rain phenomenon that depends on the ambient temperature.

$$\varphi = \frac{p_{water}}{p_{sat.\,at\,T}}. \quad \text{from Equation (6.10)}$$

The state point of p_{water} and T_{db} is no different when applied to psychrometric applications, as it was when part of a thermodynamics problem involving engines of closed pistons receiving heat and doing work. In the same way, if the superheated water, as represented by the state point, is to be cooled, then the temperature is reduced until it reaches the saturation temperature shown in Figure 6.1. This temperature is known as the wet bulb temperature. Any further attempt to cool the water vapor results in condensation, which for a layman's interpretation is called rain. Thus, the wet bulb temperature is the minimum temperature that a moist air sample can hold the amount of water per lbm of air as defined by the specific humidity, w. Since the specific humidity is defined as the ratio of water in suspension to a pound of air, the value of w can be calculated from Equation (6.9), which is repeated here:

$$\frac{M_{water}}{M_{air}} = 0.622 \times \frac{p_{water}}{14.7 - p_{water}}. \qquad (6.13)$$

The partial pressure of water is found from knowing the relative humidity and the dry bulb temperature, T_{db}.

These properties can be determined using the equations defined in the above sections. The thermodynamics relationships between these parameters have been solved and graphically displayed in a chart in Figure 6.2. This psychrometric chart will be very useful in helping to graphically display the process paths for the series of processes that moist air undergoes during an air conditioning application. It also helps to define the wet bulb and dew point temperatures.

A description of the wet bulb temperature is best left after more diagramming illustrations of previously defined parameters can be presented and understood. This diagrammatic presentation is easily demonstrated using a psychrometric chart. It may be observed that psychrometric charts are rather "busy" in their use of curves and numerical values; however, the psychometric chart is among the more useful charts in engineering because of its simplicity of use and the avoidance of time-consuming calculations. To simplify the chart, Figure 6.3 details how the new parameters of enthalpy, dew point, dry bulb temperatures, and density can be found.

In Figure 6.3, the directional arrow from the starting point of T_{db} (80°F) and relative humidity (60%) to point A shows how the wet bulb temperature is calculated. In this example, the dew point temperature is found to be 65°F. The directional arrow to point B determines the enthalpy, which is found to be 33.8 Btu/lbm$_{dry\,air}$. The directional arrow to point C determines the specific humidity, w. The specific humidity is found to be 0.013 lbm$_{water}$/lbm$_{dry\,air}$. The wet bulb temperature, T_{wb}, is found by tracing the directional arrow to point D and is found to be 70°F.

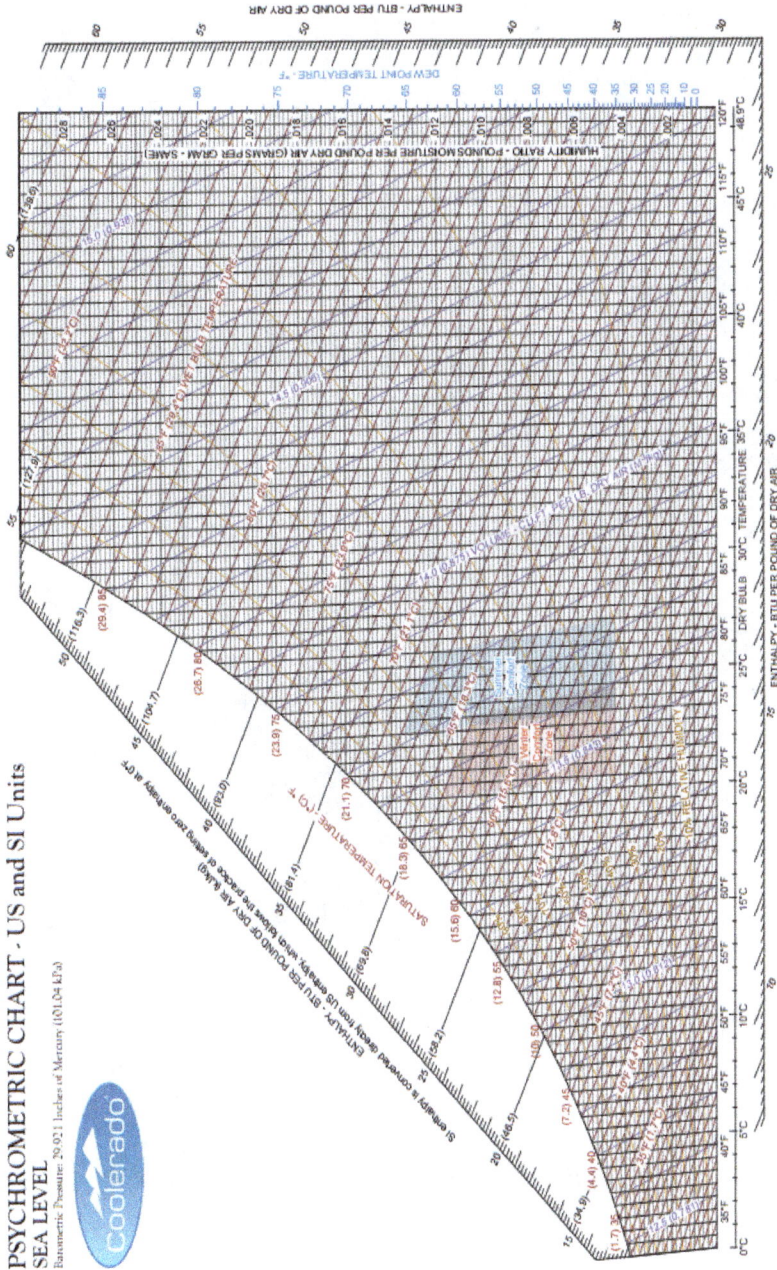

Figure 6.2. The psychrometric chart used in HVAC engineering analysis.

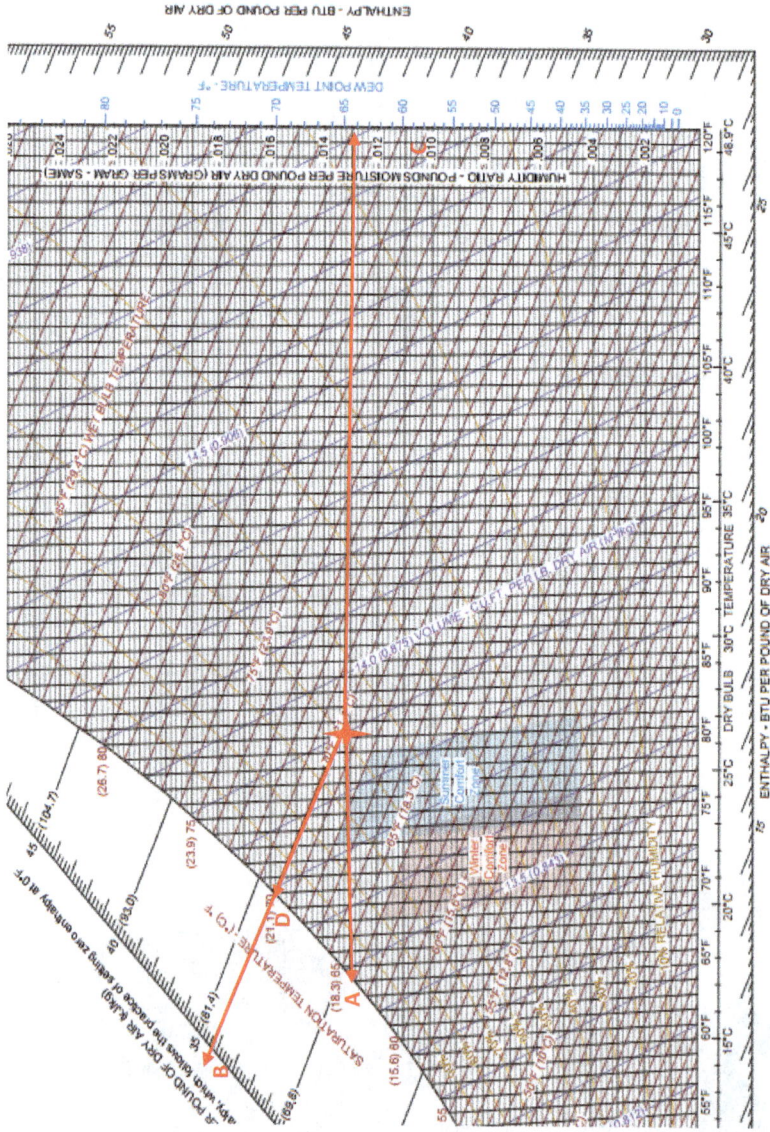

Figure 6.3. The psychrometric chart showing the process paths A, B, C, and D for determining the dew point temperature, enthalpy, specific humidity and wet bulb temperature.

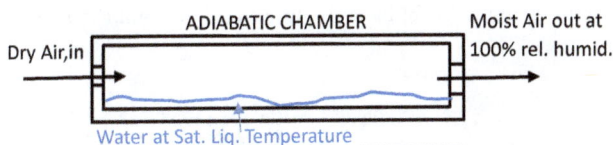

Figure 6.4. Illustration explaining how the wet bulb temperature is determined.

Knowing how to determine the wet bulb temperature is not the same as knowing what the wet bulb temperature is. The wet bulb temperature can be found by using the same sloped line that Figure 6.3 used to determine the enthalpy for the mixture. In fact, it is the enthalpy of the mixture and the condition of the mixture being "wet" that gives the reader a better understanding of what the wet bulb temperature physically represents in the thermodynamics of mixtures.

Figure 6.4 displays a shallow pan that adiabatically isolates water from the environment. A stream of dry air is entering this insulated chamber. The air that exits on the right side of the chamber evaporates the saturated liquid water in the shallow pan by using some of its sensible heat energy. As a consequence, the temperature of the air stream is reduced. This process continues until the once dry air stream is now laden with the water that was evaporated from the pan, until it cannot hold any additional water vapor in suspension. During this process, the air stream has been cooled to a saturation temperature that corresponds to the partial pressure of the steam vapor in the non-reacting mixture of water and air. The lowest temperature that can be achieved in theory is identified as the wet bulb temperature. In fact, the calculation is a little more complicated than that described if the shallow pan of water is at a temperature that is not saturated. But the physical description is nevertheless accurate.

The most accurate mathematical and thermodynamically correct formulation of the wet bulb temperature is derived by applying the First Law of Thermodynamics to the adiabatic chamber.

$$\dot{M}_{in} \times h_{in} = \left(\dot{M}_{in} + \frac{\Delta M}{\Delta t_{water}} \right) \times h_{sat.vapor\ at\ partial\ pres}$$

$$+ \left((-\Delta M) \times u_{water} \right) / \Delta time \right)_{cv}. \tag{6.14}$$

It can be shown that the assumption of the enthalpy of the inlet stream being equal to the enthalpy of the outlet stream at a relative humidity of 100% is accurate enough for most practical engineering analyses. Thus,

Table 6.3.   An example of a state point for moist air using all relevant thermodynamics properties.

Pambient=				14.7					
INPUTS				OUTPUTS		INPUT		OUTPUT	
R.H.	Tdb F	pw psia	w Lbm,v/Lbm,a	h Btu/LBm,dair	Tdew point F	Twb F	w,wet bulb Lbm,v/Lbm,a	h, iterated Btu/LBm,dair	
40%	80	0.20299	0.00890	28.95	53.54	63.6	0.01257	28.95	0.00

Equation (6.15) is useful for most applications.

$$0.24 \times T_{wb} + \omega_{wb} \times (1061 + 0.444 \times T_{wb}), out$$

$$= 0.24 \times T_{db} + \omega_{db} \times (1061 + 0.444 \times T_{db}), in, \qquad (6.15)$$

where T_{wb} and ω_{wb} are evaluated at 100% relative humidity.

The total number of properties that are commonly used in a thermodynamics analysis of moist air is now up to 10, as shown in Table 6.3, including entropy, which is not shown. However, it is still true that only two of these properties need to be known to be able to determine the others. This assumes that a single work mode is still applicable for the problem at hand, as required by the State Principle detailed in Chapter 2.

The applicability of the single work mode with the State Postulate that requires that only two independent thermodynamic properties be known to find all of the others can be easily demonstrated by studying the psychrometric chart. Pick any two parameters of the thermodynamics system. It can be observed from the chart that the intersection of these two parameters clearly defines a single point. Thus, any vertical, horizontal or inclined projection (to determine the enthalpy property value) from the intersection of two given properties on the psychrometric chart can provide the user with the corresponding value of the wet bulb and dry bulb temperatures, dew point enthalpy, or relative or specific humidity. In fact, the chart is an excellent visual guide and is faster than a computer when it comes to providing results. So much so, many HVAC industries make use of the psychrometric chart as part of their calculation blank page, so that their engineers can quickly determine the unknown properties from the two properties given in the problem.

For example, referring to Figure 6.2 and Table 6.3, given an enthalpy of 29 Btu/lbm,$_{dry air}$ and a specific humidity of 0.0126 lbm$_{water}$/lbm$_{air}$, determine all the other properties.

Now check this with the output shown in Table 6.3, which is from a spreadsheet program that the author prepared. However, because this program uses the most common pair of thermodynamics properties, T_{db} and

Relative Humidity (R.H.), as input, it will be necessary in this example to iterate the result until the output from this program matches the inputs of enthalpy and specific humidity. The reader is welcome to create a computer program similar to that which produced the output shown in Table 6.3. However, the reality is that in today's market there are many sources for psychrometric applications (apps) that can be installed across a suite of electronic devices, such as smartphones and computer tablets, that will make all these air properties easy to determine, even when given only two of the properties.

Case Study 7 provides a more detailed example of how the psychrometric chart can be applied in a real-world HVAC problem.

6.4 Combustion analysis: A common example of a reaction mixture in thermodynamics

The previous section presented the thermodynamics of non-reacting mixtures. Of equal importance to the thermodynamics analysis that involves engine analysis is a familiarity with the basics of combustion. Combustion involves the chemical oxidation of a carbon–hydrogen-based compound. The chemical equation for the oxidation of the carbon–hydrogen molecule is shown in Equation (6.16). It is clear from the equation that the carbon, hydrogen and oxygen molecules are balanced with respect to their masses (or their moles, from the left to the right of the equation). The net result of this oxidation is the exothermic release of energy at extremely high temperatures. The final temperature after combustion is a function of not only the chemical nature of the carbon–hydrogen fuel, but also the ratio of air (or oxygen) to fuel that is being consumed.

Carbon–hydrogen fuels with the formulation $C_n H_m$, where "n" and "m" refer to the number of carbon and hydrogen atoms combined to produce the molecule, are typically represented by Equation (6.16)

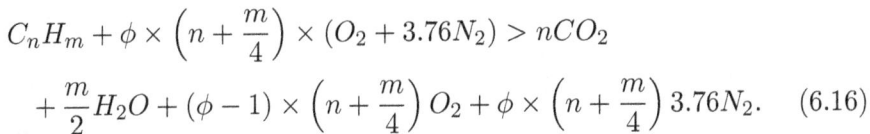

$$C_n H_m + \phi \times \left(n + \frac{m}{4}\right) \times (O_2 + 3.76 N_2) > n CO_2$$

$$+ \frac{m}{2} H_2O + (\phi - 1) \times \left(n + \frac{m}{4}\right) O_2 + \phi \times \left(n + \frac{m}{4}\right) 3.76 N_2. \quad (6.16)$$

This equation includes excess oxygen that might have been admitted to the combustion process. It is also evident from the equation that nitrogen is used, and it appears on the right and left sides of the equation without having either added to or absorbed energy. The amount of excess oxygen (or air) that is used in the combustion is represented by the parameter, ϕ. The nitrogen is in the chemical equation only because it is already present in the air. Clearly, this is true because air is literally freely available in the atmosphere. Unfortunately, it is also perhaps true that the products of combustion are literally free to be "dumped" into the same environment, having been transformed into carbon dioxide, water and oxygen (if there was an excess of air at the beginning of the process). The carbon dioxide,[1] nitrogen, and certainly the oxygen and water vapor byproducts would not have been so bad if not for the incomplete combustion processes that occur when the pressure, temperature and speed of combustion are not quite correct. The inefficient combustion results in the production of carbon monoxide (CO) and NO_x, where the "ϕ" can be 1, 2 or 3, etc., indicating a variety of unwanted and dangerous constituents of nitrogen oxide that are released into the environment during the process of combustion. These products are monitored very carefully to keep their emissions at an "acceptable" level, at least for now.

The air–fuel ratio for the combustion of the fuel is shown in Equation (6.17).

$$Air\ Fuel\ Ratio\,(AFR) = \left[\phi \times \left(n + \frac{m}{4}\right)\right] \times 4.76$$

$$\times\ 28.966/(12 \times n + 4 \times m). \qquad (6.17)$$

That is, the carbon–hydrogen molecule can contain n moles of carbon and m moles of hydrogen that have been formed into a stable or quasi-stable molecule. The molecule is ready to release some of the energy that went into the production of the molecule. The energy is released by providing a minimum amount of ignition energy that can be provided by any number of sources depending on the stability of the chemical bonds. Such sources can be as simple as operating at a temperature above a fuel's auto-ignition temperature (which is the way diesel fuel is ignited), an electric arc, a sharp impact, or sometimes even a slight vibration.

Equation (6.17) can be easily programmed to provide a quick calculation of any carbon–hydrogen compound. Table 6.4 demonstrates the result using one of the most common fuels: methane.

Tables 6.5(a) and 6.5(b) provide a list of commonly available carbon–hydrogen molecules and the amount of heat that is released per lbm from the chemical bonds. The amount of heat that is released, also known as

---

[1]Carbon dioxide is recognized as a contributor to environmental temperature changes due to its ability to absorb solar energy rather than transmit that solar energy to the ground. However, carbon dioxide is also necessary for plant growth, and as such, with reasonable controls to its emission, it provides food for the Earth's growing population.

Table 6.4. Example of a calculation of the air–fuel mass ratio and exhaust flow rate and temperature.

	CnHm:					
ϕ	200%	34.5	=Air-Fuel Mass Ratio			
n	1	21,000	High Heating Value [Btu/lbm]			
m	4	2110	F ;	C 1155	=Adiabatic Flame [F]	
Fuel Available Energy=	2,480	kWe; Qfuel		8 MMBtu/hr	2,485	=kWthermal
Air partial pres.=	4.98%					
Plant Eff=	0.97					
Exh. Mass Flow Rate	1.435E+04	Lbm/h;Lb/s	3.99	1.81		
Fuel Flow Rate	416	Lbm/h;scfm=	163.47	188.9		
		acfm=	460.20			

Table 6.5. (a) Carbon–hydrogen-based fuels and High Heating Values (L_{HHV}), and (b) heat rates and densities of biofuels.

(a)

Fuel name	L_{HHV} [Btu/lbm]	L_{LHV} [Btu/lbm]	Adiabatic flame temp. [R]
Methane	23,880	21,520	5,550
Ethane	22,320	20,430	4,020
Propylene	21,050	19,700	5,658
Propane	21,660	19,950	5,568
Butane	21,300	19,670	4,038
Pentane	21,090	19,510	—
Methanol	9,760	8,570	—
Ethanol	13,610	11,930	—
Hydrogen	61,000	51,600	5,280

(b)

BIO FUEL HEAT CONTENTS

ref. Godfrey Boyle; Renewable Energy GJ/metric ton	Btu/lbm	Density (Kg/m^3)	lBm/ft^3
Green Wood	6	2585	1166.7	72.7
Air Dried Wood	15	6464	600.0	37.4
Oven Dried Wood	18	7756	500.0	31.2
Charcoal	30	12927	300.0	18.7
Paper	17	7325	529.4	33.0
Dung (dried)	16	6895	250.0	15.6
Grass (fresh cut)	4	1724	750.0	46.8
Straw (fresh cut)	15	6464	100.0	6.2
Sugar Cane residue	17	7325	588.2	36.7
Domestic refuse	9	3878	166.7	10.4
Commercial Waste	16	6895		
Oil	42	18098	809.5	50.5
Coal	28	12065	1785.7	111.3
Natural Gas	55	23700	0.7273	0.045

the Latent Heat or High Heating Value (HHV), is usually one of the most important parameters used in a thermodynamics analysis that involves heat engines. As shown in Chapters 4 and 5, a typical output of the heat engine analysis is the determination of the amount of heat that is needed to produce a desired amount of heat engine power. With this amount of heat known, usually in the units of measure of Btu/h or kW, the amount of fuel (lbm/h or kg/s) needed to fuel the heat engine can be determined using the simple equation:

$$\dot{M}_{fuel} = \frac{\dot{Q}_{heat\,input}}{L_{high\,heating\,vlue}}$$

It is important to note the differences in the heat rates as defined by L_{HHV} and L_{LHV}. The HHV of a fuel is determined from a bomb calorimeter experiment, wherein a small amount of the fuel is burned to 100% completion. The amount of heat released is determined by measuring the change in the water bath in which the bomb calorimeter has been placed. However, the temperature of the products of combustion is typically cooled until the water bath temperature reaches equilibrium, and this is typically below the temperature that the water content of the combustion products is in a liquid state. Thus, the heat of condensation has been absorbed by the water. The amount of heat released therefore includes this heat of condensation. This quantity is defined as the HHV of the fuel.

A similar experiment could have been done with a fluid that was pressurized water and thus remains liquid at higher temperatures. For example, liquid water pressurized to 135 psia will not boil until the water reaches 350°F. In this experiment, the water products of combustion will remain a vapor as the pressurized water bath is heated by the heat released. At this elevated temperature and at the partial pressure of the water vapor in the combustion mixture, the dew point temperature of the mixture is below the final equilibrium temperature of the combustion products after it has released its heat energy into the pressurized water bath. Thus, the water vapor does not reach its condensation temperature and therefore does not release its heat of condensation. The amount of heat energy measured in this experiment is less than the HHV by the amount of condensing heat that could have been released.

For most engine and heat transfer equipment in the United States, the HHV is used to define the efficiency of the product. However, European engineering and business industries use the Low Heating Value (LHV) to define the efficiency of the component, which leads to the efficiency appearing

5 to 10% higher. It is certainly a fair engineering question to also ask whether the efficiency of a device is based on its HHV or LHV.

The science of chemistry can be used to determine how the carbon–hydrogen single molecule is broken up into several other constituent molecules, with the net results being a release in thermal energy. The release of thermal energy from the chemical oxidation process is used as the thermal energy input to heat engines. Fortunately, the oxidation process is virtually instantaneous. The key word here is "virtually" or "almost" instantaneous, in that the speed of the process is sufficiently fast to enable the use of positive displacement, reciprocating piston machines that can operate at very high speeds of 8,000 to 10,000 rpm as opposed to continuous combustion processes used in gas turbine engines and fueled combustors. The only limitations of the reciprocating piston engine are the possible excessive stresses that may be involved with the piston mass being accelerated and de-accelerated at each revolution.

The chemical balancing of the equation when the single carbon–hydrogen molecule is decomposed into several common chemical molecules is straight-forward and only depends on accounting for the mass of any one constituent, in that it has to be the same before and after the oxidation process.

For example, the equation for CH_4 (or methane) can be shown to be decomposed into three molecules — CO_2, H_2o and O_2 — IF there is an excess of oxygen in the original mixture in the attempt to oxidize the methane. That is, there is strictly a specific amount of oxygen that is required to only oxidize all the methane. If any more (or less) oxygen molecules are present than this specific amount, the difference will either appear in the products of the oxidation or will cause less of the fuel (known as unburned hydrocarbons) to appear in the products of combustion.

6.5 Combustion thermodynamics — LITE

The chemical thermodynamics of combustion presented in the previous section can be greatly simplified. This simplification has two basic advantages: (1) it enables a much clearer understanding of the basic thermodynamics of the combustion process based entirely on the First Law of Thermodynamics, and (2) the simpler equations enable their immediate use in programming models of the thermodynamics in a combustion process, particularly as it may be programmed in spreadsheets.

Starting with the major result of the more detailed chemical-thermodynamics described in the previous section, Table 6.5 provides a listing of the chemicals that can provide an exothermic reaction (a release of heat energy) when allowed to oxidize. Of most importance is the amount of energy released per unit mass (or per unit volume). For example, natural gas will release approximately $1,010\,\mathrm{Btu/ft}^3$ or $23,000\,\mathrm{Btu/lbm}$ when chemically oxidized. The initiation of the oxidation process may need a small or large amount of energy that can be provided by the engine or thermodynamics process that uses the released energy. Such thermodynamics processes include heat engines as well as the rather straightforward process of water boiling to a saturated vapor state. These processes thus provide saturated steam for industrial heating processes. Applications of industrial heating can run the gamut of simply heating a building to providing pressurized steam for operating machinery. Certainly, one of the most basic Rankine Cycle heat engine systems utilizes pressurized water from a boiler to power a steam turbine, and thus ultimately generate mechanical and then electrical power.

By applying a control volume (CV) around the combustion process that has fuel, the oxidizer entering the CV and the heated products of combustion exiting the CV, the First Law of Thermodynamics can be written for the fuel oxidation with the equation:

$$\dot{Q}_{heat\,released} = \dot{M}_{fuel} \times L_{HHV}, \tag{6.18a}$$

wherein the singular characteristic of a particular fuel, HHV, has a very prominent part in the equation.

The heat released in turn can be equated to heating the fuel and air input into the combustion process, resulting in the products of combustion being heated to a combustion temperature, $T_{flame\,temperature}$, relative to the initial ambient temperature of the fuel and the oxidizers using the equation:

$$\dot{Q} = \dot{M}_{products} \times C_{p,average} \times \Delta(T_{adiabatic\,flame\,temp.} - T_{ambient}). \tag{6.18b}$$

It must be noted that the specific heat, $C_{p,average}$, is assumed to be a constant in this equation and is determined to be the average C_p across the temperature range from the ambient to the adiabatic flame temperature. This assumption allows for the T_{flame} temperature to be easily calculated and accurate to within a few $^\circ$R or K. The more precise calculation would need to consider the nature of the gases that make-up the combustion products, the partial volumes of those gases, and the maximum combustion temperature that is achieved as a result of the release of the chemical energy. This later temperature is called the Adiabatic Flame Temperature. However, when the adiabatic flame temperature is the parameter that is unknown, then it can only be determined using an iterative calculation. For eample, an adiabatic temperature is assumed to be able to determine the enthalpies of combustion for each of the constituent gases from the enthalpy tables of each gas.

The First Law is then applied using the high heating value (Lhhv) for the fuel to determine the total enethapy change of the products of combustion gases. This enthalpy must match the total enthalpy of the constituent gases that was determined based on the assumed adiabatic flame temperature. If it is not, then a slightly different adiabatic flame temperature is assumed and the iteration continues. Alternatively, equating Equations (6.18a) and (6.18b) and dividing by the mass flow rate of fuel results in Equation (6.19).

$$\dot{M}_{fuel} \times L_{hhv} = \dot{M}_{products} \times C_{p,average} \times \Delta(T_{adiabatic} - T_{ambient}),$$

$$\dot{M}_{total\,products} = \dot{M}_{air} + \dot{M}_{fuel},$$

$$\frac{\dot{M}_{total\,products}}{\dot{M}_{fuel}} = \frac{\dot{M}_{air}}{\dot{M}_{fuel}} + 1.$$

Combining equations and solving for $T_{flame\,temperature}$:

$$L_{hhv} = (1 + AFR) \times C_{p,average} \times \Delta(T_{adiabatic} - T_{ambient}),$$

$$T_{flame\,temperature} = \frac{L_{hhv}}{1 + AFR} \times C_{p,average} + T_{ambient}. \qquad (6.19)$$

The ratio of the mass of oxidizer to the mass of fuel can be expressed as the oxidizer to fuel ratio (OFR). However, because air, a mixture of 21% oxygen and 79% nitrogen, is the most common source of the oxygen that is needed for combustion in internal combustion engines as well as gas turbines, the more common term to use in place of the OFR is the air–fuel ratio (AFR).

From a previous chemical-thermodynamics discussion of the combustion process, the reader will recall that the AFR is determined from the balanced chemical Equation (6.17) by dividing the number of moles of air consumed with respect to one mole of the fuel. Care is taken to be sure that the correct number of moles of air is used in the equation. For example, from Equation (6.18), the AFR is given as:

$$Air\,Fuel\,Ratio\,(AFR) = \left[\phi \times \left(n + \frac{m}{4}\right)\right] \times 4.76$$

$$\times 28.966/(12 \times n + 4 \times m). \qquad (6.20)$$

This equation is sometimes misunderstood because it seems that the AFR is much higher than what the OFR would require. It must be remembered that in air, the oxidizer is oxygen, and oxygen only makes up 21% of the air, and that the balance 79% of air comes along with the air charge is nitrogen, and that the nitrogen does not react chemically with the fuel, though it does take up space in the cylinder, combustor or whatever device the fuel is held in during the oxidation process.

To further understand the high value of the AFR, the reader is asked to try this "thought experiment". Starting with Equation (6.16) and letting $\phi = 1$, n = 1 and m = 4 for methane, the resulting simplified equation for the combustion of methane is:

$$1 \times C_n H_m + 1 \times \left(1 + \frac{4}{4} + 0\right) \times (O_2 + 3.76N_2)$$

$$> CO_2 + \frac{4A}{2}H_2O + xO_2 + 1 \times \left(1 + \frac{4}{4} + x\right) 3.76N_2,$$

$$C_1 H_4 + (2) \times (O_2 + 3.76N_2) = CO_2 + 2H_2O + (2)3.76N_2.$$

Therefore, it takes only 2 moles of oxygen (or air), as shown above, to completely burn (or chemically oxidize) 1 mole of methane.

Next, think about how to get 1 mole of oxygen from the air that comes "free", except that the "free" oxygen carries with it, very large amounts of nitrogen. Specifically, for every mole of oxygen needed (and you need 2 of them), 3.79 moles of nitrogen must be "taken along for the ride". It may also help the reader to better understand this relationship between the elements in the air mixture by replacing the words "moles of" with "parts of". Thus, for every 1 part of oxygen, the reader must contend with almost 4 parts of another substance, which when taken together has a molecular weight of 28.966. The total amount of moles of oxygen and nitrogen (i.e., air) that is needed is the sum of 1 mole oxygen plus 3.79 moles of nitrogen — 4.79 moles in total. The ratio of moles of air to moles of fuel is therefore 4.79:1. However, usually it is customary to change this mole ratio to "lbm Air to lbm fuel". The result is observed to be the same as the more general formula shown in Equation (6.17).

Alternatively, start with the known ratio of 0.21 parts of oxygen, plus 0.79 parts of nitrogen, BUT understand that these are partial volumes and not mass fractions. That is, $1\,\text{ft}^3$ of air contains $0.21\,\text{ft}^3$ of oxygen and $0.79\,\text{ft}^3$ of nitrogen. By using the Perfect Gas Law, $0.21\,\text{ft}^3$ of oxygen has a mass of 0.0177 lbm and $0.79\,\text{ft}^3$ of nitrogen has a mass of 0.0583 lbm. Add these two values up and the reader will get 0.076 lbm for $1\,\text{ft}^3$ of air, which everyone should recognize as being the density of air. However, the amount of air needed for burning 4 lbm of oxygen with 1 lbm of methane is:

lbm air per lbm of methane = 4 lbm oxygen/lbm methane × 0.076 lbm air/ 0.0177 lbm, oxygen in that air equals 17.2 lbm air/lbm methane.

With the AFR value known from Equation (6.20), solve for the combustion temperature, $T_{combustion}$, as shown in Equation (6.19). From Equation (6.19), if the AFR is large, then the $T_{combustion}$ temperature is reduced compared to when the AFR is at its stoichiometric value. This is reasonable, in that any excess air above the required amount of oxidizer to consume the fuel and release the chemical heat energy will cause the products of combustion to reduce in temperature. When the AFR is equal to the stoichiometric AFR, the $T_{combustion}$ temperature is called the adiabatic flame temperature. The adiabatic flame temperature is the highest temperature that can be achieved for a fuel with a known chemical latent heat of formation. By adjusting the AFR, it then becomes possible to protect the combustor and all the other devices downstream from the combustion process from being heated above the temperature limits of the materials used to construct these devices.

With the AFR known and the T_{flame} calculated from Equation (6.19), the T_{flame} from the combustion process can be determined and then used in a thermodynamic analysis that requires a heat input process. However, it is first necessary to determine the fuel rate from which the mass flow rate of the products of combustion can be determined. The fuel rate is easily known, since it is usually determined by either the customer or the application as shown in Equation (6.18a).

An example of this calculation is the best way to demonstrate the utility of the equations developed so far in this section. In the following case study, it is recommended that the reader programs these relationships into a spreadsheet model, as described at the beginning of this section. For that purpose, the reader's attention is focused on the spreadsheet output shown in Figure 6.5.

The very concise program shown in Figure 6.5 uses Equations (6.18) and (6.19). The chemical equation of the fuel is expressed by the values of n and m as shown in Cells Z128 and Z129. Equation (6.18) is used to determine the mole fraction of the products of combustion shown in Column Z, starting with Cell Z140. This mole fraction can be used to determine the mass fraction as given in Equation (6.21). This is displayed in Column AB, starting with Cell AB140.

$$Mass\,fraction_n = \frac{Mass_n}{\sum Mass_n}. \tag{6.21}$$

The relative air ratio parameter, ϕ, is typed into the spreadsheet using Cell Z127 and given the value of 200% in this example.

The energy released into the power plant, combustion furnace or engine, depending on the efficiency type, is typed into Cell Z132. As shown in Figure 6.5, a value of 97% is entered, indicating that the fuel burning system is likely to be a combustion burner for a furnace or boiler. This is based on the realization that 97% of the fuel consumed will be made available in the form of heat energy at a temperature of 2,110°F (Cell A129) based on a

	X	Y	Z	AA	AB	AC	AD	AE
125								
126			CmHn:					
127		ϕ	200%		34.5 =Air-Fuel Mass Ratio			
128		m	1	21,000	High Heating Value [Btu/lbm]			
129		n	4	2110 F	;	C 1155	Adiabatic Flm.	
130	Power Plant Rating=		2,480	kWe; Qfuel		8 MMBtu/hr;	2,485	
131	Air partial pres.=		4.98%					
132	Plant Eff=		0.97					
133	Exh. Mass Flow Rate		1.435E+04	Lbm/h;Lb/s	3.99	1.81		
134	Fuel Flow Rate		416	Lbm/h;scfm	163.47	188.9		
135				acfm=	460.20			
136		acfm=	15,504		141.36			
137		scfm=	3,133					
138								
139			Mole Frac.=		Mass Fraction - OR - Mass Fraction			
140	CO2 partial pres.=		4.98%	2.19	0.078	0.001353	0.053	
141	O2 partial pres.=		9.96%	3.19	0.113	0.00299	0.116	
142	H2O partial pres.=		9.96%	1.79	0.064	0.001685	0.065	
143	N2 partial pres.=		75.10%	21.03	0.746	0.019735	0.766	
144		Σ	100.00%	28.2			0.026	

Figure 6.5.   Excel example of using the combustion and air–fuel ratio (AFR) equations in Equations (6.18) and (6.19).

fuel heat content of 21,000 Btu/lbm (Cell AA128). The exhaust flow rate is then based on Equation (6.22), which was determined from the AFR and the rating of the heat source, in this case, a combustion system for a furnace or boiler with an efficiency of 97%, as given in Cell Z130 (or $2,480\,\text{kW}_{\text{thermal}}$).

$$\dot{M}_{exhaust} = AFR \times \left(\frac{\dot{Q}_{inout}}{L_{HHV}} \right). \tag{6.22}$$

6.5.1   Example 6.1: Launch vehicle fuel pump and power combustion systems

Aerospace launch vehicles present an interesting combustion requirement that is rarely, if ever, seen on terrestrial applications. Auxiliary power is produced with a combustion turbine system that uses more fuel than is necessary for stoichiometric combustion. Thus, the AFR is smaller than the typical AFR in the stoichiometric combustion of fuel. This occurs in launch vehicles where there is a need for a power combustion turbine to provide power to a fuel pump. In this application, the combustion takes place with liquid oxygen (LOX) as the oxidant, and liquid hydrogen as the fuel. Aerospace launch vehicle engines require that these fuels be pumped in their cryogenic liquid form from the fuel tanks into the combustion system and then to the engine nozzles. A combustion turbine is used to provide the pumping power for the oxidant and fuel pumps. More current prototype designs for launch vehicle propulsion systems are considering using methane as the fuel, in place of hydrogen. However, these combustion processes use much more methane than is required for the methane to be stoichiometrically oxidized with the oxidant. Thus, to increase turbine power, the "unburned" methane fuel is used as a working fluid mixed with the combustion products resulting from the chemical reaction produced between the oxidant and the methane fuel. The mixture's temperature and pressure are then used to determine the enthalpy-pressure relationships for the mixture. The expansion process then proceeds as it would for any expanded fluid expanded in a turbine. Unfortunately, the high pressure and high temperature of the mixture comprising methane and the combustion products, are not accurately modeled using the Perfect Gas Law. The analysis for the properties of the fluid's pressure-temperature-enthalpy must be checked using experimentation before it can be confidently applied to the design of a critical part of the launch vehicle.

However, this combustion process lends itself to an interesting exercise in using the information and methodologies presented above for determining the mixture's specific heat based on the mole and mass fractions derived from the basic combustion chemical equation. As an example of the combustion analysis for an excess fuel-combustion, expansion process, consider the following analysis.

The equation for this special combustion process that uses only stored oxygen and methane is shown in Equation (6.23). It is noted that the methane and oxygen are stored on the launch vehicle in their liquid state and are thus at temperatures of -200 to $-300°F$. The liquid fuel and oxidant are vaporized by the heat from the combustion engine nozzle; a process that cools the nozzle walls in order to survive the very hot combustion temperatures developed in the engines of the launch vehicle.

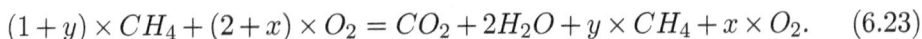

$$(1+y) \times CH_4 + (2+x) \times O_2 = CO_2 + 2H_2O + y \times CH_4 + x \times O_2. \quad (6.23)$$

In Equation (6.23), the variable "y" represents the extra amount of methane fuel that is mixed with the products of combustion in order to increase the fluid mass flow rate into the expansion turbine. This extra fuel is not burned with the limited oxidant and therefore does not provide heat energy to the system. In fact, the unburned methane actually serves to cool the products of combustion as well as increase the overall flow rate of the products of combustion into the expansion turbine. The variable "x" represents the amount of oxygen needed to oxidize (i.e., burn or combust) the methane fuel. Normally, the value of this variable would be greater than 1 in order to cool the products of combustion to temperatures that can be managed by the material used to construct the combustor or by the systems installed downstream of the combustor. However, the value of "x" for this application is zero because the combustion should be at stoichiometric conditions. That is, there should only be just enough oxidant (O_2) to be consumed by a minimum of CH_4 fuel, and this will enable the necessary combustion mixture temperature and flow rate into the expansion turbine to produce the desired power output from the turbine.

The power output from the turbine is calculated using the perfect gas relationships between the four constituent gases that form the products of combustion. The products of combustion are: CH_4, O_2, CO_2 and H_2O, and their mole and mass mixtures need to be determined based on the chemical formula shown in Equation (6.23) and the values chosen for parameters "x" and "y".

Figure 6.6 illustrates the output page from a calculation for the methane fluid-combustion turbine modeled for this worked example. It is noted that the bold "boxed" cells are the inputs to the model. For the case shown in

		Cp	Cv				
	H2O	0.560	0.447				
	CO2	0.299	0.253		(k-1)/k	0.132	
	CH4	1.039	0.910		Pr^(k-1)/k	0.644	
	O2	0.000	0.000		Turbine Eff.=	0.62	
	Cp & Cv=	0.856	0.743	Btu/Lbm/R	Texh.calc.d=	1477	R
	k=	1.152				27.77778	
	L,hhv, CH4=	20,000	Btu/Lbm		Pi	1250	psia
	AFR=	4.000			Po	45	psia
	Texh.calc.d	1435	F	1435	Mass Flow Rate=	29.75	Lbm/s
		1895	R		CH4 Mass Flow Rate=	22.75	Lbm/s
	Error=	0.0%			O2 Mass Flow Rate=	7.00	Lbm/s
					Tin,used	1895	R
		1435			Tout,used=	1477	R
	(1+y)CH4 + (2+x)O2 = 2H2O + CO2 + xO2 +yCH4				Power [kw]=	11219	
					Power [hp]=	15045	
	O2,x=	0					
	CH4,y=	12					
	Products of Combustion						
	Mole Fraction		Mass Fraction				
H2O	2	36	13.2%				
CO2	1	44	16.2%				
CH4	12	192	70.6%				
O2	0	0	0.0%				
		272	100.0%				

Figure 6.6.   Excel output for the methane combustion turbine example.

Figure 6.6, the value of "x" is zero, but the value of "y" is given as 12 to produce the desired 15,000 hp (11,219 kW) of power.

The oxidant (O_2) is given as 7.0 lbm/s. The fuel (CH_4) flow rate is calculated to be 22.75 lbm/s based on Equation (6.23) and the mass ratio of oxidant and fuel, as shown in the following equation:

$$\frac{\dot{M}_{CH4}}{\dot{M}_{O2}} = OxyFuelRatio\,(OFR) = \frac{(1+y) \times (12+4)}{(2+x) \times (16+16)}.$$

In order to determine the temperature of the products of combustion, it is necessary to determine the specific heat, C_p, of the mixture at the combustion's temperature and pressure. It is also necessary to determine the specific heat, C_v, for the products of combustion in order to determine the specific heat ratio, k, which will be used in the adiabatic, turbine power Equation (6.24):

$$Power = \dot{M}_{combustion\,total} \times C_p \times T_{in} \times \left(1 - \left(\frac{P_{out}}{P_{in}}\right)^{\frac{(k-1)}{k}}\right) \times \eta_t. \quad (6.24)$$

It is noted that the equation of the adiabatic turbine equation includes a turbine efficiency, η_t which is used to reduce the ideal, 100% efficient expansion to a more reasonable estimate of the turbine power that will actually be produced.

The moles for each of the constituents in the products of combustion given in Equation (6.23) are as follows:

H2O	2
CO2	1
CH4	12
O2	0

Thus, the mole fraction is determined by taking the number of moles of each constituent and dividing it by the total number of moles in the products of combustion (which, in this case, is 15).

		Mole Fraction
H2O	2	13%
CO2	1	7%
CH4	12	80%
O2	0	0%
	15	1

Table 6.7.   Conversion from mole fraction to mass fraction.

	No. of moles		Mass fraction
H2O	2	36	13.2%
CO2	1	44	16.2%
CH4	12	192	70.6%
O2	0	0	0.0%
		272	100.0%

The mass fraction for this mixture is determined by using the number of moles of each constituent and multiplying the mole quantity by the molecular weight of that constituent, before dividing each constituent mass by the total mass, as shown in Table 6.7. It is noted that the "total mass" is not really the combined mass, but rather a means of determining the mass fraction of each constituent, IF that constituent has the number of moles shown in the process. An important fact used in these calculations is the relative amounts of constituent that were present and known by the chemical equation. The chemical equation represents a perfect reaction and does not account for CO or CO_x, or any other transient chemical formulation that could occur from the products of combustion.

The values of the mass fraction can then be used with the C_p and C_v properties of each constituent at the temperature and pressures of the mixture. But herein lies an interesting dilemma. In order to determine the mixture temperature, it is necessary to know the specific heat, C_p, of each constituent from which the mixture temperature can be found. But the temperature of the mixture is unknown. Therefore, how can the specific heats, C_p and C_v, be determined? The solution to this dilemma is to make an initial guess of the mixture temperature for the purpose of calculating the specific heat, C_p, and then calculating the mixture temperature by using Equation (6.25a). That initial guess will be an input to the combustion model and is illustrated in the bold bordered, yellow cell in Figure 6.6.

$$h_{mixture} = h_{H_2O} \times 0.132 + h_{CO_2} \times .162 + h_{CH_4} \times .706$$

or:

$$Cp_{mixture} = Cp_{H_2O} \times 0.132 + Cp_{CO_2} \times .162 + Cp_{CH_4} \times .706$$

and:

$$Cv_{mixture} = Cv_{H_2O} \times 0.132 + Cv_{CO_2} \times .162 + Cv_{CH_4} \times .706.$$

And then with C_p and C_v: $k = \frac{C_p}{C_v}$.

The analysis continues with the calculation of the flame temperature of the mixture using the flow rate of the excess methane and oxidant, as shown in Equation (16.19).

$$Latent_{hhv} \times \dot{M}_{CH4\,fuel} = (\dot{M}_{ch4} + \dot{M}_{O2}) \times C_p \times (T_{flame} - T_\infty), \quad (6.25a)$$

where $\dot{M}_{CH4\,fuel} = \dot{M}_{O2}/OxyFuelRatio$.

Recall that only the stoichiometric amount of fuel is to be combusted with the oxidant.

Solving for T_{flame}

$$\frac{Latent_{hhv} \times \dot{M}_{CH4\,fuel}}{(\dot{M}_{ch4} + \dot{M}_{O2}) \times C_p} + T_\infty = T_{flame}. \quad (6.25b)$$

The value for the calculated flame temperature is now used as the next iteration for the temperature in the determination of the specific heats, C_p and C_v.

6.6 Some carbon dioxide facts: A global warming gas

At the time of writing of this textbook, carbon dioxide gas (CO_2) has become a cause of concern to environmentalists for being one of the largest contributing gases to global warming. Carbon dioxide causes ultraviolet radiation from the Sun to be captured by the layer of air that surrounds Earth. Such effects can be demonstrated in controlled laboratory experiments. Earth can be analyzed as a Closed Mass System. Clearly, there is no means of "$\cdots$ raising a weight in a gravity field" with respect to the CV that is placed around the earth as a system of interest for thermodynamics analysis. Without the generation of work or power being performed, ultraviolet energy enters the air and is stored by the air mass. This results in a rise in the temperature of this air mass, unless it can be radiated away into space or convected and/or conducted into land and water masses on Earth's surface. Storms and other forms of small or large air movements help distribute this energy throughout the planet. It is also true that humans exhale CO_2, as do all other living creatures on Earth. Trees and other forms of vegetation, as well as the oceans, absorb CO_2 and thus help balance the production of CO_2. However, the chemical dynamics is extraordinarily complex and accurate models depend on algorithms that are based on the laws of thermodynamics as well as Newton's laws of motion, which also require coefficients (that are continually updated) to be used by matching prior reports of CO_2 content in the air, Earth's temperatures, ice and snow land mass sizes, and the chemical analysis of the ocean and other water masses as a function of time. It is also necessary to factor in the occurrences of naturally occurring disasters, such as forest fires, earthquakes and volcanic eruptions, to correctly model the

(a) ▪ 20 lbm CO_2/gal (liquid fuel)

▪ 2 lbm CO_2/kWh (coal combustion)

▪ 12 lbm CO_2/10^5 Btu (natural gas combustion)

▪ 25 lbm CO_2/therm (wood combustion)

(b) ▪ 20 lbm CO_2/gal (liquid fuel) = 15 lbm CO_2/therm

▪ 2 lbm CO_2/kWh (coal combustion) = 24 lbm CO_2/therm

▪ 12 lbm CO_2/105 Btu = 12 lbm/therm (natural gas combustion)

▪ 25 lbm CO_2/therm (wood combustion)

(c) ▪ U.S. = 20 metric tons/yr per person

▪ Russia = 11 metric tons/yr per person

▪ China = 2.5 metric tons/yr per person

▪ India = 1 metric tons/yr per person

▪ Canada = 17 metric tons/yr per person

▪ European Community Nations ~10 metric tons/yr per person

(d) ▪ US = 20 metric tons/yr per person = 6 billion metric tons/yr

▪ Russia = 11 metric tons/yr per person = 1.7 billion metric tons/yr

▪ China = 2.5 metric tons/yr per person = 3 billion metric tons/yr

▪ India = 1 metric tons/yr per person = 1.1 billion metric tons/yr

▪ Canada = 17 metric tons/yr per person = 0.6 billion metric tons/yr

▪ European Community Nations ~10 metric tons/yr per person = 5 billion metric tons/yr

(e) ▪ Before the Industrial Revolution (circa 1750's), CO_2 levels were approximately 280 ppm

▪ In 2008, the level was approximately 387 ppm

▪ In 2007, 500,000 acres of California forests were consumed by fire, releasing 22 million tons of CO_2 in 6 days

▪ About ½ stays in the air and the rest goes… to no one knows where!

▪ CO_2 "sinks" (i.e., absorbers) include: forests (really all plant life) and oceans

▪ Human respiration accounts for only 2% of the world's CO_2 emissions

Figure 6.7. (a) CO_2 Production based on common units of measures for the fuel. (b) CO_2 Production based on the same therm heat unit (1 therm = 10^5 Btu). (c) Per Person Emissions of CO_2 by Country. (d) Net Total Emissions of CO_2 by Country. (e) 30 billion tons/year of CO_2 enters the air from the burning of fossil fuels.

effect that the combustion of fossil fuels has had (and will have) on the environment. The predictions of the effects of CO_2 on global temperatures are then based on the extrapolation of the thermos-dynamic-chemical model into the future. There is also the necessity to utilize stochastic modeling to determine these coefficients. The prediction of global temperature change must depend not only on the accuracy of algorithms, but more crucially on the coefficients used with these algorithms. A quote often used in annual business reports to protect companies from being labeled as fraudulent may also be appropriately applied here: "[that] past results are no indication of future performance".

Proponents of dire consequences to global climates due to man-made influences should also consider an observation provided by author Gavin Menzies in *1434: The Year a Magnificent Chinese Fleet Sailed to Italy and Ignited the Renaissance*. While this book may have nothing to do with global warming, it provides a theory that Chinese navigators discovered the continent that we now know as North America. How did the Chinese navigate to this "New World"? According to the author, the navigation included a passage across the North Pole leading to the North American continent. How was this possible with wooden ships? Menzies wrote: "According to the Dutch meteorological office, there were three exceptionally warm winters in the 1420s, which could have melted the Artic sea ice". Of course, proponents to the argument that links man-made effects to global climate change will focus on the words "could have" in Menzies' sentence and not be deterred in their advocacy.

There certainly should not be an argument concerning relevant facts, shown in Figures 6.7(a) through (e), about the production of CO_2, with the expectation that the reader will use this information in any future analysis that becomes necessary.

Case Studies

Case Study 1

The Thermodynamics of an Espresso Coffee Pot

This first case study in applied thermodynamics is somewhat different from the norm in that it may not at first appear to be a significant application of thermodynamics. However, as will be observed in the subsequent case studies, the applications that will be reviewed provide a good sample of what may be encountered in real-world engineering thermodynamics problems and require the tools of thermodynamics that have been presented earlier in this textbook. Examples of typical 21st century applications of thermodynamics include heat engines used as primary power generation systems, heat recovery cycles that can increase the efficiency of a prime mover engine, and the production of a chemical commodities. Besides these specific applications of thermodynamics, any invention that transfers heat, does work, or has work done to it, or that has its temperature increased or decreased should be properly analyzed by applying thermodynamics to the system.

The espresso coffee maker, the subject of this case study, shown in Figure C1.1 is one such invention that transfers heat while performing work on a liquid to produce a product: espresso coffee that inhabitants of 90% of the homes in Europe drink every day. The question that can be answered with the application of thermodynamics is: how does the heating of water, in a machine with no moving parts, produce the rich, flavorful elixir we know as espresso coffee?

If you happen to own a coffee maker like the one in Figure C1.1, your reward for following this case study will be to enjoy the coffee produced for the purpose of carrying out the analysis.

Espresso coffee pots, like the one shown in Figure C1.1, come in varying sizes. Depending on the models' individual capacities, typical espresso coffee makers can accommodate the making of 2, 4, 6, or as many 12 cups of espresso at any one time.

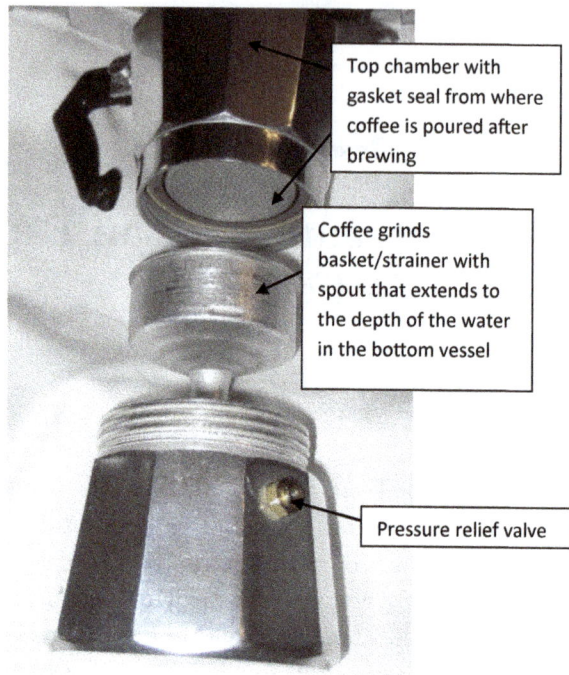

Figure C1.1.   An expresso coffee maker designed for private domestic use.

The espresso coffee pot may be treated as a closed system once the pot is filled with water, the coffee basket/strainer filled with coffee, and the top tightly screwed onto a bottom compartment that has been filled with water. In the figure, a pressure relief valve and a gasket on the top chamber that keeps a tight seal between the top and bottom halves, are shown. The heated fluid (in this case, water) is forced through the coffee strainer via the inverted spout that is submerged in the water. When heat is applied to the bottom of the pot for approximately 5 to 7 minutes per 6 oz. sized pot, the finished product — espresso coffee — is found in the top chamber.

For readers who are actively checking the author's claim that all thermodynamics systems are generally composed of any combination of the seven mechanical elements listed in Chapter 3, the espresso coffee maker can be classified as an evaporative heat exchanger, though not all of the water is evaporated in the process of making coffee.

This case study answers the following questions:

C1.1   Question 1

Provide a concise written description of the heating process. It is suggested that the description be accompanied by a sketch of the process on a P (or T) versus v (specific volume) graph. Try to sketch what you expect to be the shape of the curve reflecting the temperature change of the pot and water versus time for the entire process. Also indicate what could happen if the source of heat is not turned off after the coffee has been produced. What is the motive force that moves the cold water to the top chamber?

C1.1.1 *Analysis*

The coffee pot uses a very simple, yet ingenious method of producing coffee via the physical contact between the hot water and the finely ground coffee. The coffee must be grounded very finely (approximately $200\,\mu m$) to expose most of the coffee surface to the highest possible temperature of water that can be safely produced.

The ingenuity of the design is in how it produces coffee by vaporizing a small amount of liquid water to a much lower density saturated water vapor to safely push the saturated liquid water through the coffee grinds. In this manner, the water is heated to a safe maximum temperature of slightly above 212°F (about 100°C), while ensuring that the temperature never exceeds boiling temperature — assuming, of course, that the attentive user turns-off the energy source after the espresso coffee has been brewed. It has been observed that only 0.06% of the total water inventory needs to be boiled into saturated vapor for this method to work. This is illustrated in Figure C1.2. This design feature minimizes the energy input required to virtually only what is needed to heat the water from room temperature to boiling temperature. This evaporated mass is a very low percentage of the

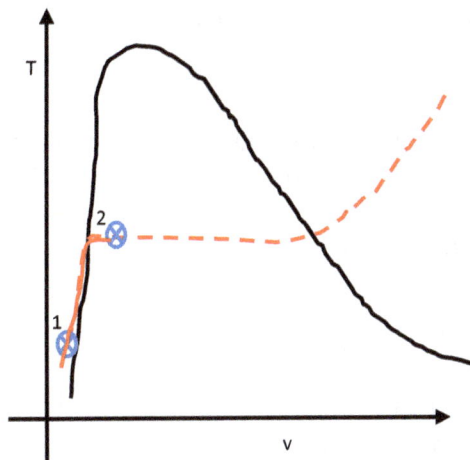

Figure C1.2. The T-v diagram of the water system (not drawn to scale) illustrating the initial water state point 1 and the final state point 2 shown as a quality of 0.06%.

initial water mass, and this percentage is referred to as the quality of the water in thermodynamic terms.

The amount of water vaporized is determined by knowing the volume of the bottom chamber and dividing this volume by the specific volume of the saturated water vapor at the boiling temperature, as shown in the numerator of this equation:

$$Quality = \frac{\left(\frac{V_{bottom\ chamber}}{v_{sat.vapor}}\right)}{M_{initial}}$$

The boiling temperature is also increased in a rather simple manner by increasing the packing density of the fresh coffee grinds in the coffee basket. By increasing the packing density, less ground coffee beans are needed to achieve the same rich coffee elixir.

Thus, methods for reducing the machine's energy consumption are observed to already be part of the existing design. It is clear that the Italian engineers who first designed this method of making coffee were as ingenious in their use of thermodynamics and heat transfer as was Master Brunelleschi in his design for the largest, free-standing dome that can still be seen and explored in Santa Maria della Fiori, Florence, Italy.

C1.2   Question 2

Using either the dimensions of the pot shown in Table C1.1 OR the dimensions measured from an actual espresso coffee pot, determine the mass of the water that is needed to fill the pot AND the mass of the coffee pot. Assuming that it is made using stainless steel (SST), you may assume that the wall thickness of the pot is 0.125 inches.

C1.2.1   *Analysis*

The mass of the water is simply the product of the volume (ft^3) of the bottom of the coffee pot and the density of the water. To perform this thermodynamics analysis, a pot with a diameter of 2.5 inches and a height of 3 inches was used. Thus, the volume of the bottom section of the coffee pot is $0.0085\,\text{ft}^3$ and the total mass of the water needed to fill the pot is 0.5 lbm or approximately 6 oz.

C1.3   Question 3: Find the Rate of Temperature change of the coffee pot as a function of time, $\frac{\partial T}{\partial t}$

Assume that the electric or gas heat input to the metal coffee pot is 200 Btu/h. NOTE: consider also that as the pot is being heated, it is losing heat energy to the room at a rate that is calculated using the equation with temperature, T in rankine $[R]$ units:

$\dot{Q}_{convection+radiation}$

$$= 0.5 \times (T_{pot+water} - T_{ambient}) + 5 \times 10^{-9} \times (T_{pot+water}^4 + T_{ambient}^4)$$

Determine the rate of the temperature change of the pot and water treated as a single (common) temperature. In other words, determine the transient temperature of the pot and water as a function of time.

C1.3.1 *Analysis*

In order to perform this analysis, a numerical solution to the nonlinear differential equation was used. The differential equation is derived from the closed mass application of the First Law of Thermodynamics and is given below:

$$\{\dot{Q}_{heat\,in} + \dot{Q}_{convection\,out} + \dot{Q}_{radiation\,out}\}$$

$$= [M \times Cp_{pot} + M \times Cp_{water}] \times \frac{(T_{pot\,at\,t+1} - T_{pot\,at\,t})}{\Delta time}$$

$$+ \Delta M_{water} \times \frac{g_g}{g_c} \times \Delta z, \qquad (C1.1)$$

where:

$Q_{convection\,loss}$ [Btu/h] $= 0.5 \times (T_{pot} - 520)$;

Radiation Loss $= K_{coef.} \times [(T_{pot})_{pot}^4 - (T_{ambient})_{amb}^4]$.

T_{pot} is the temperature of the pot at any time and $K_{coef.} = 5 \times 10^{-9}$.

Equation (C1.1) is used to solve for the parameter: $\frac{(T_{pot\,at\,t+1} - T_{pot\,at\,t})}{\Delta time}$.

This parameter is then used to determine the next temperature of the coffee pot by using the equation:

$$T_{temp.\,pot\,after} = T_{temp.\,pot\,before} + \left[\frac{(T_{pot\,at\,t+1} - T_{pot\,at\,t})}{\Delta time}\right] \times \Delta time_{increment}.$$

Using Excel, a numerical analysis was performed to determine the temperature change of the coffee pot as a function of time. The inputs for this analysis are shown in Table C1.1. It is also noted that the heat rate (Btu/h) needed to be increased to 1,205 Btu/h and not 200 Btu/h as first assumed. This was based on the reasonable engineering assumption that the espresso coffee should be ready in less than 10 minutes and not take 30 minutes or more. The results of this numerical analysis are graphically displayed as a temperature transient in Figure C1.3.

Table C1.1. An example of the inputs used in the modeling of the espresso coffee pot to determine the transient temperature as a function of time.

Dia.=	2.5 inch
Height=	3 inch
Wall Thickness=	0.125 inch
Pot Metal Density=	489 Lbm/ft^3

6.38136

Water mass=	0.53	Lbm
Vol.=	0.0085	ft^3
Pot Mass=	0.833	Lbm
Specific Heat metal=	0.12	Btu/Lbm/F
Water Specific Heat	1.0	Btu/Lbm/F
Milk Specific Heat	0.9	Btu/Lbm/F
Milk Temp=	45	F
Time Increment=	1.5	sec.s
Q,conv. & Qrad. heat loss=	0.5	Btu/hr/F	5.00E-09 Btu/hr/R^4
Tambient=	60	F
DQ/Dt), heater=	1205	Btu/hr

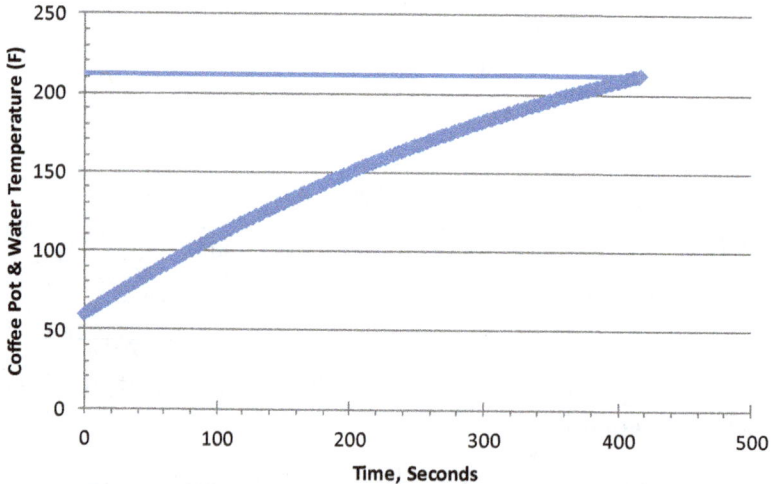

Figure C1.3. Temperature transient of a steel espresso coffee pot (capacity: 2 cups or 6 oz.) at heat rate of 1200 Btu/h.

This result was based on the assumption that the temperature of the coffee pot and the water were kept the same throughout the transient heating process. This assumption is not very realistic and is proposed here only to facilitate the analysis while using all of the necessary thermodynamic concepts that would ever be needed to analyze a multi-temperature system. A multi-temperature system only requires much more "book keeping" of each temperature but not any more thermodynamic principles. The independent variable in this analysis was the heat rate input into the pot.

C1.4   Question 4

How long does it take for the pot to reach boiling temperature? Assume that the boiling pressure is: $P_{atmosphere} + (\Delta P = .05$ psi for the coffee grinds $+.025$ psi for the inverted water spout).

C1.4.1   *Analysis*

From Figure C1.3, it may be observed that the time taken to reach the boiling temperature of 212°F from the point of heating is approximately 8 minutes.

As an extension, a comparison between the theoretical heat rate input and the theoretical heat rate required to heat the water alone can be performed using the numerical model previously derived. The time taken for the water in Figure C1.3 to reach boiling point was used as an input for the numerical model based on the heating time measured in an experiment with a 6 oz. coffee pot. The numerical model was then used to determine the theoretical heat rate from the heat source that was required to heat the water to this temperature, with the heat losses due to convection and radiation being taken into account.

C1.5   Question 5

How much energy is used to heat the water during the 8-minute process to achieve a the temperature of 212°F required to produce the coffee? NOTE: Be sure to account for the increase in the center of gravity of the water as it migrates upward from the bottom chamber to the top chamber.

C1.5.1   *Analysis*

Table C1.2 provides a summary of the analysis conducted to determine the energy and energy rate (power) required to heat the water and the coffee pot. With the specified duration of 8 min of heating, the energy and rate of energy to heat only the water is calculated to be 81 Btu and 702 Btu/h, respectively. The calculated rate of energy of 702 Btu/h may be compared to the heat rate of 1,205 Btu/h (as calculated in Table C1.1) required for heating the pot and water to 212°F, which was determined via numerical

Table C1.2.   Summary of analysis determining power
needed to heat the water and coffee pot.

THERMODYNAMIC ANALYSIS			
Specific Volume, sat. vapor=		26.71	ft³/Lbm
M.vapor=		0.00031906	Lbm
Quality, x=		0.060%	
CLOSED MASS First Law Eqn.:			
$\Sigma Q = \Delta U + \Delta P.E.$			
Q [Btu],water= 81		Q,pot=	15
DQ/Dt)water= 702		DQ/Dt)pot=	131
Coffee Pot Overall Eff.=	58%		
DQ/Dt), heater [Btu/hr]=		833	
Desired Coffee Temp.=		110	
Milk Mass [Lbm]=		0.11	

analysis in Question 3. This represents an overall 58% energy efficiency of the coffee pot.

In performing this thermodynamics analysis, attention should also be given to the amount of vapor mass, 3.19e-4 lbm, that must be produced to push the saturated liquid water into the top section. This represents a quality of 0.06% (=0.000319 lbm/0.5 lbm).

C1.6   Question 6

Imagine that you wish to make a cup of cappuccino (i.e., an espresso with milk). Assume that the milk (with an initial temperature of 45°F) must be heated in order to avoid the milk-and-espresso-mixture's temperature falling below the minimum acceptable temperature (for a cappuccino) of 110°F when the milk is added — what then, is the maximum amount of milk that can be added to the espresso without having the coffee's temperature fall below 110°F? Assume that the milk will be heated to 185°F and the coffee has cooled slightly to 200°F after being poured from the coffee pot.

C1.6.1   *Analysis*

Although it is the custom in Italy to only drink cappuccinos at breakfast, it is not uncommon in the United States for people to have a cup of cappuccino at almost any time during the day. In order to achieve the widely-preferred caramel color of cappuccino, a considerable amount of milk has to be heated and added to a very small (2–3 oz.) cup of espresso coffee. If the milk is not heated, the addition of the cold milk would cause the temperature of the mixture to fall below the acceptable norm of 110°F. The amount of cold milk that needs to be heated to make two cups of cappuccino is approximately 0.1 lbm.

Point of Information: The addition of milk to a hot cup of espresso involves an irreversible loss of energy which, unfortunately, runs contrary to the interests of the Department of Energy (DOE). A calculation of the irreversibility (or exergy destruction) indicates that as much as 0.01 kWh of energy availability is destroyed per 6 oz. pot of coffee. This amounts to 6 kWh per adult per year; assuming that 9 oz. of espresso is consumed each day.

C1.7   Question 7

You may have noticed that barristers tend to compact the coffee grounds before using them. What is the consequence of compacting the coffee grounds? Hint: See Question 4.

C1.7.1   *Analysis*

Compaction causes an increase in the pressure drop across the coffee grounds. This in turn causes the pressure of the water at the bottom of the coffee pot to increase in order to overcome the total pressure drop through the coffee grinds and the two spouts. As a consequence, the increase in water pressure causes the saturated water temperature to increase in accordance with the values stated in the saturated temperature table. The increase in water temperature results in more of the espresso beans dissolving in the water. Thus, the espresso coffee becomes even richer in taste and appearance when more coffee can be squeezed into the basket.

Organic Rankine Cycle Heat Recovery and Power Generation System

C2.1 Introduction and background

The thermodynamics term "cogeneration" has a very clear meaning to the engineering analyst. This term is used to indicate the utilization of a single source of primary fuel (or energy) for electric power generation, which is the primary purpose of a heat engine. But then there is also an interest in recovering the waste heat that is rejected by the engine and using this heat for other purposes, such as boiling water to create a low-pressure steam for industrial use. A typical heat engine, regardless of whether it uses an Otto, Diesel or Brayton Cycle, will burn fossil fuel to generate mechanical or electrical power with the help of an electric generator connected to the output shaft of the engine. Usually, there is no attempt to recover both the low and high temperature waste heat rejected from the jacket and the exhaust gas of the engine, respectively. This results in a lowered overall first law efficiency of the engine. This contradicts the objective of any engineering design: which should be to increase the first law efficiency of the prime mover. For that reason, additional thermodynamics cycles may be applied not just to burn fuel, but to also recover the otherwise wasted heat energy.

The high temperature waste heat from the exhaust gases is among the first heat energy streams to be considered economical to recover, as the high temperature energy has a potential second law efficiency that is high enough to warrant attention. For example, the temperature of the exhaust gas produced by most internal combustion engines or gas turbines is typically above 540°C. Based on the exergy analysis presented in Chapter 5 and assuming that the 540°C temperature can be reduced to 140°C, the Carnot efficiency is as high as 50% $\left(= 1 - \frac{(300\,\text{K})\ln\left(\frac{140+273}{540+273}\right)}{(140-540)} \right)$. The reader should be able to calculate the potential first law efficiency of a heat engine cycle by applying

Carnot's equation to the temperature stream. As noted in Chapter 5, using the given information on the inlet and exhaust temperature, the Carnot efficiency of an ideal heat engine with a finite waste heat stream that has its temperature change as heat is recovered from the waste heat stream was shown to be:

$$Carnot\ Eff. = 1 - \frac{T_{amb} \ln\left(\frac{T_{out}}{T_{in}}\right)}{(T_{out} - T_{in})}.$$

The term "bottoming cycle" is used in reference to the cycle involved in the recovery of waste heat energy from a prime mover's primary power cycle. The most common thermodynamics cycle used in the recovery of waste heat from the exhaust stream is the Rankine Cycle. The exhaust gas temperatures from heat engines using Otto, Diesel, or Brayton Cycles are typically less than 1,000°F. Rankine Cycles, which typically use water as the working fluid, are not particularly effective at these low temperatures. However, it is important to understand that the primary focus for a waste heat recovery cycle is not the efficiency of the cycle that might be attained with a water fluid, but rather the magnitude of the power that is recoverable from the waste heat. That is, the discharge temperature of the waste heat stream must be drawn to be as low as possible, to achieve maximum power generation and heat recovery. However, the very large latent heat of vaporization of water results in a pinch point problem that causes the discharge temperature to be higher than desired. This effect is detailed in Chapter 4.

A fluid with a lower heat of evaporation is often used to enable the discharge temperature to be lowered to a minimum temperature of approximately 300°F. This lower temperature limit is a constraint imposed by the psychrometrics of the waste heat stream that determines the behavior of the mixed gases, such as carbon dioxide (CO_2), water, and nitrogen, that constitute the exhaust gas stream. A drop in temperature to below 300°F usually causes the water vapor and nitrogen gas in the products of combustion to condense and form nitric acid. Nitric acid is known to damage exhaust gas pipes. For that reason, dropping the temperature of the exhaust gas below 300°F is regarded as an engineering limitation. For that reason, the working fluid for the Rankine Cycle is typically one of the common refrigerants used in a heat pump cycle, such as R245fa or R123a. More exotic fluids such as siloxanes are sometimes used. In call cases, the choice to enable the maximum recovery of waste heat energy is sometimes made at the expense of cycle efficiency, because the objective in waste heat recovery is to maximize the generation of useful power.

The thermodynamics modeling of a Rankine or Brayton Cycle used to recover waste heat energy from an electric utility's prime mover is no different than when the cycle was the primary power generation cycle. For this reason, the programming of a Rankine or Brayton cycle, as outlined in Chapter 3, presents no additional thermodynamics.

C2.2   Examples

The following are examples of several waste heat recovery systems that are suitable for an industrial energy user. They were prepared in the form of a report to the sales engineer from an engineering department that was tasked to assess the potential of waste heat from an industrial source of energy.

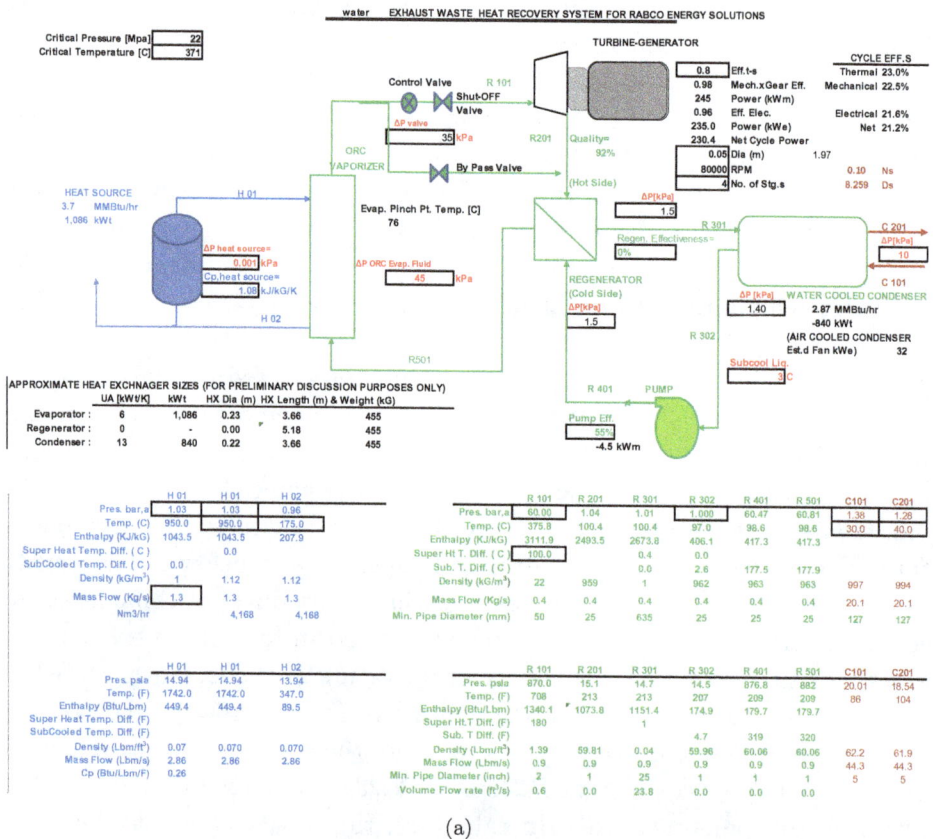

water   EXHAUST WASTE HEAT RECOVERY SYSTEM FOR RABCO ENERGY SOLUTIONS

Critical Pressure [Mpa] 22
Critical Temperature [C] 371

TURBINE-GENERATOR

CYCLE EFF.S

Control Valve R 101
Shut-OFF Valve
ΔP valve 35 kPa   R201 Quality= 92%

0.8	Eff. t-s
0.98	Mech.xGear Eff.
245	Power (kWm)
0.96	Eff. Elec.
235.0	Power (kWe)
230.4	Net Cycle Power

Thermal 23.0%
Mechanical 22.5%
Electrical 21.6%
Net 21.2%

ORC VAPORIZER
By Pass Valve

0.05 Dia (m) 1.97
80000 RPM
4 No. of Stg.s
0.10 Ns
8.259 Ds

HEAT SOURCE
3.7 MMBtu/hr
1,086 kWt
H 01

(Hot Side) ΔP[kPa] 1.5

Evap. Pinch Pt. Temp. [C] 76
ΔP heat source= 0.001 kPa
Cp.heat source= 1.08 kJ/kG/K
ΔP ORC Evap. Fluid 45 kPa

R 301

Regen. E.ffectiveness= 0%

C 201 ΔP[kPa] 10

REGENERATOR (Cold Side) ΔP[kPa] 1.5

ΔP [kPa] 1.40 WATER COOLED CONDENSER
C 101
2.87 MMBtu/hr
-840 kWt
(AIR COOLED CONDENSER Est.d Fan kWe) 32

H 02
R 302
Subcool Liq. 3C
R501

R 401 PUMP
Pump Eff. 55%
-4.5 kWm

APPROXIMATE HEAT EXCHNAGER SIZES (FOR PRELIMINARY DISCUSSION PURPOSES ONLY)

	UA [kWt/K]	kWt	HX Dia (m)	HX Length (m) & Weight (kG)	
Evaporator:	6	1,086	0.23	3.66	455
Regenerator:	0	-	0.00	5.18	455
Condenser:	13	840	0.22	3.66	455

	H 01	H 01	H 02		R 101	R 201	R 301	R 302	R 401	R 501	C101	C201
Pres. bar,a	1.03	1.03	0.96		60.00	1.04	1.01	1.000	60.47	60.81	1.38	1.28
Temp. (C)	950.0	950.0	175.0		375.8	100.4	100.4	97.0	98.6	98.6	30.0	40.0
Enthalpy (KJ/kG)	1043.5	1043.5	207.9		3111.9	2493.5	2673.8	406.1	417.3	417.3		
Super Heat Temp. Diff. (C)		0.0			100.0			0.4	0.0			
SubCooled Temp. Diff. (C)	0.0						0.0	2.6	177.5	177.9		
Density (kG/m³)	1	1.12	1.12		22	959	1	962	963	963	997	994
Mass Flow (Kg/s)	1.3	1.3	1.3		0.4	0.4	0.4	0.4	0.4	0.4	20.1	20.1
Nm3/hr		4,168	4,168									
Min. Pipe Diameter (mm)					50	25	635	25	25	25	127	127

	H 01	H 01	H 02		R 101	R 201	R 301	R 302	R 401	R 501	C101	C201
Pres. psia	14.94	14.94	13.94		870.0	15.1	14.7	14.5	876.8	882	20.01	18.54
Temp. (F)	1742.0	1742.0	347.0		708	213	213	207	209	209	86	104
Enthalpy (Btu/Lbm)	449.4	449.4	89.5		1340.1	1073.8	1151.4	174.9	179.7	179.7		
Super Heat Temp. Diff. (F)					180				1			
Sub. T Diff. (F)								4.7	319	320		
Density (Lbm/ft³)	0.07	0.070	0.070		1.39	59.81	0.04	59.98	60.06	60.06	62.2	61.9
Mass Flow (Lbm/s)	2.86	2.86	2.86		0.9	0.9	0.9	0.9	0.9	0.9	44.3	44.3
Cp (Btu/Lbm/F)	0.26											
Min. Pipe Diameter (inch)					2	1	25	1	1	1	5	5
Volume Flow rate (ft³/s)					0.6	0.0	23.8	0.0	0.0	0.0		

(a)

Figure C2.1.   (a) Steam Rankine Cycle generating 235 kWe gross (230 kWe Net) with 1 atm turbine discharge, and (b) steam Rankine Cycle generating 256 kWe gross (251 kWe Net) with 0.5 atm turbine discharge.

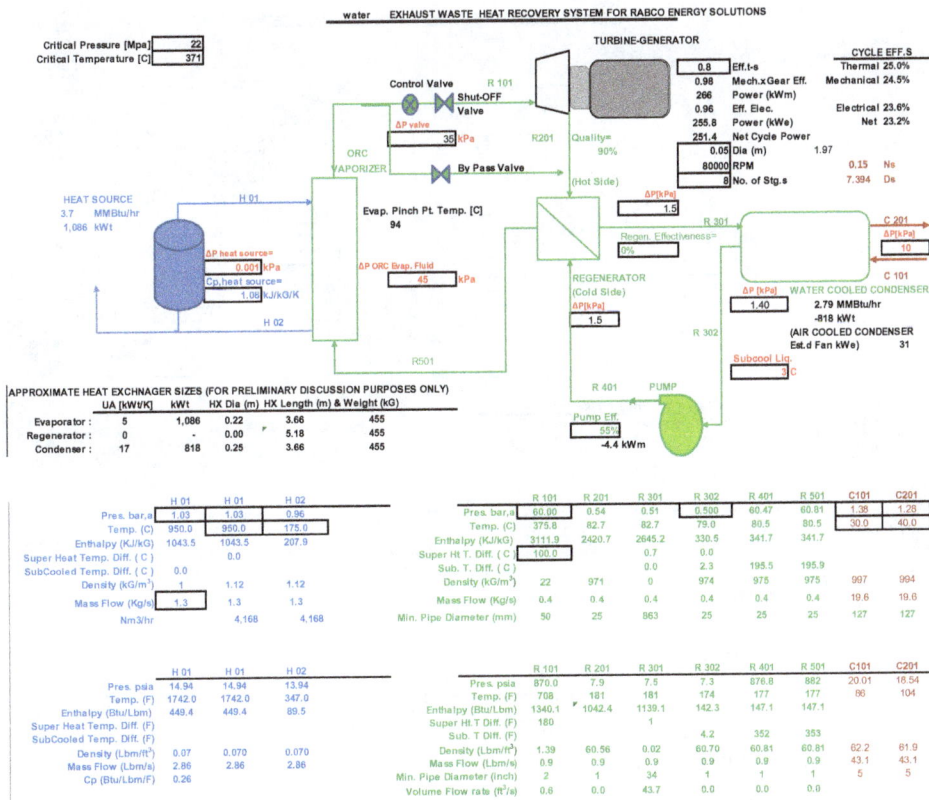

water   EXHAUST WASTE HEAT RECOVERY SYSTEM FOR RABCO ENERGY SOLUTIONS

| Critical Pressure [Mpa] | 22 |
| Critical Temperature [C] | 371 |

TURBINE-GENERATOR

CYCLE EFF.S

0.8	Eff.t-s	Thermal 25.0%
0.98	Mech.x Gear Eff.	Mechanical 24.5%
266	Power (kWm)	
0.96	Eff. Elec.	Electrical 23.6%
255.8	Power (kWe)	Net 23.2%
251.4	Net Cycle Power	
0.05	Dia (m)	1.97
80000	RPM	0.15 Ns
8	No. of Stg.s	7.394 Ds

Control Valve   R 101
Shut-OFF Valve
ΔP valve   35 kPa   R201   Quality= 90%
ORC VAPORIZER   By Pass Valve   (Hot Side)

HEAT SOURCE   H 01
3.7 MMBtu/hr
1,086 kWt

Evap. Pinch Pt. Temp. [C]   94

ΔP heat source= 0.001 kPa
Cp,heat source= 1.08 kJ/kG/K

ΔP ORC Evap. Fluid   45 kPa

H 02

R501

ΔP[kPa] 1.5

R 301
Regen. Effectiveness= 9%

REGENERATOR (Cold Side)
ΔP[kPa] 1.5

R 302

C 201
ΔP[kPa] 10

C 101
ΔP [kPa]   WATER COOLED CONDENSER
1.40   2.79 MMBtu/hr
-818 kWt
(AIR COOLED CONDENSER
Est.d Fan kWe)   31

Subcool Liq. 3 C

R 401   PUMP
Pump Eff. 55%
-4.4 kWm

APPROXIMATE HEAT EXCHANGER SIZES (FOR PRELIMINARY DISCUSSION PURPOSES ONLY)

	UA [kWt/K]	kWt	HX Dia (m)	HX Length (m) & Weight (kG)	
Evaporator :	5	1,086	0.22	3.66	455
Regenerator :	0	-	0.00	5.18	455
Condenser :	17	818	0.25	3.66	455

	H 01	H 01	H 02
Pres. bar,a	1.03	1.03	0.96
Temp. (C)	950.0	950.0	175.0
Enthalpy (KJ/kG)	1043.5	1043.5	207.9
Super Heat Temp. Diff. (C)		0.0	
SubCooled Temp. Diff. (C)	0.0		
Density (kG/m³)	1	1.12	1.12
Mass Flow (Kg/s)	1.3	1.3	1.3
Nm3/hr		4,168	4,168

	R 101	R 201	R 301	R 302	R 401	R 501	C101	C201
Pres. bar,a	60.00	0.54	0.51	0.500	60.47	60.81	1.38	1.28
Temp. (C)	375.8	82.7	82.7	79.0	80.5	80.5	30.0	40.0
Enthalpy (KJ/kG)	3111.9	2420.7	2645.2	330.5	341.7	341.7		
Super Ht T. Diff. (C)	100.0		0.7	0.0				
Sub. T. Diff. (C)			0.0	2.3	195.5	195.9		
Density (kG/m³)	22	971	0	974	975	975	997	994
Mass Flow (Kg/s)	0.4	0.4	0.4	0.4	0.4	0.4	19.6	19.6
Min. Pipe Diameter (mm)	50	25	863	25	25	25	127	127

	H 01	H 01	H 02
Pres. psia	14.94	14.94	13.94
Temp. (F)	1742.0	1742.0	347.0
Enthalpy (Btu/Lbm)	449.4	449.4	89.5
Super Heat Temp. Diff. (F)			
SubCooled Temp. Diff. (F)			
Density (Lbm/ft³)	0.07	0.070	0.070
Mass Flow (Lbm/s)	2.86	2.86	2.86
Cp (Btu/Lbm/F)	0.26		

	R 101	R 201	R 301	R 302	R 401	R 501	C101	C201
Pres. psia	870.0	7.9	7.5	7.3	876.8	882	20.01	18.54
Temp. (F)	708	181	181	174	177	177	86	104
Enthalpy (Btu/Lbm)	1340.1	1042.4	1139.1	142.3	147.1	147.1		
Super Ht.T Diff. (F)	180		1					
Sub. T Diff. (F)				4.2	352	353		
Density (Lbm/ft³)	1.39	60.56	0.02	60.70	60.81	60.81	62.2	61.9
Mass Flow (Lbm/s)	0.9	0.9	0.9	0.9	0.9	0.9	43.1	43.1
Min. Pipe Diameter (inch)	2	1	34	1	1	1	5	5
Volume Flow rate (ft³/s)	0.6	0.0	43.7	0.0	0.0	0.0		

(b)

Figure C2.1.   (*Continued*)

C2.2.1   *Example C2.1*

A thermodynamics analysis was completed for Waste Heat Energy Systems, Inc., using the waste heat specification provided by the client. The specifications included an exhaust gas, heat source temperature (950°C) and flow rate (1.3 kg/s). The exhaust gas was emitted by an internal combustion diesel engine prime mover. An ambient air coolant temperature of 30°C was also specified.

Using this information, three heat recovery systems are presented in the following case study: A steam Rankine Cycle system (Figures C2.1(a) and (b)), an Organic Rankine Cycle (ORC System) using R245fa as the working fluid (Figure C2.2), and an air-Brayton Cycle, with and without an ORC as a bottoming cycle (Figures C2.3(a) and (b)).

The air-Brayton cycle is similar to a gas turbine engine, except that the fuel combustor is replaced with a heat recovery heat exchanger. An

r245fa   EXHAUST WASTE HEAT RECOVERY SYSTEM FROM HEATTREAT FURNACE

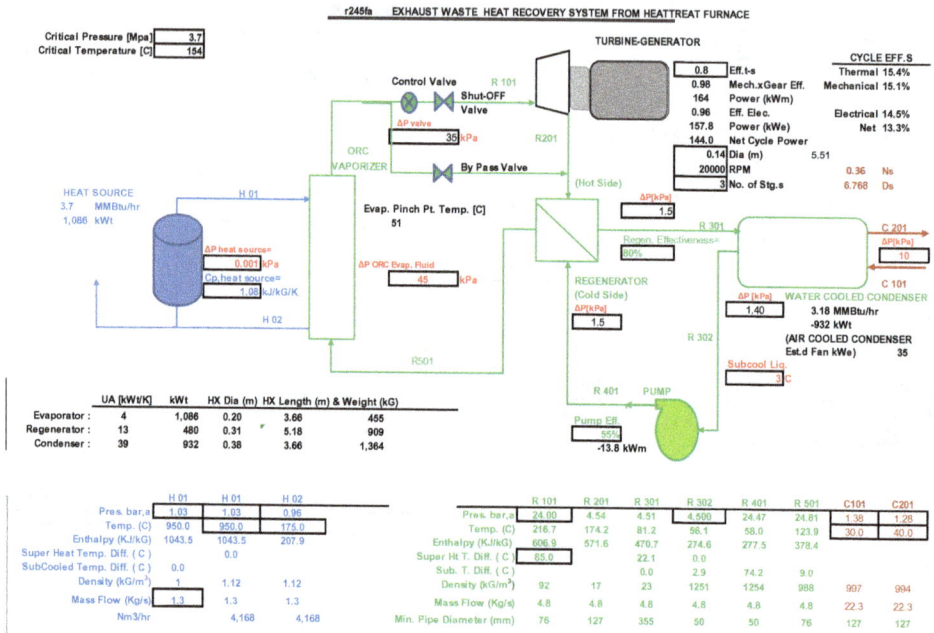

UA [kWt/K]	kWt	HX Dia (m)	HX Length (m) & Weight (kG)	
Evaporator :	4	1,086	0.20	3.66	455
Regenerator :	13	480	0.31	5.18	909
Condenser :	39	932	0.38	3.66	1,364

	H 01	H 01	H 02
Pres. bar,a	1.03	1.03	0.96
Temp. (C)	950.0	950.0	175.0
Enthalpy (KJ/kG)	1043.5	1043.5	207.9
Super Heat Temp. Diff. (C)	0.0		
SubCooled Temp. Diff. (C)	0.0		
Density (kG/m³)	1	1.12	1.12
Mass Flow (Kg/s)	1.3	1.3	1.3
Nm3/hr		4,168	4,168

	R 101	R 201	R 301	R 302	R 401	R 501	C101	C201
Pres. bar,a	24.00	4.54	4.51	4.500	24.47	24.81	1.38	1.28
Temp. (C)	216.7	174.2	81.2	56.1	58.0	123.9	30.0	40.0
Enthalpy (KJ/kG)	606.9	571.6	470.7	274.6	277.5	378.4		
Super Ht T. Diff. (C)	85.0		22.1					
Sub. T. Diff. (C)			0.0	0.0				
Density (kG/m³)	92	17	23	1251	1254	988	997	994
Mass Flow (Kg/s)	4.8	4.8	4.8	4.8	4.8	4.8	22.3	22.3
Min. Pipe Diameter (mm)	76	127	355	50	50	76	127	127

Figure C2.2.   R245fa ORC System generating 160 kWe gross power and 144 kWe net.

important consequence of this cycle is the availability of a waste heat stream from the air-Brayton Cycle that can be used for additional cogeneration heating, or as a lower temperature heat source to a secondary (Rankine) cycle. In this case study, two options were considered for the secondary or bottoming cycle: (1) steam generation using the hot air discharged from the air-Brayton Cycle, and (2) the use of an ORC System, serving as a bottoming cycle to the air-Brayton Cycle.

The results of the analysis indicate that the steam Rankine Cycle can generate 235 kWe of gross power from the waste heat source if the condensing pressure for the steam is 1 bar,a, or 256 kWe gross power, for a condensing pressure of 0.5 bar,a. The air-Brayton cycle can produce 188 kWe of gross power. In addition to the electric power produced, the air-Brayton cycle also generates waste heat steam which has considerable thermal energy potential. This waste heat could be used to generate 6,440 lbm/h of 30 psig steam. Alternatively, the exhaust emitted from the air-Brayton regenerator may be used as a source of heated pre-combustion air for a nearby, onsite gas-fired heater. An air-Brayton Cycle with an ORC serving as the bottoming cycle recovering the heat energy from the hot air rejected by the air-Brayton Cycle can generate 220 kWe of power. The ORC System can produce 160 kWe as the single heat recovery cycle applied to the waste heat.

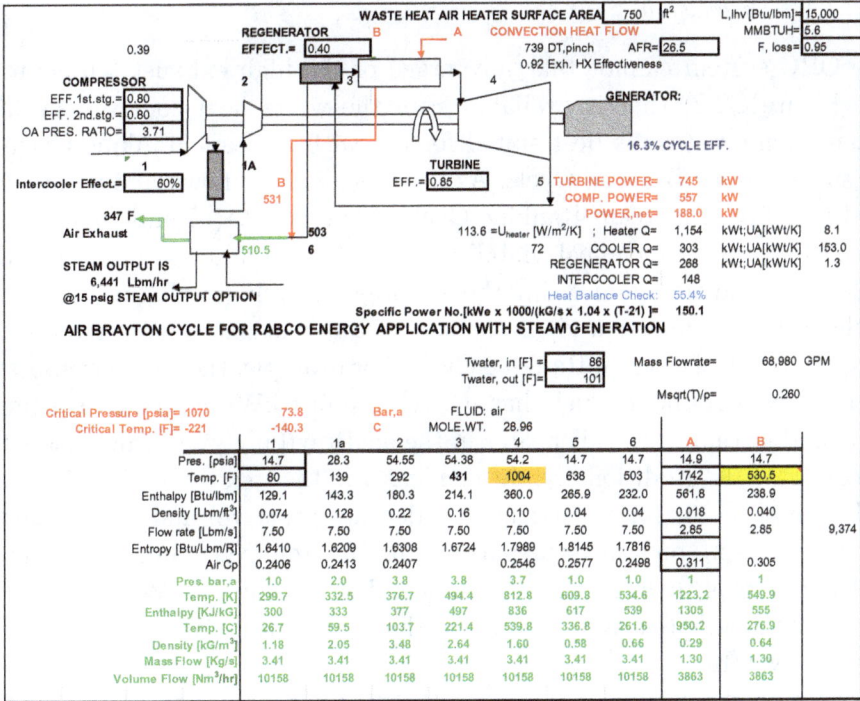

WASTE HEAT AIR HEATER SURFACE AREA 750 ft² L,lhv [Btu/lbm]= 15,000
MMBTUH= 5.6

REGENERATOR B A CONVECTION HEAT FLOW
0.39 EFFECT.= 0.40 739 DT,pinch AFR= 26.5 F, loss= 0.95
0.92 Exh. HX Effectiveness

COMPRESSOR
EFF.1st.stg.= 0.80
EFF. 2nd.stg.= 0.80
OA PRES. RATIO= 3.71

GENERATOR:
16.3% CYCLE EFF.

Intercooler Effect.= 60%

347 F
Air Exhaust

STEAM OUTPUT IS
6,441 Lbm/hr
@15 psig STEAM OUTPUT OPTION

TURBINE
EFF.= 0.85

TURBINE POWER= 745 kW
COMP. POWER= 557 kW
POWER,net= 188.0 kW
113.6 =U_heater [W/m²/K] ; Heater Q= 1,154 kWt;UA[kWt/K] 8.1
72 COOLER Q= 303 kWt;UA[kWt/K] 153.0
REGENERATOR Q= 268 kWt;UA[kWt/K] 1.3
INTERCOOLER Q= 148
Heat Balance Check: 55.4%
Specific Power No.[kWe x 1000/(kG/s x 1.04 x (T-21)]= 150.1

AIR BRAYTON CYCLE FOR RABCO ENERGY APPLICATION WITH STEAM GENERATION

Twater, in [F] = 86 Mass Flowrate= 68,980 GPM
Twater, out [F]= 101

Msqrt(T)/p= 0.260

		73.8	Bar,a	FLUID: air					
Critical Pressure [psia]= 1070		73.8	Bar,a						
Critical Temp. [F]= -221		-140.3	C	MOLE.WT. 28.96					
	1	1a	2	3	4	5	6	A	B
Pres. [psia]	14.7	28.3	54.55	54.38	54.2	14.7	14.7	14.9	14.7
Temp. [F]	80	139	292	431	1004	638	503	1742	530.5
Enthalpy [Btu/lbm]	129.1	143.3	180.3	214.1	360.0	265.9	232.0	561.8	238.9
Density [Lbm/ft³]	0.074	0.128	0.22	0.16	0.10	0.04	0.04	0.018	0.040
Flow rate [Lbm/s]	7.50	7.50	7.50	7.50	7.50	7.50	7.50	2.85	2.85	9,374
Entropy [Btu/Lbm/R]	1.6410	1.6209	1.6308	1.6724	1.7989	1.8145	1.7816		
Air Cp	0.2406	0.2413	0.2407		0.2546	0.2577	0.2498	0.311	0.305
Pres. bar,a	1.0	2.0	3.8	3.8	3.7	1.0	1.0	1	1
Temp. [K]	299.7	332.5	376.7	494.4	812.8	609.8	534.6	1223.2	549.9
Enthalpy [KJ/kG]	300	333	377	497	836	617	539	1305	555
Temp. [C]	26.7	59.5	103.7	221.4	539.8	336.8	261.6	950.2	276.9
Density [kG/m³]	1.18	2.05	3.48	2.64	1.60	0.58	0.66	0.29	0.64
Mass Flow [Kg/s]	3.41	3.41	3.41	3.41	3.41	3.41	3.41	1.30	1.30
Volume Flow [Nm³/hr]	10158	10158	10158	10158	10158	10158	10158	3863	3863

(a)

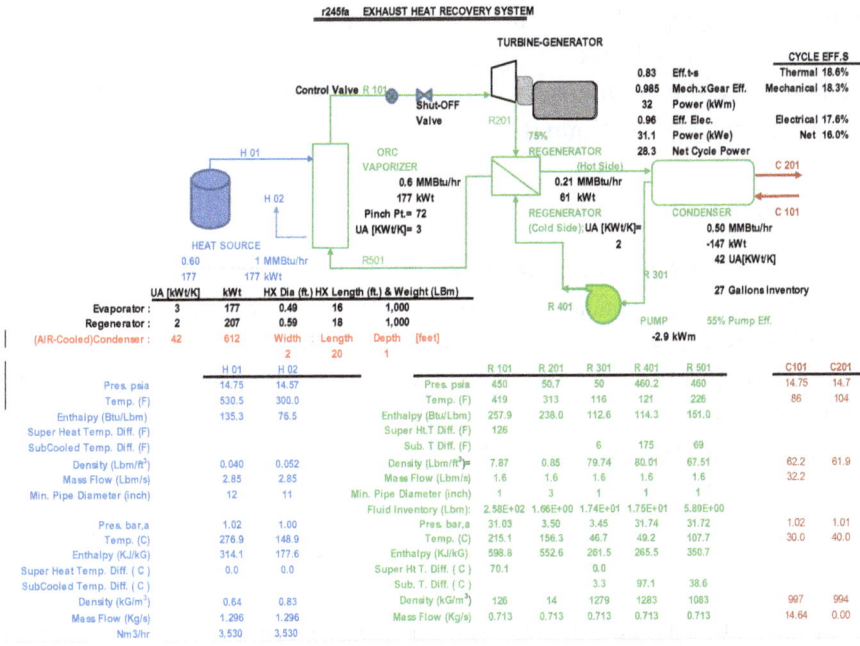

r245fa   EXHAUST HEAT RECOVERY SYSTEM

TURBINE-GENERATOR

Control Valve R 101
Shut-OFF Valve

CYCLE EFF.S
0.83 Eff.t-s Thermal 18.6%
0.985 Mech.xGear Eff. Mechanical 18.3%
32 Power (kWm)
0.96 Eff. Elec. Electrical 17.6%
31.1 Power (kWe) Net 16.0%
28.3 Net Cycle Power

ORC VAPORIZER
0.6 MMBtu/hr
177 kWt
Pinch Pt.= 72
UA [KWt/K]= 3

REGENERATOR (Hot Side)
0.21 MMBtu/hr
61 kWt
REGENERATOR (Cold Side): UA [KWt/K]= 2

HEAT SOURCE
0.60
177
UA [kWt/K] kWt HX Dia (ft.) HX Length (ft.) & Weight (LBm)
Evaporator : 3 177 0.49 16 1,000
Regenerator : 2 207 0.59 18 1,000
(AIR-Cooled)Condenser : 42 612 Width : Length : Depth [feet]
  2 20 1

CONDENSER
0.50 MMBtu/hr
-147 kWt
42 UA[KWt/K]

27 Gallons inventory
PUMP 55% Pump Eff.
-2.9 kWm

	H 01	H 02
Pres. psia	14.75	14.57
Temp. (F)	530.5	300.0
Enthalpy (Btu/Lbm)	135.3	76.5
Super Heat Temp. Diff. (F)		
SubCooled Temp. Diff. (F)		
Density (Lbm/ft³)	0.040	0.052
Mass Flow (Lbm/s)	2.85	2.85
Min. Pipe Diameter (inch)	12	11
Pres. bar,a	1.02	1.00
Temp. (C)	276.9	148.9
Enthalpy (KJ/kG)	314.1	177.6
Super Heat Temp. Diff. (C)	0.0	0.0
SubCooled Temp. Diff. (C)		
Density (kG/m³)	0.64	0.83
Mass Flow (Kg/s)	1.296	1.296
Nm3/hr	3.530	3.530

	R 101	R 201	R 301	R 401	R 501		C101	C201
Pres. psia	450	50.7	50	460.2	460		14.75	14.7
Temp. (F)	419	313	116	121	226		86	104
Enthalpy (Btu/Lbm)	257.9	238.0	112.6	114.3	151.0			
Super Ht.T Diff. (F)	126							
Sub. T Diff. (F)			6	175	69			
Density (Lbm/ft³)=	7.87	0.85	79.74	80.01	67.51		62.2	61.9
Mass Flow (Lbm/s)	1.6	1.6	1.6	1.6	1.6		32.2	
Min. Pipe Diameter (inch)	1	3	1	1	1			
Fluid Inventory (Lbm):	2.58E+02	1.66E+00	1.74E+01	1.75E+01	5.89E+00			
Pres. bar,a	31.03	3.50	3.45	31.74	31.72		1.02	1.01
Temp. (C)	215.1	156.3	46.7	49.2	107.7		30.0	40.0
Enthalpy (KJ/kG)	598.8	552.6	261.5	265.5	350.7			
Super Ht T. Diff. (C)	70.1		0.0					
Sub. T. Diff. (C)			3.3	97.1	38.6			
Density (kG/m³)	126	14	1279	1283	1083		997	994
Mass Flow (Kg/s)	0.713	0.713	0.713	0.713	0.713		14.64	0.00

(b)

Figure C2.3. (a) Air-Brayton Cycle generating 188 kW net power and 6,900 lbm steam at 15 psig, and (b) ORC R245fa system used as a bottoming cycle to the air-Brayton system.

C2.2.1.1　*Discussion*

The ORC system cannot take advantage of the high exhaust temperature (see Figure C2.2) that is available from the waste heat source. The high exhaust temperature is best suited for use with a steam Rankine Cycle or the suggested air-Brayton Cycle. As much is already known about the technical viability of a steam Rankine Cycle (Figures C2.1(a) and (b)), further elaboration on its technical viability will not be conducted here. There is considerable advantage in using the air-Brayton cycle (see Figure C2.3) due to the reduced number and/or the size of the heat exchangers required as compared with the steam Rankine Cycle. For example, the steam evaporator for the steam Rankine Cycle has a UA[1] size of 6 kWt/K, which is smaller compared to that of the air heater in the air Brayton Cycle, which has a UA size of 8 kWt/K. Additionally, the condenser in the steam Rankine Cycle has a UA equal to 13 (for the 1 bar,a condensing pressure) and 17 for the 0.5 bar,a condensing pressure, whereas the air-Brayton cycle does not require a cooler, but requires a regenerator with a UA of only 1.3 kWt/K. If steam generation is adopted as an option for the air-Brayton Cycle, the UA size is only 4.6 kWt/K.

Based on the current study, the ORC does not appear to be competitive with respect to power output or cost, compared to a steam Rankine Cycle system. It is suggested that the air-Brayton Cycle be offered as an alternative to the steam Rankine Cycle system, as a more competitive option, particularly if either steam recovery at 15 psig or pre-combustion air heating is needed at the application site.

C2.2.2　*Example C2.2*

An industrial client in China has requested a budgetary bid for several ORC Systems that can recover heat energy from 158 kg/s of water at 140°C, cooled to 120°C. This magnitude of heat energy is sufficient to power as many as five modular 300 kW ORC systems or a single, large ORC System. The 300 kW modular system was designed and deployed for several clients by ConShuai Yuan, Associate Professor, Shanghai University of Science and Engineering, cepts NREC (CN). Thus, it is reasonable to propose the use of five, modular ORC Systems as a means of reducing the development costs associated with designing and manufacturing a new ORC system to satisfy the custom

---

[1]The UA parameter is a measure of the size of the heat exchanger. It is determined by taking the ratio of the heat transfer divided by the log mean temperature difference. A large value of UA means that a large heat exchanger is required to provide the necessary heat exchange.

requirements of a new client. The state points for the single and double ORC System components are shown in Figures C2.4 and C2.5. The interconnecting pipe sizes are also provided in Figure C2.4. Of particular relevance to any qualified quote for the Balance of Plant System (heat exchangers, feed pump, control valves, fluid reservoir-hotwell, interconnect piping and assembly) is the summary of the sizes of heat exchangers as defined by the product of the heat transfer coefficient (U) and surface area (A) to form UA. The summary for a single 300 ORC System is shown in Table C2.1. These heat exchanger sizes may be compared to those of an earlier project using a modular 300 kWe ORC system (see Figure C2.4(b)), which used heat exchangers approximately half the size needed for a 300 kWe ORC system that has a higher heat source temperature, and hence, a higher cycle efficiency is achieved. The size and the amount of heat transfer doubles when the larger ORC System is used with two, modular 300 turbine-generator units (TGUs), as shown in Figure C2.5.

The same potential client in China has requested a bid for the manufacturing of two alternative ORC systems which can produce 1.0 and 2.5 MWe from heat sources of 105 and 250 kg/s of water respectively. The state points for the major ORC components for the 1 and 2.5 MWe systems are shown in Figures C2.6 and C2.7. A summary of the specifications for the heat exchangers and the feed pump is given in Tables C2.2 and C2.3.

The thermodynamic analysis summarized in Figures C2.4 through C2.7, and Tables C2.2 and C2.3, can constitute the complete and sufficient technical proposal to the client. A careful review of the figures indicates that the client has everything that is needed to know about the proposed energy recovery system, assuming, of course, that the reader is a trained engineer, familiar with thermodynamic cycles. A summary of the cycle state points shown in the tables provides sufficient and necessary information to contact the various manufacturers of the feed pump and heat exchangers. From the reader's perspective, these cycles may appear to be complicated but certainly should also appear to be familiar. The cycle is easily recognized as a regenerative Rankine Cycle heat engine; the details of which are covered in Chapters 3 and 5. Very recognizable are the use of only the four basic mechanical engineering components: the turbine, pump, heat exchanger and valve, to effectively produce a heat engine that can recover the waste heat from a manufacturing process.

Each of the cycles are identical and required only one cycle to be programmed in order to have the same inputs of energy in the form of the exhaust gas flow rate and temperature to be used with only changes made to the turbine efficiency based on the working fluid flow rate, the operating

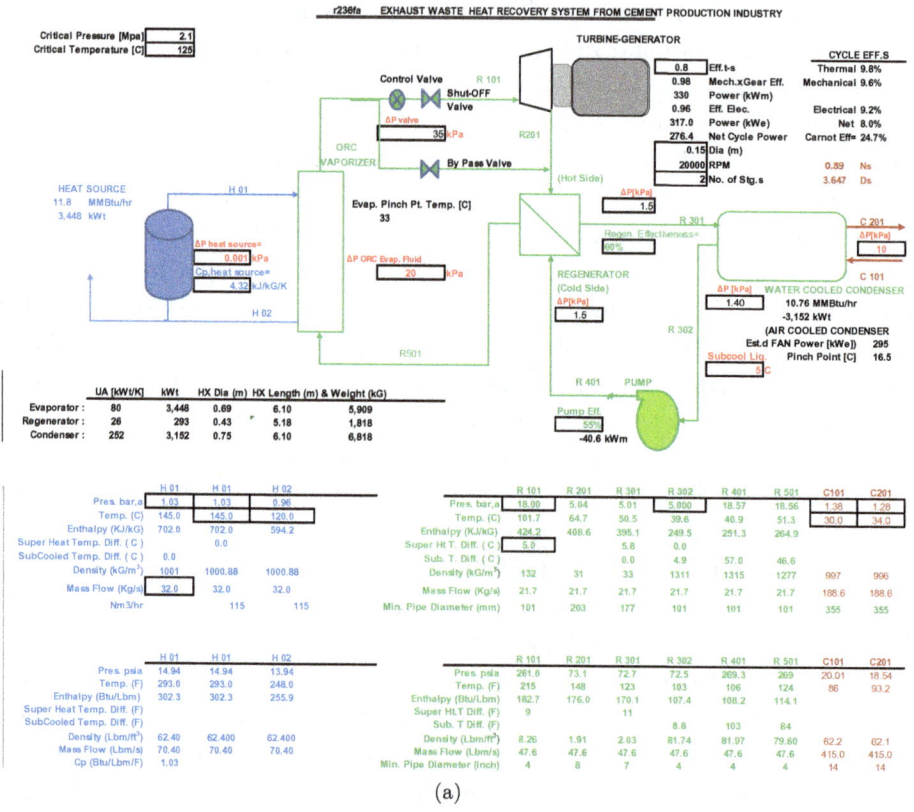

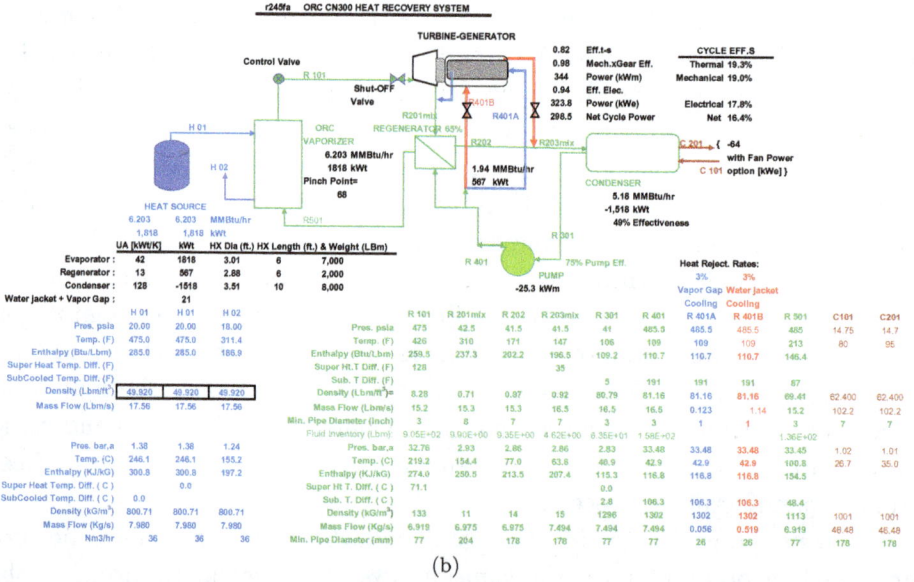

Figure C2.4.   (a) 300 kW ORC with R236fa and (b) modular 300 kW ORC shown for comparison of heat exchanger sizing; the modular ORC uses heat exchangers that were approximately half the size needed for the suggested 300 kWe ORC System.

r236fa EXHAUST WASTE HEAT RECOVERY SYSTEM FROM CEMENT PRODUCTION INDUSTRY

Critical Pressure [Mpa] 2.1
Critical Temperature [C] 125

TURBINE-GENERATOR

		CYCLE EFF.S
0.8	Eff. t-s	Thermal 9.8%
0.98	Mech.xGear Eff.	Mechanical 9.6%
660	Power (kWm)	
0.96	Eff. Elec.	Electrical 9.2%
634.0	Power (kWe)	Net 8.0%
552.7	Net Cycle Power	Carnot Eff= 24.7%
0.15	Dia (m)	
20000	RPM	2.17 Ns
4	No. of Stg.s	2.151 Ds

Control Valve R 101
Shut-OFF Valve
ΔP valve 35 kPa   R201

ORC VAPORIZER   By Pass Valve   (Hot Side)   ΔPt[kPa] 1.5   R 301

HEAT SOURCE H 01
23.5 MMBtu/hr
6.896 kWt

Evap. Pinch Pt. Temp. [C] 33

ΔP heat source= 0.001 kPa
Cp.heat source= 4.32 kJ/kG/K
ΔP ORC Evap. Fluid 20 kPa
H 02
R501

Regen. Effectiveness 60%   C 201   ΔP[kPa] 10   C 101

REGENERATOR (Cold Side)
ΔP[kPa] 1.5
R 302

ΔP [kPa] 1.40   WATER COOLED CONDENSER
21.51 MMBtu/hr
-6,303 kWt
(AIR COOLED CONDENSER
Est.d FAN Power [kWe]) 591
Subcool Liq.   Pinch Point [C] 16.5
5 C

R 401 PUMP
Pump Eff 55%
-81.3 kWm

	UA [kWt/K]	kWt	HX Dia (m)	HX Length (m)	& Weight (kG)
Evaporator :	160	6,896	0.97	6.10	11,364
Regenerator :	51	586	0.60	5.18	3,636
Condenser :	503	6,303	1.06	6.10	13,182

	H 01	H 01	H 02		R 101	R 201	R 301	R 302	R 401	R 501	C101	C201
Pres. bar,a	1.03	1.03	0.96	Pres. bar,a	18.00	5.04	5.01	5.000	18.57	18.96	1.38	1.28
Temp. (C)	145.0	145.0	120.0	Temp. (C)	101.7	64.7	50.5	39.6	40.9	51.3	30.0	34.0
Enthalpy (KJ/kG)	702.0	702.0	594.2	Enthalpy (KJ/kG)	424.2	408.6	395.1	249.5	251.3	264.9		
Super Heat Temp. Diff. (C)		0.0		Super Ht T. Diff. (C)	5.0		5.8	0.0				
SubCooled Temp. Diff. (C)	0.0			Sub. T. Diff. (C)			0.0	4.9	57.0	46.6		
Density (kG/m³)	1001	1000.88	1000.88	Density (kG/m³)	132	31	33	1311	1315	1277	997	996
Mass Flow (Kg/s)	64.0	64.0	64.0	Mass Flow (Kg/s)	43.3	43.3	43.3	43.3	43.3	43.3	377.3	377.3
Nm3/hr		230	230	Min. Pipe Diameter (mm)	152	279	254	152	152	152	482	482

	H 01	H 01	H 02		R 101	R 201	R 301	R 302	R 401	R 501	C101	C201
Pres. psia	14.94	14.94	13.94	Pres. psia	261.0	73.1	72.7	72.5	269.3	269	20.01	18.54
Temp. (F)	293.0	293.0	248.0	Temp. (F)	215	148	123	103	106	124	86	93.2
Enthalpy (Btu/Lbm)	302.3	302.3	255.9	Enthalpy (Btu/Lbm)	182.7	176.0	170.1	107.4	108.2	114.1		
Super Heat Temp. Diff. (F)				Super Ht T. Diff. (F)	9		11					
SubCooled Temp. Diff. (F)				Sub. T Diff. (F)				8.8	103	84		
Density (Lbm/ft³)	62.40	62.400	62.400	Density (Lbm/ft³)	8.26	1.91	2.03	81.74	81.97	79.60	62.2	62.1
Mass Flow (Lbm/s)	140.80	140.80	140.80	Mass Flow (Lbm/s)	95.3	95.3	95.3	95.3	95.3	95.3	830.0	830.0
Cp (Btu/Lbm/F)	1.03			Min. Pipe Diameter (inch)	6	11	10	6	6	6	19	19

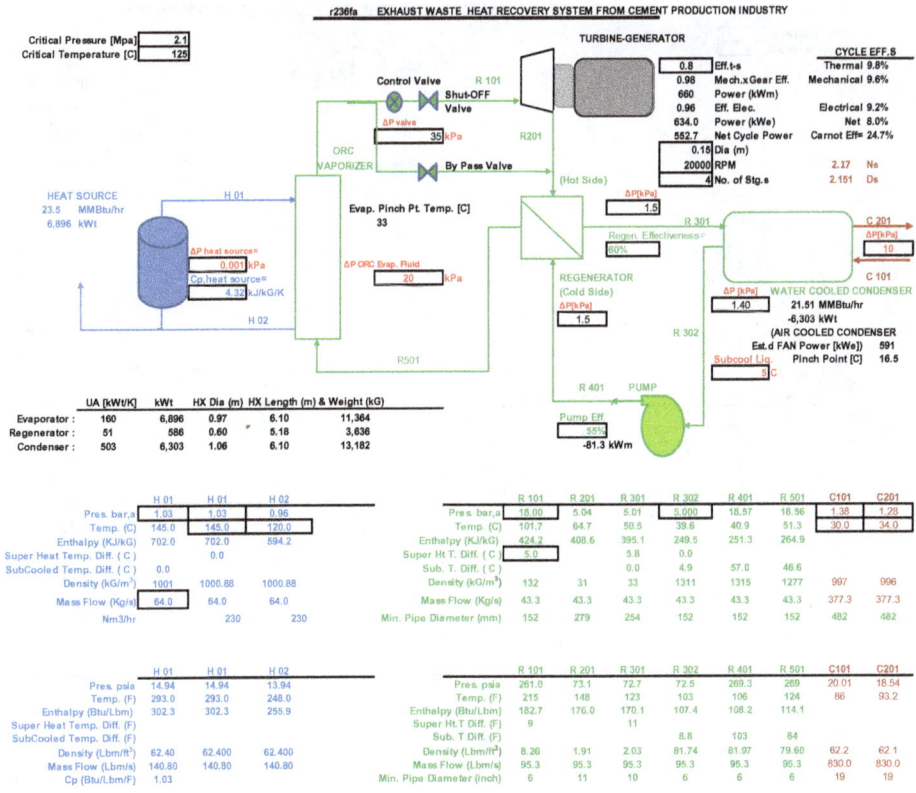

Figure C2.5. Double 300 Turbine-Generator Unit used with a single, larger Balance of Plant.

Table C2.1. Summary of Heat Exchanger Sizes.

	UA [kWt/K]	kWt	HX Dia (m)	HX Length (m) &	Weight (kG)
Evaporator:	80	3,448	0.69	6.10	5,909
Regenerator:	26	293	0.47	4.25	1,818
Condenser:	252	3,152	0.75	6.10	6,818

pressures and the type of working fluid that is used in the cycle. For the cycles shown, two different fluids are used: R245fa and R236fa. Each of these fluids requires a different turbine inlet (high) pressure and turbine outlet (low) pressure with the consequence that the flow rates of the two working fluids are different. The choice of the operating pressures must also consider the operating temperatures that are constrained by the temperature of the waste heat and the available condenser coolant water or air.

In order to fairly compare the performance of these fluids, the first obvious design point is to maximize the net power output of the cycle. What is not

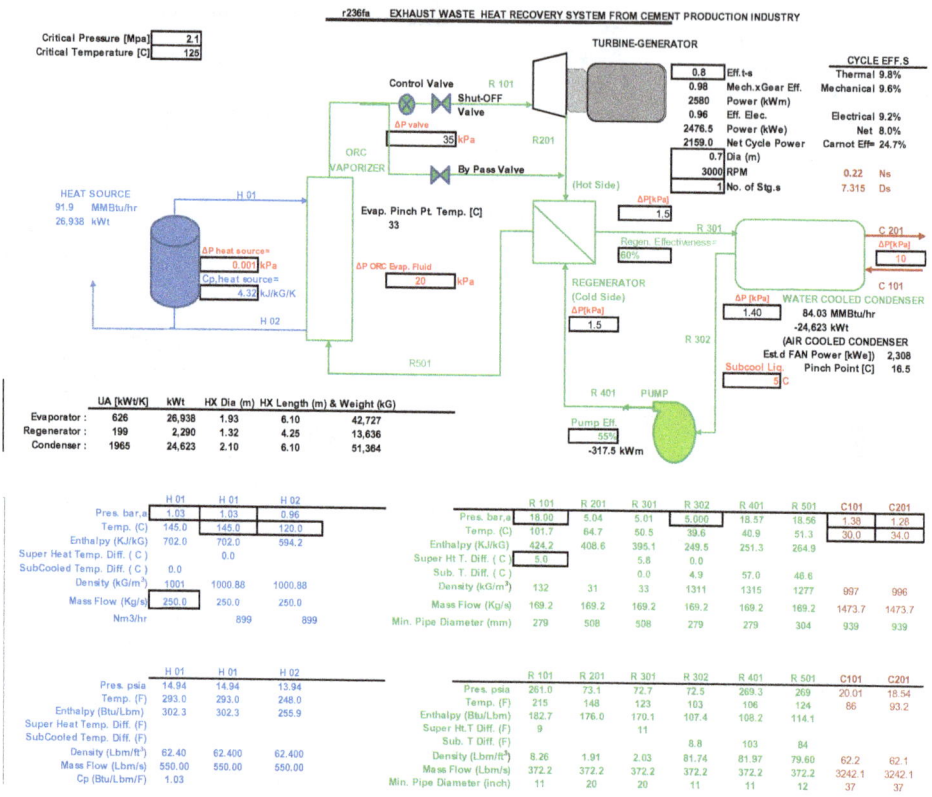

Figure C2.6. State points for ORC using R236fa fluid to generate 2.5 MWe.

as obvious is the need to keep the size of the heat exchangers to a reasonable level. For this purpose, the heat exchanger parameter, UA [kW/K] is used along with the ratio: UA/kW. The parameter UA was first introduced in Chapter 5 and is displayed for each of the heat exchangers in the table shown in each of the figures. An example of this table is shown in Table C2.1. The parameter UA is the ratio of the heat transfer that occurs in the heat exchanger to the log mean temperature difference between the hot and cold fluid streams. The U parameter in the product UA refers to the overall heat transfer coefficient, which is dependent on the fluid properties, speed, and geometry of the heat exchanger. The physical size of the heat exchanger is based on the amount of area, A, required by the heat exchanger to transfer the amount of heat transfer, $\dot{Q}$, required by the application. Thus, the area, A, is determined from knowing the UA and dividing it by the heat transfer coefficient, U, that can be calculated for each application. The sum total of the UA or — better — the sum of the individual surface areas required

r236fa   EXHAUST WASTE HEAT RECOVERY SYSTEM FROM CEMENT PRODUCTION INDUSTRY

| Critical Pressure [Mpa] | 2.1 |
| Critical Temperature [C] | 125 |

TURBINE-GENERATOR

0.8	Eff.t-s
0.98	Mech.xGear Eff.
1083	Power (kWm)
0.96	Eff. Elec.
1040.1	Power (kWe)
906.8	Net Cycle Power
0.5	Dia (m)
7500	RPM
1	No. of Stg.s

CYCLE EFF.S
Thermal 9.8%
Mechanical 9.6%

Electrical 9.2%
Net 8.0%
Carnot Eff= 24.7%

0.35   Ns
8.062   Ds

Control Valve   R 101
Shut-OFF Valve
ΔP valve   35 kPa   R201

ORC VAPORIZER   By Pass Valve

(Hot Side)   ΔP[kPa]   1.5

HEAT SOURCE   H 01
38.6 MMBtu/hr
11,314 kWt

Evap. Pinch Pt. Temp. [C]   33

ΔP heat source= 0.001 kPa
Cp,heat source= 4.32 kJ/kG/K

ΔP ORC Evap. Fluid   20 kPa

H 02

R301   Regen. Effectiveness= 60%

C 201   ΔP[kPa]   10

REGENERATOR (Cold Side)
ΔP[kPa]   1.5

C 101
ΔP [kPa]   WATER COOLED CONDENSER
1.40   35.29 MMBtu/hr
-10,342 kWt
(AIR COOLED CONDENSER
Est.d FAN Power [kWe])   969
Subcool Liq.   Pinch Point [C]   16.5
3 C

R302

R501   R 401   PUMP

Pump Eff.
55%
-133.4 kWm

	UA [kWt/K]	kWt	HX Dia (m)	HX Length (m) & Weight (kG)	
Evaporator :	263	11,314	1.25	6.10	18,182
Regenerator :	84	962	0.85	4.25	5,909
Condenser :	825	10,342	1.36	6.10	21,818

	H 01	H 01	H 02
Pres. bar,a	1.03	1.03	0.96
Temp. (C)	145.0	145.0	120.0
Enthalpy (KJ/kG)	702.0	702.0	594.2
Super Heat Temp. Diff. (C)	0.0		
SubCooled Temp. Diff. (C)	0.0		
Density (kG/m³)	1001	1000.88	1000.88
Mass Flow (Kg/s)	105.0	105.0	105.0
Nm3/hr		378	378

	R 101	R 201	R 301	R 302	R 401	R 501	C101	C201
Pres. bar,a	18.00	5.04	5.01	5.000	18.57	18.56	1.36	1.28
Temp. (C)	101.7	64.7	50.5	39.6	40.9	51.3	30.0	34.0
Enthalpy (KJ/kG)	424.2	408.6	395.1	249.5	251.3	264.9		
Super Ht T. Diff. (C)	5.0		5.8	0.0				
Sub. T. Diff. (C)			0.0	4.9	57.0	46.6		
Density (kG/m³)	132	31	33	1311	1315	1277	997	996
Mass Flow (Kg/s)	71.0	71.0	71.0	71.0	71.0	71.0	618.9	618.9
Min. Pipe Diameter (mm)	177	330	330	203	203	203	609	609

	H 01	H 01	H 02
Pres. psia	14.94	14.94	13.94
Temp. (F)	293.0	293.0	248.0
Enthalpy (Btu/Lbm)	302.3	302.3	255.9
Super Heat Temp. Diff. (F)			
SubCooled Temp. Diff. (F)			
Density (Lbm/ft³)	62.40	62.400	62.400
Mass Flow (Lbm/s)	231.00	231.00	231.00
Cp (Btu/Lbm/F)	1.03		

	R 101	R 201	R 301	R 302	R 401	R 501	C101	C201
Pres. psia	261.0	73.1	72.7	72.5	269.3	269	20.01	18.54
Temp. (F)	215	148	123	103	106	124	86	93.2
Enthalpy (Btu/Lbm)	182.7	176.0	170.1	107.4	108.2	114.1		
Super Ht.T Diff. (F)	9		11					
Sub. T Diff. (F)			8.8	103	84			
Density (Lbm/ft³)	8.26	1.91	2.03	81.74	81.97	79.60	62.2	62.1
Mass Flow (Lbm/s)	156.3	156.3	156.3	156.3	156.3	156.3	1361.7	1361.7
Min. Pipe Diameter (inch)	7	13	13	8	8	8	24	24

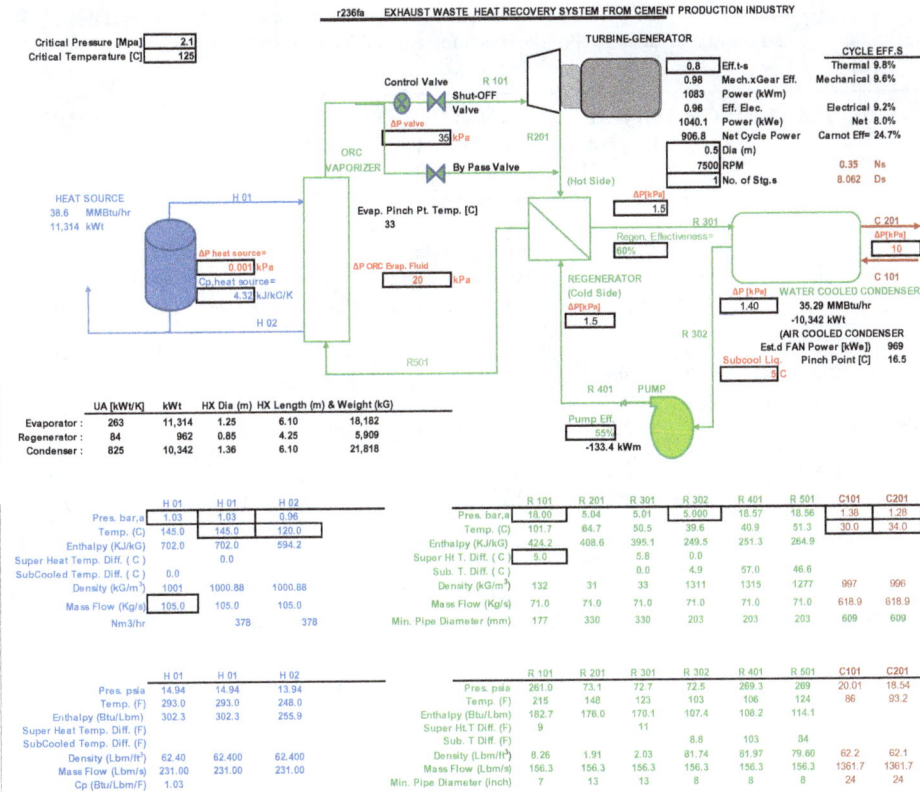

Figure C2.7.   State points for ORC using R236fa fluid to generate 1.0 MWe.

for each heat exchanger can represent how large and how expensive the heat exchangers will be, and ultimately determine the overall cost per kW ($/kW) for the overall system. The summary data for the UA for each of the heat exchangers are shown in each of the figures of the cycle analysis. The value of each exchanger's UA is dependent on the different turbine operating pressures, which affect the overall net power generated by the cycle. An iterative process of changing the cycle operating conditions can quickly discern the cycle that has the minimum surface area for the maximum net power generated and thus enable a quick estimate for the minimum cost per kW ($/kW).

For example, for the R236fa cycle (Figure C2.4(a)) and the R245fa cycle (Figure C2.4(b)) the value for the UA per kW is 1.2 and 0.39, while producing 276 kW and 144 kW, respectively. A study of the steam Rankine Cycle systems shown in Figure C2.1 indicates a UA/kW ratio of less than $0.09\,kW_{thermal}/K - kW_{mech}$. Clearly from a cost per kW perspective, the

Table C2.2.   (a) Summary of the heat exchanger and feed pump specifications for the 2.5 MWe ORC, and (b) summary of heat exchanger sizes for the 2.5 MWe ORC System.

(a)

ORC HEAT EXCHANGER DESIGN AND FEED PUMP SPECIFICATIONS

Turbine Power (kWm)=1083

Gen. Power (kWe)=1040

NET System (kWe)=907

	EVAPORATOR		Recuperator		Condenser		r236fa Feed Pump	
	Hot Exh.Gas	Cold r236fa	Hot r236fa	Cold r236fa	Hot r236fa	Cold Water	Design Point	Mfg.r's Spec.
Temp, in (F)	293	124	148	106	123	86	103	106
Sat. Liq Temp (F)	290	208			112			
Sat. vap. Temp (F)	267	207			112			
Temp,out (F)	248	215	123	124	103	93.2	108.2	114
Pres. in (psia)	14.9	263.94	73.1	269.31	73.90	20.01	72.5	69
Pres. out (psia)	13.9	261	72.7	269.09	72.50	18.54	269.3	296
Flow Rate (Lbm/s)	231.00	156.3	156.3	156.3	156.3	1361.7	156.3	187.6
Density (Lbm/ft^3)			1.91	81.97	79.19	83.83	81.72	81.33
Viscosity (Lbf-s/ft^2)			2.60E-07	5.05E-06	2.47E-07	2.32E-07	4.98E-06	4.87E-06
Temp, in (C)	145	51	65	41	50	30	40	41
Sat. Liq Temp (C)	143	98			45			
Sat. vap. Temp (C)	131	97			45			
Temp,out (C)	120	102	50	51	40	34	42	45
Pres. in (bar,a)	1.03	18.20	5.04	18.57	5.10	1.38	5	5
Pres. out (bar,a)	0.96	18.00	5.01	18.56	5.00	1.28	19	20
Flow Rate (Kg/s)	105.00	71.05	71.05	71.05	71.05	618.94	71	85
Density (kg/m^3)			31	1315	1270	1345	1311	1304
Viscosity (N-s/m^2)			1.24E-05	2.42E-04	1.18E-05	1.11E-05	2.38E-04	2.33E-04

(b)

	UA [kWt/K]	kWt	HX Dia (m)	HX Length (m)&	Weight (kG)
Evaporator:	626	26,938	1.93	6.10	42,727
Regenerator:	199	2,290	1.32	4.25	13,636
Condenser:	1965	24,623	2.10	6.10	51,364

optimization of a Rankine Cycle is not dependent entirely on the net power output but should also consider the UA per kW (UA/kW), a parameter that begins to identify the overall cost per kW for the power recovery system.

    An additional parameter to always to consider when optimizing the cycle performance is the overall, First Law Thermodynamics efficiency of the cycle. While high cycle efficiency is important for the "prime mover" heat engine

Table C2.3.   Heat exchanger specifications for the 1 MWe ORC System.

Turbine Power (kWm)=1083
Gen. Power (kWe)=1040
NET System (kWe)=907

| | EVAPORATOR | | Recuperator | | Condenser | | r236fa Feed Pump | |
	Hot Water	Cold r236fa	Hot r236fa	Cold r236fa	Hot r236fa	Cold Water	Design Point	Mfg.r's Spec.
Temp, in (F)	293	124	148	106	123	86	103	106
Sat. Liq Temp (F)	290	208			112			
Sat. vap. Temp (F)	267	207			112			
Temp, out (F)	248	215	123	124	103	93.2	108.2	114
Pres. in (psia)	14.9	263.94	73.1	269.31	73.90	20.01	72.5	69
Pres. out (psia)	13.9	261	72.7	269.09	72.50	18.54	269.3	296
Flow Rate (Lbm/s)	231.00	156.3	156.3	156.3	156.3	1361.7	156.3	187.6
Density (Lbm/ft^3)			1.91	81.97	79.19	83.83	81.72	81.33
Viscosity (Lbf-s/ft^2)			2.60E-07	5.05E-06	2.47E-07	2.32E-07	4.98E-06	4.87E-06
Temp, in (C)	145	51	65	41	50	30	40	41
Sat. Liq Temp (C)	143	98			45			
Sat. vap. Temp (C)	131	97			45			
Temp, out (C)	120	102	50	51	40	34	42	45
Pres. in (bar,a)	1.03	18.20	5.04	18.57	5.10	1.38	5	5
Pres. out (bar,a)	0.96	18.00	5.01	18.56	5.00	1.28	19	20
Flow Rate (Kg/s)	105.00	71.05	71.05	71.05	71.05	618.94	71	85
Density (kg/m^3)			31	1315	1270	1345	1311	1304
Viscosity) (N-s/m^2)			1.24E-05	2.42E-04	1.18E-05	1.11E-05	2.38E-04	2.33E-04

that is consuming the primary fuel supply, for a heat engine that is operating with heat input from a waste heat stream, the more important consideration should be given to the net amount of power that can be generated rather than the net efficiency of the waste heat recovery of the heat engine. In many instances, the objective of increasing the cycle efficiency requires that the temperature profile of the heat source maintain a high discharge temperature in order to have the Rankine Cycle's operating pressure and temperature to be as high as is practical. However, a high discharge temperature results in less waste heat recovery.

The lesson here is clear: A careful review of the performance of the Rankine Cycle, the primary cycle of choice in this Case Study, or any heat engine, must be done with a strong understanding of all of the parameters that affect the cycle's thermodynamics performance, with an appreciation that

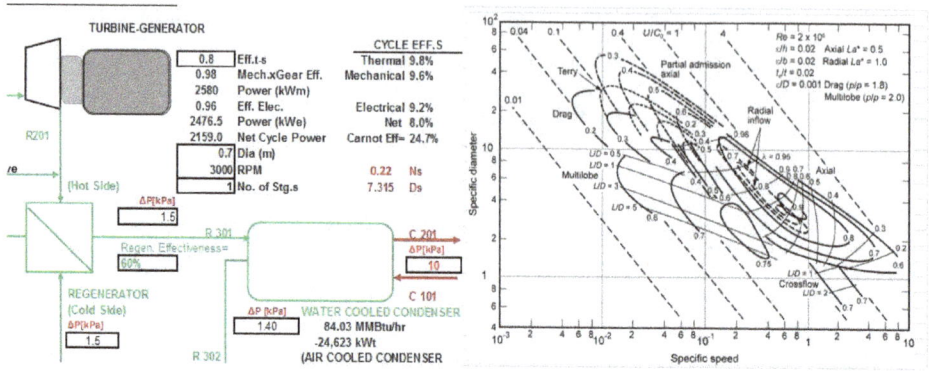

(a)

(b)

Figure C2.8.   (a) Charts and specifications/drawing of a radial inflow turbine, and (b) Photo of a radial inflow turbine (3,000 rpm; 0.7 M dia.). (Courtesy of concepts NREC, Ltd.)

high performance may come with a higher cost and overall larger size of the final system.

Case Study 3

Total Recoverable Energy Cycle (TREC): An Advanced Thermodynamics Cycle to Simultaneously Recover Low- and High-Grade Waste Energy

C3.1 Background to advanced waste heat recovery systems

This case study identifies another cycle which is unique in its ability to recover both low- and high-grade waste heat energy from a single prime mover energy source. Low-grade energy is defined here as an energy source with a temperature below 250°F (120°C). On the other hand, high-grade energy is defined as an energy source with a temperature above 250°F. The advanced cycle has been named the Total Recoverable Energy Cycle (TREC) to recognize this potential for a near 100% waste heat recovery from a single source of both low- and high-grade waste heat. TREC is an advanced hybrid binary Organic Rankine Cycle (ORC). This hybrid-binary cycle uses a single working fluid at different operating pressures and can provide more power with fewer heat exchangers and turbine-generator units (TGUs) than if two regular heat recovery ORCs were used to separately recover the low- and high-grade waste heat energy. TREC is also designed to enable either the high- or low-pressure circuits to be partially-loaded or stopped without requiring a complete shutdown of the system. This enables some power to be continually generated even when the heat source has been reduced.

In a typical ORC System, a TGU generates electrical power by expanding the pressurized, organic working fluid that has been evaporated and superheated in an evaporator using the high-grade waste heat from the prime mover. Usually, only the high-grade heat is recovered in the high-pressure evaporator. After the fluid has expanded through the turbine to produce useful mechanical or electrical power, it is then condensed and pumped through a regenerator and/or feed heater, before returning to the evaporator in a

345

SOURCES OF ENERGY FOR COMBINED
HEAT AND POWER GENERATION

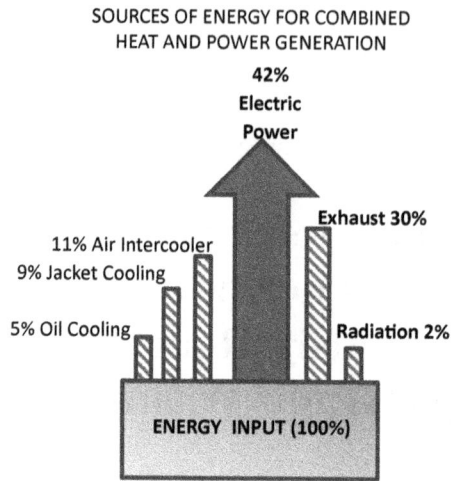

Figure C3.1.   Illustration of the fraction of waste heat available from the burning of fuel.

closed loop. The heat recovery process is an ongoing process that will happen as long as waste heat energy that is otherwise rejected into the atmosphere is produced by the prime mover. The TREC system is intended for application in cases in which the low-grade waste heat's temperatures fall below the temperature considered viable for conventional water/steam cycles.

Dual streams of waste heat are readily available from reciprocating prime movers in which the heat rejected by the water jacket and, separately, the heat rejected by the lube oil heater are typically below 200°F, while the exhaust gas waste heat has a much higher temperature, typically above 700°F. Figure C3.1, derived from a Caterpillar Engine Manufacturer's website, illustrates the fraction of waste heat that becomes available from the burning of fuel. This same fraction of waste energy is common across (>2,000 kWe) internal combustion engines used in cogeneration applications, regardless of engine manufacturer.

Gas turbines (GTs) and older fossil-fuel electric power plants using steam turbines also provide both low- and high-grade waste energy streams. Table C3.1 provides a summary of typical sources of waste heat from prime movers.[1] Table C3.2 quantifies the ratio of the low- and high-grade heat energy rejected by the prime mover per unit of electric power generated. The low-grade heat energy in these scenarios is often ignored or simply used for low-grade heating where temperatures below 40°C are required, achieving

---

[1]See www.energymanagementtraining.com/bee_draft_codes/best_practices-manual_cogen.pdf.

Table C3.1.   Summary of typical sources of waste heat from prime movers.

Cogenerateion technology uses different combinations of power and heat producing equipment, which are numerous. Most widely used combinations are mentioned below.

i.   **Stream turbine & fired boiler based cogeneration system**

Boiler	Steam turbine
Coal/Lignite fired plant	Back-pressure steam turbine
Liquid Fuel fired plant	Extraction & condensing steam turbine
Natural gas fired plant	Extraction & back-pressure steam turbine
Bagasse/Husk fired plant

ii.   **Gas turbine based cogeneration system**

Gas turbine generator	Waste heat recovery
Natural gas fired plant	Steam generation in unfired/supplementary
Liquid fuel fired plant	fired/fully fired waste heat recovery boiler
Utilisation of steam directly in process
Utilisation of steam for power generation
From steam turbine generator
[Cogeneration-cum-combined cycle]
Absorption Chiller [CHP System]
Utilisation of heat for direct heating

iii.   **Reciprocating engine based cogeneration system**

Reciprocating engine	Waste heat recovery
Liquid fuel fired plant	Steam generation in unfired/supplementary
Natural gas fired plant	fired waste heat recovery boiler
Absorption Chiller [CHP System]
Utilsation of steam directly in process
Utilsation of heat for direct heating

Table C3.2.   Heat-to-power ratio for typical prime movers.

Cogeneration system	Heat-to-power ratio (kW_{th}/kW_e)	Power output (as percent) of fuel input)	Overall efficiency %
Back-pressure steam turbine	4.0–14.3	14–28	84–92
Extraction-condensing steam turbine	2.0–10	22–40	60–80
Gas turbine	1.3–2.0	24–35	70–85
Combined cycle (Gas plus steam turbine)	1.0–1.7	34–40	69–83
Reciprocating engine	1.1–2.5	33–53	75–85

the corresponding overall efficiency shown in Table C3.2. For example, the recovery of heat energy from the lube oil that cools the bearings on a gas turbine is often ignored even though exhaust gas heat energy is — now more than ever — being recovered to improve the overall energy efficiency of the gas turbine, because the temperature of the oil rarely exceeds 60°C.

Low and high-grade energy streams are also prevalent in many energy-intensive industrial processes. However, the ratio of required heat to the required electric power for industrial processes varies across industries, as also shown in Table C3.2. If the industry heat-to-power ratio is less than what is available in the prime mover, as shown in Table C3.2, the potential overall energy efficiency of the prime mover used in the industrial process is inherently compromised, and the maximum overall efficiency cannot be achieved. For example, most industrial processes require both high-pressure steam above 100 psig and also low-pressure steam at 15 psig or less. The high-pressure steam is produced in a fired boiler while the low-pressure steam is usually produced from the same high-pressure boiler but only after having the pressure reduced to 15 psig in a Pressure Reducing Valve (PRV). The low-pressure steam is then delivered to various processes that are limited to 15 psig (250°F or less). It is interesting to note that more steam is utilized at 15 psig for safety purposes. In the United States and elsewhere in most developed countries, there are safety regulations that require a Boiler Operator-Engineer to continuously monitor any source of steam that produces steam above 15 psig. As another example, most natural gas pipeline compressor stations utilize internal combustion and, in some limited locations, gas turbine engines to drive the compressors because they can be fueled by the natural gas pressurized by the compressors. In both of these applications (the industrial processes requiring a fired boiler and natural pipeline compressor stations) there is considerable waste heat that can be recovered from three distinct sources: (1) The high-grade engine exhaust heat at temperatures that are typically above 1,000°F (540°C); (2) the engine jacket's low-grade waste energy at temperatures less than 220°F (110°C); and (3) the pressurized, moderate temperature of the air at the compressor discharge that is usually less than 400°F (200°C). Despite the advantages of recovering the compressor discharge heat, it is very rare that there's an attempt to recover this energy for electric power generation due to the complexity of the necessary waste heat recovery system. At present, there is no single heat engine

Table C3.3. Typical heat-to-power ratios for energy intensive industries.

Industry	Minimum	Maximum	Average
Breweries	1.1	4.5	3.1
Pharmaceuticals	1.5	2.5	2.0
Fertilizer	0.8	3.0	2.0
Food	0.8	2.5	1.2
Paper	1.5	2.5	1.9

cycle that can effectively and simultaneously recover waste energy from such waste heat sources that have distinctly different temperatures from the same prime mover source.

The TREC system uses a single ORC system that can recover almost 100% of the waste heat generated by the prime mover. TREC utilizes a single turbine-generator unit, which captures the primary, higher-enthalpy heat flow, as well as one or more lower-enthalpy secondary heat flows to produce turbine output power. Thus, using the TREC system, a single turbine generator can generate power using two sources of waste heat energy: the prime mover's exhaust gas and the heat lost through the water jacket (and/or lube oil) — both heat sources of 500 to 900°F in temperature. This reduces capital, operating and maintenance costs by eliminating the need to employ different turbine generators for each heat source. Additionally, heat recovery from the "low" temperature heat source may provide more supplemental power compared to the recovery of the waste heat from the higher temperature heat source. This increases the profitability of the cogeneration system given the recovery of low-temperature energy which would otherwise be wasted.

In the following case study, comparisons will be made between the TREC system and two state-of-the-art, Rankine Cycles in terms of the amount of power that may be recovered using each cycle. One option consists of a basic Rankine Cycle which uses the low-grade heat to preheat the liquid organic fluid entering the vapor generator. Such a cycle is equivalent to a pre-heater in a typical Rankine Cycle. Like the TREC system, this option combines the recovery of low- and high-grade energy in a single cycle. However, this cycle is rarely used for enhanced energy recovery because it is unable to completely recover all the low-grade heat generated by the prime mover, and thus does not offer substantial improvement in power generation.

The second state-of-the-art ORC system is commonly used to recover only high-grade waste energy, usually from the exhaust gas heat of an internal combustion or GT engine prime mover. For comparison purposes, this system will be considered as the baseline ORC system.

C3.2 The TREC system concept

The Total Recoverable Energy Cycle (TREC) enables 100% energy recovery from the waste heat generated by the exhaust, water jacket and lube oil. Figures C3.2(a) and (b) show the TREC cycle diagrams required for heat recovery from Caterpillar (CAT) and Wartsila prime movers. The TREC systems of Figures C3.2(a) and (b) use a single, multi-stage (3,600 rpm),

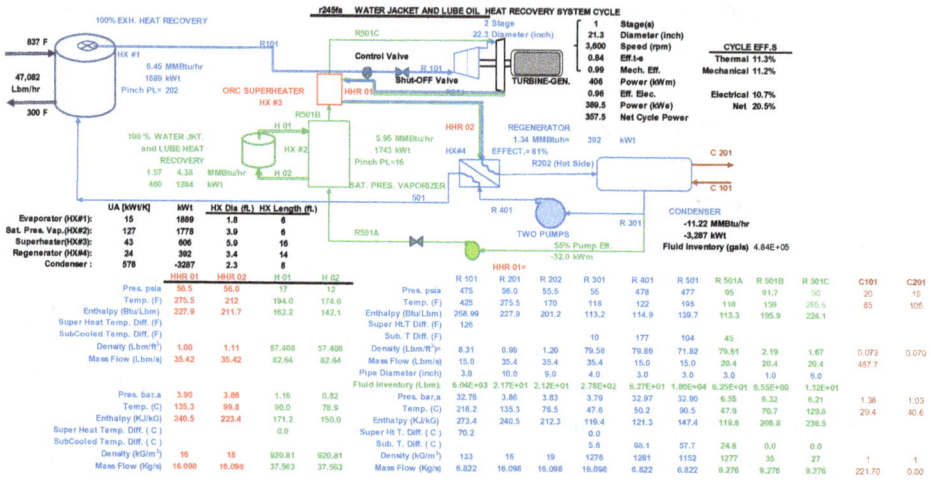

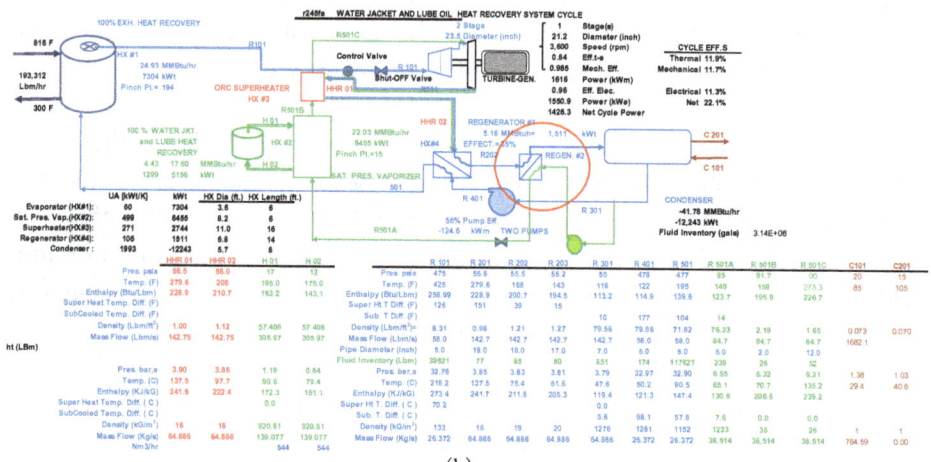

Figure C3.2.   (a) State point pressures, temperatures and flow rates for a TREC system applied to a 4,000-kWe CAT continuous duty engine; (b) State point pressures, temperatures and flow rates for the advanced TREC system applied to an 8,300-kWe Wartsila (20V34DF) continuous duty engine.

dual-entry turbine which allows the high pressure from the exhaust gas waste heat evaporator and the low pressure from the water jacket and lube oil heat recovery to enter the turbine. The second stage of the two-stage turbine utilizes the combined flow rate from the two heat sources. The discharge of the single turbine is then used to superheat the low-pressure working fluid which has been vaporized into a saturated vapor condition by the water jacket and lube oil waste heat. The heat from the turbine is used to superheat this

Table C3.4. Performance comparison of a TREC system vs. state-of-the-art cycles.

ELEC [kWe] per Wartsila Duef Fuel Engine	TGU NET ELEC [kWe] per CAT Engine	TGU NET CYCLE IMPROVE EFF. (%)	(%) IMPROVED NET PRIME POWER	WARTSILA DF 20V34DF ENGINE TURBINE-GENERATOR DESIGN TYPE
1551	1,426	14.5%	17.9%	CN'S TREC SYSTEM: A Hybrid-Binary Rankine Cycle for 100(%) Ht. Recov. with a Single, Multi-Stage Rotor with a High and Low Pres. Enry
1510	1,293	3.8%	16.2%	Modified ORC with Water Jacket Preheating PRC System with 22% Water and Lube Heat Recovery and 100% Exh. Heat Recov.
1384	1,246	~	15.6%	BASELINE ORC System with only Exhaust Heat recovery

fluid stream, thus raising its temperature and increasing the efficiency of the cycle. The result is approximately 390 kWe per CAT engine, which represents a net ORC cycle efficiency of 10.4% for a 9% overall electric power improvement in the power output of the CAT prime mover. This corresponds to a reasonable second law efficiency (defined as the ratio of the actual cycle efficiency to the Carnot maximum efficiency for the heat source and sink) of 35%.

C3.2.1 *Comparison 1: TREC system vs Wartsila 20V34DF engine: Energy recovery*

An advanced TREC system uses a second regenerator, as shown in Figure C3.1, to recover additional heat from the exhaust of the dual pressure turbine. The recovered energy is used to preheat the lower pressure working fluid used to recover the low-grade energy from the engine or industrial process. Table C3.4 provides another comparison of the net power output of the TREC system, compared to state-of-the-art cycles that can recover some (but not all) or none of the low-grade energy. For the purpose of this comparison, a Wartsila 20V34DF engine that is commercially available and produces approximately 8.4 MW of power in a dual-fuel model has been used. Such engines are used for "24/7" or "round-the-clock" continuous power generation. Engines like the Wartsila, Caterpillar, Solar and other similar internal combustion or gas turbine engines rated 2 MW and above are used as Distributed Power Generation systems in remote locations that are too distant from utility-size power plants for efficient power distribution.

Table C3.5.   Specifications of potential energy recovery turbines.

CN300 MODULE*		CN2500	
No. Stages of Stages	1	No. Stages of Stages	1
Avg. Blade Height (in)	0.74	Avg. Blade Height (in)	0.17
Avg. Wheel Hub Diameter (in)	3.71	Avg. Wheel Hub Diameter (in)	19.64
Avg. Wheel Tip Diameter (in)	5.19	Avg. Wheel Tip Diameter (in)	19.97
RPM	20,000	RPM	3,600
Est. Axial Thrust (lbf)	122.4	Est. Axial Thrust (lbf)	1937.7
*ref.: Concepts NREC, LTD		*ref.: Concepts NREC, LTD	
ORC Turbine module		ORC Turbine module	

TREC is unique in that it enables the highest energy recovery from 100% of the low- and high-grade waste heat produced by the prime mover using a multi-stage, dual-entry single turbine. Although this comes at the expense of a decreased efficiency compared to a basic ORC system which can only recover the high-grade energy, the total amount of recoverable power is greater when the TREC system is used instead of a basic ORC system. The hybrid cycle achieves this by using most of the high temperature energy from the discharge of the turbine to superheat the low-grade energy ORC circuit.

A preliminary turbine analysis indicates that a 3-stage turbine, having an overall pressure ratio of 8.2:1 and a diameter of 20 to 23 inches, can accommodate the power and working fluid flow rate (36 lbm/s) from an ORC system applied to recover the waste heat from the CAT engine. The first two stages would utilize 43% of the flow rate and have an overall pressure ratio of 5:1. The multi-stage, dual-pressure entry turbine uses an oil buffer, shaft seal and hydro-dynamic bearing design. The specifications of two viable turbine designs at 20,000 rpm and at 3,600 rpm are given in Table C3.5.

C3.2.2   *Comparison 2: TREC system vs basic Rankine Cycle — Energy recovery*

For comparison with the TREC system, an alternate "current practice" cycle which also recovers low- and high-grade waste energy from a prime mover using a single ORC system is shown in Figure C3.3. This "current practice cycle" shares a similar configuration with that of a basic Rankine Cycle, shown in Figure C3.4, except that the low-grade heat recovered from the water jacket/lube oil is used to preheat the high-pressure working fluid that is to be vaporized by the exhaust gas. This preheating by the low-grade waste heat replaces a part of the function of the regenerator in a basic Rankine

r245fa   EXHAUST HEAT RECOVERY SYSTEM CYCLE WITH WATER JACKET + LUBE HEAT RECOVERY

TURBINE-GENERATOR

		CYCLE EFF.S
0.84	Eff.t-s	Thermal 17.4%
0.98	Mech.xGear Eff.	Mechanical 17.0%
383	Power (kWm)	
0.96	Eff. Elec.	Electrical 16.3%
367.4	Power (kWe)	Net 14.8%
332.0	Net Cycle Power	

Control Valve R 101
Shut-OFF Valve   R201

ORC VAPORIZER
6.45 MMBtu/hr
1889 kWt
Pinch Point= 44   R502

80%
REGENERATOR   R202 (Hot Side)
1.91 MMBtu/hr
561 kWt
REGENERATOR (Cold Side)

C 201

CONDENSER   C 101
6.46 MMBtu/hr
-1,893 kWt

EXH. HEAT SOURCE
6.45 MMBtu/hr
1,889 kWt

Water Jacket plus Lube Ht. recovery   H 04   R 401

PUMP   55% Pump Eff.
-35.4 kWm

R 301

	UA [kWt/K]	kWt
Evaporator :	20	1889
Regenerator :	24	561
Water Jacket & Lube OilHeat :	24	359
Condenser :	568	-1893

	H 01	H 02	H 03	H 04		R 101	R 201	R 202	R 301	R 401	R 501	R 502	C101	C201
Pres. psia	15	15	20	19	Pres. psia	475	55.5	55.2	55	478.2	478	478	14.75	14.7
Temp. (F)	837.0	300.0	194	175	Temp. (F)	425	319	208	122	126	180	256.4	85	115
Enthalpy (Btu/Lbm)	213.4	76.5	162.0	142.5	Enthalpy (Btu/Lbm)	259.1	239.2	210.7	114.5	116.3	134.5	163.0		
Super Heat Temp. Diff. (F)					Super Ht.T Diff. (F)	127								
SubCooled Temp. Diff. (F)					Sub. T Diff. (F)				6	173	119	43		
Density (Lbm/ft³)	57.408	57.408	62.4	62.4	Density (Lbm/ft³)=	8.31	0.92	1.10	79.16	79.45	73.62	62.42	0.073	0.069
Mass Flow (Lbm/s)	13.08	13.08	18.51	18.51	Mass Flow (Lbm/s)	18.6	18.6	18.6	18.6	18.6	18.6	18.6	249.2	249.2
Pres. bar.a	1.02	1.01	1.38	1.31	Pres. bar.a	32.76	3.83	3.81	3.79	32.98	32.98	32.97	1.02	1.01
Temp. (C)	447.2	148.9	90.0	79.2	Temp. (C)	218.4	159.3	97.7	49.9	52.5	82.3	124.7	29.4	46.1
Enthalpy (KJ/kG)	225.3	80.7	171.0	150.4	Enthalpy (KJ/kG)	273.5	252.5	222.4	120.8	122.7	141.9	172.1		
Super Heat Temp. Diff. (C)	0.0				Super Ht T. Diff. (C)	70.3				0.0				
SubCooled Temp. Diff. (C)					Sub. T. Diff. (C)				3.3	95.9	66.1	23.7		
Density (kG/m³)	920.81	920.81	1000.88	1000.88	Density (kG/m³)	133	15	18	1270	1274	1181	1001	1	1
Mass Flow (Kg/s)	5.945	5.945	8.414	8.414	Mass Flow (Kg/s)	8.475	8.475	8.475	8.475	8.475	8.475	8.475	113.29	113.29
Nm3/hr	23	23	30	30										

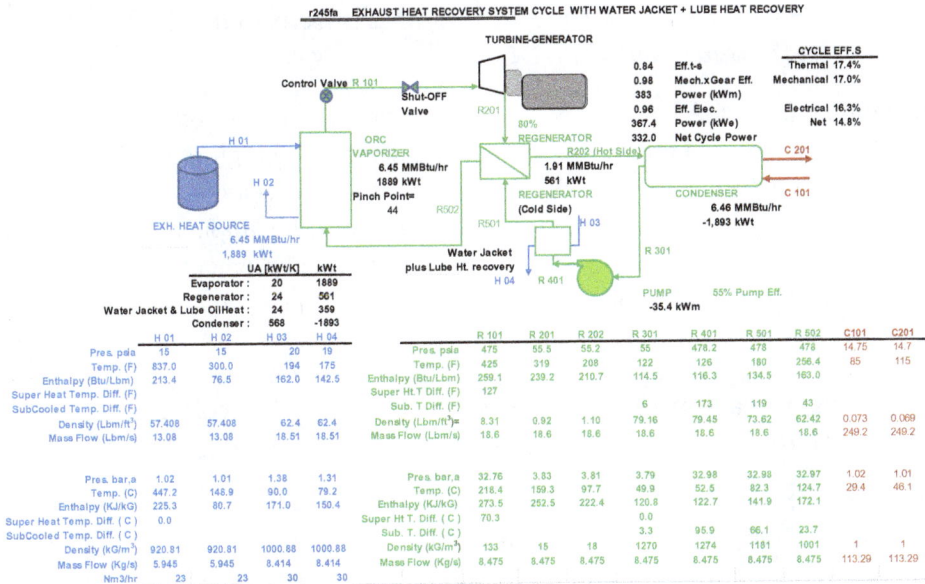

Figure C3.3.   An optional cycle for the partial recovery of low-grade heat energy and 100% recovery of the high-grade energy from the prime mover.

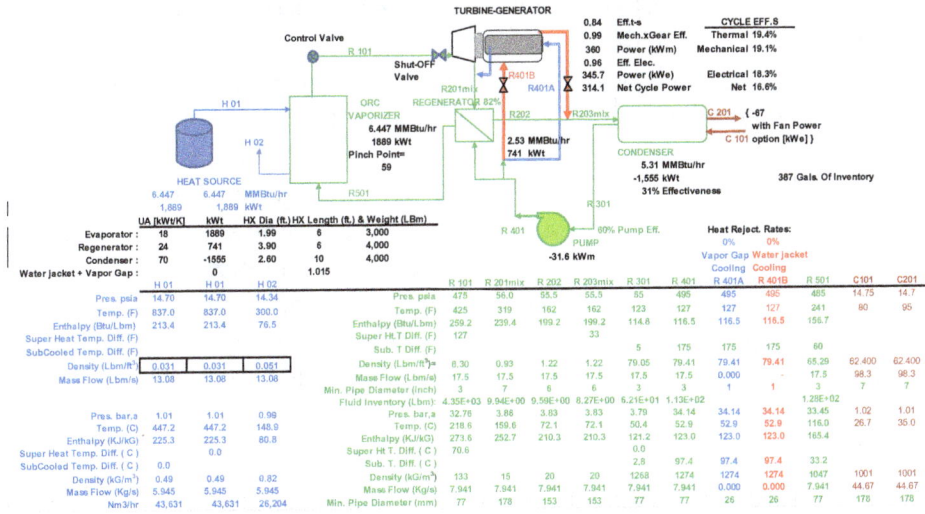

TURBINE-GENERATOR

		CYCLE EFF.S
0.84	Eff.t-s	Thermal 19.4%
0.99	Mech.xGear Eff.	Mechanical 19.1%
360	Power (kWm)	
0.96	Eff. Elec.	Electrical 18.3%
345.7	Power (kWe)	Net 16.6%
314.1	Net Cycle Power	

Control Valve   R 101
Shut-OFF Valve   R201mix   R401B   R401A

ORC VAPORIZER
6.447 MMBtu/hr
1889 kWt
Pinch Point= 59

REGENERATOR 82%   R202   R203mix

2.53 MMBtu/hr
741 kWt
CONDENSER
5.31 MMBtu/hr
-1,555 kWt
31% Effectiveness

C 201   { -67 with Fan Power
C 101 option [kWe] }
387 Gals. Of Inventory

HEAT SOURCE
6.447   6.447   MMBtu/hr
1,889   1,889   kWt

R 501   R 301   R 401

69% Pump Eff.
PUMP
-31.6 kWm

Heat Reject. Rates:
0%   0%
Vapor Gap Water jacket
Cooling Cooling

	UA [kWt/K]	kWt	HX Dia (ft.)	HX Length (ft.) & Weight (LBm)
Evaporator :	18	1889	1.99	6	3,000
Regenerator :	24	741	3.90	6	4,000
Condenser :	70	-1555	2.60	10	4,000
Water jacket + Vapor Gap :		0		1.015

	H 01	H 01	H 02		R 101	R 201mix	R 202	R 203mix	R 301	R 401	R 401A	R 401B	R 501	C101	C201
Pres. psia	14.70	14.70	14.34	Pres. psia	475	56.0	55.5	55.5	55	495	495	495	485	14.75	14.7
Temp. (F)	837.0	837.0	300.0	Temp. (F)	425	319	162	162	123	127	127	127	241	80	95
Enthalpy (Btu/Lbm)	213.4	213.4	76.5	Enthalpy (Btu/Lbm)	259.2	239.4	199.2	199.2	114.6	116.5	116.5	116.5	158.7		
Super Heat Temp. Diff. (F)				Super Ht.T Diff. (F)	127			33							
SubCooled Temp. Diff. (F)				Sub. T Diff (F)					5	175	175	175	60		
Density (Lbm/ft³)	0.031	0.031	0.051	Density (Lbm/ft³)=	8.30	0.93	1.22	1.22	79.05	79.41	79.41	79.41	65.29	62.400	62.400
Mass Flow (Lbm/s)	13.08	13.08	13.08	Mass Flow (Lbm/s)	17.5	17.5	17.5	17.5	17.5	0.000		17.5	98.3	98.3
				Min. Pipe Diameter (inch)	3	7	6	6	3	3	1	1	3	7	7
				Fluid Inventory (Lbm):	4.35E+03	9.94E+00	9.59E+00	8.27E+00	6.21E+01	1.13E+02			1.28E+02		
Pres. bar.a	1.01	1.01	0.98	Pres. bar.a	32.76	3.86	3.83	3.83	3.79	34.14	34.14	34.14	33.45	1.02	1.01
Temp. (C)	447.2	447.2	148.9	Temp. (C)	218.6	159.6	72.1	72.1	50.4	52.9	52.9	52.9	116.0	26.7	35.0
Enthalpy (KJ/kG)	225.3	225.3	80.8	Enthalpy (KJ/kG)	273.6	252.7	210.3	210.3	121.2	123.0	123.0	123.0	165.4		
Super Heat Temp. Diff. (C)	0.0			Super Ht T. Diff. (C)	70.6				0.0						
SubCooled Temp. Diff. (C)	0.0			Sub. T. Diff. (C)					2.8	97.4	97.4	97.4	33.2		
Density (kG/m³)	0.49	0.49	0.82	Density (kG/m³)	133	15	20	20	1268	1274	1274	1274	1047	1001	1001
Mass Flow (Kg/s)	5.945	5.945	5.945	Mass Flow (Kg/s)	7.941	7.941	7.941	7.941	7.941	7.941	0.000	0.000	7.941	44.67	44.67
Nm3/hr	43,631	43,631	26,204	Min. Pipe Diameter (mm)	77	178	153	153	77	77		26	26	77	178	178

Figure C3.4.   The baseline ORC system.

Cycle system. However, this cycle is unable to completely capture all of the low-grade energy. Although this cycle enables the complete recovery of the exhaust gas waste heat, only 25% of the water jacket/lube oil waste heat is recovered. Thus, it is not as efficient as the proposed TREC system.

C3.2.3   *Summary and conclusion: Maximizing engine efficiency with TREC*

In summary, the CAT prime mover is very efficient and its high output power of 4 MWe makes it suitable for continuous duty. The high efficiency results in less rejected heat in the exhaust gas or engine jacket compared to other engines. However, the analysis of the TREC system applied to this CAT engine reveals an impressive 9% performance improvement of this prime mover's output power when paired with the TREC system.

C3.3   Application of TREC system in MTU Detroit Diesel engines

This case study continues with an analysis of a smaller engine that is used for "mobile" applications. The heavy-duty, long-haul transport (i.e. mobile) market and the more typical stationary applications of engine-generator sets that generate power for isolated communities (such as "island" states or nations), are ideal for applying the TREC system. For this case study, an MTU 445 hp Detroit Diesel engine was analyzed. The MTU Detroit Diesel engine can be used in a Heavy Mobile Transport (HEMTT) military vehicle, as shown in Figure C3.5.

The feasibility analysis using the TREC system with the larger Detroit Diesel engine indicates a total power recovery of 37.5 kW of mechanical shaft power (36 kWe) from the exhaust, water jacket and lube waste heat (see Figure C3.6). This contrasts with a power recovery of 27 kWe in a case whereby only the exhaust gas heat is recovered from the same engine. Based on the results of the feasibility analysis, the TREC system demonstrates a 30% higher power output compared to that of the conventional state-of-the-art ORC system and provides an 8.4% improvement in power recovery from the Detroit Diesel engine. This translates into an 8% improvement in fuel use amongst heavy duty vehicles of the same weight.

Table C3.6 summarizes the results of the preliminary analysis comparing the output power from the application of the various heat recovery systems, shown in Figures C3.2(b), C3.3, and C3.4. Figures C3.3 and C3.4 show two alternatives to the proposed TREC system applied to these engines using a modification to the basic ORC system. Figure C3.4 shows a more typical or basic ORC system much like what is discussed in Case Study 2. The specifications for the CAT engine are given in Figure C3.7. The three cycles are listed in order of the amount of power produced from the same low- and high-grade waste heat source; from the highest amount produced to the lowest

Heavy Expanded Mobility Tactical Truck

A HEMTT loaded up and ready to go on a mission in Iraq.

Type	8x8 off-road cargo truck
Place of origin	United States
	Service history
In service	1982-Present [1]
Used by	U.S. Army [1]
	Production history

Engine	MTU Detroit Diesel 12.1 liter 445 hp (332 kW)
Transmission	Allison 4500SP/5-speed automatic
Suspension	Hendrickson w/equalizing beam
Ground clearance	24 in (610 mm)
Fuel capacity	155 US gal (587 l)
Operational range	400 mi (644 km)
Speed	62 mph (100 km/h)

Figure C3.5. A possible military application of the TREC system used to recovery waste heat from the MTU Detroit Diesel engine that improves the fuel economy of the Detroit Diesel engine by 8–10% as detailed in Figure C3.6.

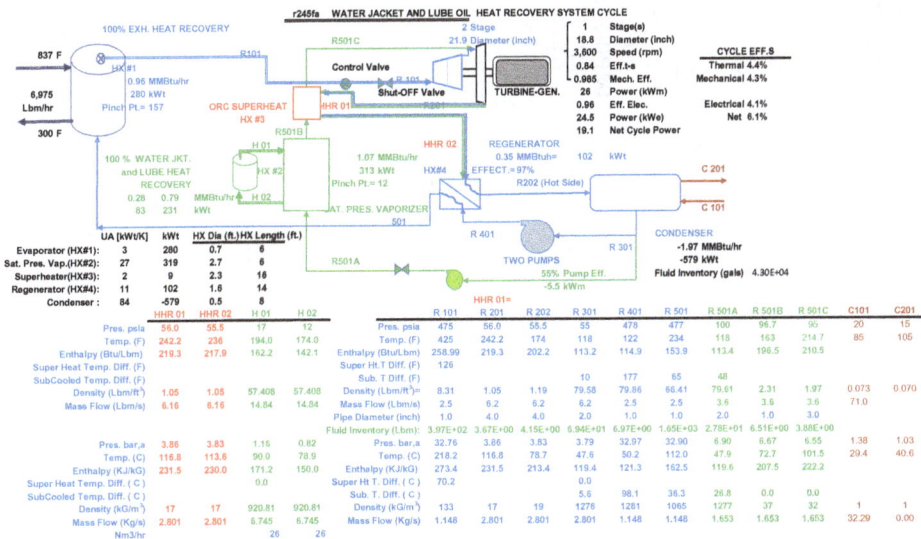

Figure C3.6. State point diagram for the TREC system applied to the Detroit Diesel engine used in the Military Heavy-Duty Transport Vehicle.

Table C3.6. Summary of ORC System Analysis ($T_{exhaust} = 300°F$).

REFERENCE	TFU GROSS ELEC [kWe] per CAT Engine	TGU NET* ELEC [kWe] per CAT Engine	TGU NET CYCLE IMPROVE. EFF. (%)	% IMPROVED NET PRIME POWER	TREC SYSTEM TURBINE-GENERATOR DESIGN CHARACTERISTICS USED TO RECOVERY WASTE HEAT FROM BIO-GAS CG3260-16
Figure 2	390	357	14%	8.9%	TREC SYSTEM: A Hybrid-Binary Rankine Cycle for 100% Ht. Recov. with a Single, Multi-Stage Rotor with a High and Low Pres. Entry & one. Regen.
Figure 3	367	332	8%	8.3%	Conventional State-of-the-Art Modification of an ORC to Provide Some Low Grade Heat Recovery used to preheat ORC Fluid entering the evaporator
Figure 4	346	314	~	7.9%	BASELINE ORC System with only Exhaust Heat recovery

*TGU NET=TGU$_{electric}$-Feed Pump parasytic

Only for reference

Technical data
4000 kWel; 4160 V, 60 Hz; Natural gas, MN = 80

CATERPILLAR®

Design conditions		
Comb. air temperature / rel. Humidity:	[°F] / [%]	77 / 60
Altitude:	[ft]	328
Exhaust temp. after heat exchanger:	[°F]	248
NOₓ Emission (tolerance - 8%):	[g/bhph]	0,94

Fuel gas data: [2]		
Methane number:	[-]	80
Lower calorific value:	[BTU/ft³]	985,2
Gas density:	[lb/ft³]	0,05
Standard gas:	Natural gas, MN = 80	

Genset:		
Engine:	CG260-16	
Speed:	[1/min]	900
Configuration / number of cylinders:	[-]	V / 16
Bore / Stroke / Displacement:	[in] / [in] / [in³]	10,2 / 12,6 / 16589
Compression ratio:	[-]	12
Mean piston speed:	[ft/s]	31
Mean lube oil consumption at full load:	[lb/hr]	2,6
Engine-management-system:	[-]	TEM EVO

Generator:	Marelli MJH 800 LA8	
Voltage / voltage range / cos Phi:	[V] / [%] / [ ]	4160 / ±10 / 1
Speed / frequency:	[1/min] / [Hz]	900 / 60

Energy balance				
Load:	[%]	100	75	50
Electrical power COP acc. ISO 8528-1:	[kW]	4000	3000	2000
Engine jacket water heat:	[MBTU/hr±8%]	4381	3173	2238
Intercooler LT heat:	[MBTU/hr±8%]	1235	877	600
Lube oil heat:	[MBTU/hr±8%]	1569	1215	1010
Exhaust heat with temp. after heat exchanger:	[MBTU/hr±8%]	7301	6114	4698
Exhaust temperature:	[°F]	837	892	945
Exhaust mass flow, wet:	[lb/hr]	47082	36809	25373
Combustion mass air flow - ISO 3046/1:	[lb/hr]	45506	34690	24496
Radiation heat engine / generator:	[MBTU/hr±8%]	751 / 348	723 / 304	710 / 273
Fuel consumption:	[MBTU/hr +5%]	31242	24173	17434
Electrical / thermal efficiency:	[%]	43,7 / 42,4	42,3 / 43,4	39,1 / 45,6
Total efficiency:	[%]	86,1	85,8	84,7

System parameters [1]		
Ventilation air flow (comb. air incl.) with ΔT = 15K	[lb/hr]	260400
Combustion air temperature minimum / design:	[°F]	41 / 77
Exhaust back pressure from / to:	[inWC]	12 / 20
Maximum pressure loss in front of air cleaner:	[inWC]	2
Zero-pressure gas control unit selectable from / to: [2]	[inWC]	8 / 120
Pre-pressure gas control unit selectable from / to: [2]	[psi]	7 / 145
Air bottle, volume / pressure	[ft³] / [psi]	71 / 435
Starter motor:	[ft³/s] / [psi]	28 / 232
Lube oil content engine / base frame:	[gal/(US)]	489 / -
Dry weight engine / genset:	[lb]	54873 / 113428

Cooling system		
Glycol content engine jacket water / intercooler:	[% Vol.]	0 / 35
Water volume engine jacket / intercooler:	[gal/(US)]	151 / 13,5
KVS / Cv value engine jacket water / intercooler:	[ft³/h]	3284 / 2189
Jacket water coolant temperature in / out:	[°F]	174 / 194
Intercooler coolant temperature in / out:	[°F]	104 / 115
Engine jacket water flow rate from / to:	[gpm]	396 / 484
Water flow rate engine jacket / intercooler:	[gpm]	454 / 286
Water pressure loss engine jacket water / intercooler:	[psi]	17 / 17
Lube oil temp. engine inlet max. / lube oil flow rate:	[°F] / [gpm]	176 / 498

1) See also "Layout of power plants": 2) See also Techn. Circular 0199-99-3017

CG260-16-E-00-04160-M-X

Engine noise level	Octave band centre frequency								Sum level
	63	125	250	500	1000	2000	4000	8000	(distance 1 meter)
Exhaust noise [dB(lin)]	131	124	122	121	117	115	113	109	123 dB (A) (±0,5 dB(A))
Air-borne noise [dB(lin)]	99	109	104	101	99	99	104	101	109 dB (A) (±1,0 dB(A))

PwrC_1.19_OrG Subject to technical changes , k578204, 16.04.2012

Figure C3.7. Specifications for the CAT engine.

amount. As shown in Table C3.6, TREC provides 14% more recoverable electric power than the state-of-the-art ORC system, and more than twice the recovered power than the rarely-used "current practice ORC cycle, which is unable to completely recover the low-grade heat concurrently produced with the high-grade exhaust gas heat. The application of the TREC system records a 9% improvement in power generation from the 42% fuel-efficient prime mover. The TREC system is predicted to have a proportional improvement in power generation for the older generation of continuous duty prime movers that have fuel efficiencies of less than 35%. That is, the expected power improvement is 9% × 0.42/0.35; equal to a power improvement of 10 to 11%.

C3.4   Dual entry, ORC turbine design for TREC system

The TREC system utilizes one turbine-generator unit to extract energy from multiple heat sources. The 2-stage turbine used has two fluid inlets and a single turbine discharge. This use of steam turbines with multi-pressure admissions is common in industrial steam turbines that are used in conventional Rankine Cycles. A multiple-admission turbine has a specific advantage in ORC applications, as the organic refrigerants used as working fluids in ORC turbines typically exit the turbine in a superheated state, as opposed to steam cycles where the steam usually exits the turbine in a saturated or wet state. A turbine with a superheated or dry vapor discharge avoids the erosion that can occur when the turbine's exhaust has even a small amount of liquid particles that have velocities that can exceed 500 m/s as they impact the turbine blades. Such erosion can damage the blades, which minimally reduces the turbine's efficiency and, in the worst-case scenario, destroys the turbine.

The turbine-generator unit in the TREC system consists of a multi-stage axial flow turbine, which uses nozzles to accelerate flow, and a moving rotor which converts the flow velocity into the rotational motion of a shaft. The turbine may either be a reaction or impulse turbine. A reaction turbine expands the vapor through the nozzle and rotor of a stage, while an impulse turbine expands the vapor primarily through the nozzle. By varying the levels of reaction in a turbine, a turbine designer can adjust the performance characteristics as well as the level of axial thrust in the turbine unit, which have a significant impact on the choice of bearing system.

The multi-admission turbine allows multiple heat flow streams to be utilized in a single turbine. A high-enthalpy flow may be admitted through a conventional high-pressure inlet, where the flow expands through one or more high-pressure (HP) stages. After the high-enthalpy flow is expanded to an appropriate pressure level, an admission of a secondary fluid flow at a lower pressure into the turbine is employed to introduce additional flow from a lower-enthalpy source into the turbine. This may be done by including a piping connection which directs the low-enthalpy flow through the turbine casing and into a manifold, or "collector". Following which, the low-enthalpy flow mixes with the high-enthalpy flow, and is expanded through one or more low-pressure (LP) stages before entering into an exhaust diffuser or the exhaust hood — a plenum that collects all of the flow streams exiting from the turbine blades before the combined fluid stream exits the turbine. The residual superheat in the exhaust flow may then be passed through a heat exchanger and used to add heat to the inlet or admission flows, in order to increase turbine and cycle efficiency and power output. Additional admissions can be employed in the same manner to accommodate more low-temperature heat sources. Because the turbine can be designed to operate with as many admissions and HP/LP stages as required, the turbine can be expected to produce close to the maximum amount of power available in an ORC process, yielding the potential for optimal cycle efficiency and resource utilization.

Case Study 4

Enhanced Gas Turbine with Water Injection

Along with the Rankine Cycle system, the Brayton Cycle (also known as a Gas Turbine (GT) Engine) is also a popular choice for providing propulsion and mechanical or electric power generation. The GT is a high-power density generation system used in aircraft, helicopters and stationary utility power generation. The GT (Brayton Cycle) processes form a cycle that results in a very high efficiency. Their first law efficiency can exceed 60% if the exhaust gas waste heat from the gas turbine is recovered to produce additional power, hot water or low-pressure steam. The GT engine has processes that follow the air-Brayton Cycle shown in Figure C4.1. Figure C4.2 illustrates the T-s diagram for the Brayton Cycle shown in Figure C4.1. However, it must be noted that the Brayton Cycle shown in Figure C4.1 consists of many more components than a basic GT engine.

The basic, much simpler GT engine follows the air-Brayton cycle, which was first introduced in Chapter 5. It consists of a compressor, a combustor and a turbine mounted onto a simple stub shaft which can be coupled to a mechanical system or an electric generator. The more complicated GT Brayton Cycle system shown in Figure C4.1 includes the components typically present in a basic turbine engine but here the compressor is illustrated as a multi-stage compressor with air intercoolers, plus the cycle includes a reheat turbine and heater and a recuperator or regenerator. There is also an additional option of water vapor injection which has the dual effect of increasing the power output of the engine and reducing the amount (ppm) of NO_x that may be found in the exhaust stream as a result of the combustion of fuel in the combustor.

AIR BRAYTON CYCLE SYSTEM WITH SINGLE REGENERATION and INTERCOOLED COMPRESSORS

			M,water inject [Lbm/s]=	**DRIVE TURBINE**		**REHEAT TURBINE**	
0.0000	0.0000	0.0000	1.564	1700 ft/s	0.275 Dia (m)	0.480	1,814 ft/s
Wtr.Inject. 0.0%	0.0%	0.0%		68 Ns	36,000 RPM	22,000	69 Ns
[Lbm/s]= 0.000	0.000	0.000		1.56 Ds	1 No. of Stg.s	1	1.64 Ds
Pres. Drop Losses	1.0%	**REGENERATOR**			0.69 V/Us	0.74	

INTERCOOLER EFFECT.S — **EFFECTIVENESS** 0.98

| 85% | 85% | 85% |

COMBUSTOR # 1

#1 TURBINE — EFF.= 90% — Pr= 2.73 — Pr,oa= 7.45

#2 TURBINE — 90% — 2.73

M,water inject [Lbm/s]= 1.564 — **POWER GEN. TURBINE**

GEAR BOX EFF.= 0.97
GENERATOR EFF.= 96%

DRIVE TURBINE POWER — Tsat= 167.6 — COMBUSTOR # 2 — Tsat= 192.6

| Comp.η= | 88% | 88% | 88% | 88.0% |
| Pr= | 1.682 | 1.682 | 1.682 | 1.682 |

8.00 **COMPRESSORS**

fdibella

Without Water Injection

	8		
kWe=	381	Hd Coef. (ψ)	1.04
Nstages=	1	FlowCoef.(φ)	1.00
RPM=	36,000	Ns=	0.97
Diameter [m]=	0.160	Tip Spd.ft/s=	989
Pr=	1.682		

0.296 / 0.226

Cp, heat source=	0.316 BTU/LBm/R
Cv, heat source=	0.246 BTU/LBm/R
K=	1.28

Heat Balance Check= 100%

DRIVE TURBINE POWER	1,774	kW		BRAYTON CYCLE EFF.	51.4%
POWER GEN. TURBINE	1,773	kW		CARNOT Cycle Eff.=	
AIR COMPRESSOR	1,547	kW		2nd Law Eff.=	
NET POWER	2,000	kW		NET POWER [kW]=	
Water recovery; COMB. #1 Q=	1,854	kWt;UA[kWt/K	6.0	
Air COOLER Q=	493	kWt;UA[kWt/K	19.7	
REGENERATOR Q=	3,755	kWt;UA[kWt/K	115.1	
INTERCOOLER IQ=	1,132	kWt;UA[kWt/K	9.0	
COMB. #2 Q=	1771	kWt;		
THERMAL STORAGE Q=	493	kWt;		

FLUID: air
MOLE. WT. 28.96

Critical Pressure [psia]= 548 — 37.8 Bar,a
Critical Temp. [F]= -221 — -140.6 C

Twater, in [F] = 70 — Water Mass Flow= 739 GPM
Twater, out [F]= 85

	1	1a	1b	1c	2	3	4	5	6	7	8
Pres. [psia]	14.70	24	41	68	114	113	112	41	41	14.9	14.72
Temp. [F]	80.6	86.0	87.0	87.1	187	1086	1500	1105	1500	1105	205
Enthalpy [Btu/lbm]	129.3	130.5	130.6	130.5	154.4	382.0	494.4	386.9	494.2	386.7	159.1
Density [Lbm/ft³]	0.07	0.12	0.20	0.34	0.48	0.20	0.15	0.07	0.06	0.03	0.06
Flow rate [Lbm/s]	15.64	15.64	15.64	15.64	15.64	15.64	15.64	15.64	15.64	15.64	15.64
Entropy [Btu/Lbm/R]	1.6413	1.6086	1.5738	1.5386	1.5431	1.7629	1.8280	1.8357	1.8976	1.9053	1.6909
Air Cp	0.24	0.24	0.24	0.24	0.24	0.27	0.28	0.27	0.28	0.27	0.24
Pres. bar,a	1.0	1.7	2.8	4.7	7.9	7.8	7.7	2.8	2.8	1.0	1.0
Temp. [K]	300.0	303.0	303.5	303.6	358.9	858.7	1088.6	869.0	1088.6	868.9	368.9
Enthalpy [KJ/kG]	300	303	303	303	358	887	1148	898	1148	898	370
Temp. [C]	27	30	31	31	86	586	816	596	816	596	96
Density [kG/m³]	1.18	1.94	3.23	5.38	7.64	3.15	2.46	1.13	0.90	0.41	0.96
Mass Flow [Kg/s]	7.11	7.11	7.11	7.11	7.11	7.11	7.11	7.11	7.11	7.11	7.11
Vol. Flow [Nm³/hr]	21183	21183	21183	21183	21183	21183	21183	21183	21183	21183	21183

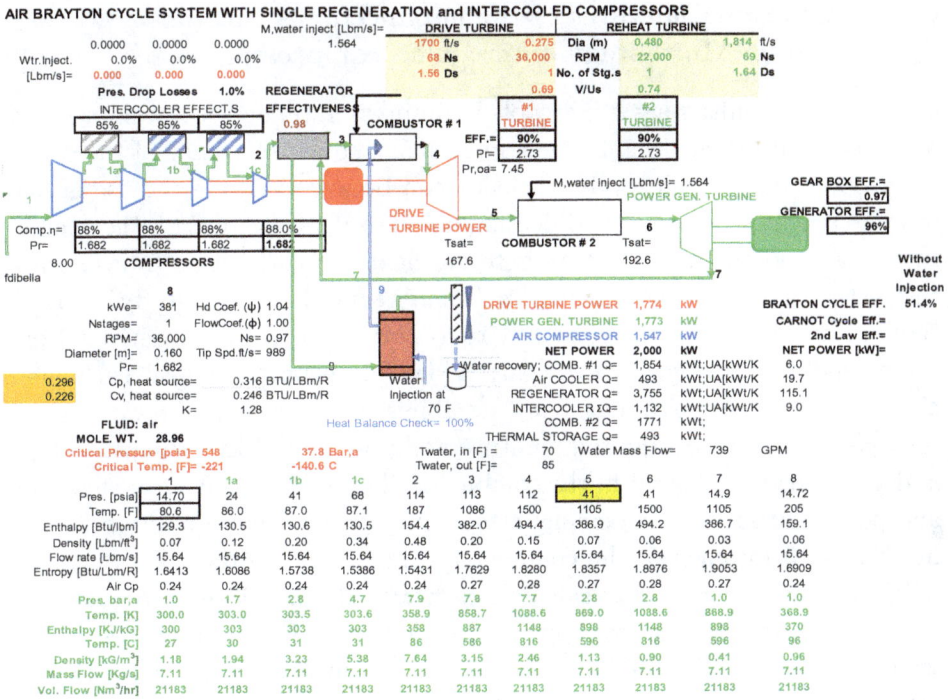

Figure C4.1. Gas turbine engine-generator based on a standard air-Brayton cycle with intercooled compressor and regenerator.

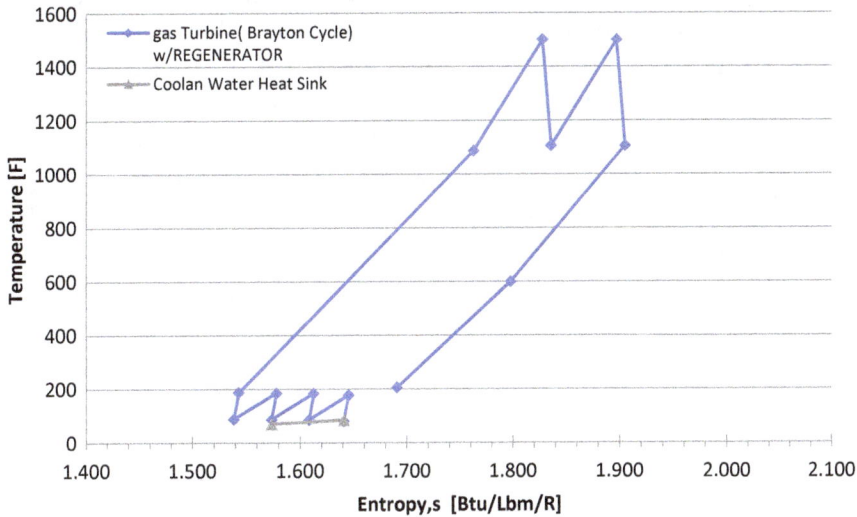

Figure C4.2. Gas Turbine (Brayton Cycle) with intercooled compressors and reheating turbine using a 1,500°F turbine inlet.

C4.1 Additional features of the enhanced gas turbine engine: Multi-stage, intercooled compressor

The use of a multi-stage, intercooled compressor reduces the compression work required for the system, thus increasing the net efficiency of the cycle. Both power output and cycle efficiency can be increased via the use of one or more reheat turbines or by injecting liquid or vapor water into the engine. Injecting vapor requires that an external heat source be used to evaporate the water that is to be injected into the engine. This case study assumes that all of the heat energy needed to vaporize the water is derived from the engine's exhaust gas waste heat. While the recuperator heat exchanger also increases the cycle's efficiency, its ability to increase the power output remains dependent on the temperature and flow rate of the exhaust stream and/or air-fuel ratio used in the combustor. These effects on cycle efficiency and power output can be demonstrated using the first law equations derived in Chapter 3 for each of the components that make up the GT cycle. The quickest way of performing calculations which demonstrate these effects is to actually construct an excel spreadsheet model of the cycle processes, like what is shown in Figure C4.1. While this author favors using Excel, any other equivalent software platform that the reader is comfortable with using or that can be easily programmed, can be used. The inputs used for the cycle analysis can be identified in Figure C4.1 to be the values within the cells with bold, black borders. All other cells contain values obtained from the application of the first law equations for each of the components: the compressor, turbine, and heat exchangers; including the thermodynamics of the water injection as a heat exchanger. The state points shown below the block diagram are the summary of the results calculated via the author's Excel model. The calculated power for the compressors and turbines, and the cycle efficiency are shown on the right side of the diagram, directly below the turbine-reheat heat exchanger-turbine module.

The reader may also have noticed that the components of the enhanced GT engine shown in Figures C4.1 and C4.2 are from among the seven major mechanical engineering components described in Chapters 3 and 5.

C4.2 Additional features of the enhanced gas turbine engine: Water injection system

Water injection has been used since the earliest phases of GT engine development. In most instances, the goal was to increase the net power output of the turbine by putting to work both the steam generated and the heat recovered from the hot exhaust gases exiting the turbine. While power augmentation remains the primary purpose of using a water injection, reducing the ppm of the GT engine's NO_x emissions has been introduced as an additional objective of stationary GT generator power generation systems. A water mass flow rate injection of as much as 20% of the air mass flow rate throughput used in the GT is possible depending on the design point selected for the regenerator effectiveness, the compressor pressure ratio and the turbine inlet temperature. It will be shown in this section that a water injection rate of 8 to 12% with respect to the compressor inlet air flow rate is more typical.

A water injection system may be considered as a heat exchanger without borders. That is, the water injection system introduces a water stream into the combustor at the same pressure that is used in the combustion process occurring between state points 3 and 4, as shown in Figure C4.1. In this Study the temperature of the injected water vapor (i.e., steam) is selected to match the regenerator discharge temperature in order to reduce the net entropy change caused by the mixing of the hot air and steam flow streams as they enter the combustor. The block diagram shown in Figure C4.3 was excerpted from Figure C4.1 to provide more details regarding the water vaporization process. Equations that suitably model the water injection process may be derived by applying the First Law of Thermodynamics to an open flow control volume around the steam boiler and the combustor.

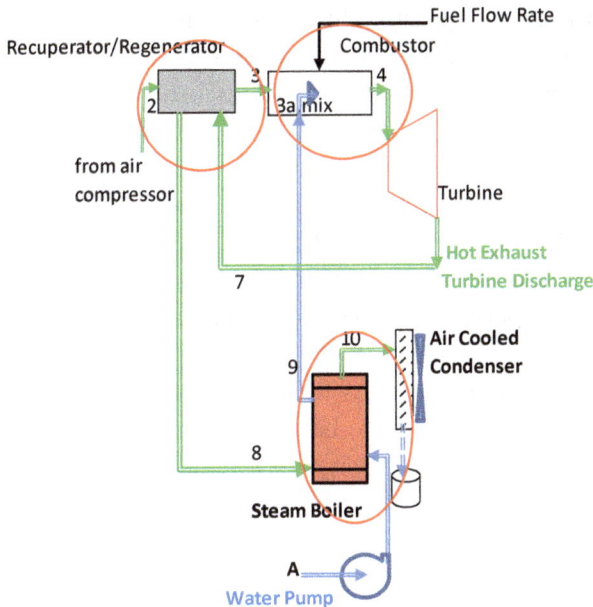

Figure C4.3. Details of the water injection system components.

Before attempting to derive equations to model the water injection process, it is necessary to understand that the introduction of a water injection system may improve the power output of the GT under some circumstances, but adversely affect cycle efficiency and the power output of the GT in other cases.

A thorough review process should be performed by the analyst regarding the specific application of water injection systems in a given thermodynamics problem, as opposed to what should now be, for the reader, a very comfortable application of the simplified equations for each of the basic mechanical engineering components of the GT cycle, i.e., the compressor, turbine and combustor and recuperator heat exchangers. Here, the necessity of this review process will be further explained using an interesting dilemma. Referring to the illustration of the water injection system in Figure C4.3, in order for the water vapor to be successfully injected into the combustor, two requirements must be met. Firstly, the pressure of the water vapor should minimally be equivalent to the pressure of the combustor. Additionally, the temperature of the water vapor should not fall below that of the recuperator exhaust at state point 3. If this temperature requirement is not met, the mixture of the two fluid streams — the air discharged from the recuperator at state point 3 and the water vapor at state point 9 — will be at an intermediate mixture temperature. An intermediate temperature has two detrimental effects on the performance of the system: (1) There's entropy production due to the mixing of the two streams of unequal temperature, and (2) The turbine requires as high a temperature as is physically possible to achieve, and the combustor will have a more difficult chore of achieving this temperature with a lower temperature entering the combustor.

A closer look at Figure C4.3 reveals that the heat obtained from the air stream exiting the recuperator at state point 8 can be used to evaporate the liquid water into a vapor, thus enabling its injection into the combustor. Enhancing recuperator heat transfer increases the cycle's efficiency and enables an increase in net power IF the same amount of fuel is used. However, it should also be understood that it is necessary to reduce the amount of heat transfer from the engine's exhaust gas waste heat to the recuperator to ensure that sufficient heat energy is available for use in increasing the temperature of the water and evaporating it into vapor.

Herein lies the dilemma: should recuperator heat transfer be reduced to maximize heat energy available for evaporating water, or should the recuperator heat transfer be enhanced in order to maximize cycle efficiency and/or power output? This is an interesting dilemma that can be resolved using a calculus approach, with the equations optimized based on the First Law of Thermodynamics, as applied to the recuperator and the water injection system, respectively. This approach requires that all of the unknowns be solved by first determining the linear equations from the First and Second Laws of Thermodynamics. By taking the derivatives of an equation that contains the desired independent variable and the desired dependent variable, the value for the dependent variable that achieves the maximum result can be determined. This was done in Chapter 5, for example, to determine the maximum pressure ratio for the GT engine if a maximum temperature ratio, equal to the temperature into the turbine divided by the ambient temperature, is known.

Alternatively, the problem may also be solved using an iterative model built on an Excel spreadsheet or generated by any other equivalent software, containing the analysis of a gas turbine engine. This technique requires that the thermodynamics of the entire engine model be prepared, with each of the individual components providing a pressure or temperature or flow rate into the component to which it is connected. When the model is complete, the analyst can change the temperatures in one of the devices, usually the turbine, and/or change the heat exchanger efficiencies (also known as effectiveness) until the desired parameter has been optimized. Usually the desired parameter is the cycle net power. This solution technique is developed in the next section of this case study.

To solve the problem with an iterative technique, it is first necessary to begin with developing the thermodynamics equations that correctly model this water injection system within the GT thermodynamics model.

The first step, as always, is to construct a control volume (CV) around the system of interest. In this analysis, the steam boiler, recuperator and combustor are the three heat exchangers of most interest in the problem-solving process. These control volumes have already been indicated in Figure C4.3. Applying the First Law of Thermodynamics for the steam boiler:

$$\dot{M}_8 \times (h_8 - h_9) = \dot{M}_{water} \times (h_9 - h_{sat.liq.}),$$

$$\dot{M}_8 \times (h_8 - h_7) = \dot{M}_2 \times (h_3 - h_2),$$

$$\text{Let}: T_3 = T_9; P_3 = P_9,$$

$$\dot{M}_{fuel} = (\dot{M}_9 + \dot{M}_3) \times (h_4 - h_3)/L_{hhv},$$

$$\dot{M}_2 \equiv \dot{M}_3,$$

$$\dot{M}_t = \dot{M}_9 + \dot{M}_3 + \dot{M}_{fuel},$$

$$\dot{M}_t = (AFR + 1) \times \dot{M}_f + \dot{M}_9,$$

$$T_4 = \frac{L_{hhv}}{\dot{M}_t \times C_p} + T_{ambient},$$

$$T_{mix,\,water+3} = (T_3 \times \dot{M}_3 + T_9 \times \dot{M}_{water,9})/(\dot{M}_3 + \dot{M}_{water,9}),$$

$$\eta_{cycle} = (Turbine\,Power - Compressor\,Power)/(\dot{M}_{fuel} \times L_{hhv}),$$

$$kW\,turbine = \dot{M}_t \times C_p \times T_4 \times \left(1 - \left(\frac{1}{Pr^{(k-1)/k}}\right)\right) \times \eta_t,$$

$$kW\,compressor = \dot{M}_2 \times C_p \times T_2 \times \left(Pr^{\frac{(k-1)}{k}} - 1\right)/\eta_c.$$

It is also of interest to consider a new equation that defines the effectiveness, ϵ, of the heat exchanger. The effectiveness of a heat exchanger is defined as the ratio of the amount of heat transfer achieved by the heat exchanger to the maximum amount of heat that could have been achieved by the heat exchanger if that heat exchanger was infinitely large. This equation has been derived in most Introductory heat transfer textbooks. For the recuperator shown in Figure C4.3, the equation is as follows:

$$\varepsilon = \frac{(T_8 - T_7)}{(T_7 - T_2)}.$$

The effectiveness equation for the boiler is:

$$\varepsilon = \frac{(T_8 - T_9)}{(T_8 - T_{sta.vapor\,water})}.$$

The insertion of these equations into an Excel spreadsheet creates a useful model of the cycle, as shown in Table C4.1, from which the cycle efficiency can be determined. This first output is used as a validation check of the programming of the equations shown above. In solving engineering problems, it is always good practice to validate a new computer program in order to avoid mistakes which can result from software errors even when the model and the equations have been correctly constructed. The solution for this problem was calculated using the specific heats and the turbine's compressor efficiencies based on air's standard properties and under conditions of perfect compression and expansion, without regeneration or recuperation; two simplifications which are always not realistic but which provide some sense of the relationship between $T_{turbine\,inlet}$, Pr and cycle efficiency. This was done in order to allow for comparisons to be made between the results

Table C4.1. Results of water injection calculation for the system in Figure C4.3.

Afr=	40		Water Injection		
T4=	1993	Turb. Eff.=	0.8		
Pr=	11	Comp. Eff.=	0.8		
Ideal Pr for T4=	11				
Regen. Effec.=	0.6	Tambient=	60		
Cp,2=	0.246				
Cp,3	0.261	L,hhv [Btu/Lbm]=	21000		
Cp,4,effective=	0.322		503.3	928.9	0.3213
Cp,7	0.303				
Cp,8-10=	0.310		0.310	598.87	555.49

{ 1-(1/Pr)^((k-1)/k) }=	0.496				
{ Pr^((k-1)/k) - 1 }=	0.984				
State Pt.	Temp.	Enthalpy			
2	699.6	281.4			
3	921.3	338.6	**Heat Balance Check:**		
503.3 3,a	750.0	505.0	Fuel Input [kw]=		
44.1 4	2075		4.923		
7	1069.2		0.00%	4.923	
8	929.3		Steam Boiler		
9,sat	364.4	1209.2	1.91		
ΔT, superheat=	20		1.65%	1.88	
ΔT pinch=	20		Regenerator		
10	404.4		0.5117		
M2=	32 Lbm/hr		0.00%	0.5117	
0.842 Mfuel=	0.80 Lbm/hr		Combustor		
Ma=	7.47 Lbm/hr		4.92		
MT=	40.3 Lbm/hr		-0.23%	4.93	
Ideal Cycle Eff. without Recup.=	49.6%		Net Power		
Actual Cycle Eff.=	41.6%		2.120		
Net Power (kW)=	2.1		0.00%	2.120	

obtained from the model versus the textbook solution for the theoretical cycle efficiency for a Brayton Cycle employed as a GT engine without regeneration. The ideal Brayton Cycle efficiency is based on the Equation (C4.1a) and assumes that the compressor and turbine efficiency is 100%.

$$\eta = 1 - \frac{1}{Pr^{\frac{(k-1)}{k}}}. \qquad (C4.1a)$$

Interestingly, there exists a similar equation which determines the optimum pressure ratio for the GT based on the ambient and inlet turbine temperatures. When using these equations, do remember that only absolute temperatures (°R or °K) should be used.

The equation for determining the proper pressure ratio based on ambient and turbine inlet temperature is:

$$Pr = \left(\frac{T_4}{T_{ambient}}\right)^{k/(2(k-1))}. \tag{C4.1b}$$

These two equations are very helpful because the results from these two simple equations can be used to validate the cycle efficiency that has been determined by the computer model. For example, start by picking an air–fuel ratio (AFR). This selection provides a turbine inlet temperature (T_4) using the equation shown below for the temperature T4, as a function of AFR and the high heating value of the fuel, L_{hhv}:

$$T_4 = \frac{L_{hhv}}{(Afr + 1) \times C_p} + T_{ambient},$$

A value for AFR is input into the spreadsheet model in Table C4.1 using the "boxed" cell. With the temperature, T_4, now calculated by the model, the overall pressure ratio (Pr) can be determined from the equation for Pr (Equation (C4.1b)) and manually typed into the thermodynamics model. The value of "$C_{p,4}$ effective", which accounts for the input of heat energy in the combustor only after the recuperator and the water injection heat and/or mass flow rate have been taken into consideration, is also input into the appropriate bordered cell.

Before proceeding to the presentation of the results obtained by combining these equations, it is necessary to first conduct an analysis of the combustor. This is necessary because of the mixing of the two different fluid streams which do not react: the heated air from the recuperator and the steam fluid flow rate from the steam generator. A straightforward analysis can be conducted by utilizing the First Law of Thermodynamics for an open flow system. For this analysis, the CV has been constructed such that it is breached both by the individual inlet streams, as well as by the mixing of the two streams. The addition of heat from the combustion of the fuel has not been considered in this model, so as to be able to determine the temperature of the mixed fluid stream that is to be heated in the combustion system. An example of this CV is shown in Figure C4.4. The resulting outlet enthalpy is simply the sum of the product of the mass flow rates and the enthalpies of each of the inlet streams. The outlet enthalpy is of the mixture that consists of water and air. The first law equation is as shown:

$$\dot{M}_{air,3} \times h_{air,3} + \dot{M}_{sat.steam,a} \times h_{sat.v} = (\dot{M}_{air,3} + \dot{M}_{stav.steam}) \times h_{mix}. \tag{C4.2}$$

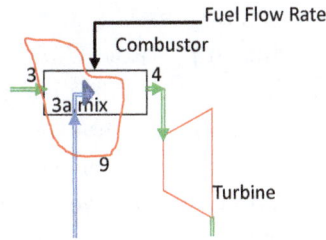

Figure C4.4.   Control volume for the thermodynamics analysis of the combustor and turbine.

The enthalpy, h_{mix}, can be calculated simply by adding the terms on the left side of the equation and dividing this total by the sum of the two flow rates. This enthalpy value can then be used to calculate the temperature of the mixture. The temperature for a mixture with a given enthalpy is not one that can be "looked up" in a textbook. Rather, it can be easily determined by an iteration technique that starts with an input of the mixture's temperature until the sum of the enthalpies of each of the constituent parts of the fluid stream equals the enthalpy solved using the equation.

However, the reader is reminded of the psychrometric analysis of an air-water mixture presented in Chapter 6. The mixing of the hot air and saturated steam flow rates is in fact an air-water mixture, although at much more elevated pressures and temperatures than normally observed in typical environmental moist air applications. Although the thermodynamics presented in Chapter 3 can be applied for this mixing of the two fluid streams, due consideration has to be given to the elevated temperature and pressure conditions. First, the partial pressure of the steam in the mixture that is being maintained at higher than atmospheric pressure must be determined. This can be done by using the same equation for the specific humidity, w, shown for the first time in Chapter 6. Recall the equation for specific humidity:

$$\omega \left[\frac{lbm_{water}}{lbm_{air}} \right] = 0.622 \times P_w/(P_T - P_w).$$

This equation can be solved for the partial pressure of water in the mixture, P_w.

The partial pressure of water in the mixture can then be used to determine the enthalpy at the common mixture temperature, $T_{3,a}$. It may be helpful to think about what is happening during the mixing of the two fluid streams when the higher temperature air (fluid stream, 3) comes in contact with the saturated steam vapor (fluid stream, 3a). Due to the fact that the temperature of the hot air is typically higher than that of the saturated steam, the hot air is cooled slightly as it superheats the saturated steam. The result of the analysis is shown in Table C4.1.

It may also be observed that the solution displays an energy balance (or 'heat balance check', as it is also known) for the components working as a complete system in a cyclic process. This is recommended to determine if the model that has been prepared should even proceed to the next step of a validation analysis. An energy balance proceeds by identifying the power and heat transfer that crosses the CV boundary. The sign convention noted in Chapter 2 must be maintained; That is, that power into the system is negative and power out is positive. Heat transfer into the system is assigned a positive sign and heat transfer out of the system is assigned a negative. The sum of these energy terms must be equal to zero for the energy into and out of the system to be balanced. For this system, the energy balance for the system shown in Figure C4.4 indicates an energy balance of almost 100%, as shown in Table C4.2; That is, the energy into and out of the system are different by only 0.4%; an acceptable difference that is likely accounted for by the precision of the iteration used to achieve the energy balance.

The validation analysis continues by comparing the results from the computer model with the published and accepted values for cycle efficiency. As shown in Table C4.3, the efficiency of the cycle calculated with the model is 0.496. By using Equation (C4.1), it is easy to show that the efficiency of the perfect air-Brayton Cycle with a Pr equal to 11 is also 0.496.

With the validation analysis of the water injection part of the computer model successfully completed, the program can now be used to validate the entire GT (Brayton Cycle) thermodynamics model using an independent resource.

Figure C4.5 provides a validation of the thermodynamics model of a generic GT without a water injection system, but with regeneration and intercooling devices — a method based on a technical paper presented by Dohmen, Schnitzler and Benra (2011) at the 8th International Conference on Heat Transfer, Fluid Mechanics and Thermodynamics. The computer model

Table C4.2.   Results of the validation calculation for water injection case study.

Fuel Input [kw] =	4.92
1.235 Exhaust Heat [kWt] =	4.72
Air inlet Heat [kWt] =	1.17
Water Inlet Heat [kWt] =	0.74
Net Power from Energy Balance	2.11
Net Power [kW] =	2.12
	−0.5%

Table C4.3.   Output from the water
injection case study.

Afr=	40
T4=	1993
Pr=	11
Ideal Pr for T4=	11
Regen. Effec.=	0.6
Cp,2=	0.246
Cp,3	0.261
Cp,4,effective=	0.322
Cp,7	0.303
Cp,8-10=	0.310

{ 1-(1/Pr)^((k-1)/k) }=	0.496
{ Pr^((k-1)/k) - 1 }=	0.984

	State Pt.	Temp.
	2	699.6
	3	921.3
503.3	3,a	750.0
44.1	4	2075
	7	1069.2
	8	929.3
	9,sat	364.4
ΔT, superheat=		20
ΔT pinch=		20
	10	404.4

M2=	32	Lbm/hr
0.842 Mfuel=	0.80	Lbm/hr
Ma=	7.47	Lbm/hr
MT=	40.3	Lbm/hr
Ideal Cycle Eff. without Recup.=	49.6%	
Actual Cycle Eff.=	41.6%	
Net Power (kW)=	2.1	

keeps the same operating parameters as used in the paper, and matches the performance of the GT using a pressure ratio of 8 and 20. The turbine and compressor efficiencies were 90% and 88%, respectively, and the regenerator's effectiveness was 95%. The actual state points for the 1,573 K and 1,173 K inlet turbine temperature are highlighted with red 'star' markers. The efficiency drops with the decrease in turbine inlet temperature (59% to 52%) and a slight reduction in the intercooler effectiveness (70% to 50%).

Next, a calculation of the GT Model with a water injection system is carried out, with the assumption that the recuperator is 60% efficient and that the temperature of the superheated water exiting the evaporator is equal to the temperature exiting the recuperator; for this case study the

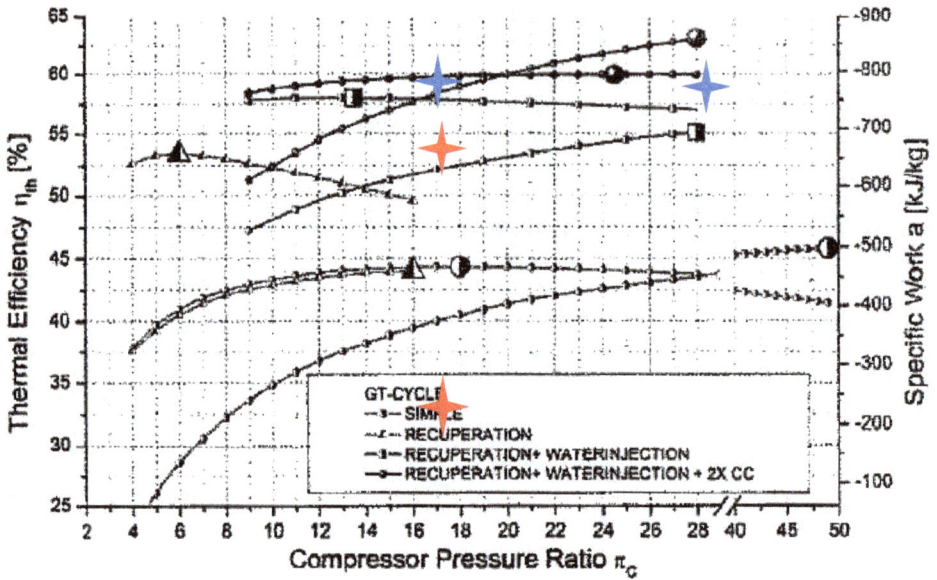

Figure C4.5. A comparison of a simple cycle vs. a recuperated cycle with cooling by water injection and reheating by sequential combustion, as presented in the paper. The blue markers reveal the results of the GT model developed for this study with intercooled compressors, as shown in Figure C4.2. The stars identify the calculation of cycle efficiency with the GT model developed for this Case Study; with and without water injection as shown in Figures C4.7 and C4.8.

water vapor temperature is 908°F (487°C). The complete state points for the cycle are shown in Figure C4.6.

Having validated the thermodynamics of the simple GT model, the model was used to include the use of water injection to increase the cycle efficiency and thus enable more power if the combustion power input is maintained. The GT cycle shown in Figure C4.7 used a water injection rate of 10%, with respect to the compressor inlet flow rate. The result is a cycle efficiency of 54% compared to a cycle efficiency of only 31% if water injection is not used.

The temperature profiles shown in Figures C4.8 through C4.10, as well as the outputs from the gas turbine cycle thermodynamics model prepared using Excel (Figures C4.6 and C4.7) provide the basis for the following observations from the thermodynamics analysis that has been completed:

1. There is very little significant increase in cycle efficiency between a Pr of 8 and 12. For that reason, the cycle analysis proceeded with only a Pr of 8:1.

2. The turbine and compressor efficiency figures taken from Professor Benra's technical paper were 90% and 88%, respectively, with a regenerator effectiveness of 95%. The cycle efficiency for a regenerative GT

AIR BRAYTON CYCLE SYSTEM WITH SINGLE REGENERATION and WATER INJECTION ANALYSIS

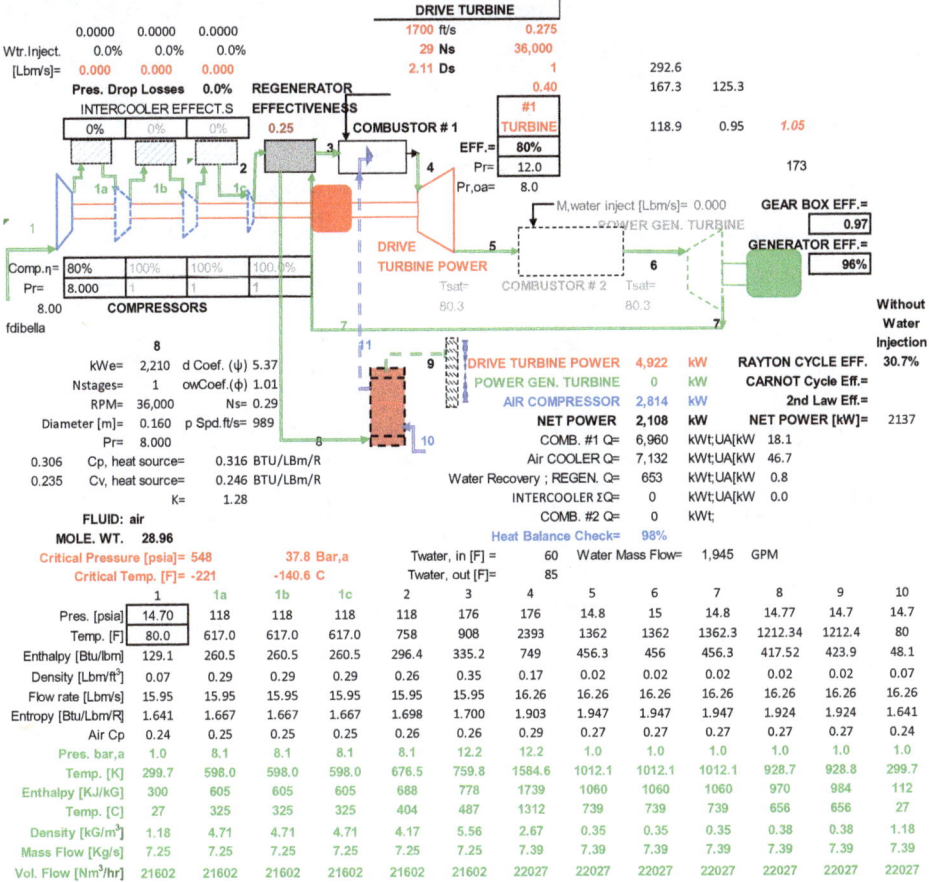

DRIVE TURBINE

	0.0000	0.0000	0.0000		1700 ft/s		0.275			
Wtr.Inject.	0.0%	0.0%	0.0%		29 Ns		36,000			
[Lbm/s]=	0.000	0.000	0.000		2.11 Ds		1	292.6		

Pres. Drop Losses 0.0% REGENERATOR 0.40 167.3 125.3
INTERCOOLER EFFECT.S EFFECTIVENESS #1
0% 0% 0% 0.25 COMBUSTOR # 1 TURBINE 118.9 0.95 *1.05*

EFF.= 80%
Pr= 12.0 173
Pr,oa= 8.0

M,water inject [Lbm/s]= 0.000 GEAR BOX EFF.=
POWER GEN. TURBINE 0.97

Comp.η= 80% 100% 100% 100.0% DRIVE GENERATOR EFF.=
Pr= 8.000 TURBINE POWER 96%
8.00 COMPRESSORS Tsat= COMBUSTOR # 2 Tsat=
fdibella 80.3 80.3

Without Water Injection

8
kWe= 2,210 d Coef. (ψ) 5.37 DRIVE TURBINE POWER 4,922 kW RAYTON CYCLE EFF. 30.7%
Nstages= 1 owCoef.(φ) 1.01 POWER GEN. TURBINE 0 kW CARNOT Cycle Eff.=
RPM= 36,000 Ns= 0.29 AIR COMPRESSOR 2,814 kW 2nd Law Eff.=
Diameter [m]= 0.160 p Spd.ft/s= 989 NET POWER 2,108 kW NET POWER [kW]= 2137
Pr= 8.000 COMB. #1 Q= 6,960 kWt;UA[kW 18.1
0.306 Cp, heat source= 0.316 BTU/LBm/R Air COOLER Q= 7,132 kWt;UA[kW 46.7
0.235 Cv, heat source= 0.246 BTU/LBm/R Water Recovery ; REGEN. Q= 653 kWt;UA[kW 0.8
K= 1.28 INTERCOOLER ΣQ= 0 kWt;UA[kW 0.0
FLUID: air COMB. #2 Q= 0 kWt;
MOLE. WT. 28.96 Heat Balance Check= 98%
Critical Pressure [psia]= 548 37.8 Bar,a Twater, in [F] = 60 Water Mass Flow= 1,945 GPM
Critical Temp. [F]= -221 -140.6 C Twater, out [F]= 85

	1	1a	1b	1c	2	3	4	5	6	7	8	9	10
Pres. [psia]	14.70	118	118	118	118	176	176	14.8	15	14.8	14.77	14.7	14.7
Temp. [F]	80.0	617.0	617.0	617.0	758	908	2393	1362	1362	1362.3	1212.34	1212.4	80
Enthalpy [Btu/lbm]	129.1	260.5	260.5	260.5	296.4	335.2	749	456.3	456	456.3	417.52	423.9	48.1
Density [Lbm/ft³]	0.07	0.29	0.29	0.29	0.26	0.35	0.17	0.02	0.02	0.02	0.02	0.02	0.07
Flow rate [Lbm/s]	15.95	15.95	15.95	15.95	15.95	15.95	16.26	16.26	16.26	16.26	16.26	16.26	16.26
Entropy [Btu/Lbm/R]	1.641	1.667	1.667	1.667	1.698	1.700	1.903	1.947	1.947	1.947	1.924	1.924	1.641
Air Cp	0.24	0.25	0.25	0.25	0.26	0.26	0.29	0.27	0.27	0.27	0.27	0.27	0.24
Pres. bar,a	1.0	8.1	8.1	8.1	8.1	12.2	12.2	1.0	1.0	1.0	1.0	1.0	1.0
Temp. [K]	299.7	598.0	598.0	598.0	676.5	759.8	1584.6	1012.1	1012.1	1012.1	928.7	928.8	299.7
Enthalpy [KJ/kG]	300	605	605	605	688	778	1739	1060	1060	1060	970	984	112
Temp. [C]	27	325	325	325	404	487	1312	739	739	739	656	656	27
Density [kG/m³]	1.18	4.71	4.71	4.71	4.17	5.56	2.67	0.35	0.35	0.35	0.38	0.38	1.18
Mass Flow [Kg/s]	7.25	7.25	7.25	7.25	7.25	7.25	7.39	7.39	7.39	7.39	7.39	7.39	7.39
Vol. Flow [Nm³/hr]	21602	21602	21602	21602	21602	21602	22027	22027	22027	22027	22027	22027	22027

Figure C4.6. Results for Gas Turbine calculation without water injection with the recuperator effectiveness equal to 68%. Note: For this analysis the compressor is not intercooled and the reheater and reheat turbine has been eliminated compared to the cycle shown in Figure C4.1.

operating with a turbine inlet temperature of 1,573 K was 59%. This efficiency drops to 52% if the turbine inlet temperature is reduced to 1,173 K. (see Figure C4.1.)

3. The cycle efficiency is reduced to 33% if the regenerator efficiency is reduced to 38%. (see Figure C4.2.)

4. In this study, it was assumed that the efficiency of the expansion turbine and the compressor were not changed as a result of the water injection. The water injection flow rate of only 10% by mass of the air flow rate through the compressor. Therefore, the power improvement of the GT due to the water injection is in direct proportion to the percentage of mass flow rate of water that can be produced by the exhaust gas heat

AIR BRAYTON CYCLE SYSTEM WITH SINGLE REGENERATION and WATER INJECTION ANALYSIS

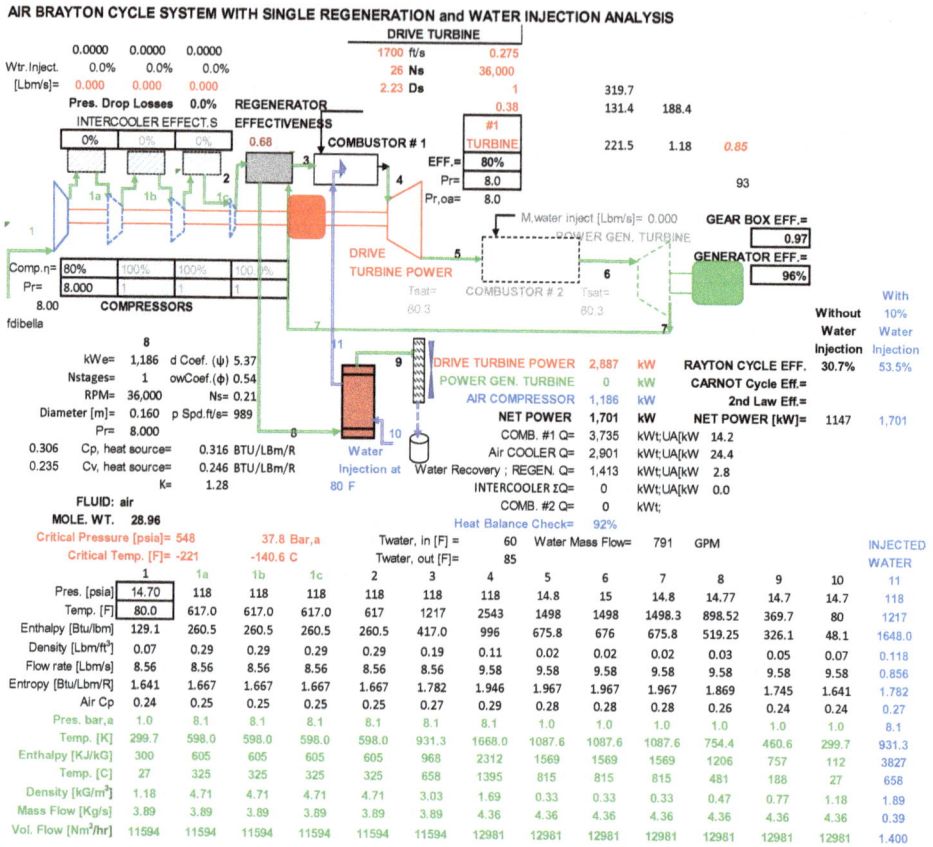

	0.0000	0.0000	0.0000		DRIVE TURBINE		
Wtr.Inject.	0.0%	0.0%	0.0%		1700 ft/s		0.275
[Lbm/s]=	0.000	0.000	0.000		26 Ns		36,000
	Pres. Drop Losses	0.0%	REGENERATOR		2.23 Ds		1	319.7
	INTERCOOLER EFFECT.S		EFFECTIVENESS				131.4	188.4

Pres. Drop Losses 0.0% REGENERATOR
INTERCOOLER EFFECT.S EFFECTIVENESS
0% 0% 0% 0.68 COMBUSTOR # 1 #1 TURBINE 221.5 1.18 *0.85*
EFF.= 80%
Pr= 8.0 93
Pr,oa= 8.0

M.water inject [Lbm/s]= 0.000 GEAR BOX EFF.=
POWER GEN. TURBINE 0.97

DRIVE
TURBINE POWER GENERATOR EFF.=
Comp.η= 80% 100% 100% 100.0% 96%
Pr= 8.000 1 1 1
8.00 COMPRESSORS Tsat= COMBUSTOR # 2 Tsat=
fdibella 80.3 80.3

| | | | | | | | | | Without | With 10% |
| | | | | | | | | | Water | Water Injection |

								RAYTON CYCLE EFF.	30.7%	53.5%

kWe= 1,186 d Coef. (ψ) 5.37 9 DRIVE TURBINE POWER 2,887 kW RAYTON CYCLE EFF. 30.7% 53.5%
Nstages= 1 owCoef.(φ) 0.54 POWER GEN. TURBINE 0 kW CARNOT Cycle Eff.=
RPM= 36,000 Ns= 0.21 AIR COMPRESSOR 1,186 kW 2nd Law Eff.=
Diameter [m]= 0.160 p Spd.ft/e= 989 NET POWER 1,701 kW NET POWER [kW]= 1147 1,701
Pr= 8.000 COMB. #1 Q= 3,735 kWt;UA[kW 14.2
0.306 Cp, heat source= 0.316 BTU/LBm/R Air COOLER Q= 2,901 kWt;UA[kW 24.4
0.235 Cv, heat source= 0.246 BTU/LBm/R Water REGEN. Q= 1,413 kWt;UA[kW 2.8
K= 1.28 Injection at Water Recovery ; INTERCOOLER ΣQ= 0 kWt;UA[kW 0.0
FLUID: air 80 F COMB. #2 Q= 0 kWt;
MOLE. WT. 28.96 Heat Balance Check= 92%

Critical Pressure [psia]= 548   37.8 Bar,a   Twater, in [F] = 60   Water Mass Flow= 791 GPM   INJECTED
Critical Temp. [F]= -221   -140.6 C   Twater, out [F]= 85   WATER

	1	1a	1b	1c	2	3	4	5	6	7	8	9	10	11
Pres. [psia]	14.70	118	118	118	118	118	118	14.8	15	14.8	14.77	14.7	14.7	118
Temp. [F]	80.0	617.0	617.0	617.0	617	1217	2543	1498	1498	1498.3	898.52	369.7	80	1217
Enthalpy [Btu/lbm]	129.1	260.5	260.5	260.5	260.5	417.0	996	675.8	676	675.8	519.25	326.1	48.1	1648.0
Density [Lbm/ft³]	0.07	0.29	0.29	0.29	0.29	0.19	0.11	0.02	0.02	0.02	0.03	0.05	0.07	0.118
Flow rate [Lbm/s]	8.56	8.56	8.56	8.56	8.56	8.56	9.58	9.58	9.58	9.58	9.58	9.58	9.58	0.856
Entropy [Btu/Lbm/R]	1.641	1.667	1.667	1.667	1.667	1.782	1.946	1.967	1.967	1.967	1.869	1.745	1.641	1.782
Air Cp	0.24	0.25	0.25	0.25	0.25	0.27	0.29	0.28	0.28	0.28	0.26	0.24	0.24	0.27
Pres. bar,a	1.0	8.1	8.1	8.1	8.1	8.1	8.1	1.0	1.0	1.0	1.0	1.0	1.0	8.1
Temp. [K]	299.7	598.0	598.0	598.0	598.0	931.3	1668.0	1087.6	1087.6	1087.6	754.4	460.6	299.7	931.3
Enthalpy [KJ/kG]	300	605	605	605	605	968	2312	1569	1569	1569	1206	757	112	3827
Temp. [C]	27	325	325	325	325	658	1395	815	815	815	481	188	27	658
Density [kG/m³]	1.18	4.71	4.71	4.71	4.71	3.03	1.69	0.33	0.33	0.33	0.47	0.77	1.18	1.89
Mass Flow [Kg/s]	3.89	3.89	3.89	3.89	3.89	3.89	4.36	4.36	4.36	4.36	4.36	4.36	4.36	0.39
Vol. Flow [Nm³/hr]	11594	11594	11594	11594	11594	11594	12981	12981	12981	12981	12981	12981	12981	1.400

Figure C4.7. Gas Turbine cycle, similar to Figure C4.6 except that it uses water injection to improve cycle efficiency.

discharged from the GT and in proportion to the increase in the cycle efficiency of the cycle due to the water injection.

5. It should be noted that only the amount of water generated from the available waste heat from the GT was used in this analysis to increase the power output of the turbine. As a consequence, the temperature of the steam that can be produced by the waste heat from the GT is limited by the exhaust gas temperature exiting the GT and the pressure ratio used in the GT cycle. It was also assumed that the temperature of the water injection is equal to the regenerator discharge temperature in order to minimize the change in entropy of the mixing air and water flow streams. Thus, the combustor inlet temperature is affected by the extent of the regeneration or recuperation used in the cycle. Figure C4.1 illustrates the exhaust gas temperatures as a function of the regenerator effectiveness. The consequence of high regenerator effectiveness resulting in a high cycle

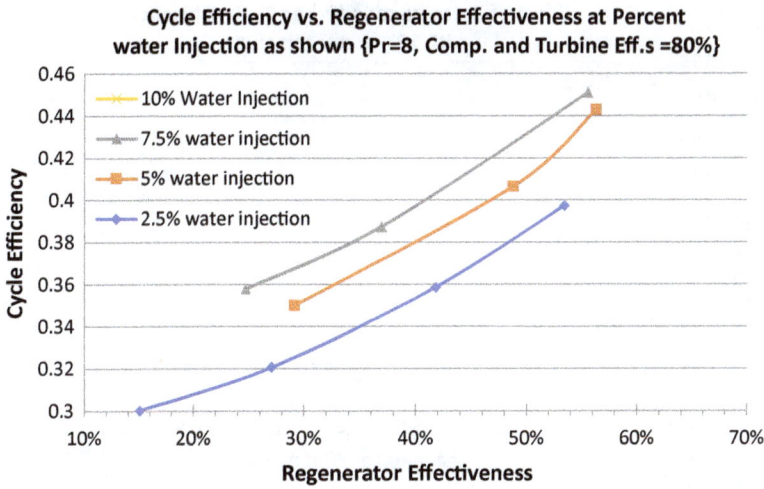

Figure C4.8. Results of the gas turbine spreadsheet model showing the effect of water injection and regenerator effectiveness on the gas turbine net cycle efficiency.

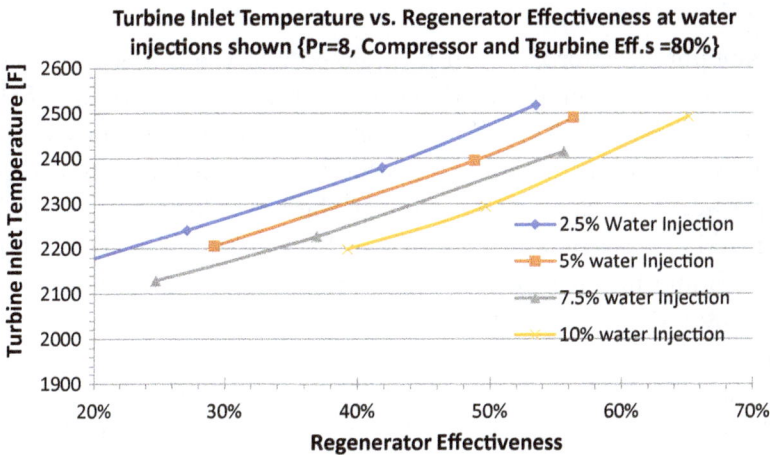

Figure C4.9. Results of the gas turbine spreadsheet model showing the required turbine inlet temperature, water injection and regenerator effectiveness required to achieve the gas turbine cycle efficiency shown in Figure C4.8.

efficiency is a low water vapor temperature derived from the exhaust waste heat that has its temperature lowered when the recuperator effectiveness is high. The need for this water vapor to be heated to the necessary turbine inlet temperature requires additional fuel to be consumed in the combustor of the GT, thus resulting in a lower overall GT cycle efficiency (see Figure C4.6). However, this drop in cycle efficiency can be prevented if the vaporization and superheating of the water to the very high turbine

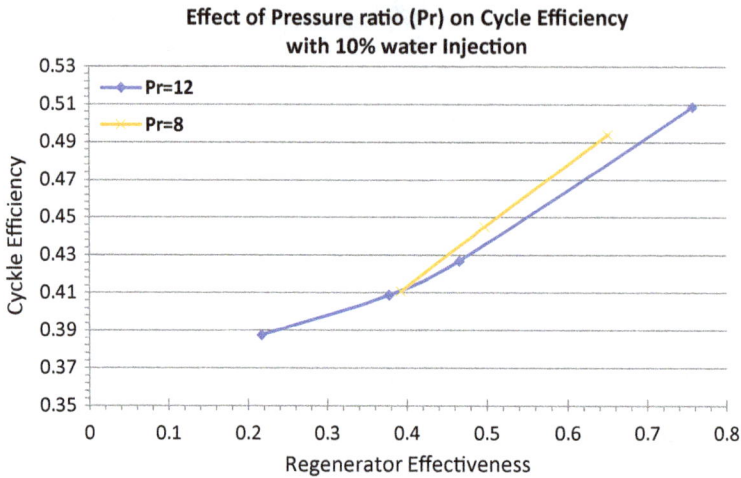

Figure C4.10.　Results of the gas turbine spreadsheet model showing the small effect that pressure ratio has on the overall efficiency of the GT with a 10% water injection.

inlet temperature can be achieved using a "free" source of waste heat energy.

6. The amount of water that can be injected into the compressors for inter-cooling purposes can also be determined to be only 1% to 2% of the flow rate of the air through the GT, on a mass basis. This may be compared to 5% to 10%, on a mass basis with respect to the air flow rate through the compressor, that can be generated by the waste heat from the GT. It is assumed in this calculation that the water is to be completely evaporated from a liquid into a vapor and be entrained in the now moist air of the compressor discharge as the combined mass flow continues to the GT combustor.

7. The analysis of the GT cycles shown in Figures C4.6 and C4.7 included a study of how the regenerator effectiveness and the turbine inlet temperature affects the cycle efficiency as the water injection is increased from 0 to 10%. The results are shown in Figures C4.8 to C4.10. The conclusion to be drawn from these last figures is that the cycle efficiency of the GT is limited by the size of the regenerator and the maximum turbine inlet temperature that is considered reasonable. For example, turbine inlet temperatures that exceed 2,500°F are not considered to be practical without using extensive turbine bade cooling and the use of exotic, high strength nickel-based steels. Similarly, regenerator effectiveness above 85% results in a non-proportional increase in heat exchange with the increase in regenerator physical size. More critical to the overall selection of the GT state points is the realization from Figures C4.9 and C4.10 that a regenerator

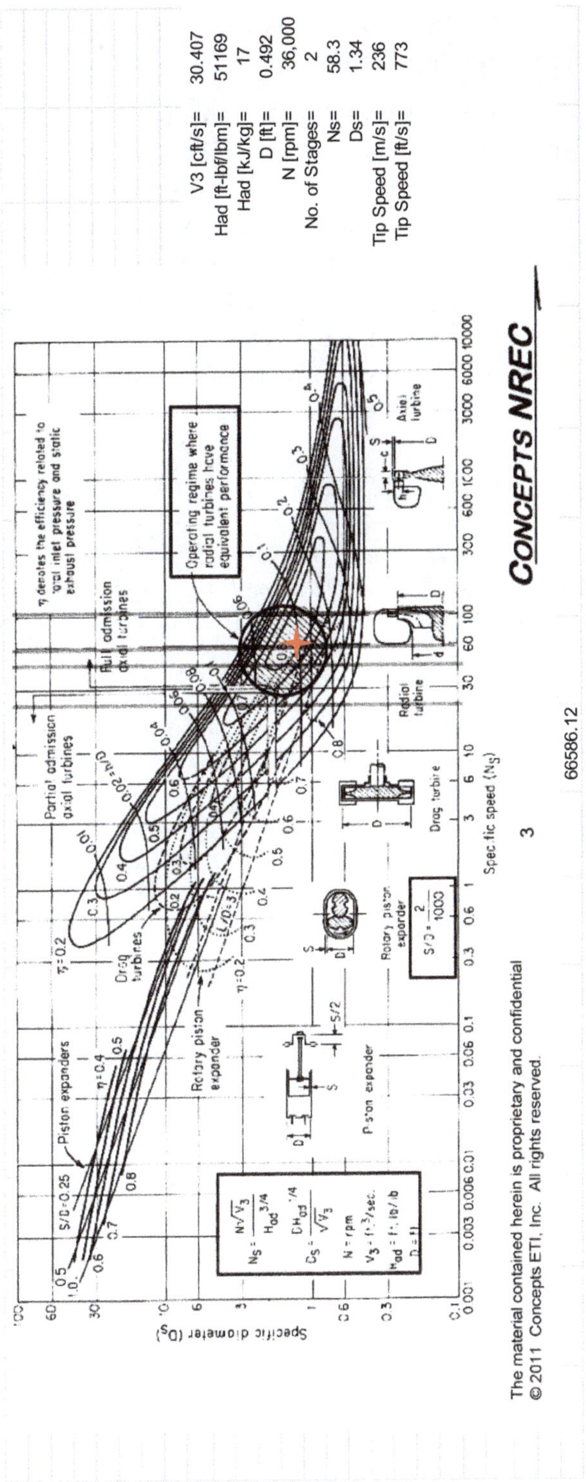

V3 [cft/s]=	30.407
Had [ft-lbf/lbm]=	51169
Had [kJ/kg]=	17
D [ft]=	0.492
N [rpm]=	36,000
No. of Stages=	2
Ns=	58.3
Ds=	1.34
Tip Speed [m/s]=	236
Tip Speed [ft/s]=	773

66586.12

Figure C4.11.   A typical Specific Diameter (Ds) and Specific Speed (Ns) Diagram for turbines.

effectiveness above 70% requires a turbine inlet temperature that exceeds the 2,500°F maximum temperature constraint.

Figure C4.11 provides a first-order analysis of the type, size and speed of the necessary turbine used in the power train of the GT. The result indicates the need for two axial turbine units, each with 2 stages, a speed of 36,000 rpm with a diameter of 0.15 m. The non-dimensional specific diameter and specific speed diagram is used to provide a first-estimate of the size, speed and type of turbine that is most suited for a particular turbine application. The formulae for the specific speed, Ns, and specific diameter, Ds, are given in the diagram. The red star identifies the intersection of Ns and Ds on the chart and the estimated efficiency.

Reference

Dohmen, H., Schnitzler, J., & Benra, F.-K. (2011). Efficiency Augmentation of Gas Turbine Cycles. *Tth International Conference on Heat Transfer, Fluid Mechanics and Thermodynamics — HEFAT2011*. Pointe Aux Piments, Mauritius.

Supercritical CO$_2$(sCO$_2$)Systems

The US Department of Energy is considering expending capital in order to promote the development of a supercritical CO$_2$(sCO$_2$) system. This has attracted many other governments and some private companies to participate in the development of the sCO$_2$ system. Echogen (owned by Dresser Rand) has developed an 8 MWe system that can operate as a waste heat recovery system for a gas turbine. Another company, Net Power, is developing the Allam Cycle. The Allam Cycle utilizes an oxygen separation process to first remove nitrogen from the air, before the remaining oxygen is utilized to burn coal. The oxygen is pressurized to 4,000 psia, and then used to oxidize coal to temperatures of 700°C (or higher). To enhance the rate of combustion, the coal used in this combustion process is first pulverized to a fine powder consistency of 70 μm–200 μm. In the absence of nitrogen, the proportion of the products of combustion is almost 100% CO$_2$. This high-pressure CO$_2$ fluid stream is expanded through a turbine before it is sequestered into the ground at pressures of 1,100 psia. This system is thus not truly a cycle but an open flow sCO$_2$ power system. With the Allam Cycle, there is potential for the United States' abundant supply of coal to be the primary fuel for the lifetime of a new power plant facility.

AnsCO$_2$ system shares the same operating principle as a Brayton Cycle. However, the sCO$_2$ system operates above the critical point of the working fluid, which is CO$_2$. Thus, the operating pressures must be above 1,070 psi ($\sim$72 bar,a) and 80°F; the critical pressure and temperature for CO$_2$, respectively. A primary benefit of the sCO$_2$ system is the ability to operate at extremely high thermal efficiencies. To attain these high efficiencies, the sCO$_2$ system must operate at very high-energy source temperatures — typically above 700°C. Such extremely high temperatures are available from nuclear or concentrated solar energy power plants. Another major benefit of a supercritical system is the relatively small size of its CO$_2$ turbine and compressor compared to steam turbines and natural gas fired, gas turbine

Reheat Cycle: On

			Eff. T1=	0.85		Tcoolant,in=	95		LT Regen=	0.92
Pr=	4		Eff.T2=	0.85		Tcoolant,out=	115			
Pr1=	2.00		LT Comp. Eff.=	0.85		Dtcool pinch=	40		HT Regen.=	0.90
Pr2=	2.00		HT Comp. Eff.=	0.85		Recomp. Fraction=	0.850		DT,heater Pinch=	100

Iterate using "SOLVER" to get Recomp. Fraction

	A	B	C	D	E	F	Fg1	Fg2	Fg3	Fh1	G	H
P[psia]	4221	2110	2110	1102	1090	1080	1070	4280	4280	4237	4237	4237
T[F]	1022	876	1022	888.90	626.00	406	155	397	607.7	713	623	862
h[Btu/Lbm]	442.8	404.6	447.1	411.7	338.9	279.9	206.6	247.6	317.0	349.6	321.9	394.6
s[Btu/Lbm/R]	0.636	0.641	0.671	0.676	0.616	0.556	0.456	0.463	0.535	0.565	0.540	0.601
Cp[Btu/Lbm/R]	0.30	0.29	0.29	0.28	0.27	0.27	0.37	0.31	0.31	0.30	0.31	0.30

Enthalpy (Btu/Lbm)

Turbine Δh, 1= 38.27   Mech. Eff.= 0.97   338.9
Turbine Δh, 2= 35.41   Gen. Eff.= 0.975   0.002   0.0%
LT Comp. Δh= 40.98
HT Comp. Δh= 69.69   O2 Main Flow [kG/s]= 352.9
S-CO2 Main Flow (Lbm/s)= 776.5

Cycle Eff.= 29.6%
Turbine Power, 1 [kW]= 31344
Turbine Power, 2 [kW]= 29002
LT Comp. Power [kW]= 28528
HT Comp. Power [kW]= 8562
Net Power (kW)= 21,994

Q, input +   Q, reheat=   Q, total   135.99
48.2   42.5   90.7   73.68
Q,cool= 62.31 CHECK 62.31

UA,super heater= 1.42E+06
UA)cooler= 1.40E+06
UA$_{HLTR}$= 1.23E+07   48.3 MWt
UA$_{HHTR}$= 1.86E+07   59.6 MWt

CSP HEAT SOURCE
Cp,solar fluid= 0.5
Q,1 (kWt)= 39,478   500 =Flowrate (Lbm/s)
Q,2 (kWt)= 82,567   1112 F

962   Turbine,1

CSP HEAT SOURCE
373 Lbm/s
1382 F

Heat Recovery HX

Low Temp. Compressor   H   Fg2   A   B   Reheat Turbine, 2

962   F; Reheat HX

Fg1   85%   F   H   D
   Low T Regen.   High T Regen.
15%   Largest Pipe ("DD") Dia. (inch)= 16.9
95 F   Fg3   G
Water Coolant   115
Fh1
High Pres. Comp.

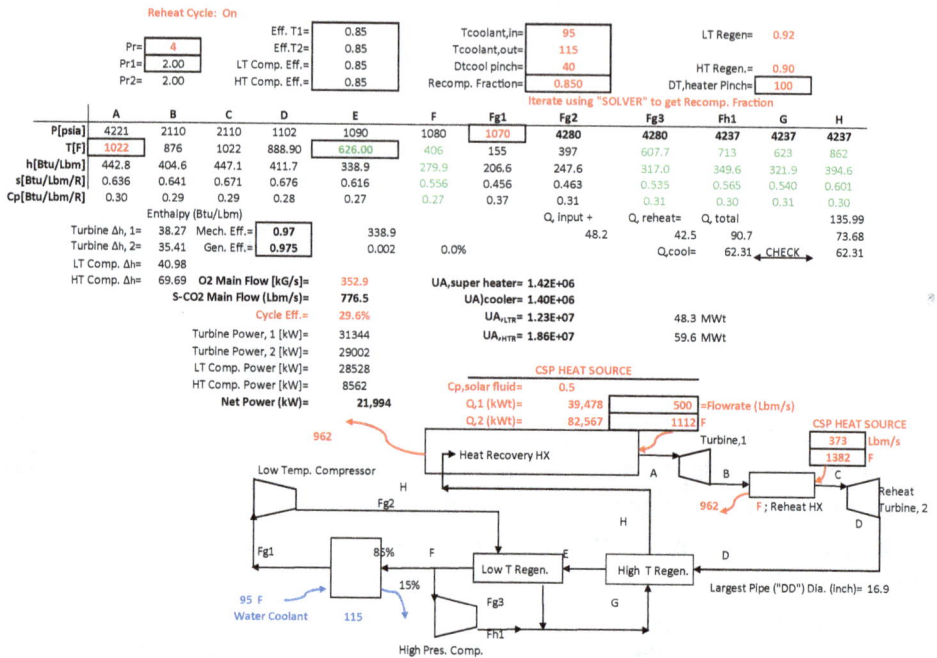

Figure C5.1.   Recompression sCO$_2$ system with a reheat turbine.

engines of comparable utility size power ratings (250 MW or higher). Unfortunately, as will be seen in this case study, the size savings available for the sCO$_2$ turbine and compressor turbomachinery is "out-weighed" by the very large regenerator and heater heat exchangers that must be utilized in the sCO$_2$ cycle to make the cycle efficient.

The components of the recompression sCO$_2$ cycle, specifically a reheater and a reheat turbine, are shown in Figure C5.1, along with the Property Table for the cycle. The recompression cycle typically uses two compressors, two recuperators, and just one turbine. Figure C5.1 illustrates the recompression sCO$_2$ cycle which includes a reheat turbine to increase the cycle efficiency. The two recuperators help to improve the efficiency of the cycle by preheating the working fluid to a very high temperature before it enters the primary sCO$_2$ heater.

The T-s diagram for the cycle is shown in Figure C5.2. The T-s diagram includes the temperature profiles of the heat source that is superimposed onto the diagram. It may be observed from Figure C5.2 that the heat source temperature profile is constrained by the temperature of the working fluid that enters the CO$_2$ heater. In order to achieve the highest cycle efficiency, the temperature of the CO$_2$ entering the heater must be as high as possible.

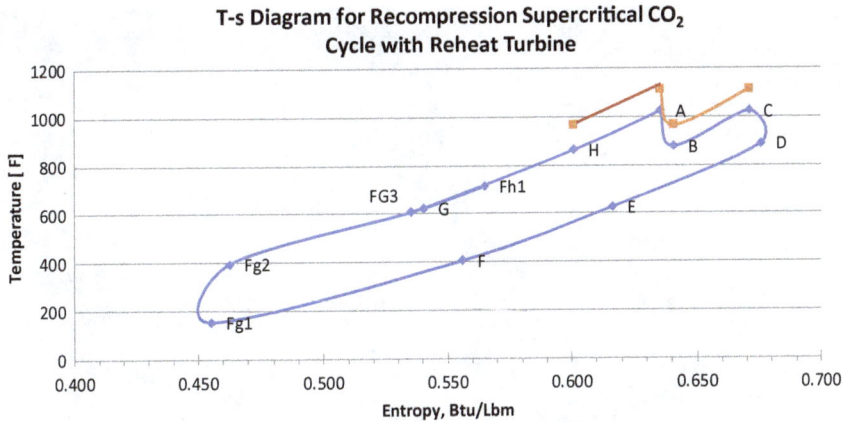

Figure C5.2.   T-s diagram for Figure C5.1's recompression system with a reheat turbine.

This high temperature imposes a constraint on the discharge temperature of the heat source. In order to generate the maximum amount of power, it is necessary to have a small change in temperature of the heat source, and a very high fluid flow rate. These requirements limit the type of heat sources that can be used for providing the heat input to the sCO_2 cycle, via the heater. For instance, despite having a high temperature, the candidate heat source should not be a waste heat stream from an industrial or prime mover engine, whereby the heat source flow rate is controlled by the nature of the industrial process and/or by the type of engine prime mover and not by an engineer. In order to recover a maximum amount of heat from the waste heat source, it is also necessary that the discharge temperature must be reduced to as close to atmospheric temperature as is practically achievable. This lower limit for this temperature is placed at approximately 300°F, in order to avoid the condensation of the water typically entrained in the hot exhaust products of combustion. The need for the heat source of an sCO_2 system to be of a high temperature while simultaneously allowing for a small change in temperature — required to recover the heat energy and a considerable amount of flow rate — can only be satisfied with a nuclear or concentrated solar power system. For these heat sources, the engineer has control of the magnitude of the heat source flow rate and the temperature difference through which this heat is recovered. In this case study, an sCO_2 system will be applied to a concentrated solar collector.

It is appropriate to present a few engineering facts concerning a concentrated solar collector system. A concentrated solar collector system, shown in Figure C5.3, includes a large area of focused solar reflectors that direct

Figure C5.3.   Concentrated solar energy heliostat and solar reflector field.

incoming solar photons onto a much smaller area, thus concentrating the solar energy. A heliostat is installed at the focal point of the focused mirror reflectors. A heliostat is a high pressure and high temperature heat exchanger designed to transfer the concentrated and hence very high temperature solar photon energy to a circulating working fluid. The heliostat is typically designed for temperatures between 2,000 and 2,500°F. The working fluid can be common heat transfer fluids such as molten nitrate (maximum operating temperature: 565°C) and molten sodium salts (maximum operating temperature: 600°C), or even gases such as air or helium (maximum temperature and pressure of 850°C, operating at 12 to 15 bar,a). Pressurized water or superheated steam could also be used. It is often necessary to design a thermal storage system that is installed alongside the solar collector field. The thermal storage system enables a continuous generation of power when solar energy is not available, or the solar energy is diminished due to weather or other natural phenomena.

C5.1 Effects of altering specific cycle conditions on the performance of the sCO$_2$ system

Assuming that a molten sodium salt heat transfer fluid is used and that it attains a temperature of 600°C and has a flow rate of 500 lbm/s, the amount of power that can be generated with this system and the system's efficiency can be calculated using the cycle conditions of the sCO$_2$ system shown in Figure C5.2. As may be observed, the flow rate of the molten salt, its temperature and specific heat are input into a computer model of the sCO$_2$ recompression system, built using Excel. For this Case Study, the reheat turbine is assumed to operate at the same inlet temperature as the first turbine, and at the same flow rate as used in the primary sCO$_2$ heater. The system can generate a net 23,655 kW from a total of 873 lbm/s (500 lbm/s + 373 lbm/s for the reheat turbine) of the 600°C molten salt heat transfer fluid. The cycle efficiency is observed to be 31%. However, note the effectiveness of the recuperators, 96% and 99%, and the size of the recuperators, as represented by the ratio of the heat transfer divided by the log mean temperature difference. This ratio is equal to the heat transfer coefficient multiplied by the total surface area of the heat exchanger or UA in units of Btu/h-°F. The two recuperators shown in Figure C5.1 are equal to 12×10^6 and 19×10^6 Btu/h $-$ F°, respectively and may serve as the basis for the following study of the effect of recuperator size on cycle efficiency.

It is interesting to determine the performance of the sCO$_2$ system with recuperators that have been reduced in size by 200% to 300%. Figure C5.4 presents this result. With a reduction in the recuperator sizes, the cycle efficiency is reduced by only 2% while maintaining a high power-output of approximately 23,500 kWe — only slightly less than before the size reduction! The larger recuperator sizes in the first cycle analysis are hardly worth the small increase in power of only 130 kWe.

It is also interesting to consider a simpler sCO$_2$ cycle as shown in Figure C5.5. In this sCO$_2$ cycle the reheater and reheat turbine has been eliminated. A three-stage, intercooled compressor is used in place of the

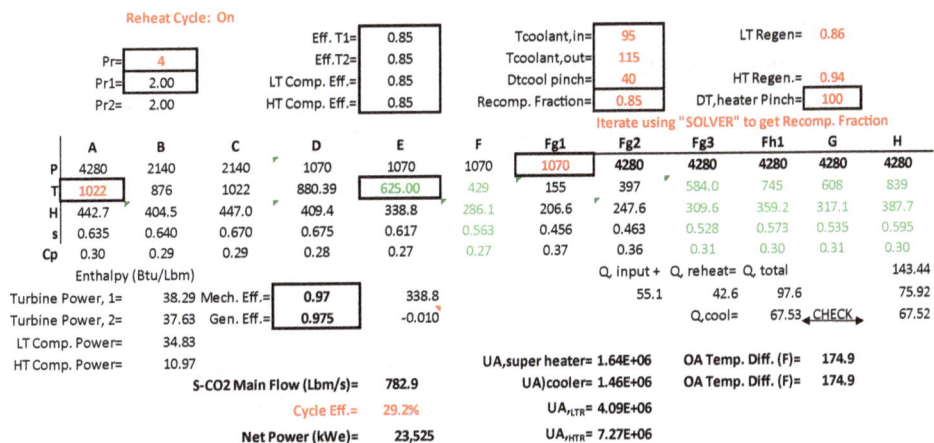

Reheat Cycle: On

	Eff. T1=	0.85		Tcoolant,in=	95		LT Regen=	0.86
Pr= 4	Eff.T2=	0.85		Tcoolant,out=	115			
Pr1= 2.00	LT Comp. Eff.=	0.85		Dtcool pinch=	40		HT Regen.=	0.94
Pr2= 2.00	HT Comp. Eff.=	0.85		Recomp. Fraction=	0.85		DT,heater Pinch=	100

Iterate using "SOLVER" to get Recomp. Fraction

	A	B	C	D	E	F	Fg1	Fg2	Fg3	Fh1	G	H
P	4280	2140	2140	1070	1070	1070	1070	4280	4280	4280	4280	4280
T	1022	876	1022	880.39	625.00	155	155	397	584.0	745	608	839
H	442.7	404.5	447.0	409.4	338.8	286.1	206.6	247.6	309.6	359.2	317.1	387.7
s	0.635	0.640	0.670	0.675	0.617	0.563	0.456	0.463	0.528	0.573	0.535	0.595
Cp	0.30	0.29	0.29	0.28	0.27	0.27	0.37	0.36	0.31	0.30	0.31	0.30

Enthalpy (Btu/Lbm)

						Q input + Q reheat= Q total		143.44
Turbine Power, 1=	38.29	Mech. Eff.=	0.97	338.8		55.1	42.6	97.6	75.92
Turbine Power, 2=	37.63	Gen. Eff.=	0.975	-0.010			Q,cool=	67.53 CHECK	67.52
LT Comp. Power=	34.83							
HT Comp. Power=	10.97					UA,super heater= 1.64E+06	OA Temp. Diff. (F)=	174.9
		S-CO2 Main Flow (Lbm/s)=	782.9			UA)cooler= 1.46E+06	OA Temp. Diff. (F)=	174.9
		Cycle Eff.=	29.2%			UA$_{rLTR}$= 4.09E+06		
		Net Power (kWe)=	23,525			UA$_{rHTR}$= 7.27E+06		

Figure C5.4. An sCO$_2$ Cycle with reduced recuperator effectiveness and thus smaller sizes compared to the cycle shown in Figure C5.1.

CONCENTRATING SOLAR APPLICATION OF CO$_2$ BRAYTON CYCLE WITH SINGLE REGEN. and 2-INTERCOOLED COMP.S

Pres. Drop Losses 1.0%
INTERCOOLERS
Clr. Effect 82% | 0% | 0%

REGENERATOR EFFECT 0.97

CO$_2$ HEATER USING CO$_2$ AS HEAT TRANSFER FLUID

108 DT,pinch

Ns 2.4E+01
Ds 2.343
Dia (m) 0.175
RPM 25000
No. of Stg.s 1

TURBINE
EFF.= 83%

Comp.n= 82% | 82% | 100% | 100%
Pr= 2.12 | 2.12 | 1.00 | 1
450
COMPRESSORS

MOTOR/GEN.
EFF.= 0.970

33.2% CYCLE EFF.
TURBINE POWER= 14,591 kW
Σ COMP. POWER= 4,394 kW
POWER,net= 10,197 kW
CO$_2$ HEATER Q= 29,760 kWt;UA[kW 398.0
CO$_2$ COOLER Q= 8,676 kWt;UA[kW 547.5
REGENERATOR Q= 33,242 kWt;UA[kW 260.1
INTERCOOLER ΣQ= 10,886 kWt; 265.5
Heat Balance Check= GOOD @ 100.00%
Heat Exchanger Factor= 71.0

Nstages/unit= 1
RPM= 25000   CO$_2$ COOLER
Dia. [m]= 0.279
Pr= 2.12
Head Coef. (ψ) 0.130
Flow Coef. (φ) 0.012   Cp, Sodium Salt Fld.= 0.304 BTU/LBm/R
Ns= 0.501   Cv, CO$_2$ Helio. Fld.= 0.235 BTU/LBm/R
K= 1.29

FLUID: carbon dioxide
MOLE. WT. 44.01
Critical Pressure [psia]= 1070   73.8 Bar,a
Critical Temp. [F]= 87   30.5 C

Twater, in [F] = 82.4   Water Mass Flow= 6,574 GPM
Twater, out [F]= 91.4

	1	1a	1b	1c	2	3	4	5	6	A	B
Pres. [psia]	1116.5	2345	4924	4875	4875	4826	4778	1128	1117	4875	4826
Temp. [F]	93.2	99.0	134.6	134.1	134	535	1004	723	151	1112	700
Enthalpy [Btu/lbm]	161.2	118.7	129.7	129.5	129.5	291.5	436.5	365.4	203.4	338.4295	213.0402
Density [Lbm/ft³]	21.03	50.89	54.17	54.11	54.11	20.63	12.56	3.93	10.60	11.839	16.693
Flow rate [Lbm/s]	194.59	194.59	194.59	194.59	194.59	194.59	194.59	194.59	194.59	225.00	225.00
Entropy [Btu/Lbm/R]	0.3752	0.2894	0.2927	0.2927	0.2927	0.5054	0.6255	0.6381	0.4490	0.6462	0.5539
CO$_2$ Cp	3.53	0.59	0.46	0.46	0.46	0.33	0.30	0.28	0.40	0.31	0.31
Pres. bar,a	77.0	161.7	339.6	336.2	336.2	332.8	329.5	77.8	77.0	336	333
Temp. [K]	307.0	310.2	330.0	329.7	329.7	552.3	813.0	657.1	339.0	873.0	644.1
Enthalpy [KJ/kG]	374	276	301	301	301	677	1014	848	472	786	495
Temp. [C]	34	37	57	57	57	279	540	384	66	600	371
Density [kG/m³]	337	816	869	868	868	331	201	63	170	190	268
Mass Flow [Kg/s]	88.45	88.45	88.45	88.45	88.45	88.45	88.45	88.45	88.45	102.27	102.27
Vol. Flow [Nm³/hr]	172531	172531	172531	172531	172531	172531	172531	172531	172531	304748	304748

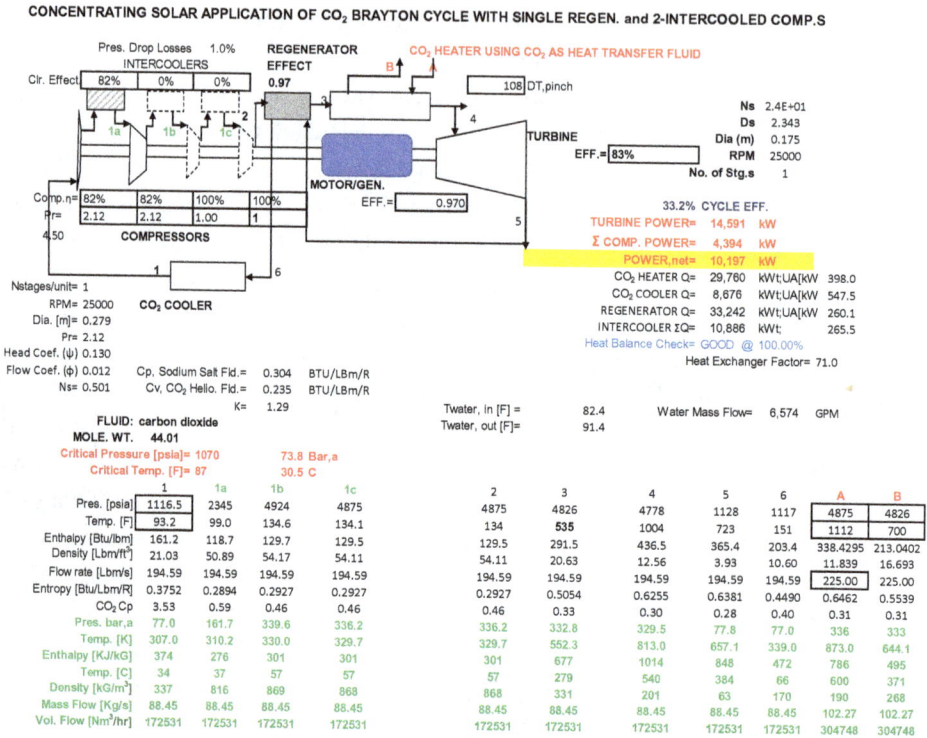

Figure C5.5. Simple sCO2 Cycle with single turbine and single, multi-stage compressor and without a reheater or reheat turbine.

high and low temperature compressors and reheater and reheat turbine as shown in Figure C5.1. Interestingly, the results demonstrate that the cycle efficiency is increased by almost 2%. In this calculation, only 225 lbm/s of heat transfer fluid flow rate was used to generate 10,000 kWe. It is noted that this system used the same recuperator size as the sCO2 cycle shown in Figure C5.5. If the heat transfer fluid flow rate were to increase and be made equal to the flow rate in the previous cycle study, the power output would be: 10, 000 kWe × (500 + 385)/225 lbm/s = 37, 111 kWe.

Lastly, as noted in Case Study No. 2, it is imperative to not always equate cycle efficiency with maximum power generation. In the case of applying an sCO2 system to heat recovery, as shown in Figures C5.6(a) and (b), a lower cycle efficiency can achieve a higher net power output. While this at first seems counter-intuitive, it is possible to achieve higher power at a lower cycle efficiency because a higher cycle efficiency is only achieved when the regeneration of the turbine exhaust is performed at high regenerative effectiveness, which ultimately translates into a higher fluid temperature entering into the cycle heater. The higher inlet temperature then constrains

(a)

Reheat Cycle: On

Pr= 4	Eff. T1= 0.85	
Pr1= 2.00	Eff. T2= 0.85	
Pr2= 2.00	LT Comp. Eff.= 0.85	
	HT Comp. Eff.= 0.85	

Tcoolant,in= 95
Tcoolant,out= 115
Dtcool pinch= 40
Recomp. Fraction= 0.850
Iterate using "SOLVER" to get Recomp. Fraction

LT Regen= 0.91
HT Regen.= 0.92
DT,heater Pinch= 318

	A	B	C	D	E	F	Fg1	Fg2	Fg3	Fh1	G	H
P[psia]	4278	2139	2139	1070	1070	1070	1070	4280	4280	4280	4280	4280
T[F]	1292	1130	1292	1134.26	635.00	410	155	397	613.6	722	630	1094
h[Btu/Lbm]	525.0	478.9	527.2	482.2	341.5	281.0	206.6	247.6	318.9	352.2	323.9	464.5
s[Btu/Lbm/R]	0.686	0.691	0.720	0.725	0.619	0.558	0.456	0.463	0.537	0.567	0.542	0.649
Cp[Btu/Lbm/R]	0.31	0.30	0.30	0.29	0.27	0.27	0.37	0.31	0.31	0.31	0.31	0.30

Enthalpy (Btu/Lbm)

Turbine Δh, 1= 46.05   Mech. Eff.= 0.97   341.5
Turbine Δh, 2= 44.98   Gen. Eff.= 0.975   0.003   0.0%
LT Comp. Δh= 40.98
HT Comp. Δh= 71.21   O2 Main Flow [kG/s]= 113.0   UA,super heater= 2.23E+05
S-CO2 Main Flow (Lbm/s)= 248.6   UA)cooler= 4.51E+05
Cycle Eff.= 39.6%   UA,LTR= 3.24E+06   15.9 MWt
Turbine Power, 1 [kW]= 12074   UA,HTR= 7.29E+06   36.9 MWt
Turbine Power, 2 [kW]= 11793
LT Comp. Power [kW]= 9132
HT Comp. Power [kW]= 2801
Net Power (kW)= 11,286

Q, input + 60.5   Q, reheat= 48.2   Q, total 108.7
Q,cool= 63.16 CHECK 63.16
154.19
91.03
63.16

CSP HEAT SOURCE
Cp,solar fluid= 0.5
Q,1 (kWt)= 15,850   500 =Flowrate (Lbm/s)
Q,2 (kWt)= 11,317   1472 F

CSP HEAT SOURCE
357 Lbm/s
1472 F

1412   Turbine,1
Low Temp. Compressor
Heat Recovery HX
1412   F ; Reheat HX   Reheat Turbine, 2
H
Fg2
Fg1   85%   F
15%
Low T Regen.   High T Regen.
95 F   Water Coolant   115
Fg3
Fh1
High Pres. Comp.
Largest Pipe ("DD") Dia. (Inch)= 10.6

(b)

Reheat Cycle: On

Pr= 4	Eff. T1= 0.85	
Pr1= 2.00	Eff.T2= 0.85	
Pr2= 2.00	LT Comp. Eff.= 0.85	
	HT Comp. Eff.= 0.85	

Tcoolant,in= 95
Tcoolant,out= 115
Dtcool pinch= 40
Recomp. Fraction= 0.850
Iterate using "SOLVER" to get Recomp. Fraction

LT Regen= 0.84
HT Regen.= 0.85
DT,heater Pinch= 200

	A	B	C	D	E	F	Fg1	Fg2	Fg3	Fh1	G	H
P[psia]	4278	2139	2139	1070	1070	1070	1070	4280	4280	4280	4280	4280
T[F]	1292	1130	1292	1134.26	735.00	446	155	397	680.9	766	694	1068
h[Btu/Lbm]	525.0	478.9	527.2	482.2	368.9	290.7	206.6	247.6	339.6	365.5	343.5	456.7
s[Btu/Lbm/R]	0.686	0.691	0.720	0.725	0.643	0.569	0.456	0.463	0.556	0.578	0.559	0.644
Cp[Btu/Lbm/R]	0.31	0.30	0.30	0.29	0.28	0.27	0.37	0.30	0.31	0.30	0.31	0.30

Enthalpy (Btu/Lbm)

Turbine Δh, 1= 46.05   Mech. Eff.= 0.97   368.9
Turbine Δh, 2= 44.98   Gen. Eff.= 0.975   0.069   0.0%
LT Comp. Δh= 40.98
HT Comp. Δh= 74.83   S-CO2 Main Flow [kG/s]= 339.4   UA,super heater= 9.66E+05
S-CO2 Main Flow (Lbm/s)= 746.6   UA)cooler= 1.42E+06
Cycle Eff.= 36.5%   UA,LTR= 4.07E+06   61.6 MWt
Turbine Power, 1 [kW]= 36266   UA,HTR= 5.77E+06   89.2 MWt
Turbine Power, 2 [kW]= 35420
LT Comp. Power [kW]= 27430
HT Comp. Power [kW]= 8840
Net Power (kW)= 33,496

Q, input + 68.3   Q, reheat= 48.2   Q, total 116.5
Q,cool= 71.44 CHECK
162.54
91.03
71.51

CSP HEAT SOURCE
Cp,solar fluid= 0.5
Q,1 (kWt)= 53,752   500 =Flowrate (Lbm/s)
Q,2 (kWt)= 38,379   1472 F

CSP HEAT SOURCE
357 Lbm/s
1472 F

1268   Turbine,1
Low Temp. Compressor
Heat Recovery HX
1268   F ; Reheat HX   Reheat Turbine, 2
H
Fg2
Fg1   85%   F
15%
Low T Regen.   High T Regen.
95 F   Water Coolant   115
Fg3
Fh1
High Pres. Comp.
Largest Pipe ("DD") Dia. (Inch)= 18.3

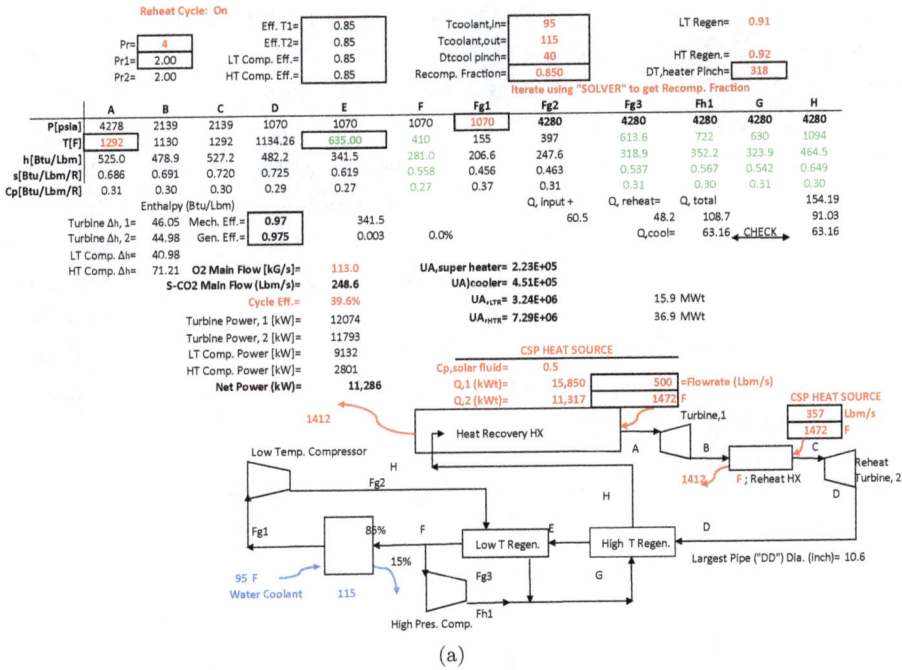

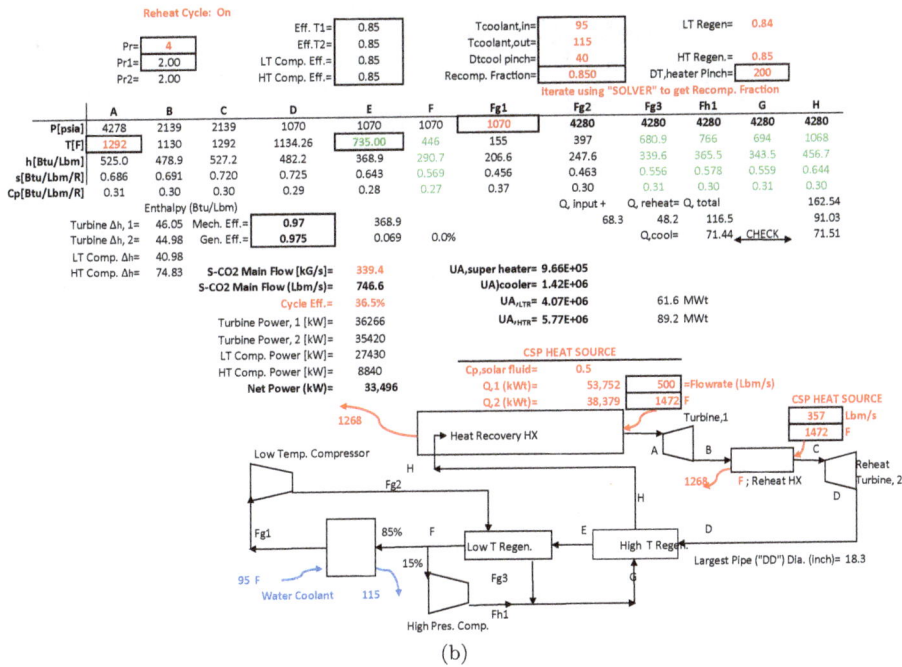

Figure C5.6.   sCO₂ Recompression Cycle with high cycle efficiency due to high fluid inlet temperature into the system heater at state point "H", resulting in lower net power generation than the cycle shown in Figure C5.6(b); (b) Almost identical cycle conditions to Figure C5.6(a) but with a lower temperature at state point "H", resulting in higher net power generation than Figure C5.6(a), but at lower cycle efficiency.

the temperature that the heat source can be reduced to, assuming that a reasonable "pinch point", i.e., minimum acceptable approach temperature between the exhaust heat discharge temperature and the fluid inlet temperature to the evaporator, can be maintained. To demonstrate this, attention is drawn to Figure C5.6(a) wherein the temperature at state point "H" is 1,086°F, while the same state point "H" is only 855°F in Figure C5.6(b). The consequence is that the exhaust gas heat source discharge temperature from the evaporator in Figure C5.6(a) can only be reduced to 1,404°F compared to 955°F in Figure C5.6(b). While this results in a lower cycle efficiency for the cycle shown in Figure C5.6(b), with more heat to recover at that lower cycle efficiency, the net power generated is higher than that generated by the cycle in Figure C5.6(a).

Case Study 6

Heat Recovery from Reciprocating Engine for Ramjet Power Augmentation

Energy recovery from waste heat sources is an essential part of energy conservation and reduction in CO_2. Most heat recovery systems draw from industrial processes that produce an end-product with the heat input or from land-based engines which generate electric or mechanical power. In this case study, a rather unique heat recovery system will be presented. Let us consider a heat recovery process for an engine that is not a land-based engine, but a vintage aircraft engine. Heat from the exhaust gas from the aircraft engine can be recovered and used to increase the propulsion of the aircraft. This very unique heat recovery application can be analyzed using thermodynamics principles, and can subsequently be used in deciding if an old engine can be altered to produce more thrust by adding an auxiliary thermodynamics-based machine, which for this application is identified as a ramjet engine.

The Wright R-1280 reciprocating engine has been used in numerous aircraft, including the classic DC-10 aircraft. A photo of the Wright R-1280 engine is shown in Figure C6.1. The engineering specifications for this engine are in Table C6.1.

The engine must produce a large amount of mechanical power to turn the propeller at a very high speed. The propeller is a form of turbomachinery which can accelerate the speed of the incoming air to a much higher one. The propulsion force that moves the aircraft forward is produced by increasing the velocity of the air mass flow rate with respect to the aircraft speed. The basic equation for determining the propulsion force is shown in Equation (C6.1). It is noted that the ΔV_x shown in the equation is the difference in velocity between the air from the engine driven propeller and the speed of the aircraft.

$$Thrust\,Force_x = \frac{\dot{M} \times \Delta V_x}{g_c}. \qquad (C6.1)$$

Figure C6.1.   A Wright R-1280 engine with bore = 6.125″, stroke = 6.875″; 9 cylinders; 790 kW blower pressure ratio = 7 : 1 to 10:1, rpm = 2200, 0.43 lbm/hp-h.

Table C6.1.   Engineering specifications for the Wright R-1280 engine.

Specifications	
Wright Cyclone R-1820-G2	
Date	1931
Cylinders	9
Configuration	Single-row, Air-cooled radial
Horsepower	1,060 hp (790 kw)
R.P.M.	2,200
Bore and Stroke	6.125 in. (155.6 mm)×6.875 in. (174 mm)
Displacement	1,823 cu. in. (29.88 liters)
Weight	1,184 lbs. (537 kg)

The Wright R-1820 propulsion engine produces a large amount of waste heat energy; approximately 30–40% of the energy input by the fuel engine is rejected in the exhaust gas from the engine. How can one recover this waste heat energy to produce additional, useful thrust? A possible method involves the use of a nozzle. A nozzle is one of the seven basic mechanical devices discussed in Chapter 3. A nozzle is a "no moving parts" device which produces thrust by converting the thermal and pressure (i.e., mechanical potential) energy at its inlet into kinetic energy at its exit. For a nozzle to recover an aircraft engine's waste heat, it must first increase the pressure of the fluid beyond atmospheric pressure. A compressor is usually what is used to increase the pressure of a gas. However, a gas compressor is relatively complex and heavy, and consequently not the best solution for improving the thrust of the Wright engine. An alternative method involves the use of a diffuser, which is another of the seven basic mechanical devices. A diffuser operates in virtually the opposite way a nozzle works: it converts high kinetic energy into higher pressure. The high kinetic energy is obtained by utilizing the aircraft's readily available very high flight speed. The higher pressure intake air can then be heated by the engine's high temperature exhaust gas — the waste heat energy. This open cycle is thus very similar to a Brayton Cycle in its thermodynamics operation.

A device that incorporates both the diffuser and nozzle functions, along with a means of combusting the fuel that is injected between these two devices is called a ramjet engine. The ramjet is thus a device with no moving parts, substituting the diffuser for the compressor which eliminates the need for a turbine to produce the power for the compressor. The absence of these heavy and more complex parts come with a compromise however; for a ramjet engine to operate, air with a very high initial velocity needs to enter the diffuser. In other words, the ramjet engine must already be in motion before it can produce the needed thrust from the nozzle.

C6.1 Integration of a ramjet engine with a reciprocating engine

The integration of a ramjet engine with a reciprocating engine can maximize the use of the discharge pressure that is produced by the Otto Cycle. More importantly, such a combination allows the use of waste heat from the engine's exhaust in place of the energy that normally would be produced by the burning of fuel in the combustor. The Otto Cycle is the thermodynamics basis for the reciprocating engine's power production. As observed in the analysis results for the Wright R-1280 engine (shown in Figure C6.2), the discharge pressure at state point 5 is 223 psia. It is important to note that the engine is supercharged and not turbocharged. Supercharging the ambient air at an altitude of 30,000 ft is necessary to increase the inlet pressure of the air into the engine. The supercharged engine draws in ambient air and pressurizes it by using an air compressor. The power for the compressor is provided by a direct, mechanical coupling to the engine's drive shaft. The

WRIGHT CYCLONE R-1920 ENGINE (OTTO) CYCLE

NOTE: *ALL 'BOLD' ENTRIES ARE INPUT PARAMETERS;*
ALL OTHER ENTRIES ARE DEPENDENT RESULTS

Vr=	**6.45**	**1.37**	=Compression Index (n, 1-2)
Vc=	**1**	**1.39**	=Expansion Index (n, 4-5)
k=	1.41	Cp=	**0.24** BTU/LBm/R
MOLE. Wt.=	28.966 LBm/LBmol	Cv=	**0.17** BTU/LBm/R

SUPERCHARGER

YES

PROPERTY TABLE

	P (psia)	T(R)	v (ft^3/LBm)		
1	**67.9**	**605**	3.30	**585**	=AMB.TEMP,R
2	872.9	1206	0.51	**14.7**	=AMB. PRES.
3	2968.0	**4100**	0.51	**0.82**	=Cyl. Cooling coef.
4	2968.0	4100	0.51	657	=Cyl. Mix Temp. (T1),R
5	222.4	1982	3.30		

9.7 Pin
520 Tin
7 Pr
0.85 Charger Eff.
943
0.8 Intercooler Effect.

0.281 =(n-1)/n , expansion

PROCESS TABLE (units: BTU/LBm)

605 F

	HEAT	INTERN. ENERGY	WORK		
1-2	-9	102.1	-111.3	**EFF=**	**0.85**
2-3	492	492.0	0.0	Delta H=	215.6 Btu/Lbm
3-4	0	0.0	0.0	Texh.out=	1083.3 R
4-5	12	-360.1	372.4	=	623.3 F
5-1	-234	-234.1	0.0	TURBO. BYPASS=	**100%**
				Tmix=	1982 R
SUM	261.0	0.0	261.0	**COMPRESSOR**	
			258.0	**EFF.=**	**0.8**

TURBO.ACTIVATED? **YES**

TURBO. EXPANDER

CYCLE	0.54	EFF. with	Pout =	14.7 psia
EFF.	0.52	LOSSES	0.261 @ Tout =	585.0 R
ENG.G EFF. CORR.=	0.505	BSFC(Lbm/hr/hp)=	0.425	125.0 F
CARNOT Eff.	85%	Intercooler Effectiveness=	**70%**	585 R; Air cooler out
Exergy in Exhaust=	183.9	Closed Mass Exergy=	149.1	-53.2

VOLUMETRIC EFF.=	**0.98**	0.0136 secs./cycle
2 OR 4 STROKE ENGINE?	**4**	
ENGINE SPEED=	**2200** rpm	**ENGINE SIZING**
ENGINE DISP.=	**1819** in.^3	**9** No.CYL.S
29801	**30** Liters	6.13 BORE
ENGINE POWER HP	KWm	6.85 STROKE
1042.3	777.2 777.2 Kwe	

No. of Engines

1

ENGINE TORQUE   2488 Ft-LBf
BRAKE MEAN EFFEC. PRES. (BMEP)=   255 psia
PISTON SPEED=   2512 ft/min;m/s   12.8
HEAT CONTENT=   1030 btu/ft^3
Air/Fuel Ratio=   46.5   2.7 PHI
Mair=   20625 LBm/Hr.; cf 4583
Mfuel=   443 LBm/Hr.; ga53.1

Qfuel=	**2973 kWt**
40% **Qwtr.+oil=**	**1189 kWt**
5% **Qradiation**	**149 kWt**
68% **Qexh.=**	**2036 kWt**

Figure C6.2. Thermodynamics cycle analysis of the Wright R-1280 engine (see Figure C6.1 and Table C6.1).

supercharger is not to be confused with the operation of a turbocharger, which is more common in today's consumer vehicles rather than the supercharger of the 1950s. A turbocharger uses the exhaust pressure at state point 5 to drive a turbine, which in turn drives an air compressor. Because the turbocharger method for increasing the engine's inlet manifold pressure is not used with the Wright R-1280 engine, the ramjet engine can utilize all of the Wright engine's exhaust gas waste heat. The waste heat's high pressure and high temperature are directly fed into the inlet of the diffuser of the ramjet. The slightly higher pressured air is then passed through the combustor and exhaust nozzle to produce the high-speed velocity that is needed to produce the additional thrust according to Equation (C6.1). It is also possible to eliminate the diffuser at the inlet to the ramjet because the high discharge pressure at state point 5 is already comparable in magnitude to the pressure that is attainable by the typical inlet diffuser of a ramjet engine.

The results of the thermodynamics analysis of the combination of the Wright R-1280's engine with the waste heat recovery capabilities of the ramjet engine are shown in Figures C6.2, C6.3(a) and C6.3(b). Again, it is important to note that the thermodynamics analysis has been conducted by configuring one or more of the seven basic mechanical devices in a manner which produces the desired result. Thus, the analysis of the combined system is as straightforward as applying the basic first and second law equations for each of the devices. The net additional thrust force is calculated to have a value within the range of 380 to 430 lbf, depending on whether a diffuser is used to provide an additional boost to the pressure achieved at state point 5 instead of simply using the high pressure that already exists at state point 5 when the exhaust valve in the engine opens.

C6.2 Hybrid Ramjet (Hy-Ram) propulsion engine with controlled sequential ignition of energetic nanoparticles to generate compression pressure waves

The ramjet engine described here and previously in Chapter 5 is among the simplest of engines to manufacture. However, this simplicity comes at a price: for thrust to be generated, the ramjet engine requires the aircraft to be in motion at a very high velocity, to enable the diffuser to produce a significant rise in air pressure of the induced air. Thus, the engine tends to be inefficient at lower velocities. The ramjet engine has been in use since the early 1950s, during the rapid development of jet aircraft engines. Its purpose then was only to produce instantaneous, additional thrust for the aircraft, e.g., during drastic flight maneuvers to escape enemy fire. For such a purpose, engine efficiency is not a priority.

An interesting new development in the evolution of the ramjet engine aims at improving its efficiency while reducing the speed of the intake air's

RAM JET ENGINE USING WASTE HEAT RECOVERY

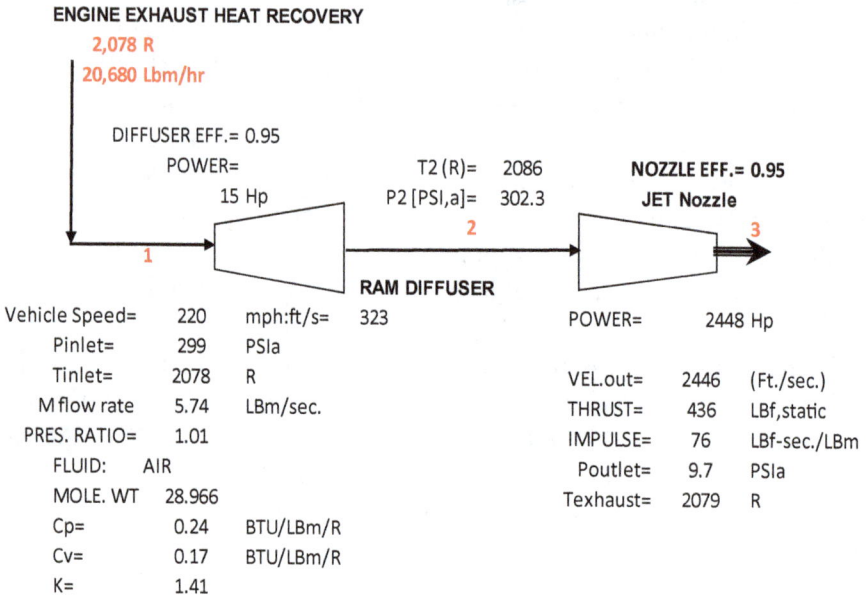

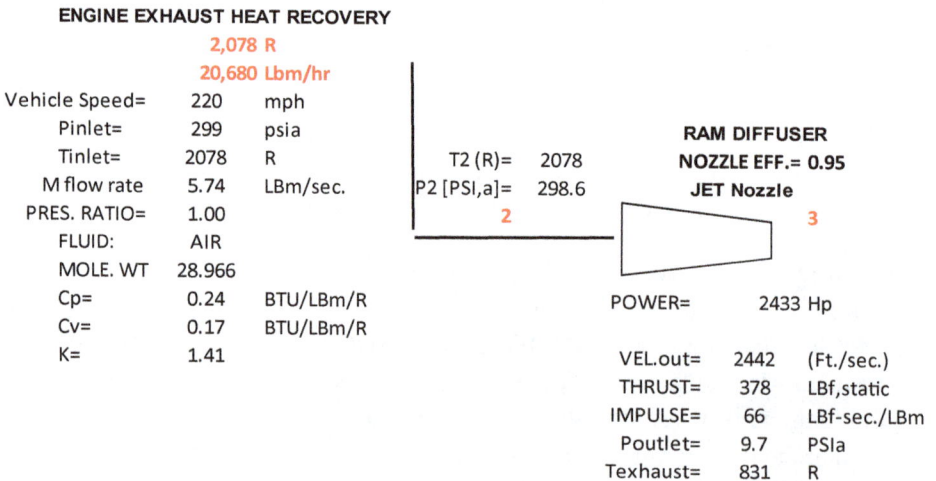

ENGINE EXHAUST HEAT RECOVERY

2,078 R
20,680 Lbm/hr

DIFFUSER EFF.= 0.95
POWER=
15 Hp

T2 (R)=   2086
P2 [PSI,a]=   302.3
2

NOZZLE EFF.= 0.95
JET Nozzle

1
RAM DIFFUSER
3

Vehicle Speed=	220	mph:ft/s=	323
Pinlet=	299	PSIa	
Tinlet=	2078	R	
M flow rate	5.74	LBm/sec.	
PRES. RATIO=	1.01		
FLUID:	AIR		
MOLE. WT	28.966		
Cp=	0.24	BTU/LBm/R	
Cv=	0.17	BTU/LBm/R	
K=	1.41		

POWER=	2448 Hp	
VEL.out=	2446	(Ft./sec.)
THRUST=	436	LBf,static
IMPULSE=	76	LBf-sec./LBm
Poutlet=	9.7	PSIa
Texhaust=	2079	R

(a)

RAM JET ENGINE USING WASTE HEAT RECOVERY

ENGINE EXHAUST HEAT RECOVERY

2,078 R
20,680 Lbm/hr

Vehicle Speed=	220	mph
Pinlet=	299	psia
Tinlet=	2078	R
M flow rate	5.74	LBm/sec.
PRES. RATIO=	1.00	
FLUID:	AIR	
MOLE. WT	28.966	
Cp=	0.24	BTU/LBm/R
Cv=	0.17	BTU/LBm/R
K=	1.41	

T2 (R)=   2078
P2 [PSI,a]=   298.6
2

RAM DIFFUSER
NOZZLE EFF.= 0.95
JET Nozzle

3

POWER=	2433 Hp	
VEL.out=	2442	(Ft./sec.)
THRUST=	378	LBf,static
IMPULSE=	66	LBf-sec./LBm
Poutlet=	9.7	PSIa
Texhaust=	831	R

(b)

Figure C6.3. Thermodynamics analyses of the ramjet waste heat recovery engine. (a) Ramjet engine with a diffuser; (b) Ramjet engine without the diffuser and using only the exhaust gas pressure and temperature at State Point 5 in Figure C6.2

velocity at which the ramjet engine is still effective in help propel the aircraft; thus making it more versatile for thrust augmentation.

The ramjet propulsion engine is the simplest aircraft propulsion system used to achieve supersonic speeds. The ramjet is essentially a gas turbine (GT) engine (which follows the Brayton Cycle) designed to eliminate the need for a turbine-compressor mechanical system. Although a GT engine enables a more controlled, stable combustion and is a more fuel-efficient propulsion system compared to the ramjet, it still requires an efficient air compressor that must be powered by an exhaust gas turbine (expander) before the hot exhaust gases can be accelerated to supersonic speeds in the GT engine's exhaust nozzle. The need for an air compressor system adds to the weight and cost of the GT engine. The simplicity of the ramjet is due to its ability to compress the combustion air by transforming the kinetic energy of the intake air stream into a stagnation pressure of approximately 2 to 6 atmospheres, without the need for a mechanical turbocompressor system. The ramjet intake air is charged to a high pressure and slow velocity, so as to enable the ramjet combustor system to continuously ignite the injected fuel. The typical combustion efficiency (~95%) of the fuel is dependent on the size of the atomized fuel, the length/diameter ratio of the fire tube, and the pressure at the inlet of the combustor. The rapid addition of heat energy in the combustor (at approximately constant pressure) increases the air temperature, and the pressurized air can be accelerated via the exhaust nozzle to increase the exhaust gas flow by 4 or 5 times, in order to propel the vehicle to supersonic speeds. However, despite the attractiveness of its simple "no moving parts" design, a significant limitation of the ramjet engine is the need for the propulsion engine to be driven at high initial speeds equivalent to or above Mach 1 — that is above the speed of sound in the local ambient air pressure an temperature — in order for the air pressure at the inlet and at the end of the diffuser section to attain stagnation pressures and temperatures suitable for sustaining the ramjet's combustor's performance.

Figure C6.4 identifies the specifications for the ramjet propulsion engine required for a typical flight mission, as it ascends to the given elevation above sea level. Figure C6.2 provides an estimate for the required pressure ratio (Pr) across the exhaust nozzle to achieve the vehicle Mach number shown. As may be observed from Figure C6.1, the specific impulse from the typical ramjet may attain magnitudes of 250 s at the prevalent local ambient conditions at 30,000 ft elevations, **only if** the combustion temperatures are extremely high, approximately 6,500 to 7,500°R (depending on the ambient conditions and ramjet size). Such high temperatures are difficult to achieve with typical aviation fuels. However, such temperatures may be achieved in

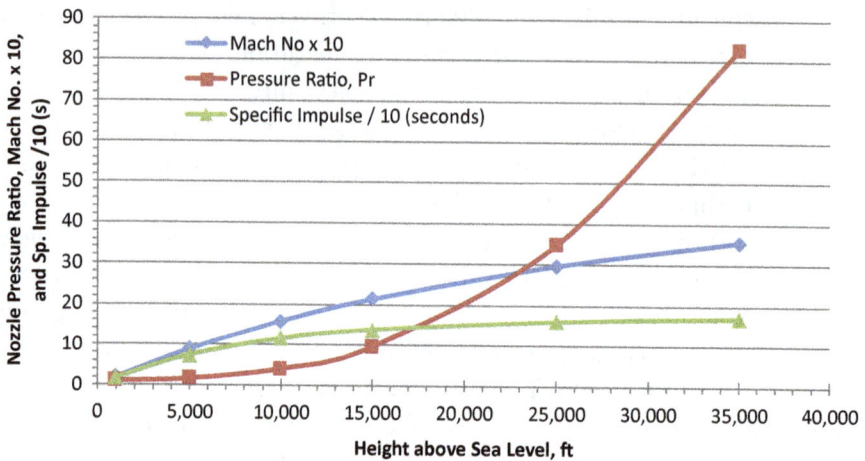

Figure C6.4.   Typical mission Mach number and Pr specification for a conventional ramjet propulsion engine as a function of height above sea level.

the proposed Hy-Ram engine design and can be effectively utilized in the Hy-Ram propulsion engine. Unlike conventional gas turbine propulsion systems, Hy-Ram systems do not contain mechanical turbo-compressor rotors which restrict combustion temperatures to approximately 3,500°R due to their metal fatigue temperatures. For this reason, it is rare for continuously-fired gas turbine propulsion systems to operate at temperatures above 3,500°R for long periods of time without adequate, expensive and complex blade cooling systems. It is also apparent from Figure C6.4 that in order to achieve Mach 2.5 and above, the Pr into the combustor with respect to the local ambient pressure (at the flight elevations shown) must be 25:1. Such a Pr is not attainable at low take-off speeds without some means of assisted propulsion. To achieve a higher Pr, a means of establishing a high enough inlet pressure to the ramjet combustor to enable the sustained combustion of fuel, without resorting to typical mechanical turbo-compressor systems that are prevalent in GT (Brayton Cycle) engines, is required. Concepts NREC (CN), a world renown manufacturer of prototype turbomachinery systems, has completed a feasibility analysis which demonstrates the ability of a unique combustion propulsion system to provide the necessary air pressure and extreme temperatures (7,500°R) required in a typical ramjet combustor system. The proposed combustion propulsion system utilizes energetic nanofuel particles which can be ignited at low ignition energies and temperatures in a sequential and planar pattern that would generate the necessary combustion pressures and temperatures for the conventional state-of-the-art ramjet combustor. This propulsion system is called the 'Hybrid-Ramjet Sequential-Pulse Dual

Table C6.2.   Energetic nano-particle combustor analysis.

Energetic Nano-particle combustor analysis		
MPH=	152	miles/hr.
Height above Sea level=	1000	ft
Nano-Particle Pres. Heat Value=	8000	BTU/LBm
Sonic Vel.=	1,114	ft/s
Pamb=	14.2	psia
Tamb=	516	R
Cp,air=	0.27	Btu/LBm/R
Stag. Temperature=	1,745	R
Combustor Exh. Temp.=	4,808	R
P/R=	71.0	
Thrust=	35,179	Lbf
Specific Impulse=	209	s

Combustion' (HR-SPDC) engine. This propulsion system operates without rotating turbomachinery. Having no moving parts allows it to maintain its simplicity as a propulsion system. This system is able to start at very low aircraft speeds (sometimes called "static" speeds) and attain the desired specific impulse of 250 s at zero altitude and STP (standard temperature and pressure)[1] ambient conditions. The design specifications for the HR-SPDC engine at takeoff are shown in Table C6.2. A comparison of the conventional ramjet propulsion engine versus the HR-SPDC on the same flight mission is shown in Figure C6.5, with combustor temperatures constrained to below 5,500°R.

The HR-SPDC engine produces the necessary ramjet combustor air pressure by using a controlled stream of energetic nanoparticles that are sequentially injected into the fire-tube to create a series of small ignitions upstream of the ramjet combustor. This firing pattern forces the combustion flame by each individual ignition downstream towards the ramjet combustor. This ignition pattern may be analogized to the process taking place in a multi-stage axial compressor that produces a high air pressure in the inlet to the combustor in a typical gas turbine propulsion engine which follows the Brayton Cycle. In the HR-SPDC, the series of small ignitions has the net effect of achieving a higher effective compression ratio, due to the more efficient combustion of the nanoparticles.

The air pressure boost from the combustion of the energetic nanoparticles would continue until the aircraft is moving at supersonic speeds, at which time the nanoparticle boost system can be shut down, and the ramjet can provide all of the necessary propulsion thrust. However, the nanoparticle

---

[1]STP is defined as air at 0°C (273.15 K, 32°F) and at the standard pressure of 1 atm.

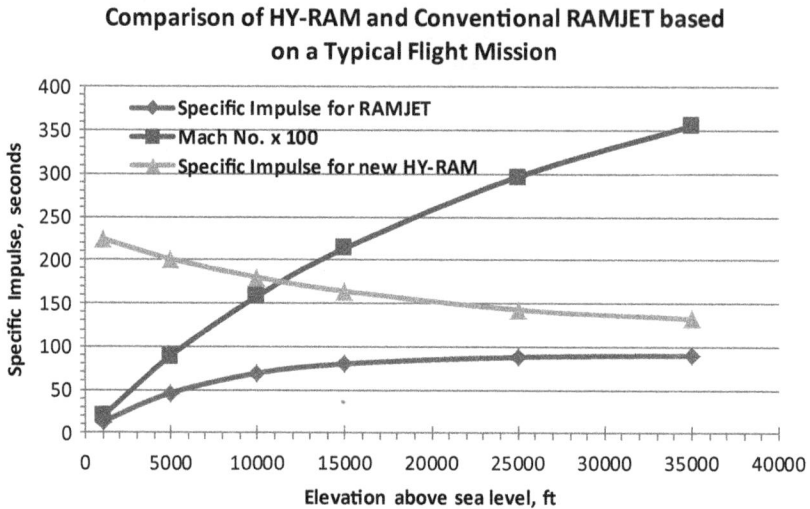

Figure C6.5. Comparison of Specific Impulse for HY-RAM and Ramjet Engines on the same mission.

combustion system can be engaged as necessary to provide auxiliary power boost when necessary to propel the aircraft to higher speeds.

C6.3   Hybrid Ramjet-Sequential Pulse Dual Combustion (HR-SPDC) engine versus a conventional Pulse Detonation Engine (PDE)

The study of Pulsed Detonation Engines (PDEs) has a long history dating from the early 1940s. More recently, PDEs have been researched as propulsion systems (either as primary or auxillary systems) for aircraft and missiles. PDEs provide thrust by combining the instantaneous ignition of individual charges of fuel while maintaining a cost-effective and lightweight system. This is in contrast to a typical reciprocating (Otto Cycle) engine or even a GT jet engine — though both engines may be fuel-efficienct, the weights of these prime movers reduce the overall engine power densities and thus decrease the flight mobility of the aircraft that use them. PDEs have been shown to be effective propulsion systems at low subsonic vehicle speeds relative to those of the ramjet engine. A study published in 2006 (Ma, Choi, & Yang, 2006) concluded that:

> "··· a Pulse Detonation Engine (PDE) out-performs its ramjet counterpart for all the flight conditions··· but the benefit decreases with increasing Mach no." and that "··· the specific impulse is 36% and 23% higher than a ramjet at Mach 1.2 and 2.1 but falls off to 3% at Mach 3.5."

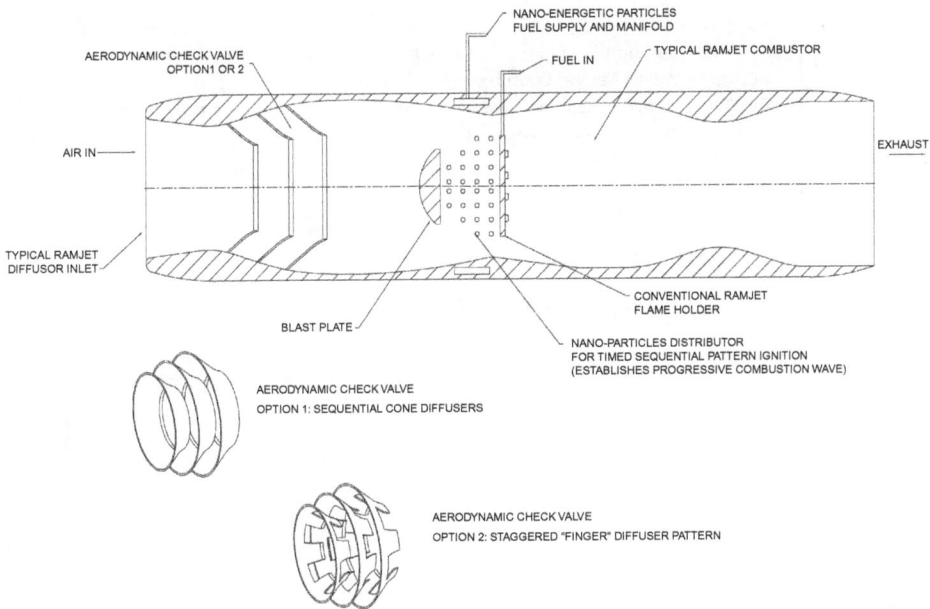

Figure C6.6. Conceptual diagram of a proposed hybrid ramjet propulsion system.

Although the rapid firing sequence (50 cycles/s and higher) of PDEs was timed to provide a continuous high-velocity streamline of exhaust, the process can be fuel inefficient due to the somewhat uncontrollable combustion of the fuel charge resulting in a larger fuel mass per ignition event. In contrast, the HR-SPDC engine utilizes a controlled ignition process. The HR-SPDC engine achieves this by utilizing the low energy ignition properties of energetic nanoparticles to provide a controlled ignition, resulting in the generation of a pattern of wave fronts which progresses down the fire tube toward the ramjet combustor, enhanced by a reflected pressure wave from a blast plate. The details of these features are explained in the next section.

The HR-SPDC propulsion system is conceptually diagrammed (not to scale) in Figure C6.6. The P-v diagram for the continuous, open-flow combustion process is identified in Figure C6.7.

The HR-SPDC propulsion engine utilizes an injection zone (Process $A_1 - A_2 - B$ in Figure C6.7) of energetic nano-energetic particles. These particles can be controlled to ignite in a sequence which will effectively create a compression wave sufficient in magnitude and frequency to provide the conventional ramjet engine with compressed air. This enables the ramjet combustor to provide supersonic speed, starting at a low or zero speed.

The HR-SPDC engine can be viewed as a combination of a PDE and a ramjet engine, but with the advantage of higher fuel efficiency while retaining

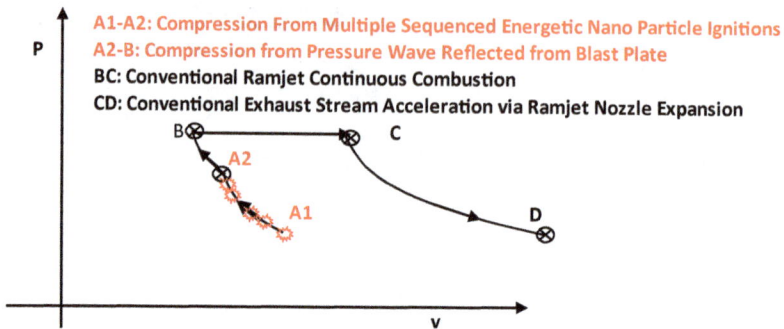

Figure C6.7.   Dual pressure combustion, continuous open-flow process from controlled sequential energetic nanoparticle ignition.

the mechanical simplicity of the ramjet. As mentioned earlier, the mechanical simplicity is due to the absence of moving parts in the flame tube. The higher efficiency is due to the fact that the combustion process utilized is an open-cycle version of the combination of two of the most efficient thermodynamics cycles: the dual-cycle diesel and the air–Brayton Cycle thermodynamics processes, as shown in Figure C6.7. The HR-SPDC engine will also enable the conventional ramjet engine to perform better at cruise speeds, due to its ability to provide a higher combustor inlet pressure and temperature.

While the efficiency of the HR-SPDC engine versus that of the conventional ramjet engine is not easily assessed — because the HR-SPDC may be operated as a static takeoff or lift engine while the ramjet cannot — comparison can still be made with some assumptions in place. Figure C6.5 provides a comparison with the proper assumptions made for air–fuel ratios, combustion efficiencies as well as aircraft height and speed.

The energetic nanoparticle combustor (the combustor part of the HR-SPDC) is also expected to be more efficient than the conventional axial compressor in a GT engine in achieving the desired pressure ratio for the main combustor. The controlled, sequential ignition of the energetic nanoparticles enables the pre-ramjet combustion zone to achieve high compression air pressures with a larger number of small ignitions. Therefore, smaller standing compression waves are continuously generated in a standing wave pattern (and in the limit, achieving a shock wave) progressing down the ramjet fire tube (from left to right, with respect to the fire tube shown in Figure C6.6), starting from the right side of the blast plate. The progressive wave is fortified by the continuous self-ignition (using the combustion temperature) of the energetic nanoparticles from the temperature generated by the progressing wave front. The combustion of the energetic nanoparticles is known to result in adiabatic flame temperatures of 6,500°R, and the timed sequential injection of the nanoparticles adds energy to effectively "tune" the combustion waves, so as to enhance the resulting pressure. When the sequential ignition pattern has completed its firing pattern from the front to the rear end, the firing starts again at the inlet plane. The fuel ignition in the proposed HR-SPDC is continuous and controlled, unlike in PDEs where a "pulse" ignition causes a detonation wave and produces the necessary propulsion thrust via the nozzle. The smaller, planar compression waves, when summed together, achieve the same total pressure as may be achieved by a single detonation of a larger fuel charge.

An additional significant innovation is claimed for the proposed hybrid ramjet engine: the ability for the energy in the progressive wave front to be enhanced by the additive effect of the pressure waves that have been reflected off the blast plate. This energy then moves from left to right along the fire tube to provide a secondary (albeit smaller) compression (see Process $A_2 - B$ referenced in Figure C6.7) of the charged air entering the ramjet combustor.

The timed series of sequential ignitions of the energetic nanoparticles and its effect on improving the compression of the inducted air-charge may be analogized to be similar to a multistaged axial compressor that continually compresses the air charge before the combustor, and achieves a higher, effective compression ratio. The timed, progressive detonation of the planar layers of energetic nanoparticles will not only provide a more efficient, complete combustion of the fuel, but when it is in series with the ramjet combustion, it will provide a cycle efficiency that is comparable to a Brayton Cycle with a high compression efficiency.

Another analogy of the proposed contributory effects of the staged, sequential ignition of the energetic nanoparticles can be made with the compression of air caused by the conservation of momentum principle employed in a wave rotor compression device. In a wave rotor, the progressive compression of the air charge is due to the rapid opening and closing of small air passages causing the air to be compressed by the impulse forces generated by the rapid change of momentum in the air. The success of this wave rotor to produce sufficient pressure is well known to researchers and the industry, and has been offered as a means of augmenting the compressed air charge entering a gas turbine combustor (see Nalim & Karem (2003)). One unfortunate attribute is the need for the wave rotor to be rotated at high speeds, thus making it an unacceptable choice as an alternative to the ramjet engine, for which the goals are to be simple and have "no moving parts". The need for a high rotational speed decreases the engine's reliability and increases its cost and weight.

In the HR-SPDC engine, the need to provide a robust propulsion engine to enable a continuous, open-flow process, is aided by another innovation: the aerodynamic-based, air flow "check valve". As indicated previously, unlike the PDE, the proposed HR-SPDC process does not resort to a pulsed ignition in order to instantaneously establish a combustion pressure wave. It does, however, need to maintain and control the air flow direction during start up and the throttling-up or down in power. This can be effectively accomplished using an aerodynamic check valve, as conceptualized in Figure C6.6 where two options are shown for the aerodynamic check valve — Option 1: The conical staged check valve; and Option 2: the staged check valve with openings to increase turbalence. The basic design concept for the aerodynamic check valve is to establish a high ΔP across the check valve to prevent the reverse flow of air in the fire tube (i.e. the air flow rate moving backwards through the air intake). The two options shown in Figure C6.6 can provide the excessive back pressure to prevent reverse air flow. The use of the "no moving parts" aerodynamic check valve eliminates the need for the conventional spring-loaded flapper valves commonly used with PDEs. The elimination of these mechanical flappers and the instantaneous "pulsed" ignition of the fuel reduces the noise and exhaust emissions from the HR-SPDC engine as compared to the PDE engine.

C6.4   Description and research summary of energetic nanoparticles materials[2]

As elaborated in previous sections, the ability to energize the nanoparticles is an essential part of the working principle of the HR-SDPC engine. The feasibility of using energetic nano particles as a source of controlled combustion is supported in part by previous, independent work that has been done at Northeastern University by Prof. Latika Menon and colleagues (see Apperson *et al.* (2007)). The independent work only concerned research in the properties of the energetic nanoparticles. The concept of using these particles in a controlled combustion was considered novel.

Energetic materials are a class of substances that store energy chemically, and when ignited, undergo an exothermic thermal reaction without the need for an external substance such as oxygen. The fabrication of such energetic

---

[2]Study and work in progress in 2015 by Professor Latika Menon of Northeastern University. The following description is taken from a joint proposal to the Department of Defence.

materials traditionally involves a packaged mixture of oxidizer and fuel material (e.g. black powder), or the combination of both oxidizer and fuel material into one molecule (e.g. trinitrotoluene). These materials have applications in defense systems, where they are used to produce weapons, explosives, propellants, etc. (Menon, *et al.*, 2004; Menon, Aurongzeb, Patibandla, & Ram, 2006; Miziolek, Karna, Mauro, & Vaia, 2005). Some major considerations for the successful weaponization of energetic materials include energy density, rate of energy release, long-term storage stability, and sensitivity to unwanted initiation. In recent years, it has been found that nanoscale energetic materials have the potential for increased energy release, stability, sensitivity and mechanical properties compared to conventional energetic materials (Miziolek A., 2002). Simply stated, nanoenergetic materials can store higher amounts of energy than conventional energetic materials, which can be used to maximize the lethality of the weapons produced.

Currently, three new kinds of nanoenergetic materials are being researched: metastable intermolecular composites (MICs), sol-gel nanocomposites, and thin film multilayer films (Prakash, McCormick, & Zachariah, 2005; Wilson & Kim, 2005; Blobaum, Reiss, Plitzko, & Weihs, 2003; Aurongzeb, *et al.*, 2003; Ma, Thompson, Clevenger, & Tu, 1990; Tillotson, *et al.*, 2001; Gash, Tillotson, Satcher Jr., Hrubesh, & Simpson, 2001; Myagkov, Zhigalov, Bykova, & Mal'tsev , 1998) MIC materials based on mixed nanoclusters of oxidizer and fuel can release nearly twice the energy of the best monomolecular energetic materials (Prakash, McCormick, & Zachariah, 2005; Wilson & Kim, 2005). In the MIC formulation, maximum energy density can be achieved through a complete balance between the oxidizer and fuel material. However, due to the granular nature of MIC, reaction kinetics are largely controlled by mass transport rates between reactants. Fuel- or oxidizer-containing domains limit the mass transport, thereby decreasing the efficiency of the thermite reaction in MICs. The extreme energy densities postulated in nanocomposites may thus be difficult to reach in such systems.

Sol-gel chemistry involves chemical reactions in solution to produce nanometer-sized particles immersed in a solid network (Tillotson, *et al.*, 2001; Gash, Tillotson, Satcher Jr., Hrubesh, & Simpson, 2001). Such structures are macroscopically uniform, due to the small particle size and the small interparticle separations. In both the granular mixtures and sol-gel materials, the distribution of the particles are random. The randomness in interparticle separation can inhibit self-sustaining processes by locally separating the fuel and oxidizer. Sol-gel reactants often have organic impurities that make up about 10% of the sample mass (Tillotson, *et al.*, 2001). This results in reduced energy release. Thin film technology allows the fabrication of multilayered foils consisting of alternating layers of oxidizer and fuel material, forming sol-gel nanocomposites (Blobaum, Reiss, Plitzko, & Weihs, 2003; Ma, Thompson, Clevenger, & Tu, 1990; Aurongzeb, *et al.*, 2003; Myagkov, Zhigalov, Bykova, & Mal'tsev, 1998). Such foils provide large, regular planar interfaces and close contact between the oxidizer and fuel reactants(Aurongzeb, *et al.*, 2003). They are nanoscaled in one dimension and the energy release proceeds through a surface reaction between the oxidizer and the fuel material.

C6.5 Modeling of energetic materials

Early models of solid-solid systems focused on powder-based systems relevant to applications such as propellants, pyrotechnics and explosives (Subrahmanyam & Vijayakumar, 1992). Notably, based on the results from an early self-propagating high-temperature synthesis (SHS) work, Armstrong and Koszykowski predicted the steady flame front velocity for the solid-solid propellants. More recent models are discussed in Suvaci, Simkovich & Messing (2004), Granier *et al.* (2003) and Gaus *et al.* (1999). All these models (early and recent) consist of an energy balance equation, coupled with other equations that account for inter-species diffusion and transformation constituting a strongly exothermic heat source.

References

Apperson, S., Shende, R. V., Subramanian, S., Tappmeyer, D., Gangopadhyay, S., Chen, Z., & Kapoor, D. (2007). Generation of fast propagating combustion and shock waves with copper oxide/aluminum nanothermite composites. *Applied Physics Letters, 91*(24). doi:https://doi.org/10.1063/1.2787972

Aurongzeb, D., Holtz, M., Daugherty, M., Berg, J. M., Chandola, A., Yun, J. H., & Temkin, H. (2003). Influence of nanocrystal growth kinetics on interface roughness in nickel–aluminum multilayers. *Applied Physics Letters, 83*(26). doi:https://doi.org/10.1063/1.1637155

Besnoin, E., Cerutti, S., Knio, O. M., & Weihs, T. P. (2002). Effect of reactant and product melting on self-propagating reactions in multilayer foils. *Journal of Applied Physics, 92*(9). doi:https://doi.org/10.1063/1.1509840

Blobaum, K. J., Reiss, M. E., Plitzko, J. M., & Weihs, T. P. (2003). Deposition and characterization of a self-propagating CuOx/Al thermite reaction in a multilayer foil geometry. *Journal of Applied Physics, 94*(5). doi:https://doi.org/10.1063/1.1598296

Eisenreich, N., Fietzek, H., Juez-Lorenzo, M. d., Kolarik, V., Koleczko, A., & Weiser, V. (2004). On the Mechanism of Low Temperature Oxidation for Aluminum Particles down to the Nano-Scale. *Propellants Explosives Pyrotechnics, 29*(3), 137–145. doi:https://doi.org/10.1002/prep.200400045

Gash, A. E., Tillotson, T. M., Satcher Jr., J. H., Hrubesh, L. W., & Simpson, R. L. (2001). New sol–gel synthetic route to transition and main-group metal oxide aerogels using inorganic salt precursors. *Journal of Non-Crystalline Solids, 285*(1–3), 22–28. doi:https://doi.org/10.1016/S0022-3093(01)00427-6

Gaus, S. P., Harmer, M. P., Chan, H. M., & Caram, H. S. (1999). Controlled Firing of Reaction-Bonded Aluminum Oxide (RBAO) Ceramics: Part I, Continuum-Model Predictions. *Journal of the American Ceramic Society, 82*(4), 897–908. doi:https://doi.org/10.1111/j.1151-2916.1999.tb01851.x

Gavens, A. J., van Heerden, P. J., Mann, A. B., Reiss, M. E., & Weihs, T. P. (2000). Effect of intermixing on self-propagating exothermic reactions in Al/Ni nanolaminate foils. *Journal of Applied Physics, 87*(3). doi:https://doi.org/10.1063/1.372005

Glassman, I., & Papas, P. (1999). Combustion thermodynamics of metal-complex oxidizer mixtures. *Journal of Propulsion Power, 15*(6), 801–805. doi:https://doi.org/10.2514/2.5499

Granier, J. J., Mullen, T., & Pantoya, M. L. (2003). Nonuniform laser ignition in energetic materials. *Combustion Science and Technology, 175*(11), 1929–1951. doi:https://doi.org/10.1080/714923185

Granier, J. J., Plantier, K. B., & Pantoya, M. L. (2004). The role of the Al2O3 passivation shell surrounding nano-Al particles in the combustion synthesis of NiAl. *Journal of Materials Science, 39*(21), 6421–6431. doi:10.1023/B:JMSC.0000044879.63364.b3

Jayaraman, S., Knio, O. M., Mann, A. B., & Weihs, T. P. (1999). Numerical predictions of oscillatory combustion in reactive multilayers. *Journal of Applied Physics, 86*(2). doi:https://doi.org/10.1063/1.370807

Jayaraman, S., Mann, A. B., Reiss, M. E., Weihs, T. P., & Knio, O. M. (2001). Numerical study of the effect of heat losses on self-propagating reactions in multilayer foils. *Combustion and Flame, 124*(1–2), 178–194. doi:https://doi.org/10.1016/S0010-2180(00)00192-9

Ma, E., Thompson, C. V., Clevenger, L. A., & Tu, K. N. (1990). Self-propagating explosive reactions in Al/Ni multilayer thin films. *Applied Physics Letters, 57*(12). doi:https://doi.org/10.1063/1.103504

Ma, F., Choi , J.-Y., & Yang, V. (2006). Propulsive Performance of Airbreathing Pulse Detonation Engines. *Journal of Propulsion and Power, 22*(6), 1188–1203. doi:10.2514/1.21755

Mann, A. B., Gavens, A. J., Reiss, M. E., van Heeden, D., Bao, G., & Weihs, T. P. (1997). Modeling and characterizing the propagation velocity of exothermic reactions in multilayer foils. *Journal of Applied Physics, 82*(3). doi:https://doi.org/10.1063/1.365886

Menon, L., Aurongzeb, D., Patibandla, S., & Ram, B. K. (2006). Size dependence of energetic properties in nanowire-based energetic materials. *Journal of Applied Physics, 100*(3). doi:https://doi.org/10.1063/1.2234551

Menon, L., Patibandla, S., Ram, B. K., Shkuratov, S. I., Aurongzeb, D., & Holtz, M. (2004). Ignition studies of Al/Fe2O3 energetic nanocomposites. *Applied Physics Letters, 84*(23). doi:https://doi.org/10.1063/1.1759387

Miziolek, A. (2002). Nanoenergetics: An Emerging Technology Area of National Importance. *AMPTIAC Quarterly, 6*(1), 43–48. Retrieved from https://p2infohouse.org/ref/34/33119.pdf

Miziolek, A. W., Karna, S. P., Mauro, J. M., & Vaia, R. A. (2005). *Defense Applications of Nanomaterials* (Vol. 891 of ACS Symposium Series). Washington, D.C.: American Chemical Society.

Munir, Z. A. (1988). Synthesis of high temperature materials by self-propagating combustion methods. *American Ceramic Society Bulletin, 67*(2), 342–349.

Myagkov, V. G., Zhigalov, V. S., Bykova, L. E., & Mal'tsev , V. K. (1998). Self-propagating high-temperature synthesis and solid-phase reactions in bilayer thin films. *Technical Physics, 43*, 1189–1192. doi:https://link.springer.com/article/10.1134/1.1259177

Nalim, R., & Kerem, P. (2003). Internal Combustion Wave Rotors for Gas Turbine Engine Enhancement. In Japan Gas Turbine Society (Ed.), *Proceedings of the International Gas Turbine Congress 2003 Tokyo*. Tokyo: Gas Turbine Society of Japan.

Prakash, A., McCormick, A. V., & Zachariah, M. R. (2005). Synthesis and reactivity of a super-reactive metastable intermolecular composite formulation of Al/KMnO4. *Advanced Materials, 17*(7), 900–903. doi:10.1002/adma.200400853

Rai, A., Lee, D., Park, K., & Zachariah, M. R. (2004). Importance of Phase Change of Aluminum in Oxidation of Aluminum Nanoparticles. *The Journal of Physical Chemistry B, 108*(39), 14793–14795. doi:https://doi.org/10.1021/jp0373402

Subrahmanyam, J., & Vijayakumar, M. (1992). Self-propagating high-temperature synthesis. *Journal of Materials Science, 27*(23), 6249–6273. Retrieved from https://link.springer.com/article/10.1007/BF00576271

Suvaci, E., Simkovich, G., & Messing, G. L. (2004). The Reaction-Bonded Aluminum Oxide (RBAO) Process: II, The Solid-State Oxidation of RBAO Compacts. *Journal of the American Ceramic Society, 83*(8), 1845–1852. doi:https://doi.org/10.1111/j.1151-2916.2000.tb01480.x

Tillotson, T. M., Gash, A. E., Simpson, R. L., Hrubesh, L. W., Satcher Jr., J. H., & Poco, J. F. (2001). Nanostructured energetic materials using sol–gel methodologies. *Journal of Non-Crystalline Solids, 285*(1–3), 338–345. doi:https://doi.org/10.1016/S0022-3093(01)00477-X

Trunov, M. A., Schoenitz, M., & Dreizin, E. L. (2005). Ignition of Aluminum Powders Under Different Experimental Conditions. *Propellants, Explosives, Pyrotechnics, 30*(1), 36–43. doi:https://doi.org/10.1002/prep.200400083

Trunov, M. A., Schoenitz, M., Zhu, X., & Dreizin, E. L. (2005). Effect of polymorphic phase transformations in Al2O3 film on oxidation kinetics of aluminum powders. *Combustion and Flame, 140*(4), 310–318. doi:https://doi.org/10.1016/j.combustflame.2004.10.010

Wilson, D. E., & Kim, K. (2005). Combustion of Consolidated and Confined Metastable Intermolecular Composites. *43rd AIAA Aerospace Sciences Meeting and Exhibit*. Reno: American Institute of Aeronautics as Astronautics. doi:https://doi.org/10.2514/6.2005-275

Wu, S., Holz, D., & Claussen, N. (1993). Mechanism and kinetics of reaction-bonded aluminum oxide ceramics. *Journal of the American Ceramic Society, 76*(4), 970–980. doi:https://doi.org/10.1111/j.1151-2916.1993.tb05321.x

Case Study 7

Solar Powered Air Conditioning System Using Water as the Working Fluid

Air conditioning in a workspace is very energy intensive. A workspace can be defined as any enclosure that needs to be kept at a controlled temperature and humidity, such as a room(s) in a home, a theater, museum exhibit room, etc. Air conditioning can mean much more than just heating or cooling the workspace. It could also mean the need to maintain the space's relative humidity as well as a comfortable temperature for the contents of the workspace (we use 'contents' here as the term includes both inanimate as well as animate entities). The contents of a workspace can range anywhere from paintings and furniture in a museum to couples who are dancing in a dance hall.

According to the Clausius or Kelvin Statements of the Second Law of Thermodynamics, it was made clear in Chapter 3 that the continuous removal of heat energy from a workspace must be performed by providing power to a cooling system. The source of the power for cooling the workspace can come from any energy source, including renewable energy sources such as solar energy. There has been considerable research in the use of solar energy as the only power source needed to air condition a workspace. Solar energy can be used in a Rankine Cycle to generate power, which is required for the refrigeration system. There is also an added advantage to using only water as the refrigerant in air conditioning systems.

A thermodynamics model combining a Rankine Cycle-based power generation system and a vapor-compression cycle (also called a reverse Rankine Cycle system) is shown in Figure C7.1. In this solar powered refrigeration system, water is used as the working fluid for both the Rankine cycle and the vapor-compression refrigeration cycle. Figures C7.2(a)–(d) provide the results of the thermodynamics analysis of a solar-powered, water-based refrigeration system.

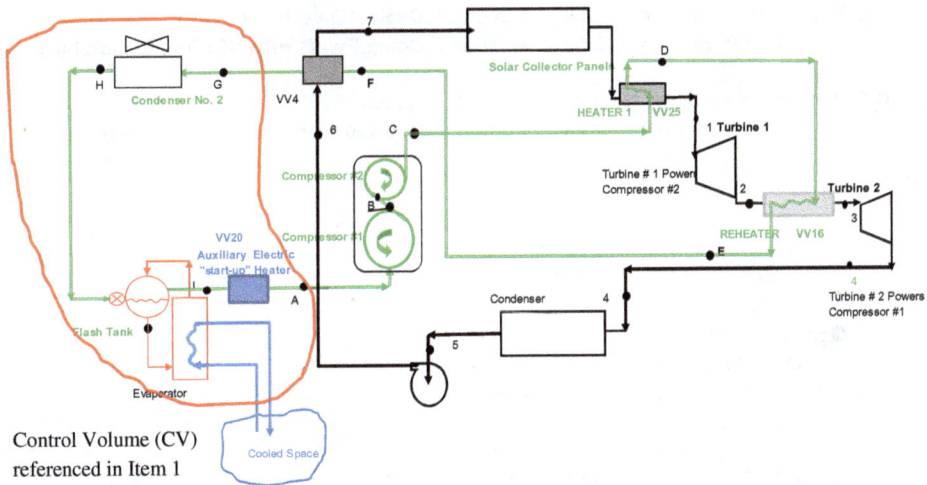

Figure C7.1. Diagram of an air conditioning system combining a Rankine cycle-based power generation system with a vapor-compression cycle.

As often mentioned in this textbook, the cycles shown in Figure C7.1 are constructed by using only one or more of the seven mechanical systems identified in detail in Chapter 3. Each of these components are well defined by the First Law of Thermodynamics which usually can be applied to each of the components by the much simpler equations of heat transfer or power equal to the product of mass flow rate and the enthalpy difference across the component. Thus, once two properties at the inlet and outlet of each component are known, the enthalpy can be determined. Following this, the work output or work input can then be determined. After this is done for each of the components, the net power and/or net cooling (i.e., the removal of energy from the workspace) can finally be determined.

A study of the innovative solar-powered, water-based air conditioning system shown in Figure C7.1 has revealed the following salient points:

1. A Control Volume placed around the workspace (shown in Figure C7.1) can be used to determine the governing equation for the fluid flow rate of the working fluid as a function of the amount of desired cooling from the workspace:

$$Refrigeration\ Cycle\ Flow\ Rate = Qcool/(h_{Flash\ tank\ out} - h_{Flash\ tank\ in}).$$

From this equation, it can be noted that the net cooling in the cycle is performed between a superheated water vapor temperature measured at the inlet of the condenser (T_G) and the saturated vapor temperature exiting the flash tank (T_A). The heat absorbed from the workspace (Q_{cool})

C.O.P. oa thermal cycle=	0.69	=Qcool. / (Qsolar+Q vv 20 elec.htr.)
C.O.P. refrig.=	9.58	=Qcool. / (Comp.Pwr#1 +Pwr#2+Q vv20 elec.htr.)

C.O.P. oa thermal cycle=	0.72	=Qcool. / Qsolar
C.O.P. room=	2.98	=Qcool. / (Q, v v20 elec. Htr.+ Σ Fan power)

Thermal Eff. # 1=	0.80		0.67	Compressor #2 Eff.
Expander #1=	0.86	kWm	0.83	Compressor #2
Expander #2=	0.69	kWm	0.71	Compressor #1
Thermal Eff. # 2=	0.82		0.68	Compressor #1 Eff.
Qcooling = Qevap.=	9.67	kWt		
Qcondenser #2=	-10.17	kWt		
Q v v4=	-0.31	kWt		
Q v v 25=	-1.18	kWt	1.18	
Q v v 16=	-0.06	kWt		
Q v v 20 (Elec. Htr.)=	0.53	kWt		
REFRIG. CYC. HT. BALANCE=	100.0%	Balanced	0.00 kWt diff.	

Solar Energy in=	13.44	kWt	
Condenser #1=	-13.44	kWt	
Feedpump Power=	0.001	kWt	
POWER CYCLE HT. BALANCE=	100.0%	Balanced	0.01 kWt. Diff.

(a)

		HX Effectiveness=	0.78		0.45		
			VV25 REHEATER		VV16 REHEATER		

	1 (amb.)	SOLAR IN [7]	SOLAR OUT [8]	TURBINE 1 IN [1]	TUR. 1 OUT [2]	TURBINE 2 IN [3]	TUR. 2 OUT [4]	FEED-WATER IN [6]
P (mPa)=		71.9	65.0	65.0	23.4	23.4	9.1	71.9
T Deg.C =		54.6	89.3	195.9	114.1	119.9	44.0	41.2
h (kJ/Kg)		168.4	0.0	2112.6	1997.7	2005.8	1913.5	127.1
Quality (x)=		N/A	N/A	N/A	erheated	perheated	verheated v	N/A
Flow rate (kg/hr)=		20.0	20.0	20.0	20.0	20.0	20.0	20.0
P (psia)=		10.4	9.4	9.4	3.4	3.4	1.3	10.4
T (F)=		130.3	193	385	237	248	111	106
h (Btu/Lbm)=		98.38		1234.27	1167.15	1171.91	1117.963	74.27
Quality (x)=		N/A	N/A	N/A	erheated	perheated	verheated v	N/A
Flow Rate (Lbm/s)=		0.0122	0.0122	0.0122	0.0122	0.0122	0.0122	0.0122
Vol. Flow (ft³/min)=								

(b)

Figure C7.2.   (a) The output page of the thermodynamics model of the cycle shown in Figure C7.1, including all of the power ratings for the major components and the cycle COP; (b) State points for the Rankine Cycle portion of the air conditioning system shown in Figure C7.1; (c) State points for the Vapor Compression (Refrigeration) Cycle portion of the air conditioning system shown in Figure C7.1; (d) State points (continued) for the Vapor Compression (Refrigeration) Cycle portion of the air conditioning system shown in Figure C7.1.

	COMP. 1 IN [A]	COMP. 1 OUT [B]	COMP. 2 OUT [C]	HEATER OUT [D]	REHEATER OUT [E]	OR CONDENSER #2 OUT [F]	IN	FLASH TANK IN [H]	FLASH TANK OUT
Cond. No. 2 Effectiveness=				0.63	HX Effectiveness= 0.49			CYCLE VAPOR SIDE	
					FEED-WATER OUT				
P (mPa)=	2.0	3.9	7.2	7.2	7.2	7.2	7.2	7.0	2.0
T Deg.C =	86.0	177.0	280.5	131.4	123.6	83.6	83.6	39.0	17.7
h (kJ/Kg)	1960.1	2087.8	2237.0	2023.2	2012.3	1956.3	1956.3	120.4	1865.3
Quality (x)=	N/A	N/A	N/A	N/A	N/A	0.0	0.0	0.0	1.0
Flow rate (kg/hr)=	14.7	14.7	14.7	14.7	14.7	14.7	14.7	14.7	14.7
P (psia)=	0.3	0.6	1.0	1.0	1.0	1.0	1.0	1.0	0.3
T (F)=	186.8	350.6	536.9	268.5	254.5	182.6	182.6	102.3	63.8
h (Btu/Lbm)=	1145.2	1219.8	1307.0	1182.1	1175.7	1143.0	1143.0	70.3	1089.8
Quality (x)=	N/A	N/A	N/A	N/A	N/A	0.0	0.0	0.0	1.0
Flow Rate (Lbm/s)=									
Vol. Flow (ft³/min)=									

(c)

	EVAP. Sat. Liq IN	EVAP. Sat. Vap. OUT	AIR OUT = ROOM IN	ROOM AIR OUT	EVAP. AIR IN
	EVAP. SIDE		Evap. Effect.= 0.63		
			AIR SIDE OF EVAP.		
					15
P (mPa)=	2.1	2.0	101.4	101.4	101.4
T Deg.C =	18.2	17.7	20.0	23.3	23.3
h (kJ/Kg)	56.2	1865.3			
Quality (x)=	0.0	1.0			
Flow rate (kg/hr)=	14.2	14.2	10416.7	10416.7	10416.7
P (psia)=	0.3	0.3	14.7	14.7	14.7
T (F)=	64.8	63.8	68.0	74.0	74.0
h (Btu/Lbm)=	32.8	1089.8			
Quality (x)=	0.0	1.0			
Flow Rate (Lbm/s)=			6.4	6.4	6.4
Vol. Flow (ft³/min)=			5304.8	5304.8	5304.8

(b)

Figure C7.2. (*Continued*)

is then used to determine the air flow rate from the workspace based on the equation:

$$Qcool = Room\ Air\ Mass\ Flow\ Rate$$
$$\times \{h(at\ T_{air\ out}) - h(at\ T_{air\ into\ workspace})\}$$

It may be shown that this air flow rate must be kept high enough so that the temperature of the room air flow rate leaving the cooled space

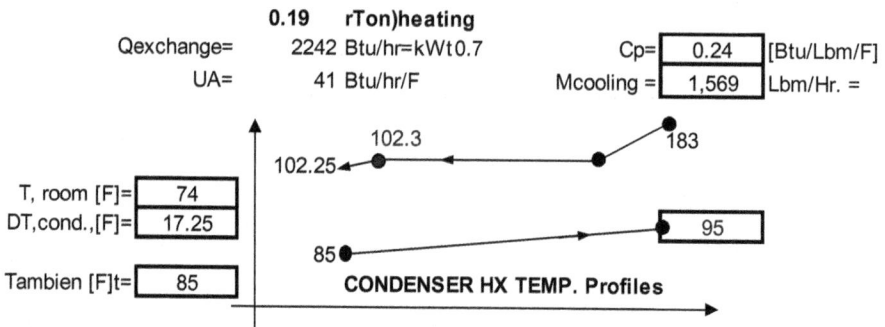

Figure C7.3.   Condenser heat exchanger temperature profiles.

does not approach and certainly cannot fall below the desired evaporative temperature of the water refrigerant in the flash tank. In the flash tank, the vapor temperature of the refrigerant in turn determines the operating suction pressure for the first stage of the compressor. The overall result is that the temperature of the room space's air flow rate that was used to condense the refrigerant cycle water before it is throttled into the flash tank can reach temperatures of over 95°F. These relationships are best illustrated by the temperature profiles for the condenser as shown in Figure C7.3.

2. A study of the operating suction for the first stage of the compressor and discharge temperature of the compressor vs. the pressure profiles (as shown in Figure C7.4) reveals that the refrigeration part of the combined cycle is a combination of a reverse Brayton Cycle (without the turbine) and a vapor recompression cycle. These two cycles differ in their fluid expansion process: a reverse Brayton Cycle involves the use of a turbine to expand the fluid in order to lower its pressure back to the that of the compressor suction. This solar-powered, water refrigeration system does not use a turbine for this pressure reduction process. The necessary reduction in the pressure at this point in the refrigeration system is brought about by the successive pressure (friction energy) losses in the three heat exchangers and interconnecting piping, and most critically in the flashing process wherein the expansion process achieves the cooling of the water refrigerant. This is a reasonable inference, given the very small pressure differential (1.1 to 0.3 psia) between the high- and low-pressure levels.

3. From Figure C7.1 it can be observed that, assuming that only the amount of heat energy input to the system via the electric heater VV20 is considered, the overall coefficient of performance (COP) of the cycle is 0.69.

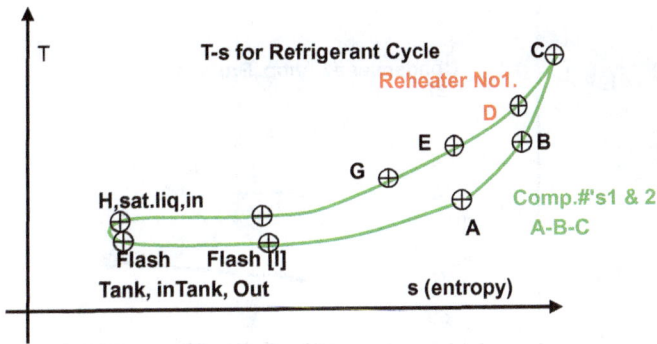

(a)

Legend	Temp. (F)	Temp. (C)	psia pressure	Btu/Lbm/F entropy	Btu/Lbm enthalpy
1	385	196	9.42	1.92	1234.3
2	237	114	3.4		
B	351	177	0.56	2.21	1219.8
I	64	18	0.29	2.08	1089.8
H	102.9	39	1.04	1.98	1106.7
I	64	18	0.29	2.08	1089.9
3	255	124	1.04	2.08	1175.7
A	187	86	0.29	2.18	1145.2
C	537	281	1.04	2.24	1307.0
D	268	131	1.04	2.09	1182.1
G	183	84	1.04	2.04	1143.0

(b)

Figure C7.4. (a) Temperature-entropy (T-s) diagram for the vapor-recompression (cooling) cycle, with emphasis on the flash expansion at "H"; (b) Associated property table.

4. It is important to note that the amount of cooling, Q_{cool}, for the system is only 9.7 kW. This has been determined by the spreadsheet model using the cycle state points shown in Figures C7.2(c) and (d).

5. Considering the operating pressure of the water as a refrigerant, a very small change in the operating pressures and temperatures may result in a very large disparity between the predicted performance and the measured performance if the system were to actually be built and tested. For example, if the temperature of the air exiting the workspace were to be changed from 69.5°F (as was used in Figure C7.3) to 72°F, the air temperature of the air flow rate from the workspace would be too high (at 144°F, as may be observed from Figure C7.4(a)) and the cycle would not be able to operate.

6. The humidity of the workspace air has not been considered in any of the analyses. However, the high temperature of the room air exiting from the cycle condenser may be useful in humidifying the air.

(a)

(b)

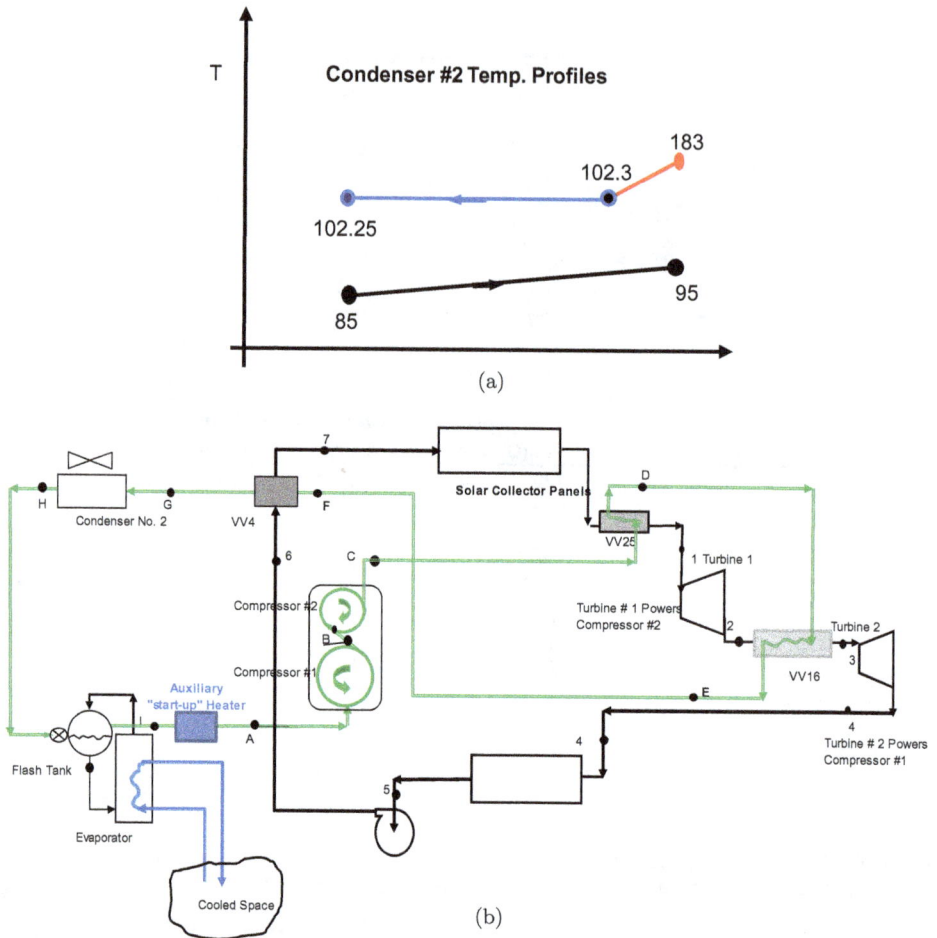

Figure C7.5.   (a) Temperature profile for Condenser 2's water vapor refrigerant and the 85°F ambient coolant; (b) A larger detail of the combined Rankine Cycle and water refrigeration cycle shown in Figure C7.1.

Case Study 8

Utilization of R134a and R1234ze for the Development of New Chiller Systems and as "Drop-In fluids" in Existing Systems

The heating, ventilation and air conditioning (HVAC) industry is rapidly proceeding toward the use of R1234ze and other earth-friendly refrigerants in the development of new refrigeration systems. The industry is also gradually abandoning the use of a conventional refrigerant, R134a, which came into prominence as a substitute for R11, and later R22, as a result of the Montreal Protocol that identifies R11 and R22 as ozone depletion gases. Table C8.1 summarizes the Ozone Depletion Potential (ODP) and the Global Warming Potential (GWP) for a list of conventional chiller refrigerants currently in use. From the results shown in Table C8.1, although R134a and the other proposed refrigerant substitutes listed are not considered fluids that contribute to the depletion of the ozone, they do however have a significant impact on global warming. Considering this impact, the adoption of R1234ze as a refrigerant for use in new refrigeration systems is being explored as a means of significantly reducing global warming emissions, while maintaining little to no effect on ozone depletion.

R1234ze(E) (trans-1,3,3,3-Tetrafluoroprop-1-ene, $CF_3CH = CHF$) is a part of the hydrofluoroolefin (HFO) family of fluorinated gases. The extremely low GWP value (less than 1) of R1234ze is its most significant physical characteristic. It is considered by the American Society of Heating, Refrigerating and Air-Conditioning Engineers (ASHRAE) as a "mildly flammable fluid" at temperatures above 30°C. The manufacturer of R1234ze also suggests careful attention to the choice of material and lubricants used with the refrigerant. Attention must be paid to the suitability of acrylics used with R1234ze under some conditions, as well as any contact the refrigerant may have with neoprene and polypropylene. Polyester oils are recommended lubricants for compressor bearings or other mechanical components used in

Table C8.1.   Comparison of contemporary refrigerant fluids with respect to GWP and ODP.

Refrigerant	Global warming potential	Ozone depletion potential
R407C	1774	0
R134A	1300	0
R410A	1924	0
R513A	572	0
R1233ZD	1	0
R1234ZE	LESS THAN 1	0

Note: R11 is given the OZP (Ozone Depletion Potential) ranking of 1 to indicate that it has the lowest contribution to ozone depletion.

chiller systems utilizing R1234ze as refrigerant. Additionally, when R1234ze is to be used as a "drop-in fluid" in a chiller system that was previously using R134a as the refrigerant, the operational guidelines of the compressor manufacturer should be closely observed.

Many manufacturers of water chillers and the manufacturers of the major component, such as Concepts NREC, are currently studying the application of R1234ze in chiller systems, with the aim of adjusting and improving on the design of compressors that must work with RF1234ze. This section presents some of the basic results obtained by comparing the chiller efficiency, also known as the coefficient of performances (COPs), of R1234ze and R134a. The analysis presented in this section consists of two independent analyses, A and B, as follows.

A. Analysis A compares the performance of each refrigerant, with an assumption that the specific speed (N_s) is theoretically optimized for a basic chiller which follows the vapor compression cycle, and a chiller system which utilizes an economizer (a cascade system). This comparison is consistent with the engineering task of developing a "new" chiller system that is specifically designed to use R1234ze. In this comparison, the efficiency of the compressor is kept at the same value of 85% and the diameter of the compressor impeller and speed are determined from a theoretical specific speed (Ns) and specific diameter (D_s) feasibility analysis. The theoretical compressor (total-to-static) adiabatic efficiency versus specific speed is shown in Figure C8.1. Figure C8.1 has been prepared based on a review of published data on the viable design efficiency for compressors based on generic radial compressor research, and compares very favorably with other independent, published compressor analyses, as noted in Figure C8.1.

Viable Compressor Efficiency (+/-5%) as a Function of (NASA) Specific Speed (NS)

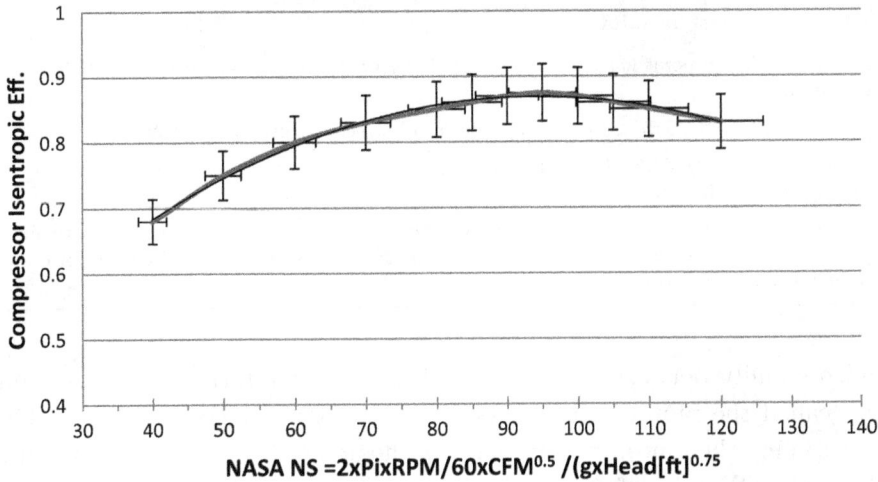

Figure C8.1.   Potential design specification for a compressor's total-to-static adiabatic efficiency versus specific speed (N_s).

B. Analysis B determines the performance of R1234ze when used in a chiller system originally designed to operate on R134a. In this case study, the existing R134a compressor speed and diameter are unchanged, and new N_s and D_s values are calculated for the R1234ze refrigerant operating at the pressure ratios that correspond to the evaporator and condenser saturation pressures for the temperatures that are fixed by the chiller application. As revealed by the use of the optimum N_s and D_s, chosen for their highest efficiency, the performance of the chiller system designed to operate on R134a is less optimal when R1234ze is used in place of R134a. This affects the compressor efficiency. Depending on the severity extent of the change in the COP due to the change in compressor efficiency, the size of the condenser may also be undersized for R1234ze. In this case, the pressure ratio across the compressor may increase, which would also decrease the COP for this "drop-in" R1234ze application.

The analysis proceeded by adopting the ASHRAE guideline for conventional HVAC chiller applications. This guideline is summarized in Table C8.2.

The analyses of the chiller systems include the use of an integrated, hermetically sealed, electric motor drive that uses the refrigerant to cool the motor jacket and internal windings. For the following analyses, the efficiency for the motor was taken to be 95%. The heat rejected from the motor is

Table C8.2.   ASHRAE guideline used in the current chiller system analyses.

Target Design Conditions

3.1 Full Load Design point:

The full load design point is based on a water chiller operating at standard ARI-550/590-2003 conditions:

Evaporator = 12.2°C/6.67°C/0.56°K approach, 0.043/sec per evaporator kW.

Condenser = 29.44°C/34.57°C/0.7°K approach, 0.054/sec per evaporator kW. Discharge piping loss less than 10 kPa.

The full load design point must comply with Ashrae 90.1 full load requirement for both the 2004 editons and 2010; of these two the 2004 has the most stringent requirement with a 6.11COP limit (0.576 kW/ton).

divided equally between the jacket and the motor internal windings, sometimes called the motor winding cooling gaps. It was also assumed that the vapor cooling the motor winding gap was heated to a superheat temperature of 25°F (14°C). The refrigerant is returned to the chiller system. Thus, the chiller system is closed with respect to maintaining its inventory of refrigerant. The performances of the following two types of chiller systems designed for use with R134a and R1234ze as refrigerants were analyzed:

1. The basic vapor-compression chiller system, consisting of four components: a one-stage compressor, condenser, throttle valve and evaporator. This vapor compression cycle has been shown in Figure C8.2(a).
2. A chiller system which includes an economizer in its design, which is often called a cascade chiller system. The cascade chiller system consists of the same components as a vapor-compression refrigeration system, with the inclusion of the following additional components: a single flash tank, and a two-stage compressor. The vapor from the flash tank is injected into the inlet of the second stage of the compressor. This system is shown in Figure C8.2(b).

C8.1   Results of the analysis for Case Study A: Designing new chillers with either R134a or R1234ze and comparison with an economizer

Tables C8.3 and C8.4 present a summary of the analyses for new R134a and R1234ze chiller systems — each specifically designed to operate on one of the two refrigerants mentioned. Thus, the 85% efficiencies for the compressor are assumed to be achievable with the speed and diameter of the compressor shown as determined by the equations for the dimensionless parameters

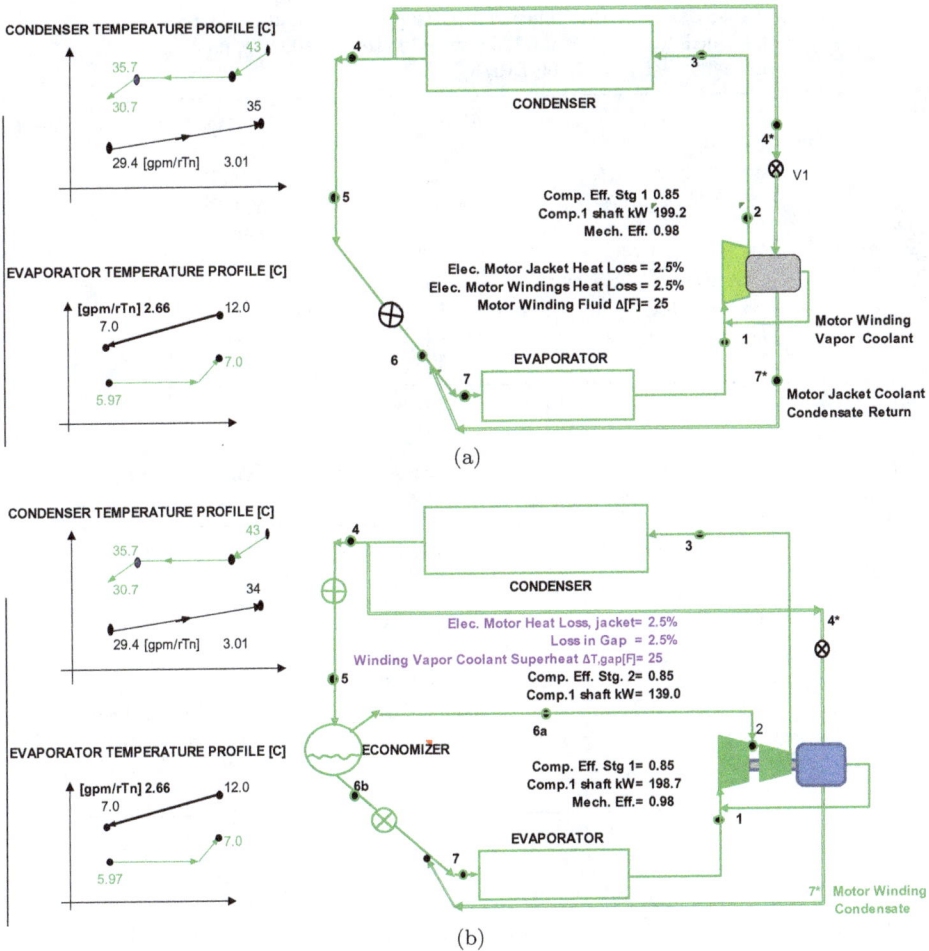

Figure C8.2. (a) Diagram of a basic vapor-compression chiller system; (b) Diagram of a cascade chiller system.

Specific Speed (Ns) and Specific Diameter (Ds) shown below.

$$N_s = \frac{RPM_{comp.} \times \sqrt{Volume_{flow}}}{\Delta Enthalpy^{0.75}}$$

$$D_s = \frac{Dia._{comp.} \times \Delta h^{0.25}}{\sqrt{Volume\ \dot{F}lowrate}}$$

The speed and diameter are based on a specific speed of 105 Ns and a specific diameter of 1.5. The comparison between the performance of a vapor-compression system utilizing R134a and another using R1234ze, as shown in Table C8.3, is sized for a cooling capacity of 400 RT. The corresponding cycle state points for Table C8.3 are shown in Figures C8.3 and

Table C8.3. Performance comparison of basic chiller systems designed for use with R134a or R1234ze at 400 RT cooling capacity. [1 RT = 12,000 Btu/h]

400 RT capacity	R134A	R1234ze
COP	6.58	6.57
$\dot{M}_1\,AND\,\dot{M}_2$ [kg/s]	8.75; 8.84	9.52; 9.61
$\dot{V}_1\,AND\,\dot{V}_2$ [m^3/s]	0.50; 0.21	0.67; 0.27
P1 [bar,a]	3.6	2.67
P3 [bar,a]; T3 [C]	9.05; 43.1	6.82; 38.3
Compressor efficiency	0.85	.85
Ns and Ds	105 and Ds = 1.5	105 and Ds = 1.5
Diameter [m]	0.22	.254
Speed [rpm]	15,500	12,600

Table C8.4. Performance comparison of cascade chiller systems designed for use with R134a or R1234ze at 700 RT cooling capacity. [1 RT = 12,000 Btu/h]

700 RT capacity	R134A	R1234ze
COP	6.79	6.80
$\dot{M}_1\,AND\,\dot{M}_2$ [kg/s]	14.4; 15.4	15.6; 16.75
$\dot{V}_1\,AND\,\dot{V}_2$ [m^3/s]	0.82; 0.52	1.09; 0.69
P1 & P2 [bar,a]	3.6 & 6.29	2.67
P3 [bar,a]; T3 [C]	9.05; 44.4	6.82; 38.4
Compressor efficiencies ($\eta1$ & $\eta2$)	0.85 & 0.85	0.85 & 0.85
Ns$_1$ & Ns$_2$ and Ds$_{1,2}$	105 & 92 and Ds$_{1,2}$ = 1.5	105 & 88 and Ds$_{1,2}$ = 1.5
Diameters, D1 & D2 [m]	0.32 & 0.23	0.37 & 0.27
RPM	8,100	6,600

C8.4. The economizer (cascade) system comparison of the same fluids is sized for a cooling capacity of 700 RT and the result is summarized in Table C8.4.

Several critical observations can be made from studying Tables C8.3 and C8.4:

1. The COPs for a basic vapor-compression system using R134a or R1234ze are very comparable.
2. The volume flow rate for the basic vapor-compression system using R1234ze is 30–35% more than the volume flow rate for the system using R134a for the same capacity. The mass flow rate for the R1234ze system is 8–9% higher than the mass flow rate for the R134a chiller.
3. The diameter for the R1234ze system compressor is 15% larger than the compressor diameter for the R134a.
4. The speed for the R1234ze system compressor is 18–20% less than that of the R134a compressor.

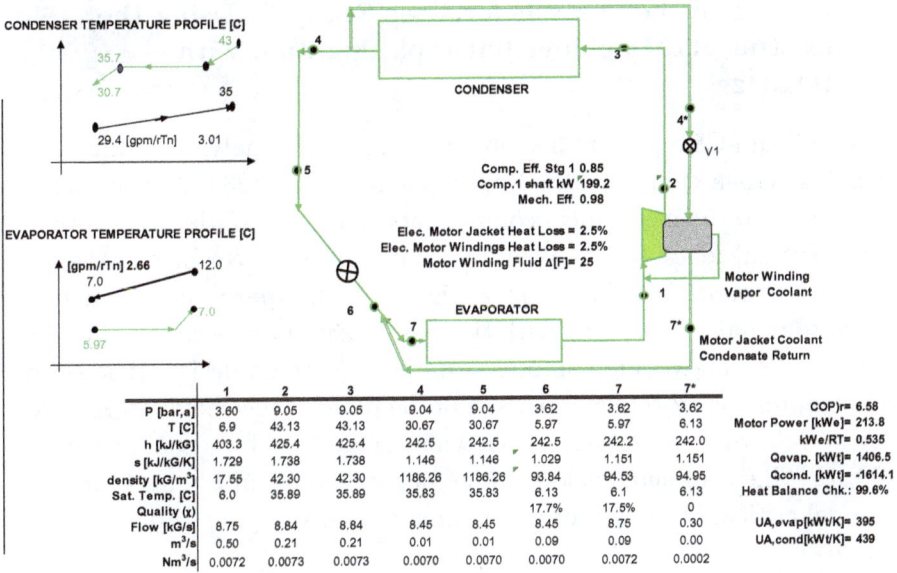

	1	2	3	4	5	6	7	7*	
P [bar,a]	3.60	9.05	9.05	9.04	9.04	3.62	3.62	3.62	COP)r= 6.58
T [C]	6.9	43.13	43.13	30.67	30.67	5.97	5.97	6.13	Motor Power [kWe]= 213.8
h [kJ/kG]	403.3	425.4	425.4	242.5	242.5	242.5	242.2	242.0	kWe/RT= 0.535
s [kJ/kG/K]	1.729	1.738	1.738	1.146	1.146	1.029	1.151	1.151	Qevap. [kWt]= 1406.5
density [kG/m³]	17.55	42.30	42.30	1186.26	1186.26	93.84	94.53	94.95	Qcond. [kWt]= -1614.1
Sat. Temp. [C]	6.0	35.89	35.89	35.83	35.83	6.13	6.1	6.13	Heat Balance Chk.: 99.6%
Quality (x)						17.7%	17.5%	0	
Flow [kG/s]	8.75	8.84	8.84	8.45	8.45	8.45	8.75	0.30	UA,evap[kWt/K]= 395
m³/s	0.50	0.21	0.21	0.01	0.01	0.09	0.09	0.00	UA,cond[kWt/K]= 439
Nm³/s	0.0072	0.0073	0.0073	0.0070	0.0070	0.0070	0.0072	0.0002	

Figure C8.3. A basic chiller system designed for use with R134a at 400 RT cooling capacity with compressor efficiency = 85%. Data used in Table C8.3 have been taken from this complete cycle.

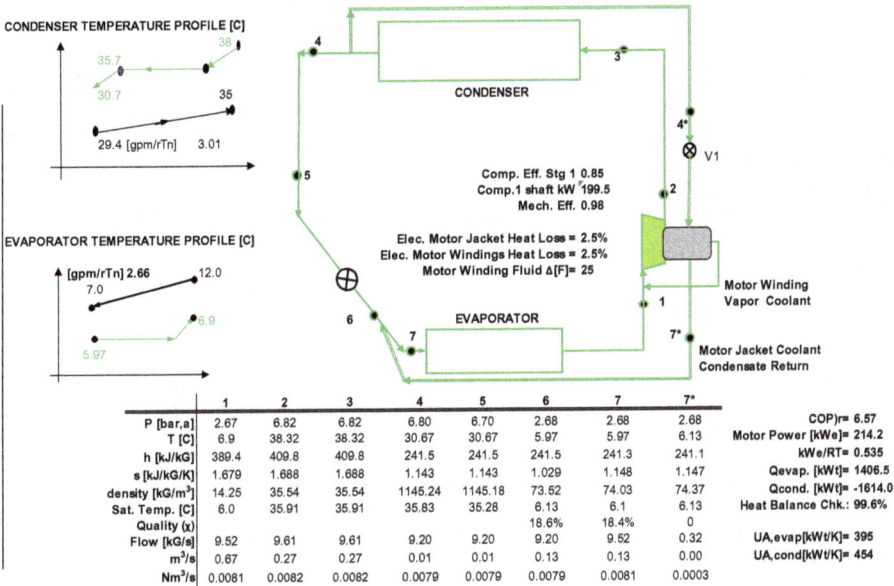

	1	2	3	4	5	6	7	7*	
P [bar,a]	2.67	6.82	6.82	6.80	6.70	2.68	2.68	2.68	COP)r= 6.57
T [C]	6.9	38.32	38.32	30.67	30.67	5.97	5.97	6.13	Motor Power [kWe]= 214.2
h [kJ/kG]	389.4	409.8	409.8	241.5	241.5	241.5	241.3	241.1	kWe/RT= 0.535
s [kJ/kG/K]	1.679	1.688	1.688	1.143	1.143	1.029	1.148	1.147	Qevap. [kWt]= 1406.5
density [kG/m³]	14.25	35.54	35.54	1145.24	1145.18	73.52	74.03	74.37	Qcond. [kWt]= -1614.0
Sat. Temp. [C]	6.0	35.91	35.91	35.83	35.28	6.13	6.1	6.13	Heat Balance Chk.: 99.6%
Quality (x)						18.6%	18.4%	0	
Flow [kG/s]	9.52	9.61	9.61	9.20	9.20	9.20	9.52	0.32	UA,evap[kWt/K]= 395
m³/s	0.67	0.27	0.27	0.01	0.01	0.13	0.13	0.00	UA,cond[kWt/K]= 454
Nm³/s	0.0081	0.0082	0.0082	0.0079	0.0079	0.0079	0.0081	0.0003	

Figure C8.4. A basic chiller system designed for use with R1234ze, at 400 RT cooling capacity, with compressor efficiency = 85%. Data used in Table C8.4 have been taken from this complete cycle.

C8.2   Results of the analysis for Case Study B: Using the existing R134a chiller but replacing fluid with R1234ze

Tables C8.5 and C8.6 present a summary of the cycle analysis assuming that an existing R134a chiller has its fluid replaced with R1234ze, with no additional change to the mechanical components. Thus, the initial 85% optimum design point efficiency for the compressor is changed according to the aerodynamic characteristics of the new refrigerant. The speed and diameter are assumed to remain unchanged with the new R1234ze replacement. Thus, the Ns and Ds for the existing compressor impellers are off-design. The simple, vapor compression system comparison of the performance between the use of R134a and the drop-in R1234ze is shown in Table C8.6 for a cooling capacity of 400 RT. The complete cycle state points are shown in Figure C8.8.

Several critical observations can be made from studying Tables C8.5 and C8.6:

Table C8.5.   Performance evaluation of basic chiller system originally designed for use with R134a, with R1234ze used as "drop-in" fluid, at 700 RT cooling capacity. [1 RT = 12,000 Btu/h]

700 RT capacity	R134A (see Figure C8.4)	R1234ze (see Figure C8.7)
COP	6.79	5.63
$\dot{M}_1$ AND $\dot{M}_2$ [kg/s]	14.4; 15.4	15.54; 16.75
$\dot{V}_1$ AND $\dot{V}_2$ [m³/s]	0.82; 0.52	1.09; 0.69
P1 & P2 [bar,a]	3.6 & 6.29	2.67 & 4.66
P3 [bar,a]; T3 [C]	9.05; 44.4	6.82; 42.5
Compressor efficiencies ($\eta 1$ & $\eta 2$)	0.85 & 0.85	0.70 & 0.74
Ns1 & Ns2 and Ds	105 & 92 and Ds = 1.5	129 & 109 and Ds = 1.29
Diameters, D1 & D2 [m]	0.32 & 0.23	.32 & .23
RPM	8,100	8,100

Table C8.6.   Performance evaluation of basic chiller system originally designed for use with R134a, with R1234ze used as "drop-in" fluid, at 400 RT cooling capacity. [1 RT = 12,000 Btu/h]

400 RT capacity	R134A (see Figure C8.9)	R1234ze (see Figure C8.12)
COP	6.58	5.7
$\dot{M}_1$ AND $\dot{M}_2$ [kg/s]	8.75; 8.84	9.52; 9.62
$\dot{V}_1$ AND $\dot{V}_2$ [m³/s]	0.50; 0.21	0.67; 0.27
P1 [bar,a]	3.6	2.67
P3 [bar, a]; T3 [C]	9.05; 43.1	6.82; 41.4
Compressor efficiency	0.85	.74
Ns and Ds	105 and 1.5	127 and 1.3
Diameter [m]	0.22	.22
Speed [rpm]	15,500	15,500

1. The COP for a drop-in of R1234ze in an R134a chiller is reduced by 13–17% for the full-load capacity.
2. The volume flow rate for the R1234ze system is 30–35% more than the R134a system for the same cooling capacity. The mass flow rate for the R1234ze system is 8–9% higher than the mass flow rate for the R134a chiller.
3. The condenser for the R1234ze chiller will be undersized by 2–3% due to the lower COP.

Regulatory authorities throughout the world continue to emphasize that changing the refrigerant in chillers from R134a to R1234ze will protect the environment from global warming. This initiative has been encouraged by industrial consensus over the last several decades, that R1234ze is a very attractive replacement fluid for conventional refrigerants, as its use has been demonstrated to have little to no contributions to global warming. The results shown in this case study indicate that new chiller units designed for use with R1234ze will be comparable in COP performance with the "older" chiller units designed for use with R134a. However, the use of R1234ze as a "drop-in fluid" into existing chiller systems that were designed for use with R134a will result in a 13–17% reduction in COP performance at the full capacity design point.

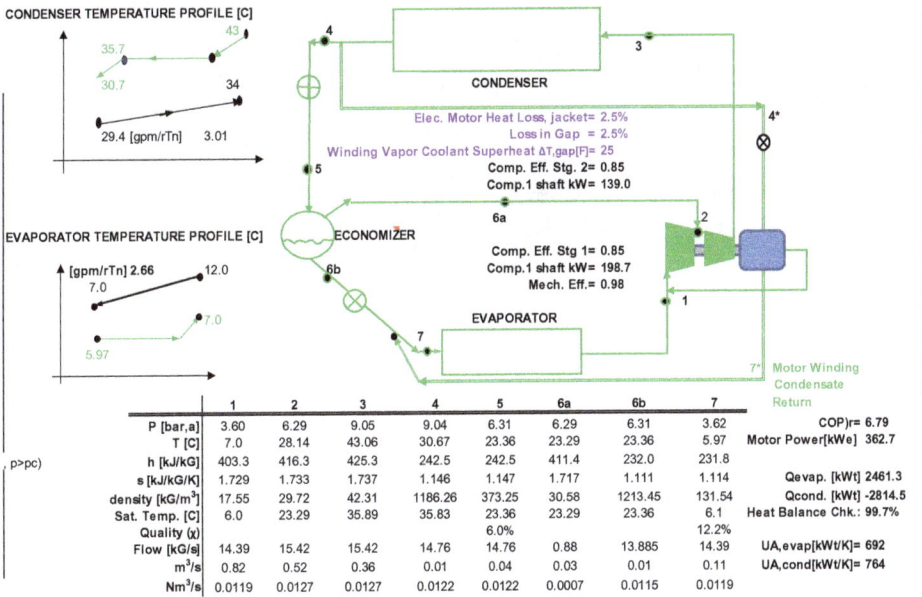

	1	2	3	4	5	6a	6b	7	
P [bar,a]	3.60	6.29	9.05	9.04	6.31	6.29	6.31	3.62	COP)r= 6.79
T [C]	7.0	28.14	43.06	30.67	23.36	23.29	23.36	5.97	Motor Power[kWe] 362.7
h [kJ/kG]	403.3	416.3	425.3	242.5	242.5	411.4	232.0	231.8	
s [kJ/kG/K]	1.729	1.733	1.737	1.146	1.147	1.717	1.111	1.114	Qevap. [kWt] 2461.3
density [kG/m³]	17.55	29.72	42.31	1186.26	373.25	30.58	1213.45	131.54	Qcond. [kWt] -2814.5
Sat. Temp. [C]	6.0	23.29	35.89	35.83	23.36	23.29	23.36	6.1	Heat Balance Chk.: 99.7%
Quality (x)					6.0%			12.2%	
Flow [kG/s]	14.39	15.42	15.42	14.76	14.76	0.88	13.885	14.39	UA,evap[kWt/K]= 692
m³/s	0.82	0.52	0.36	0.01	0.04	0.03	0.01	0.11	UA,cond[kWt/K]= 764
Nm³/s	0.0119	0.0127	0.0127	0.0122	0.0122	0.0007	0.0115	0.0119	

Figure C8.5. A cascade chiller system designed for use with R134a at 700 RT cooling capacity, with compressor efficiency = 85%.

R1234ze Fluid

CONDENSER TEMPERATURE PROFILE [C]

35.7 · 35 ·
30.7 · 34
29.4 [gpm/rTn]   3.01

CONDENSER

Elec. Motor Heat Loss, jacket= 2.5%
Motor Gap Winding Heat Loss= 2.5%
Vapor Gap Cooling Superheat [F]= 25
Comp. Eff. Stg. 2= 0.85
Comp.1 shaft kW = 142.0

4" V1

6a

EVAPORATOR TEMPERATURE PROFILE [C]

[gpm/rTn] 2.66
7.0   12.0
6.9
5.97

ECONOMIZER
6b

Comp. Eff. Stg 1= 0.85
Comp.1 shaft kW = 195.0
Mech. Eff.= 0.98

EVAPORATOR
7
7*
Condensate Return

	1	2	3	4	5	6a	6b	7	
P [bar,a]	2.67	4.66	6.82	6.80	4.67	4.66	4.67	2.68	COP)r 6.80
T [C]	6.9	25.03	38.39	30.67	23.03	22.94	23.03	5.97	Motor Power [kWe] 361.8
h [kJ/kG]	389.4	401.3	409.9	241.5	241.5	399.3	230.9	230.7	
s [kJ/kG/K]	1.679	1.684	1.688	1.143	1.144	1.677	1.108	1.110	
density [kG/m³]	14.25	24.35	35.53	1145.24	297.76	24.62	1169.99	105.31	Qevap. [kWt]= 2461.3
Sat. Temp. [C]	6.0	22.94	35.91	35.83	23.03	22.94	23.03	6.1	Qcond. [kWt]= -2812.8
Quality (y)					6.4%			12.6%	Heat Balance Chk. 99.6%
Flow [kG/s]	15.55	16.74	16.74	16.05	16.05	1.03	15.012	15.55	UA,evap[kW t/K]= 692
m³/s	1.09	0.69	0.47	0.01	0.05	0.04	0.01	0.15	UA,cond[kW t/K]= 788
Nm³/s	0.0133	0.0143	0.0143	0.0137	0.0137	0.0009	0.0128	0.0133	

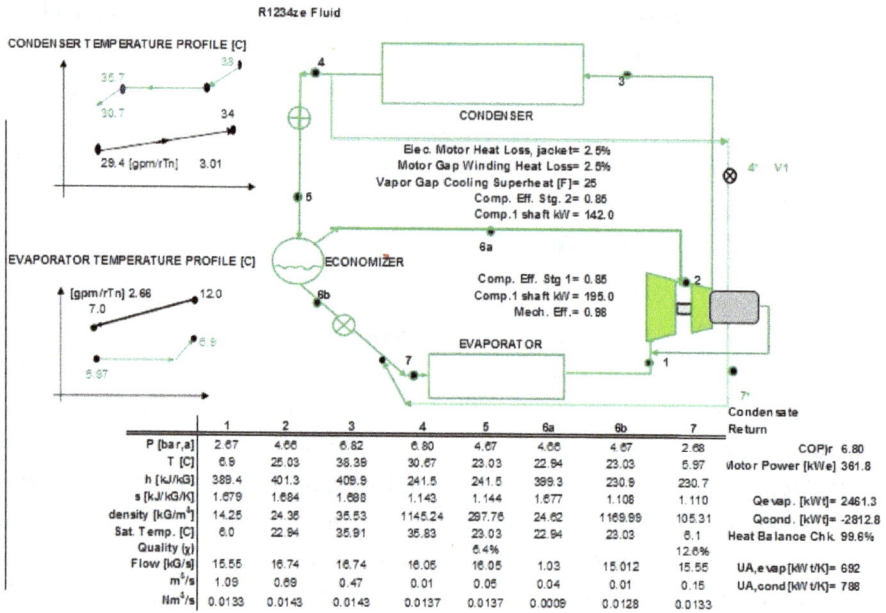

Figure C8.6.   A cascade chiller system designed for use with R1234ze at 700 RT cooling capacity, with compressor efficiency = 85%.

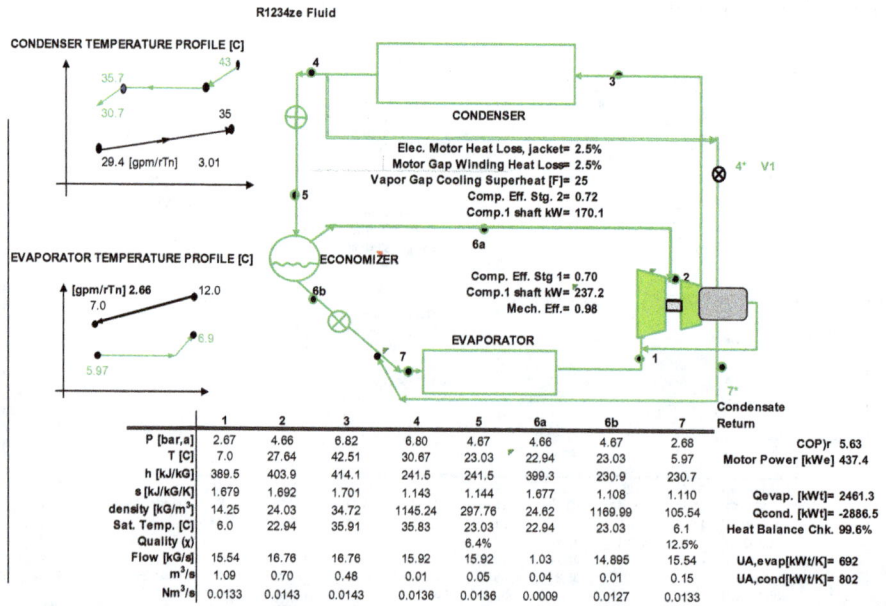

R1234ze Fluid

CONDENSER TEMPERATURE PROFILE [C]

35.7 · 43 ·
30.7 · 35
29.4 [gpm/rTn]   3.01

CONDENSER

Elec. Motor Heat Loss, jacket= 2.5%
Motor Gap Winding Heat Loss= 2.5%
Vapor Gap Cooling Superheat [F]= 25
Comp. Eff. Stg. 2= 0.72
Comp.1 shaft kW= 170.1

4" V1

6a

EVAPORATOR TEMPERATURE PROFILE [C]

[gpm/rTn] 2.66
7.0   12.0
6.9
5.97

ECONOMIZER
6b

Comp. Eff. Stg 1= 0.70
Comp.1 shaft kW= 237.2
Mech. Eff.= 0.98

EVAPORATOR
7
7*
Condensate Return

	1	2	3	4	5	6a	6b	7	
P [bar,a]	2.67	4.66	6.82	6.80	4.67	4.66	4.67	2.68	COP)r 5.63
T [C]	7.0	27.64	42.51	30.67	23.03	22.94	23.03	5.97	Motor Power [kWe] 437.4
h [kJ/kG]	389.5	403.9	414.1	241.5	241.5	399.3	230.9	230.7	
s [kJ/kG/K]	1.679	1.692	1.701	1.143	1.144	1.677	1.108	1.110	
density [kG/m³]	14.25	24.03	34.72	1145.24	297.76	24.62	1169.99	105.54	Qevap. [kWt]= 2461.3
Sat. Temp. [C]	6.0	22.94	35.91	35.83	23.03	22.94	23.03	6.1	Qcond. [kWt]= -2886.5
Quality (x)					6.4%			12.5%	Heat Balance Chk. 99.6%
Flow [kG/s]	15.54	16.76	16.76	15.92	15.92	1.03	14.895	15.54	UA,evap[kWt/K]= 692
m³/s	1.09	0.70	0.48	0.01	0.05	0.04	0.01	0.15	UA,cond[kWt/K]= 802
Nm³/s	0.0133	0.0143	0.0143	0.0136	0.0136	0.0009	0.0127	0.0133	

Figure C8.7.   A cascade chiller system originally designed for use with R134a, with R1234ze here used as a "drop-in fluid".

CONDENSER TEMPERATURE PROFILE [C]

EVAPORATOR TEMPERATURE PROFILE [C]

CONDENSER

Comp. Eff. Stg 1 0.74
Comp.1 shaft kW 229.9
Mech. Eff. 0.98

Elec. Motor Jacket Heat Loss = 2.5%
Elec. Motor Windings Heat Loss = 2.5%
Motor Winding Fluid Δ[F]= 25

400 rTons
EVAPORATOR

V1

Motor Winding
Vapor Coolant

Motor Jacket Coolant
Condensate Return

	1	2	3	4	5	6	7	7*	
P [bar,a]	2.67	6.82	6.82	6.80	6.70	2.68	2.68	2.68	COP)r= 5.70
T [C]	7.0	41.42	41.42	30.67	30.67	5.97	5.97	6.13	Motor Power [kWe]= 246.8
h [kJ/kG]	389.5	413.0	413.0	241.5	241.5	241.5	241.2	241.0	kWe/RT= 0.617
s [kJ/kG/K]	1.679	1.698	1.698	1.143	1.143	1.029	1.148	1.147	Qevap. [kWt]= 1406.5
density [kG/m³]	14.25	34.93	34.93	1145.24	1145.18	73.52	74.25	74.60	Qcond. [kWt]= -1645.8
Sat. Temp. [C]	6.0	35.91	35.91	35.83	35.28	6.13	6.1	6.13	Heat Balance Chk.: 99.5%
Quality (x)						18.6%	18.3%	0	
Flow [kG/s]	9.51	9.62	9.62	9.15	9.15	9.15	9.51	0.36	UA,evap[kWt/K]= 395
m³/s	0.67	0.28	0.28	0.01	0.01	0.12	0.13	0.00	UA,cond[kWt/K]= 460
Nm³/s	0.0081	0.0082	0.0082	0.0078	0.0078	0.0078	0.0081	0.0003	

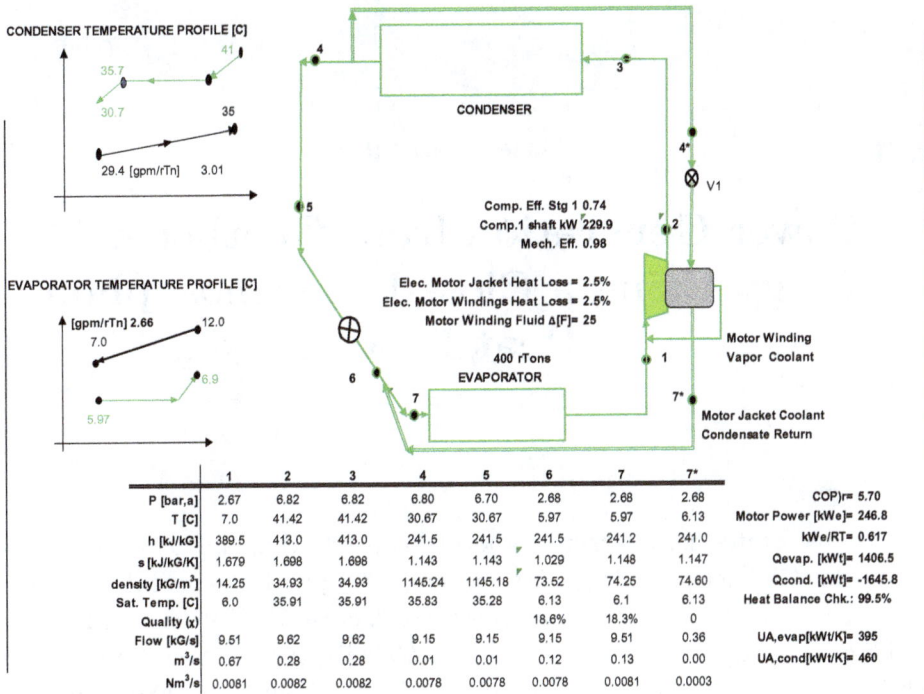

Figure C8.8. A basic chiller cycle originally designed for use with R134a with a 400 RT cooling capacity, with R1234ze here used as a "drop-in" fluid.

The complete cycle state points shown in Figures C8.3 to C8.8 can also be used to determine several important results. For example, this case study showed that a small increase of 2–3% in heat transfer would not significantly change the hot and cold temperature profiles in the condenser, based on the engineering practice of having considerable performance margins designed into the heat exchanger.

Case Study 9

Power Generation from Geothermal Energy Using a Closed Thermosyphon Heat Pipe

The United States has a larger geothermal capacity than any other nation in the world (Geothermal Energy Association, 2009); with the potential to produce 8,000–73,000 MW of geothermal power across 13 of its western states (Williams, Reed, Mariner, DeAngelo, & Galanis, 2008). According to the National Renewable Energy Laboratory (NREL) (Kutscher, 2001), there is immediate and significant potential in those lands for small-scale geothermal power development. The global market is strong for 100 to 500 kWe geothermal systems, which can serve these rural, remote areas as mini grids.

Demand is also high for systems that improve or expand the co-production of geothermal energy. Vast potential exists for the utilization of slim holes, which typically measure approximately 100 mm in diameter or less, to explore for the presence of geothermal energy potentials, in addition to the 823,000 200 + mm diameter size drilled wells in the United States as geothermal power generation units.[1] In fact, slim holes alone could potentially deliver 100 kW to 1 MWe of geothermal power per slim hole (Garg & Combs, 1997).

However, high system costs per kWh for the original hardware needed to create these small (less than 100–500 kW of power) geothermal systems are an obstacle to the utilization of this clean, sustainable energy source (Kutscher, The Status and Future of, 2000). The construction costs of power plant subsystems and production wells are high. Additionally, the operation and maintenance (O&M) of these geothermal systems are also expensive.

---

[1]Geothermal Energy Association Report: Geo101-Binder1 from www.geothermalenergy association.org

It is therefore of benefit to propose the design of a cost-effective, modular 100 kWe geothermal energy-based power-generating system that uses dry steam in place of a binary (i.e., two fluid)-type geothermal power. This system features the use of an environmentally-safe and inexpensive working fluid — water — in a closed Rankine Cycle system that is installed in a completely self-contained module that fits into the small diameter slim hole, rather than placing the components for a typically multi-MWe size system on the surface of the geothermal field. This approach is unique for geothermal applications.

This system consists of a long heat pipe that can be installed in both an existing slim (exploratory drilled) hole or larger diameter, geothermal well. It is suggested that one or more heat pipes, each rated at 100 kWe, can provide the geothermal energy user with as much power as the geothermal field can provide while reducing the per unit cost of the Rankine Cycle System. The heat pipe(s) used in this system operate(s) on a thermosyphon principle which will be explained in detail in the next section. Analysis of the performance of the thermosyphon system can be performed by numerically integrating the pressure increase of the water as it descends into the heat pipe and the pressure decrease of the lighter vapor as it rises in the heat pipe. This analysis can be performed using an Excel spreadsheet. The start of such a spreadsheet is shown in Figure C9.2. A proper sequence of calculations that must be performed on the liquid in the heat pipe as it descends, and the vapor as it ascends, has also been provided in the spreadsheet. The pressure at depth "n + 1" can be calculated by adding the incremental change in pressure due to the weight of the fluid to the pressure at depth at "n". This can be easily performed using Equation (C9.1).

$$P_{n+1} = P_n + \Delta ht. \times \frac{g_g}{g_c} \times \rho(P, T). \tag{C9.1}$$

Given that the density of a liquid changes as a function of pressure and temperature, the very large depth of the slim hole results in the pressurization of the water used as the liquid working fluid. This effect is similar to that achieved by a pump in the Rankine Cycle. After the vapor has been evaporated, the lower density of the vapor-phase fluid results in the rising of the vapor towards the surface. As the fluid rises, it also loses pressure, due to pipe friction energy. However, the net pressure drop of the vapor fluid as it rises nearer the center of the heat pipe is less than the pressure increase of the liquid that descends along the evaporator walls of the heat pipe. This pressure reduction of the vapor fluid as it rises along the center of the heat pipe may be thought as being similar to the process which occurs when a throttle valve is used to reduce the pressure in a thermodynamic process, except that, in the geothermal system, heat transfer is lost between the hot vapor fluid and the colder, surrounding environment as the fluid rises in the heat pipe. As such, the loss of pressure due to gravity can be modeled as a virtual valve in the riser tube.

The process of heat transfer into the evaporator section of the geothermal energy-based power generation system can be modelled by using the thermo-fluid-heat transfer engineering equations presented in this text. The computer model is structured to be a numerical integration of the heat transfer and the thermodynamics that is occurring along the entire geothermal well. The model proceeds by dividing the entire length of the geothermal well into "n" sections. Each section is treated as a short pipe with a small length (Δl) and diameter (d). The heat pipe diameter and the length of the heat pipe are inputs into the computer model. The amount of heat input into each of the evaporator sections of the power-generating system is determined by the products of the temperature difference between the brine and the evaporating fluid, the heat transfer coefficient, and the surface area of that section of the heat pipe:

$$\dot{Q}_{heat\,trans.\,from\,geothermal\,brine} = U_{heat\,trasn.\,coef} \times A_{section\,surface\,area} \times \Delta T$$

With the heat transfer from the geothermal brine into the working fluid properly modeled, the thermodynamics of this power generation system reverts to the analysis typically done for a Rankine Cycle system, as described in Chapter 5. That is, regardless of the method of heat transfer into the cycle, once the heat input from the geothermal brine is determined and is used to evaporate the working fluid, the thermodynamics of the Rankine Cycle representing the downhole heat recovery process is identical to that of a typical Rankine Cycle system. The illustration of the computer model on a spreadsheet used for the analysis of the geothermal energy-based power generation system is shown in Figure C9.3.

C9.1 Reduction in amount of initial capital required and installation costs

As shown in Figure C9.1, the working fluid — water — is evaporated by the heat from the geothermal brine that is flowing upward between the well wall and the smaller diameter wall of the heat pipe. Using the surface of the heat pipe as the evaporating surface for the working fluid eliminates the need for a mounted evaporator. Additionally, with the effects of gravity serving to pressurize the working fluid that is falling into the slim hole, a feed pump is no longer required.

After evaporation, the vaporized working fluid ascends to the top of the heat pipe due to its lowered density, while simultaneously undergoing a decrease in pressure due to the line pressure drop due to pipe friction. As noted above, this loss of vapor pressure can be thought of as the working fluid being passed through a "virtual" valve (the "valve" is virtual because the pressure drop occurs without an actual valve, as per the previous discussion) except that the enthalpy change in each section of the numerical model is positive due to the friction heating, and not zero as it would be if the pressure change were in fact occurring in a valve. The net effect is to slightly increase the temperature of the vapor fluid as it rises in the slim tube. The model continues with the water vapor entering the turbine located in a vault near the surface. The turbine expands the steam to 3–4 psia (0.2 to 0.28 bar,a), thus generating power. To generate power outputs in multiples of the 100 kWe system shown in Figure C9.1, several heat pipes can be installed into the geothermal field. The ascended water vapor also serves to cool the generator windings (a permanent magnet coupled to the turbine). The magnetic bearings of the turbine eliminate the need for a gearbox, lubricated bearings or a shaft seal.

C9.1.1 *Principle of the thermosyphon system*

A feasibility analysis for this Case Study has been completed that includes a computer model for the thermo-fluids analysis of a thermosyphon-based

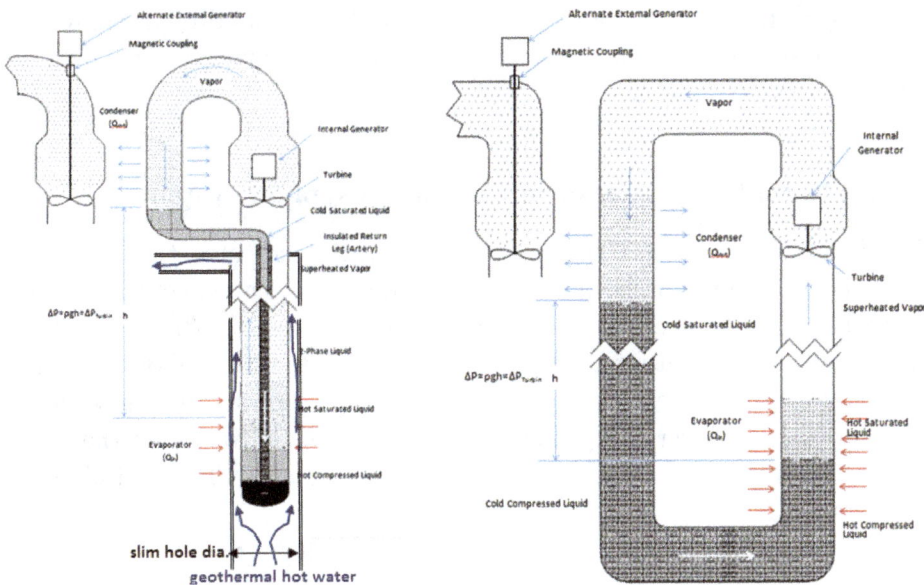

Figure C9.1. Two configurations of a thermosyphon heat pipe modular geothermal power generation system.

geothermal energy recovery concept which offers the potential for enhanced power generation in a single geothermal well, plus a reduction in manufacturing and installation costs.

A thermosyphon heat pipe, as applied to a geothermal energy source, is a heat pipe which includes an internal power generation subsystem that produces electric power in a closed system. The evaporator section of the heat pipe based thermosyphon evaporates the working fluid (water) using heat from the geothermal heat source. Following evaporation, the density of the vapor is much less than the liquid fluid that is flowing into the geothermal well and thus the vapor rises to the top of the heat pipe where it is expanded through a high-speed turbine. The expanded fluid is then condensed and drained under the effect of gravity, back to the evaporator section. A thermosyphon heat pipe is designed to operate at very large depths within the geothermal field; the column of liquid fluid that is flowing downward in the geothermal well builds pressure linearly with each increased unit of depth of the well due to the weight of the fluid. Thus, in this system, gravity serves a similar function as a feed pump of a working fluid, creating a "gravity feed pump". Two configurations of a thermosyphon heat pipe are shown in Figure C9.1, in which a controllable height of water functions as a "gravity feed pump", maintaining 65–75 psia (4.5 to 5 bar,a) in the evaporator section of the pipe positioned at the bottom of the slim hole.[2] The power module of the thermosyphon heat pipe produces as much as 100 kWe of electrical power, following a basic Rankine Cycle in a closed system. The turbine-generator system is installed either close to or on the surface, housed by a semi-closed vault that also contains the power conditioning and control systems required for the power conversion.

C9.2   Notable advantages of the thermosyphon system

The feasibility study for this case study clearly indicates that the thermosyphon heat pipe-based power generation concept can effectively recover geothermal heat energy with a temperature as low as 325°F at depths of 1,500 ft in a slim hole that is only 8 inches in diameter. Several other notable advantages of the thermosyphon system are as follows, starting with the most obvious and very important benefit: the use of water as the working fluid and not an organic fluid such as R245fa, R134a or isobutene-isopentane.

---

[2]This work was in collaboration with a colleague at Concepts NREC, Mr. Fred Becker, Director of Engineering Sales.

Rock temp. Diff. Multiplier= **2.2**

									Geo-Rock Temp. Change per Hole Depth=	0.523	F/ft		0.001
60.1 Average Down Hole Density									Vapor Temp. Change per HX length=	0.28	F/ft		0.010
3.5 Average Riser Density									Depth Increment=	10.00	ft	
282 Down Hole Volume					1.650				Surface Condensing Temperature	206	F	
132 Riser Volume (ft³)					0.261				Depth to Top of Heat Pipe Evap. Section=	850	ft		C=
17,425 Lbm total									DPm HX loss=	0.99	Down-Hole
42.09							Riser Pres. Riser Fluid	Liquid Temp Temp. of	Hole	Down Fluid	Down Liquid [Lbm/ft³]

Depth	Fluid Down Temp.	Fluid Rise Temp.	Rock Temp		Enthalpy	R236 fa	Vapor Temp.	At Depth	Solid Rock	Depth (ft)	Pressure	Density &	Vapor
0	199	433	50	269.9	0.6	168.9	269.9	433	198.8	50	0	13	60.1	0.033
10	199	433	55	270.0	0.6	168.9	270.0	433	198.84	55	10.0	13	60.1	0.033
20	199	433	60	270.0	0.6	168.9	270.0	433	198.85	60	20.0	13	60.1	0.033

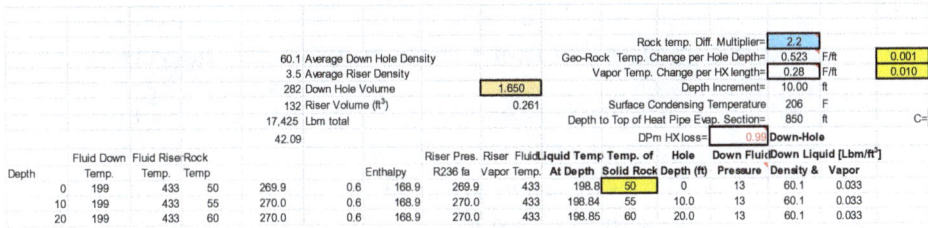

Figure C9.2.   A numerical integration technique for calculating the pressure increase of the water descending into the evaporator, at large depths of the slimhole.

From an environmental perspective, the choice of water as the working fluid for the thermosyphon-based power generation system has a significant advantage over isobutane, isopentane, or other refrigerant fluids that are conventionally used in geothermal power systems.

No external refrigerant or thermal fluid circulating pumps are required for the operation of the thermosyphon-based power generation system. The Rankine Cycle uses the weight of the fluid between the evaporator and condenser to pressurize the working fluid before it enters the turbine. Thus, the typical feed pump used in a Rankine Cycle system is not needed. This results in cost savings, not only with its initial and O&M costs, but also with the reduction in installation space required.

The evaporator of the proposed thermosyphon-based power generation system operates over a range of pressures by adjusting the height of the liquid column in the thermosyphon. This results in an increased rate of heat transfer compared to a constant pressure/temperature evaporation process, as well as an increase in the net temperature difference between the evaporating fluid and the geothermal heat source. The increase in the temperature difference between the geothermal brine heat source and the evaporating fluid will either enable the surface area of the heat pipe wall to decrease or, more importantly, enable the amount of heat input into the heat pipe to increase and thus allow an increase in flow rate of the fluid which in turn enables an increase in the amount of power that can be generated from the heat pipe.

The turbine is integrally coupled to a permanent magnet (a generator that is cooled by the working fluid), uses magnetic bearings and operates at high speeds (10,000 to 20,000 rpm). Such generators are available from the Dresser-Rand Synchrony Division or from Total Power Systems (New Castle, England).

C9.3   Discussion of results of a feasibility study conducted using a computer model for thermo-fluids analysis

For the feasibility study, it was assumed that the geological strata used in the analysis were depths as shallow as 1,500 ft and temperatures ranging from 325 to 500°F. The assumption of shallow depth may have resulted in a conservative estimate of the performance of the thermosyphon-based power generation system. The computer model for thermo-fluids analysis, used and presented in this case study, was designed to enable the above assumptions to be used in determining the pressure and temperature state points along the length of the geothermal well, and ultimately the inlet and outlet pressures and temperatures of the turbine. Using the computer model, the fluid flow rate through the turbine could also be determined. The results obtained from analysis conducted using the computer model confirmed that the physical constraints typically imposed on heat pipes — such as the flooding limit, sonic limit, heat flux and boiling limit — were not exceeded, and thus the design point fluid flow rate within the heat pipe and the desired power rating of the heat pipe can be achieved. The analysis proceeded by determining the performance of a 100 kWe system for 100 to 350°F geothermal temperatures. It is interesting to note that as much as 2.5 MWe could be generated if the heat pipe model shown in Figure C9.3 uses a geothermal temperature as high as 500°F and if the diameters of the slimehole wells are increased to 24 inches.

The temperature profiles of the water working fluid and the geothermal well from the feasibility analysis for a 100 kWe rated, thermosyphon-based power generation system are shown in Figures C9.3 and C9.4. The operating turbine's inlet pressure and temperature is 60 psia at 290°F. A preliminary design of the turbine describes a three-stage steam turbine, with stage diameters from 4 to 7 inches and operating at 20,000 rpm. The proposed thermosyphon heat pipe utilizes a hermetically-sealed, integrated, high-speed turbine-generator unit (TGU) which does not require a gear box, shaft seal or a dynamic lubrication system, for reasons previously described. The discharge pressure is 3 psia. Figure C9.4 identifies the temperature profiles of the three principal heat sources and heat sink as modeled by the heat pipe thermo-fluids model. The loci of points presented for each of these profiles is taken directly from the three columns for the brine, working fluid vapor and liquid return temperatures. It is noted that the temperature of the liquid that is returned by gravity down the heat pipe is at the saturation liquid temperature corresponding to the depth, and thus also the pressure of the fluid at the depths shown. Although the temperature may look constant, the saturation temperature does increase with an increase in the saturation pressure.

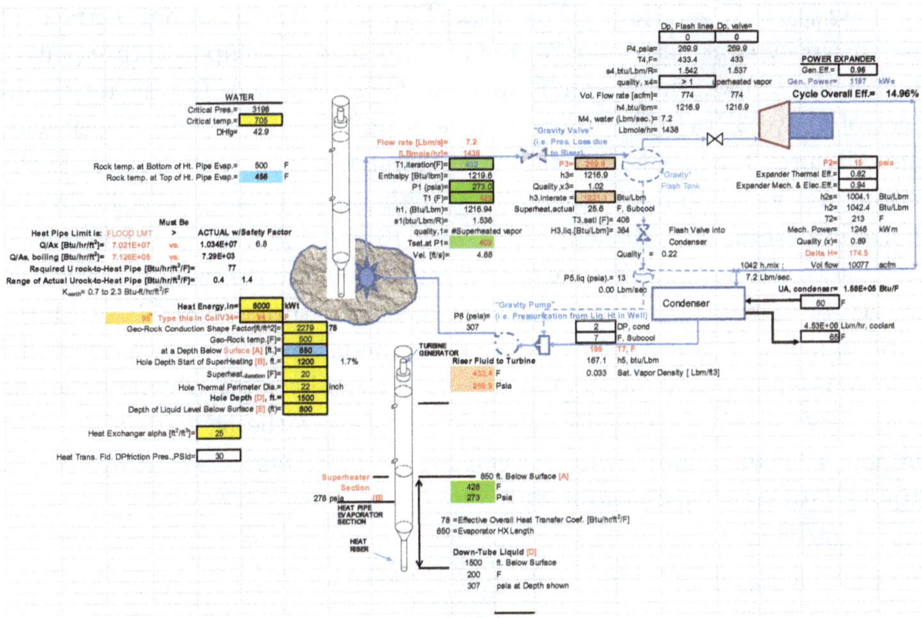
WATER
Condenser
POWER EXPANDER
Cycle Overall Eff.= 14.96%
Flash Valve into Condenser
TURBINE GENERATOR
HEAT PIPE EVAPORATOR SECTION
HEAT RISER
Down-Tube Liquid

Figure C9.3. Schematic of the Rankine Cycle as a model of the heat recovery process in a geothermal energy-based power generation system.

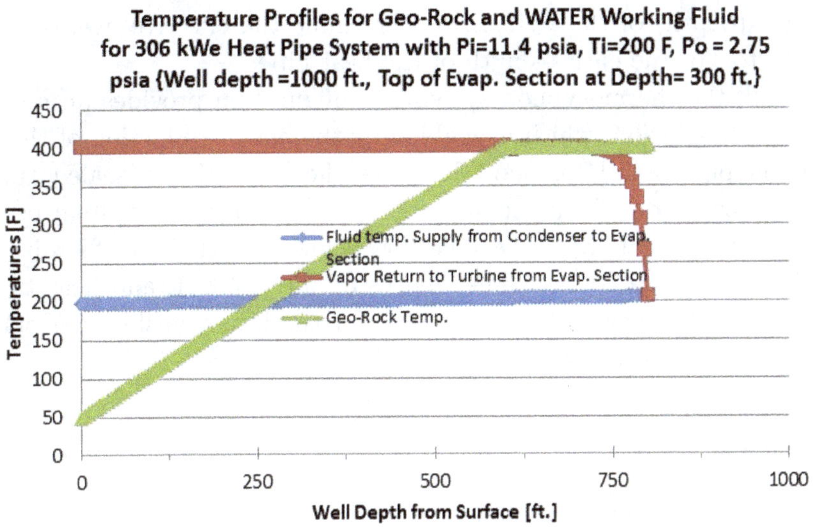
Temperature Profiles for Geo-Rock and WATER Working Fluid for 306 kWe Heat Pipe System with Pi=11.4 psia, Ti=200 F, Po = 2.75 psia {Well depth =1000 ft., Top of Evap. Section at Depth= 300 ft.}
Temperatures [F]
450
400
350
300
250
200
150
100
50
0
0
250
500
750
1000
Well Depth from Surface [ft.]
Fluid temp. Supply from Condenser to Evap. Section
Vapor Return to Turbine from Evap. Section
Geo-Rock Temp.

Figure C9.4. Temperature profiles for a case study analysis of a downhole Rankine Cycle system.

The key to the successful operation of a thermosyphon-based power generation system is an external surface of a thermosyphon heat pipe with a high heat transfer coefficient. The surface of the heat pipe serves as the evaporator section. The heat transfer coefficient is defined as:

$U_{oa} = Q_{input}/(A_{surface\,area} \times (T_{geo-mass}-T_{surface}))$. The external surface of the thermosyphon heat pipe is assumed to have a temperature equal to that of the evaporating water fluid used. Generally, a heat transfer coefficient of above $3 \mathrm{Btu/h/ft^2/°F}$ is considered unlikely for the hard rock with micro-fissures that make up the geothermal material that surrounds the geothermal well. However, the results of the analysis suggest a likelihood that a heat transfer coefficient of between 10 and 100 may be achievable, if the Hot Dry Rock contains voids or fissures that are continually served by hot geo-brine fluids or geo-magma with a temperature of 500^{+}°F. A Monte Carlo simulation was used in this thermo-fluids analysis to determine the average possible value for the thermo conduction heat transfer between the geophysical substrate and the cylindrical surface of the heat pipe based on the shape factor equation used for a pipe that is buried vertically in the ground with a thermal conductivity, k:

$$S_{shape\,factor} = \frac{2 \times \pi \times L_{Depth\,of\,well}}{\ln \frac{4 \times L}{D_{well\,diamter}}}; \; \dot{Q}_{brine\,heat\,recovery}$$

$$= S \times k \times (T_{brine} - T_{fluid\,evap.})$$

Of course, the equation for the shape factor can be integrated with respect to length (depth) of the well, L, to determine the effective average of the shape factor for the entire length of the heat pipe.

Overall, the thermosyphon system's configuration provides an economical, modular solution, which is enabled by the mass production of the thermosyphon piping and the modularity of the hermetically sealed turbine. As can be concluded from the analysis, the thermosyphon system provides useful power for heat transfer coefficients below $100 \mathrm{Btu/h/ft^2/°F}$ with optimal diameters of 12 to 18 inches, depths of only 1,500 ft, and heat transfer coefficients of less than $50 \mathrm{Btu/h/ft^2/°F}$ for the heat transfer area between the geo-mass and the evaporating section.

References

Garg, S. K., & Combs, J. (1997). Use of slim holes with liquid feedzones for geothermal reservoir assessment. *26*(2), 153–178. doi:https://doi.org/10.1016/S0375-6505(96)00038-7

Geothermal Energy Association. (2009). *U.S. Geothermal Power Production and Development Update.*

Kutscher, C. F. (2000, June 16–21). The Status and Future of. Madison, Wisconsin, USA: National Renewable Energy Laboratory. doi:NREL/CP-550-28204

Kutscher, C. F. (2001). *Small-Scale Geothermal Power Plant Field Verification Projects.* doi:NREL/CP-550-30275

Williams, C. F., Reed, M. J., Mariner, R. H., DeAngelo, J., & Galanis, S. (2008). *Assessment of moderate- and high-temperature geothermal resources of the United States: U.S. Geological Survey Fact Sheet.* U.S. Geological Survey. Retrieved from https://pubs.usgs.gov/fs/2008/3082/

Case Study 10

Applications of Mechanical Vapor Recompression

The vapor recompression cycle was reviewed in detail in Chapter 5, where it was presented as a cycle that could provide cooling. The working fluid in the vapor recompression cycle is typically a refrigerant, such as R134a or R245fa. Case Studies 7 and 8 presented several alternative cycles which could also provide cooling using either a conventional refrigerant or water as a working fluid. This case study reviews another type of vapor recompression cycle traditionally known as Mechanical Vapor Recompression (MVR). MVR often uses water in the form of a superheated vapor as the working fluid. MVR uses a steam compressor to increase the pressure of saturated steam in industrial scale applications, such as in packaged water boilers. Most packaged water boilers easily produce saturated steam at 15 psig. However, the production of steam above 15 psig usually requires the constant attention of a boiler operator, which comes at an additional expense. This higher pressure is required for the transport of steam over large distances between buildings. To address this, MVR can be applied to increase the pressure of the saturated steam without intervention from a boiler operator. As the density of the steam increases in proportion with its pressure, even a pipe with a small diameter can be used to transport a large mass flow rate of steam. Once the pressurized steam has been transported, it is often immediately reduced in pressure to 15 psig. Following the reduction in pressure, the steam is used directly for heating purposes.

The performance of a system utilizing MVR is measured by calculating the Coefficient of Performance (COP) of the system. The COP is defined by the equation:

$$C.O.P. = \frac{\dot{M} \times \Delta h_{fg} \times \eta_{compressor}}{\dot{M} \times C_p \times \left(\left(\frac{P_{out}}{P_{in}} \right)^{(k-1)/k} - 1 \right)}. \qquad (C10.1)$$

The ratio of the latent heat of condensation (Δh_{fg}) of steam to the power of the steam compressor (in kW or hp) is typically above 1.

C10.1 Practical applications of mechanical vapor recompression

C10.1.1 *Application of mechanical vapor recompression in district heating systems*

In many European and Scandinavian countries, MVR has been applied in district heating systems, which are used in the heating of buildings. Heat distribution is carried out via a closed network of piping that circulates heat in the form of hot water or steam from a central power plant to the outlying buildings. Figure C10.1 shows a diagram of a district heating system. The choice between using hot water or steam as the medium to transport the heat energy is based on the magnitude of the amount of heat energy that needs to be transferred and the cost per mile of piping with respect to design of the piping. In more modern district heating systems, hot water is generally the accepted transport medium. The hot water is carried in pipes with a smaller diameter than those required if pressurized steam were to be used as the heat transfer medium. The smaller the diameter of the pipe, the less expensive the installation costs for the district heating pipes, which are typically buried in the ground or carried in cement-lined vaults for easy maintenance. Of course, consideration must be given to the desired temperature of the heat sink. That is, if the application requires only potable water to be heated for lavatory use, the temperatures need only be 120–140°F, which is suitable for hot water, and district heating systems. Steam could also be used for district heating but there is an exergy penalty to overcome and the district heating piping would need to be at pressures usually exceeding 15 bar,a.

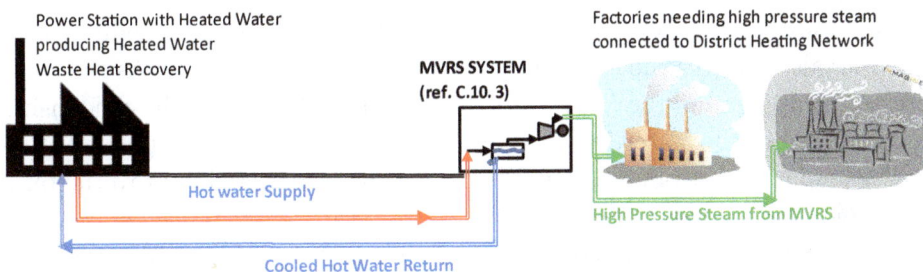

Power Station with Heated Water
producing Heated Water
Waste Heat Recovery

Factories needing high pressure steam
connected to District Heating Network

MVRS SYSTEM
(ref. C.10. 3)

Hot water Supply

High Pressure Steam from MVRS

Cooled Hot Water Return

Figure C10.1.   Application of MVR in a district heating system.

C10.1.2   *Application of mechanical vapor recompression in waste treatment*

While district heating systems use pressurized hot water that is to be flashed into water vapor as the working fluid for MVR systems (MVRSs), there are other systems that use the superheated steam vapor directly produced by a waste heat or renewable energy-based heat source as the working fluid for an MVRS; Figure C10.2(a) illustrates the use of an MVRS in such an application. Figure C10.2(b) illustrates an application of MVR in the treatment of raw waste that removes water from the sludge to recover only the solid materials.

C10.1.3   *Application of mechanical vapor recompression system for waste heat recovery*

Figures C10.3(a) and (b) illustrates an MVRS that is powered by the waste heat recovered from the condenser of a refrigeration system. The large vessel that contains the water to be evaporated serves as a flash tank which also functions as the receiver for the water entering through the flash valve.

    All the components that constitute the MVRS are among the seven basic mechanical devices first identified in Chapter 3, and subsequently applied to common thermodynamics processes as detailed in Chapter 5. The application of the First Law of Thermodynamics to each of the MVRS components is based on the basic equations developed in Chapter 3.

C10.1.4   *Application of mechanical vapor recompression in the desalination of water*

One of the most interesting and practical applications of an MVRS is the distillation of water. In this application, the high discharge temperature of the compressor allows for the evaporation of the contaminated or salt-water. This is possible because the compressor component of the MVRS serves to reduce the pressure of the water to below atmospheric pressure. This reduction in pressure also reduces the evaporating temperature of the water. As the compressor component of the MVRS pressurizes the water vapor, the condensing temperature of the water is raised and the water vapor is superheated. The outlet fluid stream from the compressor is then recirculated back to the tank that holds the contaminated water. The temperature difference between the tank temperature and the compressor discharge temperature compressor is sufficient to transfer the heat of compression to the contaminated water. This heat transfer causes the water to evaporate, leaving the solids and/or

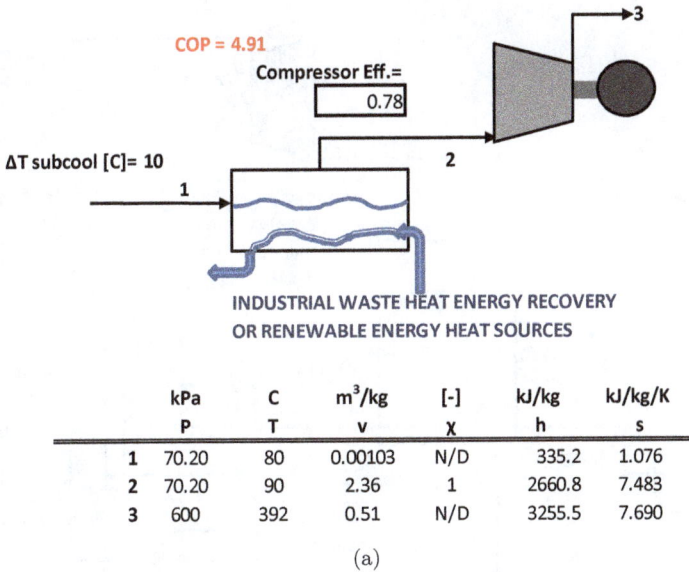

COP = 4.91

Compressor Eff.= 0.78

ΔT subcool [C]= 10

INDUSTRIAL WASTE HEAT ENERGY RECOVERY
OR RENEWABLE ENERGY HEAT SOURCES

	kPa	C	m³/kg	[-]	kJ/kg	kJ/kg/K
	P	T	v	X	h	s
1	70.20	80	0.00103	N/D	335.2	1.076
2	70.20	90	2.36	1	2660.8	7.483
3	600	392	0.51	N/D	3255.5	7.690

(a)

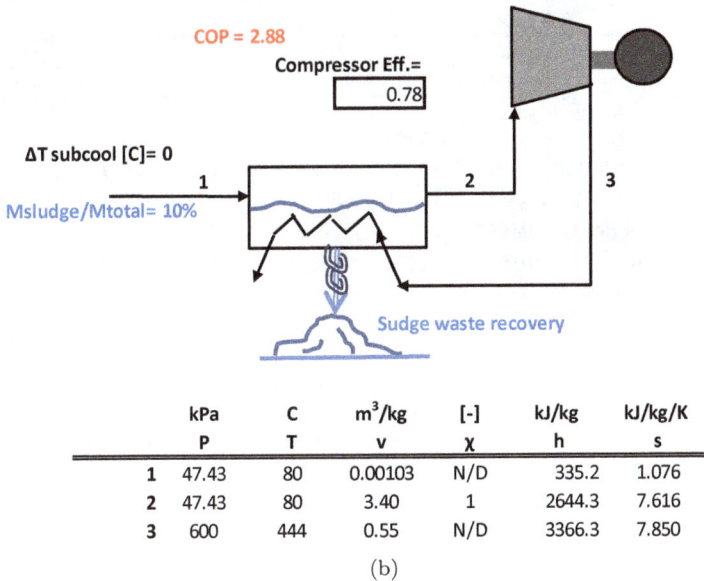

COP = 2.88

Compressor Eff.= 0.78

ΔT subcool [C]= 0

Msludge/Mtotal= 10%

Sudge waste recovery

	kPa	C	m³/kg	[-]	kJ/kg	kJ/kg/K
	P	T	v	X	h	s
1	47.43	80	0.00103	N/D	335.2	1.076
2	47.43	80	3.40	1	2644.3	7.616
3	600	444	0.55	N/D	3366.3	7.850

(b)

Figure C10.2.   (a) Application of MVRS in a waste heat recovery process to provide a compressor component with low-pressure steam; (b) Application of MVRS in the removal of water and recovery of solids from raw waste.

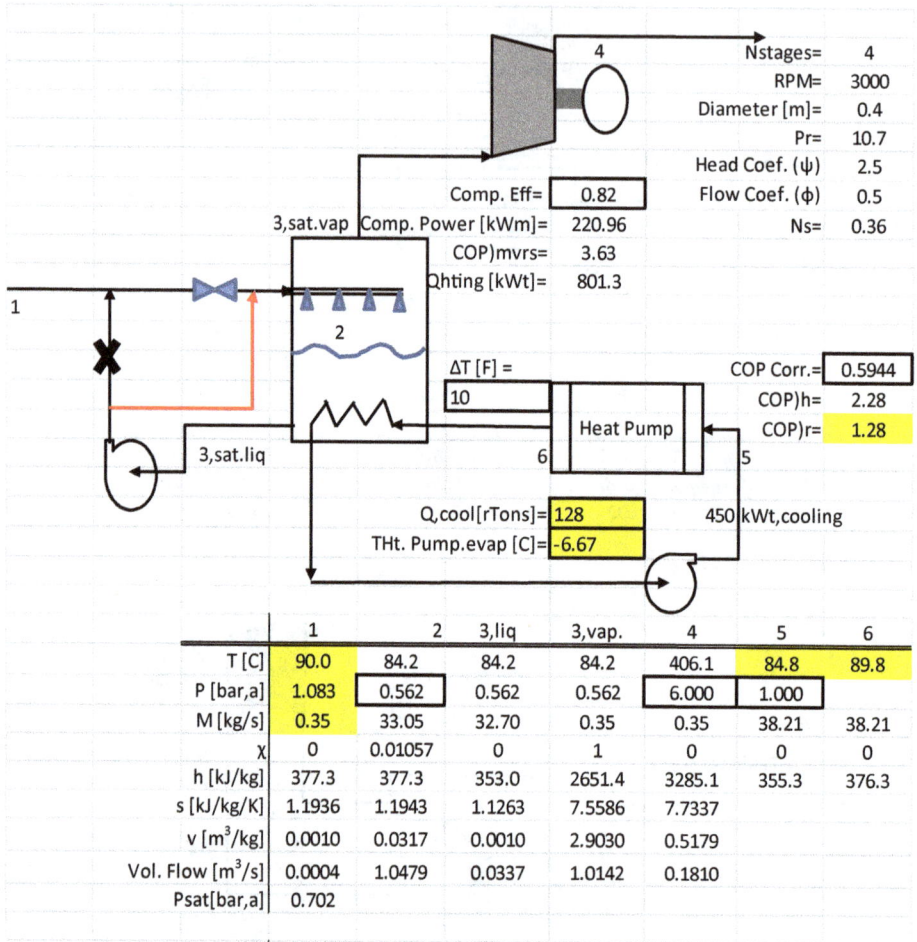

(a)

Figure C10.3. (a) MVRS powered by waste heat recovery from condenser: 4-stage steam compressor for 6 bar,a steam; (b) MVRS powered by waste heat recovery from condenser: 4-stage steam compressor for 4 bar,a steam.

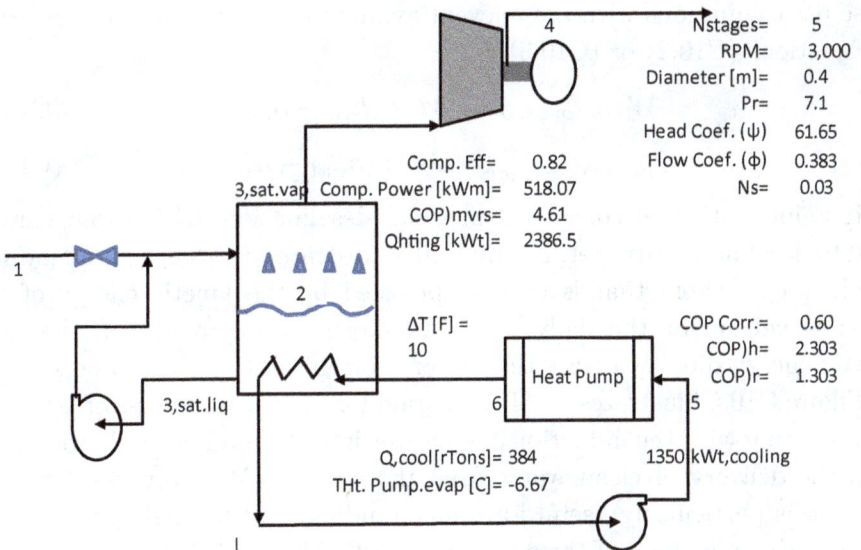

Figure C10.3. (*Continued*)

	1	2	3,liq	3,vap.	4	5	6
T [C]	90.0	84.2	84.2	84.2	339.7	84.8	89.8
P [bar,a]	1.083	0.562	0.562	0.562	4.0	1.0	0.86
M [kg/s]	1.04	98.44	97.40	1.04	1.04	113.81	113.81
X	0	0.01057	0	1	0	0	0
h [kJ/kg]	377.3	377.3	353.0	2651.4	3150.3	355.3	376.3
s [kJ/kg/K]	1.1936	7.5586	7.5586	7.5586	7.7106		
v [m³/kg]	0.0010	2.9030	2.9030	2.9030	0.7011		
Vol. Flow [m³/s]	0.0011	285.7659	282.7452	3.0207	0.7295		
Psat[bar,a]	0.702						

	1	2	3,liq	3,vap.	4	5	6
T [F]	194	183.6	183.6	183.6	643.5	184.6	193.6
P [psia]	15.7	8.14	8.14	8.14	58	14.5	12.5
M [lbm/s]	2.29	216.6	214.3	2.29	2.29	250.4	250.4
X		1.06%	0	1			
h [Btu/Lbm]	162.2	162.2	151.8	1140.1	1354.6	152.8	161.8
s [Btu/Lbm/R]	0.2851	1.8054	1.8054	1.8054	1.8417		
v [ft³/Lbm]	0.0166	46.5634	46.5634	46.5634	11.2451		
Vol. Flow [ft³/s]	0.038	10083.98	9977.39	106.59	25.74		
Psat[psia]	10.2						

(b)

Figure C10.3. (*Continued*)

minerals in the tank. This condensed, clean water is then collected and stored for use.

To determine the performance of the MVRS, it is necessary to determine the enthalpy at the inlet and outlet of each of the components. After the respective enthalpies have been determined, the basic equations detailed in Chapter 3 for each of the components can be applied. Finally, combining these basic equations with the relevant assumptions results in the derivation of Equations (C10.2) or (C10.3).

$$\dot{W}_{compressor} = \dot{M} \times (h_{in} - h_{out}), \qquad (C10.2)$$

$$\dot{Q}_{heatexchanger} = \dot{M} \times (h_{out} - h_{in}). \qquad (C10.3)$$

It is interesting to consider a modular desalination MVRS that can be used to desalinate saltwater. Figure C10.4 illustrates a concept for a modular desalination system that is entirely powered by the kinetic energy of the water wave and/or the daily tides. As shown in Figure C10.4, the tidal kinetic energy directly drives the compressor and initiates the process.

Figure C10.5 illustrates the T-s diagram for the entire process of desalination: starting with the induction of seawater into the module, and proceeding with the delivery of clean water from the cyclic, MVRS process. The T-s diagram is particularly useful for understanding, and ultimately as an aide to guide the modelling of the complex MVRS. The modelling process is also facilitated by using the equations for each of the components, as described in Chapters 3 and 5, and the property table summarizing the properties at the inlet and outlet of each component. The cycle analysis is considered complete only when it can be demonstrated that the energy input into the open system is equal to its energy output. This is easily done by using Equation (C10.2) or Equation (C10.3) for each of the components that constitute the MVRS.

The desalination system shown in Figure C10.4 will produce clean water from ocean water using marine hydrokinetic (MHK) energy as the only power source. The modular system will be sized to be compatible with the water requirements of the electrolyzer system that continues to be advanced to include the delivery of 5,000 psig hydrogen. The modular system will be partially submerged in the ocean — ultimately on a floating barge and integrated with all the other subsystems that would result in a fully operational hydrogen generation and storage system, as seen in the example in Figure C10.6.

The basic cycle for the desalination phase of the desalination-hydrogen production system is shown in the one-line diagram provided in Figure C10.4. The engineering basis for the proposed desalination system combines a proven MVRS with a novel application of a mechanically powered water cavitator that generates vapor generator using sonic shock waves to convert the mechanical energy into thermal energy. The cavitator essentially, serves as an inducer to a steam compressor by ensuring that there is sufficient water vapor moving into the compressor inlet to enable the compressor to operate efficiently. The clean water is produced by the evaporation and condensation method of desalination. The ocean water is evaporated at sub-ambient

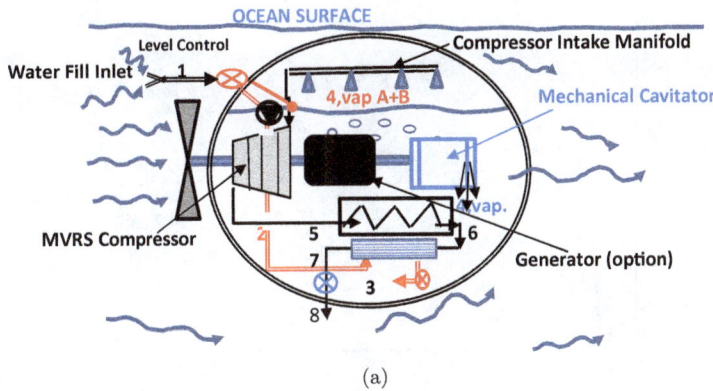

(a)

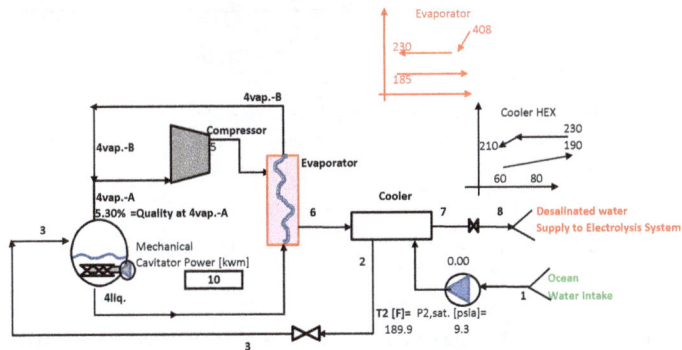

Compressor Pr 2.5
Compressor Eff. 0.7
Pump Pr 1.1
Pump Eff. 0.65
Valve Pr 2.2
Cooler Effectiveness 2%
Evaporator Pinch [F] 10
Mass Flowrate [Lbm/s] 0.2
Gals./Day 2019

Ambient Pressure [psia] 14.7
Ocean Temp. [F] 60
Ocean Depth [ft.] 5
Line Pres. Losses [%] 0.50%
SubCooling in Evap. ΔT7 20 45.2
Evap. Recovery Fraction 89.5%

Compressor Power [kWm]= 21.7
Pump Power [kW]= 0.002
C.O.P.= 6.4
0.1145

ENERGY BALANCE
Cavitator kW 10
Pump kW 0.002
Compressor Kw 21.7
Ocean Heating kw 31.8
ENERGY BALANCE 100%

	1	2	3	4liq.	4vap.-A	4vap.-B	5	6	liquid 6A	vapor 6B	7	8
P [psia]	16.9	18.6	8.5	8.42	8.42	8.38	21.0	20.9	20.9	20.9	20.84	17.0
T [F]	60	190.0	185	185	185	408.0	230.4	230.4	210	210.4		
Quality			0.50%	0%	100%	100%		11.45%	0%	100%	0.0%	#Subcooled liquid
h [Btu/Lbm]	28.1	158.2	158.2	153.3	1140.7	1140.7	1243.7	309	198.9	1157.9	178.7	178.7
v [ft3/Lbm]	0.0160	0.0160	0.2290	0.0165	#[PHFLSH	as not conv	as not conv	2.2	0.0168 has not conv		0.0167	0.0167
s [Btu/Lbm-R]	0.0556	0.2789	0.2718	0.2714	1.8027	1.8032	1.8397	0.3397	0.3397	0.3397	0.3100	0.3100
Flowrate [Lbm/s]	0.2	0.2	0.2	0.1894	0.0106	0.1894	0.2000	0.2000	0.18	0.0229	0.2000	0.2000
							0.2000			0.2000		

219.4124

T2 [F]= 189.9 P2,sat. [psia]= 9.3

Average Heat Exchanger Core Surface Geometry, Alpha (α) [ft²/ft³]
Average Ht. Trans. Coef.s

430	W/m²/K	Evaporator	76	Btu/Hr/F/ft	25
140	W/m2/K	Cooler	25	Btu/Hr/F/ft	75

APPROXIMATE HEAT EXCHANGER SIZES

	UA [kWt/K]	kWt	HX Dia (m)	HX Length (m) & Weight (kG)
Evaporator :	7	197	0.42	1.50 909
Cooler:	1	27	0.20	0.50 455

(b)

Figure C10.4. (a) 2,000 gal/day water cavitator, self-contained, desalination module, powered by *CN*'s Water Turbine with Vessel at Lower Operating Pressure; (b) Block Diagram of the components with state points.

pressure that has been created at the compressor suction. The heat energy for the evaporation is derived from the heat of condensation from the clean steam vapor that is at the compressor discharge and that has been recircuited back into the standing pool of ocean water contained in the desalination pod.

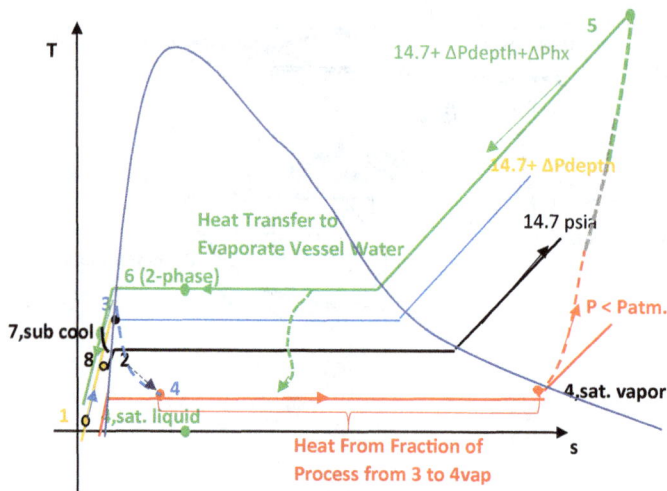

Figure C10.5.   T-s diagram for water, illustrating the thermodynamic processes used to desalinate water with a mechanical vapor recompression cycle integrated with a cavitation heater-vapor inducer.

The system shown in Figures C10.4 and C10.5 includes the following major features:

1. All the power needed for the system is recovered from the ocean's MHK energy source. However, the floating platform (or barge) can be tethered close to shore in order to facilitate the delivery of high-pressure hydrogen to the onshore hydrogen pipeline delivery system or high-pressure storage.

2. The MHK power is used to power a water vapor compressor that is used to desalinate water, based on an MVRS.

3. The clean water is further purified in a commercially available reverse osmosis system to a specification that is required by the hydrogen electrolyzer.

4. Giner ELX is developing an electrolyzer that is able to produce high-pressure hydrogen at 5,000 psig, in what will accurately be identified as an electrochemical hydrogen compressor (EHC).

5. The output from the EHC is then boosted by a compressor to a pressure of 10,000 psig: a pressure that is thought to be the most economical pressure for use in the transportation and industrial energy sectors in the US.

6. The most economical and less complicated procedure is to then deliver the high-pressure hydrogen to a nearby, onshore storage system or insert it directly into the hydrogen pipeline delivery system.

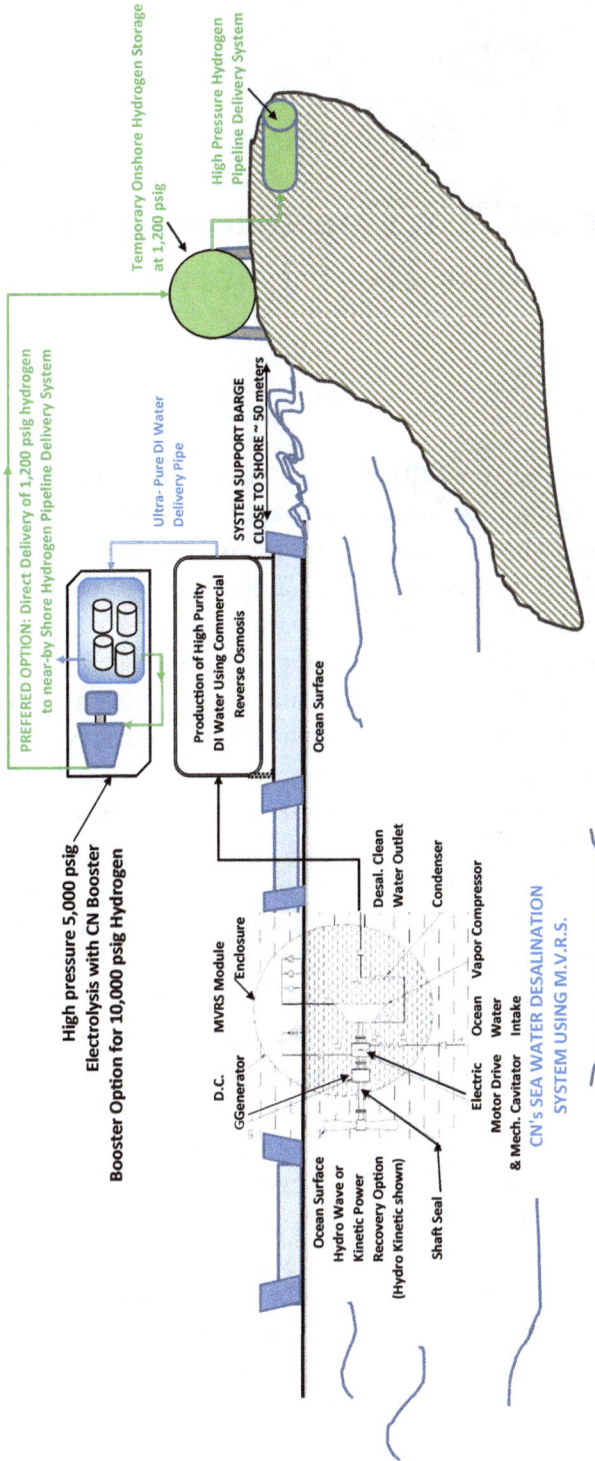

Figure C10.6. <u>Alternative</u> configuration of the proposed hydrogen production system with lower pressure hydrogen delivery to on-shore pipeline network.

Case Study 11

Desiccant Air Conditioning Using Solar Heat Energy

In the late 1970s and early 80s, there were many innovative applications of thermodynamics using renewable energy in the United States. Such activities were encouraged and almost mandated by the oil embargo during the 1970s that forced the reduction of oil consumption in USA. Although the embargo had been lifted within a year, consumer confidence in the constant supply of oil had already been eroded. The Government started and funded many programs to explore ways of reducing oil consumption in USA and to encourage renewable energy resources. A large number of energy conservation programs were initiated, including projects involving the recovery of exhaust gas waste heat from long haul diesel vehicles as described in Case Study 3.

Another project was the use of solar energy to produce air conditioning. It is very reasonable to expect that the Department of Energy (DOE) would be focused on using renewable energy to power air conditioning cycles given the need for air conditioning throughout the commercial and industrial sectors in the United States. The industries in the southern states are very much dependent on the availability of air conditioning for the comfort of the workers. Case Study 7 described a solar-powered air conditioning system involving the Rankine Cycle, which required compressors, pumps and heat exchangers. Like Case Study 7, this case study also details the use of solar energy as the heat input to system; however, the system in question completely avoids the use of compressors and pumps that are required in typical air conditioning systems that rely on the vapor compression cycle. In place of the multiple heat exchangers used in the solar-powered air conditioning system involving the Rankine Cycle, only a single, robust, atmospheric pressure air-to-air heat exchanger — called a rotary regenerator — is used in a cycle that utilizes the thermodynamic properties of moist air and a desiccant that is an absorber of moisture in a thermodynamic cycle called a Dunkie Cycle. The desiccant material coats the walls of the rotary regenerative heat exchanger.

C11.1 The working principle of the Dunkie Cycle

Using low-grade heat energy — for instance, solar energy or low temperature waste heat — for air conditioning is a matter of processing the hot, humid air, taken from the space that is to be cooled, through a sequence of state points that are best represented on a psychrometric chart. The thermodynamics of water-laden air as the subject of psychrometric analysis has been discussed in Chapter 6. One of the major thermodynamic cycles that use the processes identified in Chapter 6 and the input of only low-grade heat (i.e., requiring temperatures less than 120°F) is known as the Dunkie Cycle. Figure C11.1 shows an application of the Dunkie Cycle in an air conditioning system, with solar energy as the heat input to the system.

Figure C11.2 illustrates the processes, traced on a psychrometric chart, that make up the Dunkie, open cycle. The cycle consists of several heat exchangers, several humidifiers, a single dehumidifier and a heat source. The dehumidifier is a heat exchanger known as the rotary regenerator wheel which contains an active desiccant bed. Desiccants are materials which adsorb and desorb typically water moisture from the surrounding air. Zeotropic desiccants are most commonly used in luggage or packages that must be kept free of moisture. Desiccants often appear in the form of small pre-packaged packets that accompany fine leather products such as briefcases.

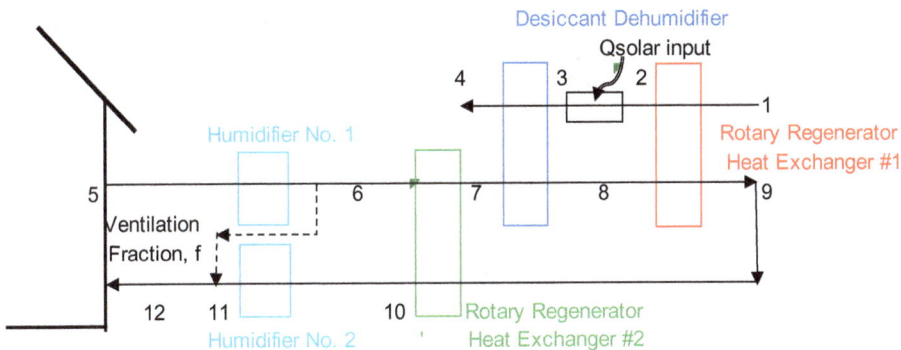

Figure C11.1. A typical Dunkie Cycle applied to an air conditioning system, with solar energy as the heat input.

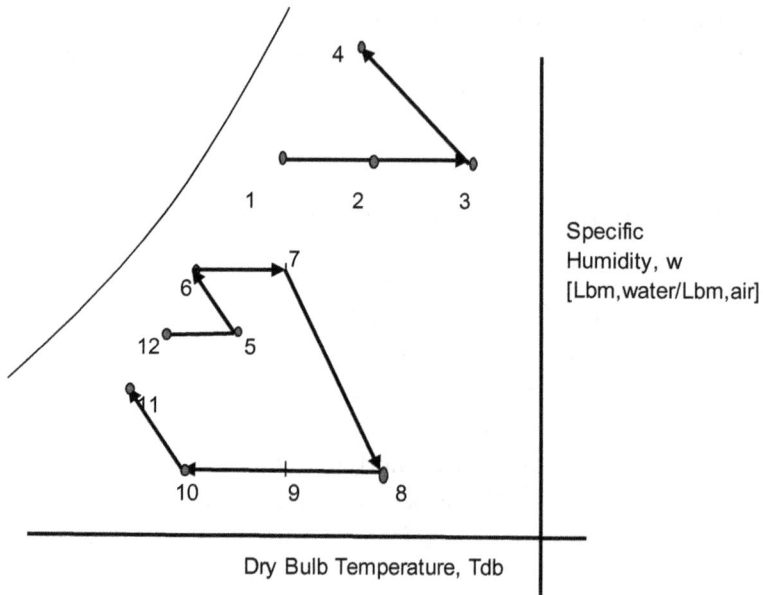

Figure C11.2. The thermodynamics processes that define the Dunkie Cycle shown in Figure C11.1.

C11.2   Thermodynamics analysis of the Dunkie Cycle

The thermodynamics processes shown in Figure C11.2, along with the descriptions of these processes provided in Chapter 6 can be used to determine the temperatures and enthalpies for each of the 12 state points shown in Figure C11.2. As discussed in the previous case studies and throughout this textbook, determination of heat transfer or work consumed (or produced) can be carried out using the enthalpy values at the inlet and outlet of the device. The components that constitute the Dunkie Cycle can be identified as the heat exchanger mechanical device; one of the seven basic mechanical systems used in all thermodynamics systems, as described in Chapter 3.

C11.2.1   *Effects of the humidifier and dehumidifier on specific (w) and relative (φ) humidity*

The humidifier and dehumidifier — and certainly the application of a desiccant material as part of the construction of the dehumidifier — require some additional clarification as to their (respective) effects on the specific humidity, w [lbm,water/lbm,air], and the dry bulb temperature at the inlet and outlet of the heat exchanger. Recalling the instruction given in Chapter 6, the humidification (i.e., the addition of water to a stream of dry air) or the dehumidification of moist air (i.e., the removal of some water content from moist air) is a matter of recognizing that the enthalpy of the humid air at the inlet of the humidifier/dehumidifier is equal to the enthalpy of the humid air at the outlet of the device. The enthalpy of the moist air is a function of the specific humidity, w, and the dry bulb temperature as determined from Equation (C11.1).

$$h = 0.24 \times T_{db} + w \times (1061 + 0.444 \times T_{db}). \qquad \text{(C11.1)}$$

The specific humidity, w [lbm,water/lbm,air], can be calculated using Equation (C11.2). As may be observed from Equation (C11.2), the specific humidity, w, is a function of the partial pressure of the water in the atmospheric air. The partial pressure of the water can be determined by knowing the relative humidity of the air, ϕ, and the saturated water vapor pressure, p_{satv}, which is only a function of the dry bulb temperature, T_{db}.

$$w = 0.622 \times \phi \times p_{satv}(T_{db})/(14.696 - \phi \times p_{satv}(T_{db})). \qquad \text{(C11.2)}$$

C11.2.2 *Defining the state points of the Dunkie Cycle*

The state principle can be applied in the analysis of the Dunkie Cycle. Using this principle, only two thermodynamics properties need to be known to determine the rest of the thermodynamic properties for each of the 12 state points that constitute the Dunkie Cycle. As indicated in Chapter 6, there are two additional common temperature properties: wet bulb (T_{wb}) and dew point temperatures (T_{dp}), as well as the relative (ϕ) and specific (w) humidity.

A property table for the analysis of a Dunkie Cycle, shown in Table C11.1, can be constructed via the application of Equations (C11.1) and (C11.2), along with the First and Second Law applied to each of the mechanical elements that constitute the cycle that can be broadly identified as "heat exchangers". The obvious difference between this property table for the Dunkie Cycle and those that have been used for the thermodynamic cycles presented in Chapters 3 and 4 as well as many of the case studies, is the inclusion of the two "new" temperatures: T_{wb}, T_{dp} and the two humidity values: specific humidity, w and relative humidity, ϕ). The dry bulb temperature, T_{db} used in this property table is identical to the typical temperature used throughout this text but without a clarifying subscript. Certainly, the enthalpy at each state point is also included in the property table. The enthalpy is necessary to apply the First Law of Thermodynamics to each heat exchanger component. The only difference here is that the enthalpy

Table C11.1. Property table showing the state points for the analysis of the Dunkie Cycle shown in Figures C11.1 and C11.2.

Heat Exchanger Effectiveness #1= 93%				
Heat Exchanger Effectiveness#2= 42%			$\Delta T_{,6-7}=$ 30	
Effectiveness of Humidifier No. 1 95%			$\Delta T_{,9-10}=$ 15	
Effectiveness of Humidifier No. 2 95%				
Effectiveness of Desiccant Dehumidifier= 85%			Cooling Capacity [Btu/h]= 12,000	
M_1/M_6 = 1.2				
M_5/M_6 = 0.7			M_5 [lbm/h]= 4,062	
Ventilation Mixture [M_6/M_{12}]= 0.3				
C.O.P.,sensible= 0.17				

		INPUTS		OUTPUTS			INPUT	OUTPUT	
		R.H.	Tdb	w	h	Tdew point	Twb	w,wet bulb	h, iterated
			F	lbm,v/lbm,a	Btu/lbm,dair	F	F	lbm,v/lbm,a	Btu/lbm,dair
Calc. #1	1	98.0%	95	0.0358	62.3	94.7	95.1	0.0367	63.3
Calc. #7	2	37.8%	128	0.0358	70.8	94.7	99.7	0.0426	71.0
Calc. #9	3	22.7%	163	0.0532	99.5	113.1	113.1	0.0651	99.5
Calc. #10	4	67.3%	121	0.0532	88.2	113.0	108.5	0.0564	88.6
Calc. #2	5	60.0%	75	0.0112	30.3	60.7	65.4	0.0134	30.3
Calc. #4	6	97.5%	66	0.0133	30.3	66.3	65.4	0.0134	30.3
Calc. #5	7	75.0%	96	0.0278	53.7	87.3	88.5	0.0296	53.8
Calc. #6	8	15.5%	138	0.0184	53.7	74.8	88.5	0.0296	53.8
Calc. #3	9	47.2%	98	0.0184	43.8	74.9	65.6	0.0135	30.4
Calc. #8	10	75.4%	83	0.0184	40.1	74.8	70.0	0.0158	34.0
Calc. #11	11	96.0%	78	0.0196	40.1	76.7	44.0	0.0061	17.1
Calc. #12	12	98.5%	69	0.0152	33.3	69.0	65.3	0.0134	30.2

includes the energy content of the water vapor, h_w, as presented in Chapter 6, and the dimensional unit for the enthalpy changed from Btu/lbm to Btu/lbm, dry air to give proper reference to the basis upon which the total enthalpy energy is based. The boxed parameters shown in Table C11.1 are inputs to the Dunkie Cycle. Temperature differences for the heat exchanger processes taking place between state points 6 to 7 and 9 to 10, are known at the onset of the analysis. It is noted that the effectiveness of the heat exchangers is calculated using Equations C11.3(a) and C11.3(b).

$$\varepsilon_1 = \frac{(T_8 - T_9)}{(T_1 - T_8)}, \tag{C11.3a}$$

$$\varepsilon_2 = \frac{(T_6 - T_7)}{(T_8 - T_6)}. \tag{C11.3b}$$

Usually, the effectiveness, ϵ_1 and ϵ_2, are used as inputs to the thermodynamics solution of the temperatures. However, the Dunkie Cycle uses the temperature differences for the heat exchanger processes, ΔT_{6-7} and ΔT_{9-10}, as inputs and then uses Equations C11.3(a) and C11.3(b) to calculate the heat exchanger effectiveness. To avoid oversizing the heat exchangers, the heat exchanger effectiveness should be limited to less than 90%.

The leftmost column and highlighted cells in Table C11.1 help to identify the sequence of the analysis of the Dunkie Cycle which consists of

12 processes, a relatively large number compared to the other thermodynamics cycles (typically heat engines or refrigeration cycles) that have been
studied in this book. The calculation of each of the individual state points
is as follows, starting with state point 1. State point 1 is clearly defined
with the input of the relative humidity and the dry bulb temperature. Equations (C11.1) and (C11.2) can be directly applied to determine the other
properties at state point 1. State points 2 and 3 must have the same specific humidity, w, as state point 1. Similarly, state points 5 and 9 have two
thermodynamics properties, T_{db} and RH, that enable the other state points
to be directly calculated using Equations (C11.1) and (C11.2) or, as always,
using a psychrometric chart.

The definition of state point 6 is based on the knowledge that the process
between state points 5 and 6 is a humidification process. The humidification
process increases the specific humidity, w, of the moist air at state point 5.
The wet bulb temperature corresponding to state point 5 places a minimum
temperature limit on state point 6. This limitation is observed more clearly
by referencing Figure C11.2. It is this limit, $T_{5,wb}$, that is used to determine the temperature, T_6, of state point 6 using Equation C11.4(a) with an
assumed value for the effectiveness of the humidifier.

$$\varepsilon_{humidifier,1} = \frac{(T_5 - T_6)}{(T_5 - T_{5,wb})}. \tag{C11.4a}$$

The same defining equation is used to determine the effectiveness of the
second humidifier shown in Figure C11.2, but with the necessary changes
in the temperature state points, in Equation (C11.4b). This effectiveness
equation is used to determine the temperature at the state point, T_{11}, with
state point 10 completely defined and a value for the effectiveness of the
humidifier provided as an input. The effectiveness must be less than 1, and
a value of 90% is commonly used for humidifiers.

$$\varepsilon_{humidifier,2} = \frac{(T_{10} - T_{11})}{(T_{10} - T_{10,wb})}. \tag{C11.4b}$$

The effectiveness of the humidifier is defined as the fraction of the total
possible temperature change of the moist air with the wet bulb temperature
set as the limit. The humidification of state points 3 to 4 follow the same
format as detailed above for temperatures, T_6 and T_{11}, with state point 3
being completely defined.

State points 7 and 8 are linked by the dehumidification process, which
is the reverse of the humidification process. Thus, state points 7 and 8
share a common enthalpy. With the properties of the state point 7

known, state point 8 can be defined using Equation (C11.4), as shown in Equation C11.4(c). This also includes solving for T_8 by an algebraic method:

$$\varepsilon_{dehumidifier,1} = \frac{(T_8 - T_7)}{(T_8 - T_{7,wb})}; \quad \text{therefore:} \quad T_8 = \frac{(T_{7,wb} - T_7)}{(1 - 1/\varepsilon)}.$$

$$(C11.4c)$$

The mixture of the two air streams at state points 6 and 11, with each stream having a different amount of water content as determined by their respective specific humidities, w_6 and w_{11}, is performed by conserving the amount of water in both streams, before applying the first law equation. Thereafter, the water content of state point 12 is determined from a mass conservation equation shown in Equation (C11.5):

$$w_{12} = \dot{M}_6 \times w_6 + \dot{M}_{11} \times w_{11}, \qquad (C11.5)$$

$$h = \dot{M}_6 \times h_6 + \dot{M}_{11} \times h_{11}. \qquad (C11.6)$$

The results of the analysis shown in Table C11.1 were calculated using Excel. Other similar spreadsheet software platforms are also capable of providing similar results. One of the attractive features of using a spreadsheet is that the analyst must continue to be a part of the solution. That is, an iteration is performed manually by the analyst to complete the cycle calculations and ensure that the energy and mass balances are maintained. For this purpose, the differently-highlighted spreadsheet cells help the analyst to easily identify the properties which need to be matched between the pair of properties that define the process. This method is particularly helpful when using a simple spreadsheet solution which only has Equations (C11.1) and (C11.2) programmed into it, with the assumption that the relative humidity and the dry bulb temperature are already known. A more sophisticated psychrometric computer model that allows for the definition of state points would simply require any two thermodynamics properties to be known in order to calculate the others.

C11.2.3 *Performing calculations to determine the COP for the Dunkie Cycle and the dry air-flow rate required*

The last calculations to be performed are to determine the COP for the Dunkie Cycle process and the magnitude of the flow rate of air through the cycle that is needed to produce the required cooling capacity [Btu/h, kW or RT (1 refrigeration ton = 12,000 Btu/h] for the system. Equations (C11.7) and (C11.8) are used in these calculations.

$$C.O.P. = \frac{(T_{12} - T_5)}{(T_2 - T_3)}, \tag{C11.7}$$

$$\dot{M}_5 = \frac{\dot{Q}_{cooling}}{(h_{12} - h_5)}. \tag{C11.8}$$

It is quickly observed in Table C11.1 that the COP is very small. A "sanity check" of this small value is necessary and, as noted in Chapter 2, recommended for all thermodynamics analyses. The procedure used to determine if this COP value is reasonable (Equation (C11.7)) is based on calculating the COP for a Carnot Refrigerator based on an Exergy or Availability analysis using the same cycle operating temperatures. An exergy analysis for an adsorption chiller system is reviewed in Chapters 4 and 5. The ideal COP for the Dunkie Cycle can be calculated using the following equation:

$$COP_{\cdot ideal} = \frac{\dot{Q}_c}{\dot{Q}_h} \times \frac{1 - \frac{T_c}{T_{solar}}}{1 - \frac{T_c}{T_{ambient}}}, \quad \text{(from Equation (4.17))}$$

where T_c refers to the cold air temperature produced by the cycle.

For the parameters shown in Table C11.1, the ideal COP is calculated to be 0.57 compared to that of the cycle, which is equal to 0.17. The second law efficiency, which is the ratio between these two values, is equal to 30%. Thus, it may be concluded that the Dunkie Cycle analysis that results in a calculated COP of 0.17 is a reasonable result compared to the ideal COP of 0.57.

Case Study 12

Carbon Dioxide Capture and Sequestration by Integrating Pressure Swing Adsorption with an Open Supercritical CO_2(sCO_2) Brayton Cycle

C12.1 Background of the concept

There has been continuous and increasing concern surrounding the need to reduce fossil fuel-generated carbon emissions that are entering the atmosphere. The COP21 Conference was successful in achieving consensus from 196 countries that the release of carbon, in the form of carbon dioxide (CO_2) and carbon monoxide (CO), must be curtailed.

Climate scientists have shown that the carbon emissions from the combustion of fossil fuels for the generation of electric power can most significantly be controlled using these three strategies:

1. *Conservation efforts, which curtail the amount of power required*
2. *A shift in power generation from fossil fuels to renewable energy sources*
3. *Efficient capture and sequestration of carbon dioxide, the byproduct gas resulting from fossil fuel combustion, particularly coal combustion*

Although the first two strategies are currently being implemented, the fact remains that fossil fuels will still be needed as sources for power generation for many years to come. This leads us to the third strategy for the resolution of the CO_2 dilemma, while maintaining reliable power generation. The USA's Department of Energy (DOE) has been at the forefront of supporting innovative approaches to enable coal combustion at high cycle efficiencies, while also capturing the CO_2 generated by power plants. Once captured, the CO_2 can then be sequestered. Considering also the high costs involved in the deployment of current CO_2 reduction techniques, the technical objective of several research programs, sponsored by DOE has also been to reduce the initial cost of systems that effectively capture, reduce, or sequester carbon dioxide.

An example of a cycle currently being developed with the support of the DOE for the removal of harmful emissions is the Allam Cycle. The Allam cycle utilizes an oxygen separation process to remove nitrogen from the air, and then uses the oxygen for combustion with pulverized coal. The oxygen is first pressurized to 4,000 psia, and then burned with coal to temperatures of 700°C. With the removal of nitrogen, the products of combustion almost fully comprise of CO_2. This high-pressure CO_2 fluid stream is expanded through a turbine before it is permanently sequestered into the ground at pressures of 1,100 psia. Rather than a closed cycle, this system should be considered as an open-flow, supercritical CO_2 (sCO_2) power-generation system.

Reduction of CO_2 emissions through sequestration is an energy-intensive process, requiring as much as 10–15% of a utility's net electric power output. To address this, Concepts NREC (CN) has developed a hybrid cycle which combines an sCO_2 Brayton Cycle with a Pressure Swing Adsorption (PSA) process.

C12.2 Working principles of the hybrid $(sCO_2)^2$ cycle

Although both the Allam system and the hybrid cycle developed by CN utilize CO_2 recovered from fuel combustion in the power plant as the working fluid in a subsequent open-flow sCO_2 cycle, there is one important distinction between them: the hybrid cycle developed by CN can be retrofitted onto existing coal-fired, power generation plants until renewable energy power generation systems become the dominant source of utility power. It is thus described in thermodynamics terms as a 'topping cycle'. CN has labeled this hybrid system "$(sCO_2)^2$".

The goal of the new $(sCO_2)^2$ system is to significantly reduce the amount of electric power required by the CO_2 sequestration compressors by approximately 30–40% of the typical power required. This case study also reveals that this new hybrid cycle could further reduce the amount of a utility's net electric power consumed by CO_2 sequestration by up to 80% if a CO_2 pressure letdown turbine is used during the underground vault-filling process. For a 500 MWe power plant, this amounts to an operational cost savings of US$80 million over a 20-year period.

C12.3   Feasibility study of proposed $(sCO_2)^2$ system

The proposed $(sCO_2)^2$ cycle is shown in Figures C12.1 and C12.2, along with performance parameters associated with the integration of a 10 mMWe sCO_2/PSA system into a 335-MWe coal-fired power plant. Figures C12.1(a) and (b) illustrate the proposed $(sCO_2)^2$ cycle, using a reheat SCO_2 turbine. Figure C12.2 provides a preliminary specification for the speed and diameter of the single stage SCO_2 turbine. Figure C12.3 illustrates the same hybrid system with its turbine inlet temperature changed from 700°C to 900°C. However, the exhaust heat temperature from the power plant's coal fired steam boilers are not changed.

When the reheat turbine is used in the cycle, the net power generated by the sCO_2 process is 7.2 MWe. This translates to a 35% reduction in the net power output required for the compression process in the sequestration of the CO_2. Even if a reheat turbine was not used, a reduction of as high as 30% in the amount of a utility's net power output required for the sequestration process is still possible. This percentage can further increase to as high as 55% if higher turbine inlet temperatures can be made available. This results in a decrease in the cost per kWe ($/kW) of the initial capital and operation and maintenance (O&M) costs of the system. The additional recovery of power using "letdown" turbines from the expansion of the CO_2 pressure as it enters the partially filled underground vaults results in an 80% reduction in the power typically used by the primary sequester compressor.

Figure C12.2(a) provides the details of the turbine's design specifications of speed and diameter. The basis for this first-order feasibility analysis is the use of standardized Ds-Ns charts as shown in Figure C12.2(b). The specific speed (Ns) and specific diameter (Ds) are dimensionless parameters that are used in the turbomachinery industry to estimate the diameter and speed of the largest stage in the turbine for turbines that have more than one stage. The Ds and Ns are defined using the isentropic enthalpy drop and volume flow rate of the fluid according Equations (C12.1) and (C12.2) below. The reader is cautioned to carefully check the units and the nomenclature used against those used in the Ds-Ns graph to ensure consistency in the use of units.

$$Ds = D \times \Delta h_s^{0.25}/\dot{V}^{0.5} \tag{C12.1}$$

$$Ns = N \times \dot{V}^{0.5}/\Delta h_s^{0.75} \tag{C12.2}$$

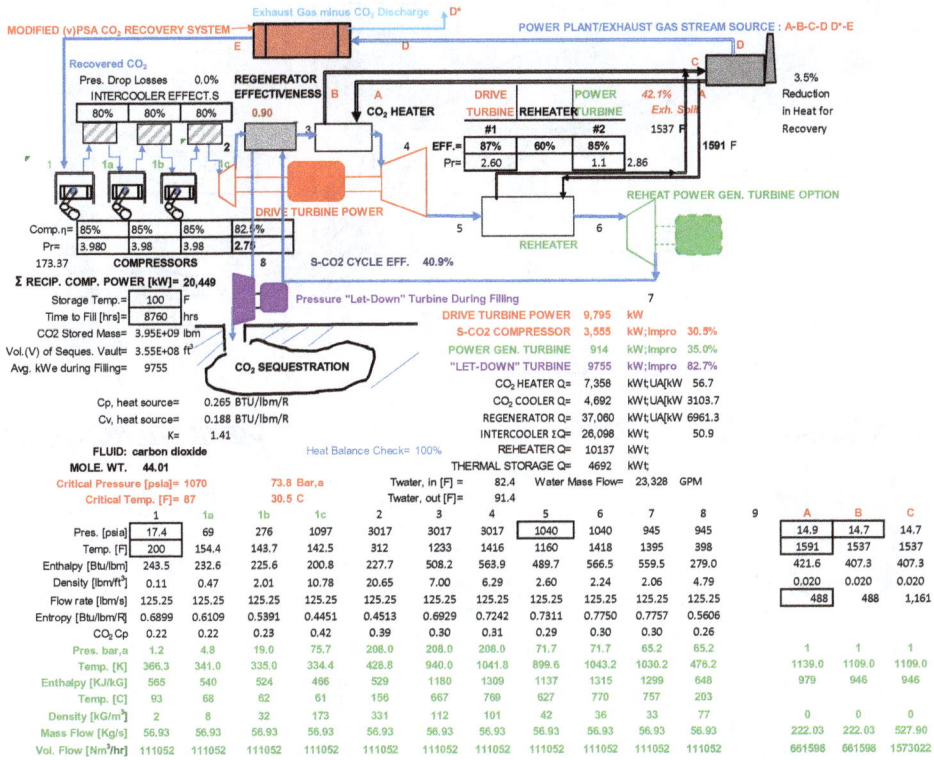

Exhaust Gas minus CO$_2$ Discharge D*

MODIFIED (v)PSA CO$_2$ RECOVERY SYSTEM

POWER PLANT/EXHAUST GAS STREAM SOURCE : A-B-C-D D*-E

Recovered CO$_2$

Pres. Drop Losses	0.0%	REGENERATOR				DRIVE	POWER	42.1%				3.5%
INTERCOOLER EFFECT.S		EFFECTIVENESS	B	A		TURBINE	REHEATER	TURBINE	Exh. Split			Reduction
80%	80%	80%	0.90	CO$_2$ HEATER		#1		#2	1537 F			in Heat for
						EFF.= 87%	60%	85%		1591 F		Recovery
						Pr= 2.60		1.1	2.86		

DRIVE TURBINE POWER

Comp.η=	85%	85%	85%	82.5%
Pr=	3.980	3.98	3.98	2.76
173.37	COMPRESSORS			

REHEATER

REHEAT POWER GEN. TURBINE OPTION

Σ RECIP. COMP. POWER [kW]= 20,449

Storage Temp.=	100	F
Time to Fill [hrs]=	8760	hrs
CO2 Stored Mass=	3.95E+09 lbm
Vol.(V) of Seques. Vault=	3.55E+08 ft³
Avg. kWe during Filling=	9755

Pressure "Let-Down" Turbine During Filling

DRIVE TURBINE POWER	9,795	kW
S-CO2 COMPRESSOR	3,555	kW; Impro 30.5%
POWER GEN. TURBINE	914	kW; Impro 35.0%
"LET-DOWN" TURBINE	9755	kW; Impro 82.7%
CO$_2$ HEATER Q=	7,358	kWt; UA[kW 56.7
CO$_2$ COOLER Q=	4,692	kWt; UA[kW 3103.7
REGENERATOR Q=	37,060	kWt; UA[kW 6961.3
INTERCOOLER ΣQ=	26,098	kWt; 50.9
REHEATER Q=	10137	kWt
THERMAL STORAGE Q=	4692	kWt

Cp, heat source=	0.265 BTU/lbm/R
Cv, heat source=	0.188 BTU/lbm/R
K=	1.41

FLUID: carbon dioxide

Heat Balance Check= 100%

MOLE. WT. 44.01

| Twater, in [F] = | 82.4 | Water Mass Flow= 23,328 GPM |
| Twater, out [F]= | 91.4 |

Critical Pressure [psia]= 1070   73.8 Bar,a

Critical Temp. [F]= 87   30.5 C

	1	1a	1b	1c	2	3	4	5	6	7	8	9	A	B	C
Pres. [psia]	17.4	69	276	1097	3017	3017	3017	1040	1040	945	945		14.9	14.7	14.7
Temp. [F]	200	154.4	143.7	142.5	312	1233	1416	1160	1418	1395	398		1591	1537	1537
Enthalpy [Btu/lbm]	243.5	232.6	225.6	200.8	227.7	508.2	563.9	489.7	566.5	559.5	279.0		421.6	407.3	407.3
Density [lbm/ft³]	0.11	0.47	2.01	10.78	20.65	7.00	6.29	2.60	2.24	2.06	4.79		0.020	0.020	0.020
Flow rate [lbm/s]	125.25	125.25	125.25	125.25	125.25	125.25	125.25	125.25	125.25	125.25	125.25		488	488	1,161
Entropy [Btu/lbm/R]	0.6899	0.6109	0.5391	0.4451	0.4513	0.6929	0.7242	0.7311	0.7750	0.7757	0.5606				
CO$_2$ Cp	0.22	0.22	0.23	0.42	0.39	0.30	0.31	0.29	0.30	0.30	0.26				
Pres. bar,a	1.2	4.8	19.0	75.7	208.0	208.0	208.0	71.7	71.7	65.2	65.2		1	1	1
Temp. [K]	366.3	341.0	335.0	334.4	428.8	940.0	1041.8	899.6	1043.2	1030.2	476.2		1139.0	1109.0	1109.0
Enthalpy [KJ/kG]	565	540	524	466	529	1180	1309	1137	1315	1299	648		979	946	946
Temp. [C]	93	68	62	61	156	667	769	627	770	757	203				
Density [kG/m³]	2	8	32	173	331	112	101	42	36	33	77		0	0	0
Mass Flow [Kg/s]	56.93	56.93	56.93	56.93	56.93	56.93	56.93	56.93	56.93	56.93	56.93		222.03	222.03	527.90
Vol. Flow [Nm³/hr]	111052	111052	111052	111052	111052	111052	111052	111052	111052	111052	111052		661598	661598	1573022

S-CO2 CYCLE EFF. 40.9%

CO$_2$ SEQUESTRATION

(a)

(s-CO$_2$)^2 Open Brayton Cycle with Heat Source and Sequestration

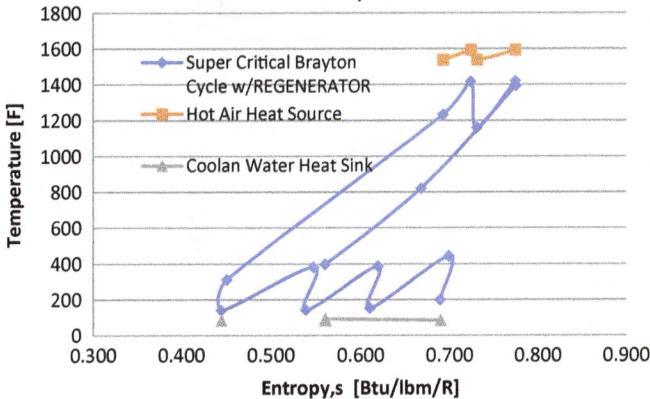

Legend:
— Super Critical Brayton Cycle w/REGENERATOR
— Hot Air Heat Source
— Coolan Water Heat Sink

Temperature [F] vs Entropy,s [Btu/lbm/R]

(b)

Figure C12.1.   (a) Proposed hybrid (sCO$_2$)^2 cycle at 700°C inlet turbine temperature, with reheat turbine for sequestration of CO$_2$ from coal-fired utility power plants; (b) T-s diagram for the sCO$_2$ cycle shown in Figure C12.2(a).

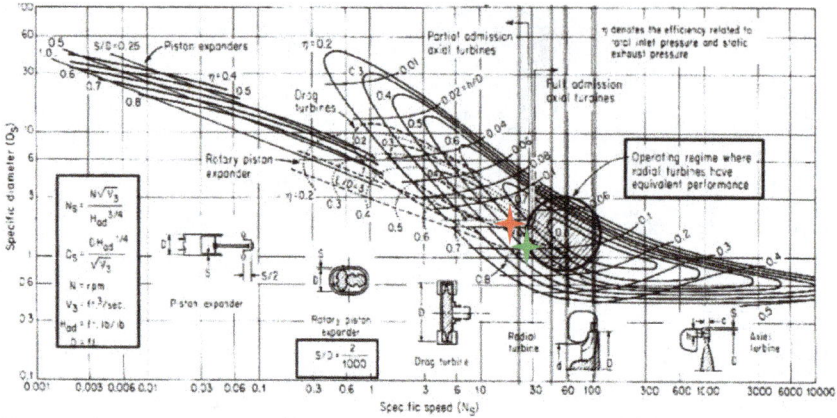

CONCEPTS NREC

(a)

TURBINE No. 2

V3 [cft/s]=	24.461	Flow Rate	3.24E+05 lbm/hr	40.92 kg/s
Had [ft-lbf/lbm]=	25158	P,in	1645 psia	11.34 Mpa
Had [kJ/kg]=	75	T,in	1330 F	994.0 K
D [ft]=	0.492	P,out	1096.67 psia	7.563 MPa
N [rpm]=	30,000	P,stage,out	1097 psia	7.563 MPa
No. of Stages=	1		85% Turbine Eff.	
Ns=	74.3	Δh,issen.=	75.2 KJ/Kg	
Ds=	1.25			2614 kW
Tip Speed [m/s]=	236	Ns=	16.51	
Tip Speed [ft/s]=	773	Ds=	2.41	

V/Us= 0.61

0.150	Dia (m)	773 ft/s
30000	RPM	
1	No. of Stg.s	

TURBINE No. 1

V3 [cft/s]=	21.341	Flow Rate	3.24E+05 lbm/hr	40.92 kg/s
Had [ft-lbf/lbm]=	37970	P,in	3126 psia	21.56 Mpa
Had [kJ/kg]=	113	T,in	1259 F	954.6 K
D [ft]=	0.492	P,out	1645 psia	11.345 MPa
N [rpm]=	40,000	P,stage,out	1645 psia	11.345 MPa
No. of Stages=	1		87% Turbine Eff.	
Ns=	67.9	Δh,issen.=	113.4 KJ/Kg	
Ds=	1.49			4038 kW
Tip Speed [m/s]=	314	Ns=	14.10	
Tip Speed [ft/s]=	1030	Ds=	3.69	

V/Us= 0.66

1564

0.150	Dia (m)	1030 ft/s
40000	RPM	
1	No. of Stg.s	

(b)

Figure C12.2.   (a) Ds-Ns diagram for turbine design; (b) Turbine design specifications for the 700°C cycle shown in Figure C12.1.

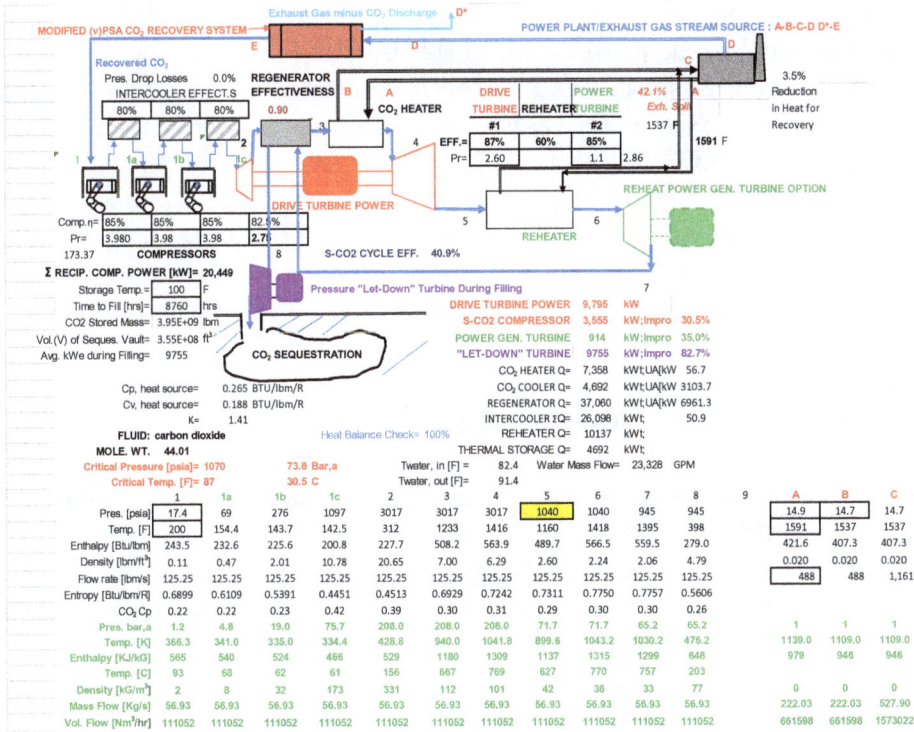

Figure C12.3.   A $(sCO_2)^2$ system with a 900°C turbine inlet temperature.

C12.4   Description of the sCO$_2$/PSA cycles in the $(sCO_2)^2$ system

The proposed new $(sCO_2)^2$ system consists of a conventional sCO$_2$ cycle coupled with a Pressure Swing Adsorption (PSA) cycle. A typical primary sCO$_2$ turbo-compressor was used, along with two additional turbines; one of which served as an sCO$_2$ reheat turbine, while the other turbine models a pressure-reduction or "letdown" turbine which can recover energy from the pressure differential between the exit of the sCO$_2$ turbine (74 bar,a) and the partial pressure in the remote ground sequestration vault until the storage volume is filled to its design pressure.

Figure C12.1(a) shows the hybrid $(sCO_2)^2$ system, optimized to capture and sequester CO$_2$ gas using a minimum of net parasitic power at the lowest cost per kW (\$/kWe), for a 235-MWe-rated power plant that burns coal as its primary fuel. A cost reduction is made possible by reducing the rating of the motor that is used to drive the CO$_2$ reciprocating compressor and electrical sub-systems that would otherwise be required to power the CO$_2$

reciprocating compressor. Note that net power generated by the sCO_2 system is proportional to the power rating of the coal-fired power plant.

The hybrid cycle first uses a PSA process to recover CO_2 from the exhaust gas stream. The recovered CO_2 is then utilized as the working fluid in an sCO_2 cycle. The discharge of the sCO_2 cycle is still approximately 1,100 psia and is readily available to be permanently sequestered into an underground geophysical vault via a vault-filling process. During the vault-filling process, the sCO_2 passes through a letdown turbine which recovers a significant portion of the energy used for compression, thus generating additional power. The filling continues until the pressure in the vault increases to the acceptable storage level.

The features of the $(sCO_2)^2$ cycle include:

- An sCO_2 cycle that uses, as a working fluid, the same CO_2 that is recovered from the utility waste heat source.
- A PSA system that removes CO_2 gas from the exhaust gas waste stream. The best choice for CO_2 adsorbents that can operate at the temperatures anticipated in the integrated supercritical cycle has been researched by several contemporary and independent chemical engineering researchers (see (Boon, *et al.*, 2014; Grande, Cavenati, & Rodrigues, 2005; Reynolds, Ebner, & Ritter, 2005; Gomes & Yee, 2002; Takamura, Narita, Aoki, Hironaka, & Uchida, 2001). There is consensus that K-promoted hydrotalcite, when used in a high-pressure, high-temperature PSA system, can effectively recover CO_2 from a utility power plant's exhaust gas at higher-than-typical operating temperatures. In addition, a novel application of a pressure letdown turbine can be integrated into the proposed hybrid system. If a pressure reduction turbine is used as part of the PSA process, a reduction of 10% of the power typically used in the vacuum pumping of the PSA sequence can be obtained.
- A reheat turbine, which is added to the cycle to improve the sCO_2 cycle's efficiency to 57%.
- Recovery of the high-pressure energy (used to generate power) from the sCO_2 turbine discharge, which occurs as the pressurized CO_2 leaves the open sCO_2 cycle and begins to fill the underground sequestration vaults. The suggested letdown turbine would be installed at the remote, sequestration site and not at the site of the coal power plant unless, of course, the sequestration ground vault is located at the power plant.

It is proposed that the $(sCO_2)^2$ cycle functions as a topping cycle which operates at the highest combustion temperatures that a power plant can generate while having only minimal impact on the continuous power generation of existing steam boilers positioned downstream from the planned primary sCO$_2$ heater. To ensure this, the temperature of the combustion exhaust gases that are returned to the power plant is only reduced by 50–100°F. The sCO$_2$ cycle efficiency is 40–45%, which is higher than that of typical older, coal-fired power plants. Therefore, using the (sCO$_2$)^2 system as a topping cycle, the power conversion efficiency of the power plant can be slightly increased in two ways:

1. Less of the power generated by coal combustion is required to offset the power required for the CO$_2$ compression process.
2. The extra coal that is burned is burned at a higher cycle efficiency.

As shown in Figure C12.1(a), approximately 40–45% of the combustion gases are diverted to the reheat heat exchanger to increase the inlet temperature of the reheat turbine, and the balance is directed to the primary sCO$_2$ turbine. This combustion heat stream is then utilized by the power plant, to generate electric power in a Rankine Cycle which includes feed water heaters and reheat turbines.

The exhaust from the power plant is then directed to the PSA system which removes CO$_2$ from the products of combustion. The PSA process used for CO$_2$ capture has also been successfully used in a variety of industrial gas processes for the removal of gases (such as hydrogen). As may be observed from Figure C12.1(a), the proposed sCO$_2$ cycle is a regenerative Brayton Cycle and not a recompression sCO$_2$ cycle, which would typically be used with waste heat of much higher temperature. The sCO$_2$ cycle utilizes the captured CO$_2$ gas from the exhaust products generated by the combustion process in the power plant.

The recovered CO$_2$ is then directed to the sequestration compressors, shown in Figure C12.1 as multi-stage, intercooled reciprocating compressors. The compression of the CO$_2$ from the atmosphere to 1,000 psia is most effectively accomplished using positive displacement (reciprocating) compressors, due to the relatively low volume flow rate and high-pressure ratio requirements for compression. Such CO$_2$ compressors are currently available for purchase and installation from many commercial vendors, without any additional research and development required.

The intercooling of the CO$_2$ between the compressor stages is strongly recommended unless there is an inadequate amount of ambient water available or a high local ambient air cooling. This intercooling improves the compression efficiency, and thus reduces the work of compression. The use of a "let down" turbine before the CO_2 stream is sequestered into the storage vault increases the energy storage density of the storage system by reducing the temperature of the CO_2 at the discharge of the "let down" turbine which consequently increases the density of the CO_2. Thus, more CO_2 mass can be stored in the same volume of the sequestration vault. In the proposed hybrid sCO_2/PSA cycle, the compressed CO_2 proceeds to the sCO_2 axial or radial compressor, where its pressure can be boosted to supercritical pressures in preparation for the sCO_2 power generation cycle.

The use of the compressed CO_2 as the working fluid in an open sCO_2 cycle is unique. This open cycle design eliminates the need for a very large sCO_2 cooler that is otherwise required in a typical closed sCO_2 cycle. Also, the open cycle design allows any CO_2 working fluid that leaks past shaft seals to be easily recovered, depressurized and returned to the inlet of the PSA process for continued service. This design feature consists of appropriately-designed labyrinth seals or equivalent less-expensive seals, thus eliminating the need for very expensive dry gas seals.

There is consensus (Steeneveldt, Berger, & Torp, 2006) that future sequestration of CO_2 will need to be done at a great distance from the source of the CO_2, given the limited number of geophysical (naturally-occurring) vaults in the ground. For this to be successful, the CO_2 would need to be compressed at its source, and then transported in pipelines. This is similar to how natural gas is transported at pressures of approximately 700 psia, from oil fields or liquid natural gas ports, to the site of demand where its pressure is reduced to local (end-of-the-line) pipeline pressure of 40 psig.

In this approach of transportation, the pipes carrying CO_2 are small in diameter, which makes them inexpensive. The CO_2 is transported at very high, supercritical pressures. Similar to the case of natural gas transportation, once the CO_2 arrives at the empty and/or partially pressurized geophysical vault, the pressure is reduced, or "let down", until the vault is filled.

In a typical CO_2 sequestration process, there is an empty hole in the ground that needs to be filled. The filling continues until the pressure in the vault increases to the acceptable storage level, or until the gas is absorbed into the ground substrate, or both. During this filling process, a pressure letdown turbine can recover approximately another 30% of power.

C12.5 Summary

The proposed sCO_2/PSA hybrid system will enable the cost-effective use of a CO_2 sequestration system in a fossil fuel power plant. The $(sCO_2)^2$ cycle operates as a "topping cycle", enabling the recovery of heat from the power plant's heat source that is at the highest combustion temperatures to provide a partial power assist to drive the commercially-available sequestration CO$_2$ compressors. Concepts NREC conducted a feasibility study which has demonstrated the ability of a 10 MWe sCO$_2$/PSA system to service a 235 MWe coal-fired power plant. The proposed hybrid system enables CO$_2$ sequestration at 1,000 psig and provides up to 80% of the necessary CO$_2$ compression power required for CO$_2$ sequestration.

References

Boon, J., Cobden, P. D., van Dijk, H. A., Hoogland, C., van Selow, E. R., & van Sint Annaland, M. (2014). Isotherm model for high-temperature, high-pressure adsorption of CO2 and H2O on K-promoted hydrotalcite. *Chemical Engineering Journal, 248,* 406–414. doi: https://doi.org/10.1016/j.cej.2014.03.056

Gomes, V. G., & Yee, K. W. (2002). Pressure swing adsorption for carbon dioxide sequestration from exhaust gases. *Separation and Purification Technology, 28*(2), 161–171. doi: https://doi.org/10.1016/S1383-5866(02)00064-3

Grande, C. A., Cavenati, S., & Rodrigues, A. E. (2005). Pressure swing adsorption for carbon dioxide sequestration. *Proceedings of the 2nd MERCOSUR Congress on Chemical Engineering.* Rio de Janeiro.

Reynolds, S. P., Ebner, A. D., & Ritter, J. A. (2005). New pressure swing adsorption cycles for carbon dioxide sequestration. *Adsorption, 11,* 531–536. doi: https://doi.org/10.1007/s10450-005-5980-x

Steeneveldt, R. L., Berger, B., & Torp, T. A. (2006). CO2 Capture and storage: Closing the Knowing–Doing gap. *Chemical Engineering Research and Design, 84*(9), 739–763. doi: https://doi.org/10.1205/cherd05049

Takamura, Y., Narita, S., Aoki, J., Hironaka, S., & Uchida, S. (2001). Evaluation of dual-bed pressure swing adsorption for CO2 recovery from boiler exhaust gas. *Separation and Purification Technology, 24*(3), 519–528. doi: https://doi.org/10.1016/S1383-5866(01)00151-4

Thermodynamics Principles Applied to the "Drinking Bird" Toy

The "Drinking Bird" that is shown in Figure C13.1 is an iconic toy which can be used to promote the understanding of several thermodynamic processes. Its intentionally comical appearance belies an appreciation for the lessons that it can teach about applied thermodynamics. A careful study of this toy through the lens of thermodynamics reveals it to be a very good example of the application of the Rankine Cycle, albeit across a very small temperature difference and with no net power produced.[1] The head chamber of the Drinking Bird serves as the condenser, while the body chamber acts as the evaporator. Although the Drinking Bird does not seem to use a feed pump and turbine, closer observation reveals that fluid is being pumped upward from the bottom chamber through a transparent glass tube via thermal expansion, and evidence of power generation is demonstrated by the continuous oscillation of the Drinking Bird. Any number of mechanical systems could be designed to utilize this oscillating motion to produce a continuous power that can wither mechanical or be transformed into electric power.

Let us break this thermodynamics system down into its critical components, before transform the Drinking Bird into a functional power generation system. It is helpful to first identify the three distinct sequences that constitute the cyclic operation of the Drinking Bird.

Figure C13.1 shows the system during the evaporating process of the fluid in the bottom chamber which causes the fluid to rise in the feed tube. The Bird's beak is made of a moisture-laden fabric which holds water the way

---

[1]It is more accurate to say that the power that is produced by this heat engine is consumed by the friction at the support, and the air drag friction that is exerted on the Bird's head and body are equal and thus cancel each other out, so as to provide no additional energy that would accelerate the system to ever higher oscillation speeds.

Figure C13.1.   The Drinking Bird shown with its head about to move downward due to the filling of the feed tube connecting the body-chamber to the head-chamber.

a washcloth holds water until it dries. The water is pulled up via capillary action through the glass tube, into the fabric-lined Bird's head. The damp fabric is used to cool the glass beak-like chamber, reducing its temperature to the wet-bulb temperature; a temperature less than the ambient temperature. A detailed explanation of wet bulb temperature is given in Chapter 6. The wet-bulb temperature is produced through the evaporation of the liquid water from the beak cloth; a process requiring energy derived from the enthalpy of the fluid inside the beak and head of the bird. The reduction in the temperature of the fluid in the beak results in a slight reduction in the pressure in the beak and head chamber which has the consequence of helping to draw up the fluid from the bottom "body" chamber and fill the top chamber.

Simultaneous to the evaporative cooling of the vapor-state fluid that is contained in the head-chamber, the body-chamber of the Drinking Bird is being heated by the ambient temperature. The oscillating movement of the bottom chamber in the warmer ambient temperature aids the heating of the working fluid inside the bottom chamber by causing an increase in the heat transfer coefficient between the ambient air and the glass bulb. The liquid fluid changes its density very quickly as it is heated due to its low vapor pressure. The low vapor pressure with respect to the ambient temperature is the

Figure C13.2.　(Left) The Drinking Bird at the moment when the bottom chamber is at its lowest point; (Middle) When the Bird's head dips into the "drinking cup". Here,the dipping of the head has caused a vapor-opening between the head-chamber and the bottom body-chamber, which enables the liquid to return by gravity to the bottom-chamber; (Right) When the sudden flow of fluid back to the body-chamber causes a force imbalance that results in the characteristic swinging of the Drinking Bird.

physical property which makes the working fluid ideal for this application. The heating thus causes the liquid to quickly rise into the tube that connects the body chamber with the head chamber. The rising liquid starts to cause the Drinking Bird to rotate in a counterclockwise direction as the weight of the rising fluid changes the balance of the tube. This is best seen in Figure C13.2. This process continues until the liquid in the bottom-chamber is depleted just enough to open a vapor fluid connection between the top and bottom chambers. The photo in Figure C13.2 (Right) shows how the weight of the working fluid that is rising in the glass tube causes the Drinking Bird to tip until it is almost horizontal. This horizontal orientation creates a vapor opening between the head-chamber and the bottom-chamber that enables the trapped vapor in the top chamber to be released into the bottom chamber, which immediately allows the liquid in the glass feed tube to drain back into the body of the Drinking Bird as shown in Figure C13.2 (Right), which causes the Drinking Bird to return to its original vertical position, due to the weight of the liquid mass that has returned to the bottom chamber. It is this sudden imbalance in forces that causes the Drinking Bird to continue to pendulum-swing when the Drinking Bird is in the up-right position. This swinging motion is very important as it helps to increase the rate of heat transfer via convection from the ambient into the liquid fluid in the bottom-chamber, while also helping to evaporatively cool the vapor fluid in the head-chamber of the Drinking Bird.

Once set into motion, the Drinking Bird appears to be unstoppable — an example of a perpetual motion machine. However, it can be easily demonstrated by simple tests that without gravity to cause the liquid to re-fill the bottom chamber, or without water in the "drinking" cup for the Bird to pick up, the swinging ("drinking") motion of the Drinking Bird will eventually stop. It can also be demonstrated that the frequency of the oscillation of the Drinking Bird is dependent on the temperature difference between the head and bottom-chambers. This temperature difference can be increased by placing the Drinking Bird toy in a very dry ambient air environment. As indicated in Chapter 6, the dry bulb temperature of dry air can be much higher than the corresponding wet-bulb temperature, depending on the relative humidity of the dry air.

It is also interesting to note that the amount of mechanical work that is produced by the Drinking Bird is exactly equal to the magnitude of the drag and kinematic friction that is exerted on the Drinking Bird. The drag force is caused by the displacement of the ambient air due to the swinging motion of the body of the Drinking Bird. The kinematic friction force is caused by the mathematical product of the weight of the body and the coefficient of friction, μ, at the point of contact with the knife-edge of the fulcrum. The fulcrum provides the point support for the Drinking Bird that enables the expected "see-saw" action of the Drinking Bird. Although very small in magnitude compared to the drag friction, the friction force is the mathematical product of the weight and the coefficient of friction between the steel knife edge and the plastic support. The energy (or power) that this system can produce can be increased by allowing more of the expanding liquid fluid from the bottom-chamber to fill the glass feed tube, causing the glass tube to be moved into a horizontal position. The counter-clockwise motion of the head–bottom chamber system can be used to produce power by using a piston (or a rack) and pinion gear system connected to the right side of the fulcrum, where it provides an opposing force that the weight of the rising liquid must overcome before the system is moved to a horizontal position. That opposing force is moved through a distance that is equal to the chord of the arc of motion and thus produces recoverable work. This is best illustrated by using an engineering representation of the Drinking Bird system as shown in Figure C13.3.

The theoretical efficiency of the Drinking Bird system can be determined from the Carnot Equation by using the temperature of the ambient, T_{amb}, and the wet bulb temperature, T_{wb}, as shown in Equation (C13.1).

$$\eta_{Carnot} = 1 - \frac{T_{wb}}{T_{ambient}}. \tag{C13.1}$$

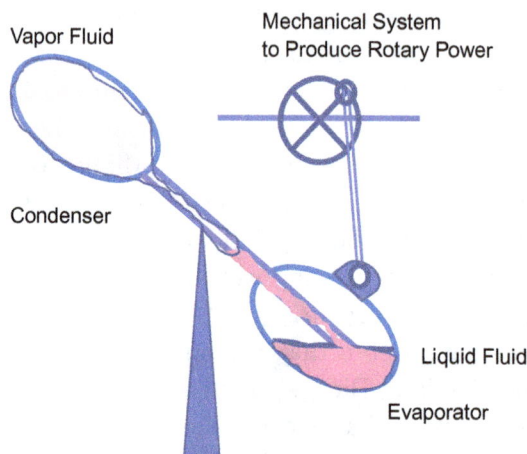

Figure C13.3. An engineering sketch of the Drinking Bird's power take-off system that provides net power.

The wet-bulb temperature is determined, as discussed in Chapter 6, by referring to the psychrometric chart shown in Figure C13.4. The methodology for determining the wet-bulb temperature using Figure C13.4 was previously described in detail in Chapter 6. For this case study, the ambient temperature is taken to be 85°F and the relative humidity to be 40%. With this state point thermodynamically well-defined and with reference to Figure C13.4, the wet-bulb temperature is found to be 67°F. Based on this, the Carnot Cycle efficiency is calculated to be 3.3%. Since this Drinking Bird system is likely to have very high mechanical losses due to the drag and contact friction as presented above, the actual cycle efficiency is likely to be less than 1%. However, the power that is produced is basically "free", as the Drinking Bird does not consume fossil fuels, and is only powered by the ambient temperature.

Using another equation shown in Equation (C13.2), the amount of energy lost to the environment can also be calculated. This equation can then be used in another equation, Equation (C13.3), to estimate the amount of friction energy consumed by the Drinking Bird. The amount of heat rejected by the condenser of this system (i.e., from the head-chamber) is simply the amount of energy needed to evaporate the amount of water removed from the drinking glass. The amount of water removed from the glass can be measured by using a graduated cylinder or a simple mass balance. The amount of heat rejected by the condenser of the system can then be found

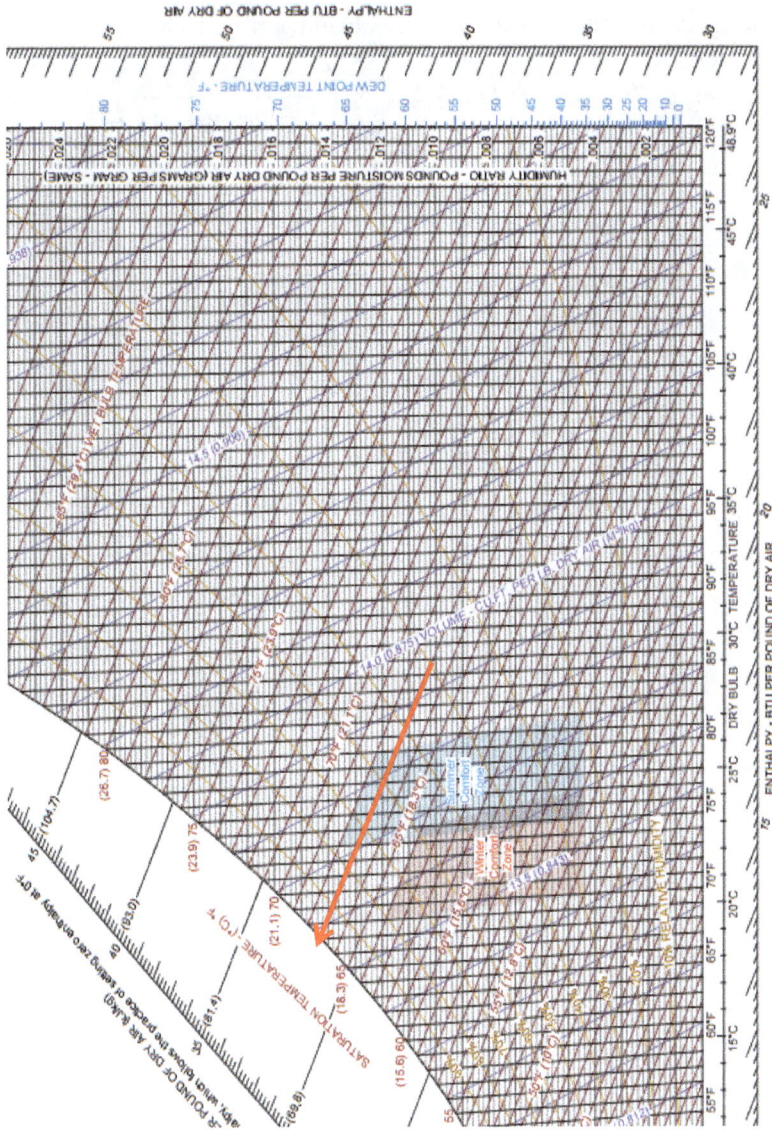

Figure C13.4. Psychrometric chart demonstrating a colder wet bulb temperature (60°F) for the head chamber compared to the drier ambient temperature (80°F) for the bottom chamber.

by Equation (C13.2).

$$Q_{condenser} = \Delta h_{fg} \times M_{evaporated\ water}. \tag{C13.2}$$

Assuming the system is not ideal and therefore operates at a fraction, f, of the Carnot efficiency, then the amount of friction energy, $E_{friction}$, consumed by the Drinking Bird system is found from Equation (C13.3).

$$Friction\ Energy\ (E_{friction}) = Q_{condenser} \times \left(\frac{\eta_{Carnot} \times f}{1 - \eta_{Carnot} \times f} \right). \tag{C13.3}$$

Refrigeration Cycle with an Ejector

The development of efficient and inexpensive air conditioning systems has been a boon to industrial development in arid and hot locations around the world. It is not an exaggeration to say that many states making up the Southern United States have been industrialized since the early 1900s due to the invention of the Carrier cooling system. The system is based on the vapor recompression system also known as a reverse Rankine Cycle system or Vapor Compression System. The vapor recompression system was reviewed in Chapter 5. Two systems are commonly used for refrigeration purposes: the basic vapor compression system and the economizer vapor compression system. These are shown in Figures C14.1 and C14.2.

An interesting modification to the basic refrigeration cycle is to substitute the vapor compressor with an ejector, as shown in Figure C14.3.

An ejector is a device with no moving parts, which can function as a small vacuum pump using only the kinetic energy (in the form of velocity) and potential energy (in the form of pressure) of a motive stream. A schematic of an ejector is shown in Figure C14.4. The high pressure, motive stream is shown entering the plane section labelled "1" (this is equivalent to the stream labelled "6c" in Figure C14.3), and the lower pressure stream that is to be suctioned enters at the section labeled "2" (this is equivalent to the stream labelled "1" in Figure C14.3). The outlet stream is a thermodynamic mixture of the motive stream and the suctioned stream that is maintained at an intermediate pressure that is shown exiting the ejector at the plane section "C" in Figure C14.4 (this is equivalent to stream "2" in Figure C14.3). The exhaust stream at "C" is at a slightly higher pressure than the stream entering the suction stream at "2". Thus, the ejector is able to draw the suction stream at "2" toward the exit at plane section "C" without the need for a vacuum pump which would require additional power input. An ejector is very often used in steam Rankine Cycles to draw out non-condensable gases (typically air) from the condenser. These non-condensable gases accumulate

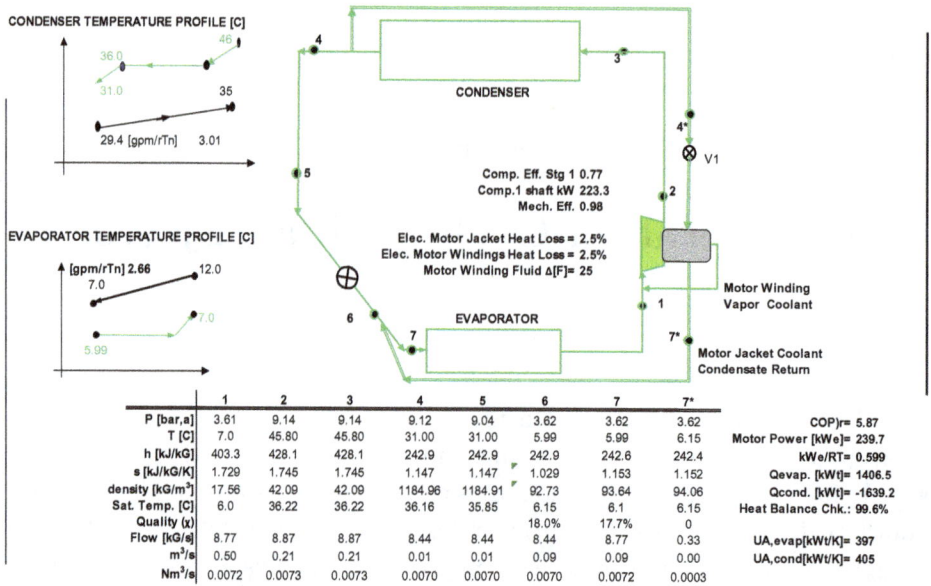

	1	2	3	4	5	6	7	7*
P [bar,a]	3.61	9.14	9.14	9.12	9.04	3.62	3.62	3.62
T [C]	7.0	45.80	45.80	31.00	31.00	5.99	5.99	6.15
h [kJ/kG]	403.3	428.1	428.1	242.9	242.9	242.9	242.6	242.4
s [kJ/kG/K]	1.729	1.745	1.745	1.147	1.147	1.029	1.153	1.152
density [kG/m³]	17.56	42.09	42.09	1184.96	1184.91	92.73	93.64	94.06
Sat. Temp. [C]	6.0	36.22	36.22	36.16	35.85	6.15	6.1	6.15
Quality (x)						18.0%	17.7%	0
Flow [kG/s]	8.77	8.87	8.87	8.44	8.44	8.44	8.77	0.33
m³/s	0.50	0.21	0.21	0.01	0.01	0.09	0.09	0.00
Nm³/s	0.0072	0.0073	0.0073	0.0070	0.0070	0.0070	0.0072	0.0003

COP)r= 5.87
Motor Power [kWe]= 239.7
kWe/RT= 0.599
Qevap. [kWt]= 1406.5
Qcond. [kWt]= -1639.2
Heat Balance Chk.: 99.6%
UA,evap[kWt/K]= 397
UA,cond[kWt/K]= 405

Figure C14.1. Refrigeration cycle that is similar to Figure 5.25, but with the refrigerant also used in the cooling of the electric motor with two cooling streams: one that cools the generator housing and the other to cool the generator windings.

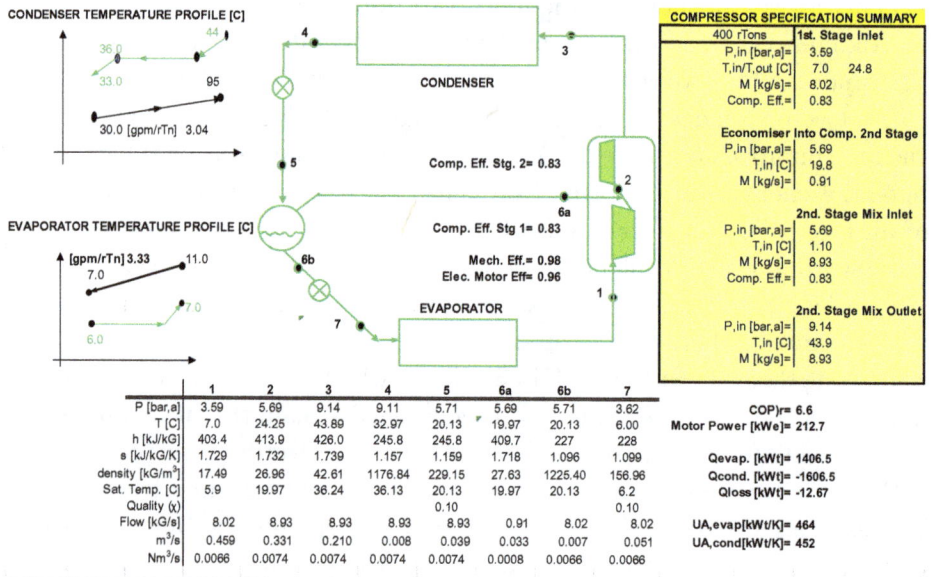

COMPRESSOR SPECIFICATION SUMMARY

400 rTons	1st. Stage Inlet	
P,in [bar,a]=	3.59	
T,in/T,out [C]=	7.0	24.8
M [kg/s]=	8.02	
Comp. Eff.=	0.83	

Economiser Into Comp. 2nd Stage
P,in [bar,a]= 5.69
T,in [C] 19.8
M [kg/s]= 0.91

2nd. Stage Mix Inlet
P,in [bar,a]= 5.69
T,in [C] 1.10
M [kg/s]= 8.93
Comp. Eff.= 0.83

2nd. Stage Mix Outlet
P,in [bar,a]= 9.14
T,in [C] 43.9
M [kg/s]= 8.93

	1	2	3	4	5	6a	6b	7
P [bar,a]	3.59	5.69	9.14	9.11	5.71	5.69	5.71	3.62
T [C]	7.0	24.25	43.89	32.97	20.13	19.97	20.13	6.00
h [kJ/kG]	403.4	413.9	426.0	245.8	245.8	409.7	227	228
s [kJ/kG/K]	1.729	1.732	1.739	1.157	1.159	1.718	1.096	1.099
density [kG/m³]	17.49	26.96	42.61	1176.84	229.15	27.63	1225.40	156.96
Sat. Temp. [C]	5.9	19.97	36.24	36.13	20.13	19.97	20.13	6.2
Quality (x)					0.10			0.10
Flow [kG/s]	8.02	8.93	8.93	8.93	8.93	0.91	8.02	8.02
m³/s	0.459	0.331	0.210	0.008	0.039	0.033	0.007	0.051
Nm³/s	0.0066	0.0074	0.0074	0.0074	0.0074	0.0008	0.0066	0.0066

COP)r= 6.6
Motor Power [kWe]= 212.7
Qevap. [kWt]= 1406.5
Qcond. [kWt]= -1606.5
Qloss [kWt]= -12.67
UA,evap[kWt/K]= 464
UA,cond[kWt/K]= 452

Figure C14.2. A Cascade (Economizer) cycle: a vapor compression refrigeration cycle using a cascade system to improve its COP.

CONDENSER TEMPERATURE PROFILE [C]

36.0 47
33.0 35
30.0 [gpm/rTn] 3.33

CONDENSER
Economiser Comp. Eff.= 0.706
Comp.1 shaft kW= 6.3

239.947164
3338.70161
3578.64877
3578.64877

ECONOMIZER

EVAPORATOR TEMPERATURE PROFILE [C]

[gpm/rTn] 2.66
7.0
6.9
8.0

12.0

Heater
Q [kWt]= 1539
26.0 C
Pump Power [kWm]=
14.6

EVAPORATOR

6a 6c
EJECTOR

3338.74
3338.70
-264.52
1747.77
1746.13
264.52

7.10
6.54
4.99
7.87% 99.5983964

COMPRESSOR SPECIFICATION SUMMARY

75 rTons	1st. Stage Inlet	
P,in [bar,a]=	3.59	
T,in/T,out [C]	7.10	30.2
M,total [kg/s];Vol. [m³/s]=	1.54	0.088
(Mgap [kg/s])=	0.0327	)
Comp. Eff.=	0.741	
	Economiser Into Comp. 2nd Stage	
P,in [bar,a]=	6.24	1.735
T,in [C]	22.6	
M,economizer,6a [kg/s]=	0.126	
1 stg.Comp.Temp.out [C]=	30.2	
	2nd. Stage Mix Inlet	
P,in [bar,a]=	6.24	
T,in [C]	29.84	
Comp. Eff.=	0.706	
	2nd. Stage Mix Outlet	
P,in [bar,a]=	9.14	1.485
T,out [C]	47.3	
M at 3 & M at 7* [kg/s]=	1.700	0.097
4*, M,jacket cool.[kg/s]=	0.129	

	1	2	3	4	5	6a	6b	6c	7
P [bar,a]	3.59	9.13	9.14	9.11	6.26	6.26	6.26	31	3.62
T [C]	7.1	125.66	111.38	32.97	23.13	23.13	23.13	171	6.00
h [kJ/kG]	403.4	510.9	503.8	245.8	245.8	411.3	231.7	544.0	231.8
s [kJ/kG/K]	1.730	1.977	1.939	1.157	1.158	1.717	1.110	1.967	1.114
density [kG/m³]	17.48	30.08	31.56	1176.84	299.00	30.44	1214.30	101.18	131.84
Sat. Temp. [C]	5.9	36.20	36.22	36.13	23.13	23.13	23.13	88.0	6.2
Quality (x)						8.0%			12.1%
Flow [kG/s]	1.54	6.54	7.10	7.10	7.10	0.57	1.54	4.99	1.54
m³/s	0.09	0.22	0.23	0.01	0.02	0.02	0.00	0.05	0.01
Nm³/s	0.0013	0.0054	0.0059	0.0059	0.0059	0.0005	0.0013	0.0041	0.0013

t value is outside limits

COP)r= 0.17
Motor Power [kWe]= 20.8

Qevap. [kWt]= 263.7
Qcond. [kWt]= -1832.5
Qheater [kWt]= 1538.9
Heat Balance Chk.: 99.9%
UA,evap[kWt/K]= 439
UA,cond[kWt/K]= 82

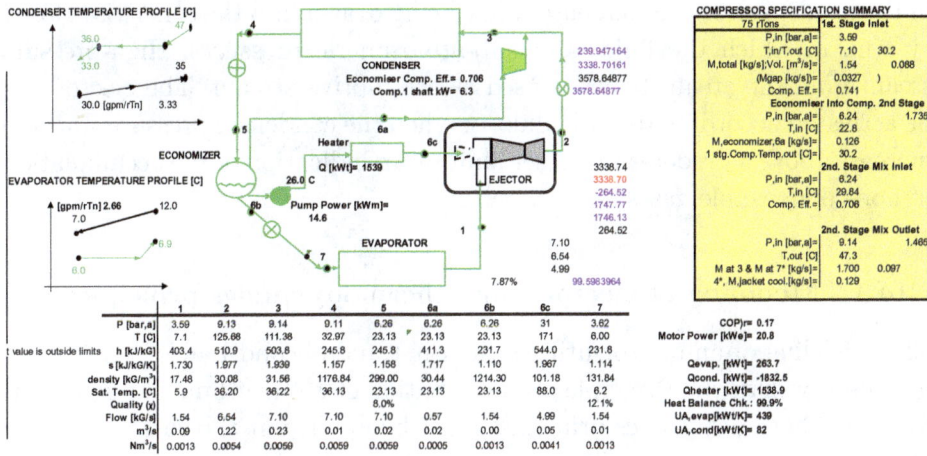

Figure C14.3. Refrigeration cycle with an ejector pump operating in parallel to the system's main compressor.

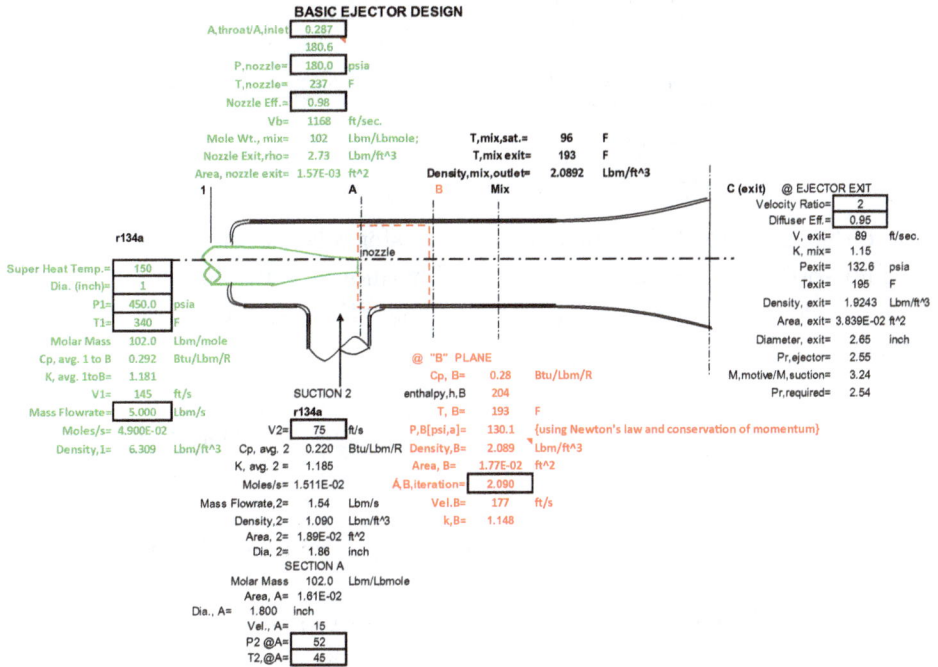

BASIC EJECTOR DESIGN

A,throat/A,inlet 0.287
180.6
P,nozzle= 180.0 psia
T,nozzle= 237 F
Nozzle Eff.= 0.98
Vb= 1168 ft/sec.
Mole Wt., mix= 102 Lbm/Lbmole;   T,mix,sat.= 96 F
Nozzle Exit,rho= 2.73 Lbm/ft^3   T,mix exit= 193 F
Area, nozzle exit= 1.57E-03 ft^2   Density,mix,outlet= 2.0892 Lbm/ft^3

1   A   B   Mix

r134a

Super Heat Temp.= 150
Dia. (inch)= 1
P1= 450.0 psia
T1= 340 F
Molar Mass 102.0 Lbm/mole
Cp, avg. 1 to B 0.292 Btu/Lbm/R
K, avg. 1toB= 1.181
V1= 145 ft/s
Mass Flowrate= 5.000 Lbm/s
Density,1= 6.309 Lbm/ft^3

@ "B" PLANE
Cp, B= 0.28 Btu/Lbm/R
enthalpy,h,B 204
T, B= 193 F
P,B[psi,a]= 130.1 (using Newton's law and conservation of momentum)
Density,B= 2.089 Lbm/ft^3
Area, B= 1.77E-02 ft^2
A,B,iteration= 2.090

SUCTION 2
r134a
V2= 75 ft/s
Cp, avg. 2 0.220 Btu/Lbm/R
K, avg. 2= 1.185
Moles/s= 1.511E-02

Mass Flowrate,2= 1.54 Lbm/s
Density,2= 1.090 Lbm/ft^3
Area, 2= 1.89E-02 ft^2
Dia, 2= 1.86 inch

Vel.B= 177 ft/s
k,B= 1.148

SECTION A
Molar Mass 102.0 Lbm/Lbmole
Area, A= 1.61E-02
Dia., A= 1.800 inch
Vel., A= 15
P2 @A= 52
T2,@A= 45

C (exit)   @ EJECTOR EXIT
Velocity Ratio= 2
Diffuser Eff.= 0.95
V, exit= 89 ft/sec.
K, mix= 1.15
Pexit= 132.6 psia
Texit= 195 F
Density, exit= 1.9243 Lbm/ft^3
Area, exit= 3.839E-02 ft^2
Diameter, exit= 2.65 inch
Pr,ejector= 2.55
M,motive/M,suction= 3.24
Pr,required= 2.54

Moles/s 4.900E-02

Figure C14.4. Thermodynamics model of ejector used in the refrigeration cycle shown in Figure C14.3.

in the headspace of the condenser after being drawn into the otherwise closed system, in which the fluid is at sub-atmospheric pressures. High-pressure steam from the steam boiler is used as the motive stream. The ejector can be activated according to a schedule, or when the condenser pressure starts to increase above a predetermined set limit — an indication of the accumulation of non-condensable gases.

C14.1 Modeling of ejector using thermodynamics principles

The modeling of an ejector utilizes the First and Second Law of Thermodynamics as well as the Principle of Conservation of Momentum. The equations for these thermodynamics principles must be applied at the correct sections at the inlet and outlet of the ejector and also within the ejector, specifically at the plane sections "A" and "B".

Recalling that the First Law is expressed as:

$$\sum \dot{Q} + \sum \dot{M} \times (h + KE + PE)_{inlets}$$

$$= \sum \dot{W} + \sum \dot{M} \times (h + KE + PE)_{outlets} + \frac{\partial U}{\partial t})_{cv}$$

$$\dot{Q} = -\sum \dot{Q}_k,$$

where the subscript "k" indicates heat transfer relative to the surroundings.

The application of the First Law of Thermodynamics for the nozzle section positioned between the plane sections "1" and "A" is given by the following equation.

$$C_p \times T_1 \times \left\{ 1 - \left(\frac{P_a}{P_1}\right)^{((k-1)/k)} \right\} \times nozzle\ eff. = \frac{(V_{out}^2 - V_{in}^2)}{2 \times g_c \times J}.$$

where J is a unit conversion factor depending on whether the units used are metric or imperial.

The Second Law of Thermodynamics for an Open Flow System may be applied to determine the total rate of entropy change for the system. That is, if the ejector is performing ideally (not a very realistic assumption) and at steady state, then the heat transfer from the ejector $(\dot{Q}_k) = 0$ *and the* $\frac{\partial S_{system}}{\partial t}$ *term* $= 0$. The isentropic efficiency of the ejector would be identified by the manufacturer such that $(\text{mxs})_{\text{exit}} > (\text{mxs})_{\text{inlet}}$.

$$+\frac{\sum \dot{Q}_k}{T_k} \geq -\left[\sum \dot{m}\, s_{\text{exit}} - \sum \dot{m}\, s_{\text{inlet}} + \frac{\partial S_{system}}{\partial t}\right].$$

The Principle of Conservation of Momentum is applied to the sections marked by planes "A" and "B", as seen in the following equation:

$$P_1 \times A_1 - P_A \times A_A = (\dot{m} \times V_A - \dot{m} \times V_1)/g_c.$$

The application of these equations in modelling the ejector is shown in Figure C14.4. The ejector spreadsheet is programmed on a second worksheet within the Excel workbook model. The cells in the ejector model (Figure C14.4) that are bordered contain the inputs required for the equations. The inputs for the ejector model are taken from the vapor compression refrigeration cycle model that is shown in Figure C14.3.

The refrigeration cycle shown in Figure C14.3 has been designed to have the ejector replace a fraction of the function of the refrigerant compressor. Due to its limited ability to provide the total compression ratio (P_{out}/P_{in}) at the total cycle flow rate, the ejector cannot fully replace the compressor. The design of the refrigeration cycle in Figure C14.3 solves this problem by having the ejector provide the total required pressure ratio but at a fraction of the full flow rate. The other components of the cycle remain identical to those of a typical vapor recompression refrigeration system, as shown in Figure C14.1, or the more advanced refrigeration cycle, identified as the Cascade (Economizer) Cycle, shown in Figure C14.2.

The complete cycle analysis shown in Figure C14.3 indicates a Coefficient of Performance (COP; which is calculated as follows: $\frac{\dot{Q}_{cooling}}{W_{compressor}}$) of 0.17 compared to a COP of 5.7 for the more conventional cycle as shown in Figure 5.25 of Chapter 5. This very large disparity in the COP performance is due to the need for a considerable amount of refrigerant fluid — 5 kg/s — to be used as the motive fluid as represented by arrow 6c in Figure C14.3. Clearly, the use of an ejector to substitute a part of the function of the typical compressor is not recommended.

However, a refrigeration system with an ejector component can still be considered as a viable alternative to the typical vapor recompression refrigeration system, if the condenser pressure was reduced to enable a lower pressure ratio (Pr) and if there was no alternative source of external power. The requirements for its consideration would be met if the ambient temperature was lower and if the cycle was needed in a very remote location, such as in an emergency isolation chamber.

However, if the ambient temperature was below 72°F, there would not be a need for an air conditioning system in the first place. In this scenario, an option would be to have the ejector refrigeration cycle operate as a heat pump instead. For this option to be viable, the evaporation temperature must not be much lower than the 6°C shown in Figure C14.3, as any lower temperature would result in a lower evaporator pressure, and consequently a higher Pr.

Case Study 15

Analysis of an Advanced Compressed Air Energy System with Continuous Onsite Power Augmentation via the Air-Brayton Cycle[1]

A thermodynamics analysis of an advanced Compressed Air Energy System (CAES) for Distributed Power Generation (DPG), which both utilizes turbomachinery for waste heat recovery from onsite prime movers and provides continuous power generation using sources of renewable energy to augment on-site power, is presented. Sources of renewable energy used by the advanced CAES for power augmentation include wind power and solar photovoltaics (PV), in the range of 1,500 to 2,500 kW. The coupling of the recuperation of waste heat energy from the exhaust gas of the existing onsite prime mover and using renewable energy resources to reheat the stored compressed air, enables the continuous generation of power, and thus increases the time that the CAES air-Brayton Cycle is available for power generation. The proposed system utilizes battery storage to maintain high-energy-density storage for the purpose of using the stored electric power only for preheating the stored compressed air and not to supply the grid. This preferred use of stored electric power for only heating purposes would eliminate costly electrical rectifying and inversion systems that are typically used to provide AC power to the electrical grid. Using the stored electrical energy for heating can help stabilize the grid's power generation. Thus, this proposed system may be thought of as a "crossover" system combining CAES technology with battery storage technology but without the need for electrical inversion and synchronizing switchgear, which makes it particularly useful in cases whereby

---

[1] Originally prepared with contributions by Dr. David Japiske, CEO of Concepts NREC, LLC, White River Junction, VT, USA.

the stored electric power is used directly as DC power at an industrial facility. The direct use of stored energy from a battery as heat input for the proposed "crossover" system may also be considered in other applications. For example, an ideal application of the advanced CAES system is in isolated DPG systems in remote sites utilizing "power islands" of renewable energy augmented by onsite fossil fuel prime movers and power generation systems. The proposed "crossover" system enables high reliability, a fast response to transient power loads and the efficient use of renewable energy, as well as heat recovery from onsite conventional prime mover systems.

C15.1 CAES: A state-of-the-art energy storage system

The prevalent energy storage system in use today is the battery energy storage system. Electric batteries have a long history of usage in storing electrical energy and were typically devoted to storing small amounts of energy. The advent of power generation using stored renewable wind and solar energy, and the growing interest in coupling this energy with utility grids has encouraged the development of much larger battery storage systems rated from 3 to 30 MW and 12 to 120 MWh.[2] Lithium-ion batteries are effective in providing spinning reserves but must use solid-state inverters to connect to the utility grid. At 250–690 Wh/l (100–265 Wh/kg), the energy density of a lithium-ion battery has made it the most popular battery for energy storage. Typically, the energy conversion and recovery efficiency for a lithium-ion battery is approximately 70%, over the (average) 1,200 lifetime cycles of the batteries.

Over the past five years, many entrepreneurial companies have been conducting both private and publicly-funded research to design and implement thermomechanical energy storage systems. A common example of thermomechanical energy storage is the CAES system. A CAES is relatively simple in concept, but complicated in design and execution. This is especially true when the CAES is scaled for use by electric utilities to store the electric energy that is available, for future power generation at a reduced cost, when the current demand for the electric power is low. Such systems use axial or reciprocating compressors, depending on the storage pressures and the

---

[2]These ranges are different. The first (3 to 30 MW) refers to power generation but the second to energy produced (MWh). CAES systems are rated according to the amount of energy that can be stored, but there's always interest in knowing how much power can be generated at any time to meet utility demands. Info source: San Diego Gas & Electric Company Proposal and Awards, April 2017.

volumetric size of the storage. For example, plants designed for large-scale compressed air energy storage in the 21st century use very deep caverns in the ground that may have storage pressures of up to 70 bar,a. Designers of early CAES systems conceptualized to operate with competitive Energy Recovery Efficiency (ERE) considered using lower storage pressures of only 13 bar,a to 20 bar,a. The lower pressures were considered due to the limited availability of commercially available compressors that could operate efficiently at the higher pressures and consequential higher temperatures. The energy recovery cycle in early CAES systems typically required the use of a heat input system such as a combustion system that is installed upstream of the pressure let down turbine. Most often, the pressure let down turbine is actually the combustion turbine of an onsite gas turbine (GT) power generation unit and the combustor of the GT serves to preheat the stored compressed air that is released from storage. The preheating of the stored, compressed air must be done, as shown in this case study, to increase the power output of the expansion turbine.

It is generally accepted that a CAES is more economical and efficient, compared to battery storage systems, over the total operating lifetime of a wind turbine or PV system. However, the CAES has a lower energy density than battery storage systems. As such, a major objective in developing the CAES is to increase the energy density (often via latent heat energy storage), and to increase the ERE of the mechanical energy storage system. The ERE is defined in Equation (C15.1):

$$E.R.E. = \frac{\sum \dot{W}_{out} \times \Delta Time_{discharging}}{\sum \dot{W}_{in} \times \Delta Time_{charging} + \dot{Q}_{reheat} \times \Delta Time_{discharging}}$$

where:

$$\dot{W}_{in} = \dot{W}_{net\ compressor} + \dot{Q}_{auxillary\ heat\ input} \times f \qquad (C15.1)$$

C15.2   Common elements of state-of-the-art designs for energy storage systems

State of the art designs for energy storage systems have several common elements. These elements are itemized here with suggestions on how to these features can be improved upon.

1. Air in a vapor phase is typically used as the storage medium for energy storage systems; although cryogenic energy storage, i.e., air that has been cryogenically liquefied to increase its energy density, has recently been

made commercially available with a prototype developed by Concepts NREC (CN). There has been considerable interest in using such phase-change material to store thermal energy in an attempt to achieve a higher specific energy (Btu/lbm or kJ/kg) by taking advantage of either the latent heat of liquification or heat of vaporization. However, the latent heat energy from changing the phase of the fluid is typically used only for low-grade heat energy storage and recovery. The exception to this is the use of molten salts with concentrated solar energy systems wherein temperatures of 600°C or slightly higher are commonplace; but only after many years of funded research by the United States Department of Energy (DOE). Using air as the storage medium in an open energy recovery system is preferable to using phase-change material as the storage medium from the point of view of cost and complexity. In open systems, the compressed air used as the storage medium can be directly released into the inlet of a GT combustor, thus avoiding some or all of the compressor power required by the GT engine. The use of stored compressed air also enables a more rapid start-up of the GT from an idle state to produce power on demand. Decreasing the time between idle speed and full speed and power is considered an important specification in enabling the utility to match fluctuations in the power demand that is experienced by the utility.

2. A very large storage reservoir, which is typically constructed from geophysical phenomena, is used; therefore, most energy storage systems are classified as 'utility-scale', and thus are very costly with respect to the purchase of components, site preparation and installation.

3. One or more compressors to compress the air that is to be stored, and one or more positive displacement or turbomachinery expanders are used to directly recover the high-pressure energy of the air working fluid. Alternatively, the compressed air is directly used in an industrial process that requires pressurized air.

4. The process of energy recovery from the stored fluid must contend with a negative pressure and temperature gradient as the storage medium exits the fixed volume storage reservoir. Thus, the reheating of the medium and some means of flow and/or pressure control is essential to enable the expander to operate most efficiently and with the maximum high specific enthalpy change through the expander. In order to preserve or increase the ERE of the storage system, heat for the reheating must either be derived from a waste heat stream or the prime mover engine must use some form of heat recuperation or regeneration. The recuperation or regeneration of heat energy involves the heat transfer between

the working fluid and a porous substrate matrix via conduction and convection. The porous substrate matrix has energy stored as sensible heat during the "charging" sequence of the energy storage cycle. The recuperator or regenerator heat exchanger that is used for this heat exchange can also be designed to include the recovery of the latent heat of vaporization if the heat exchanger is constructed with a phase change material. The phase change material would be contained in the heat exchanger tubing and the flow of hot air over and around the tubing would change the phase of the material from solid to liquid during the "charging" or heat absorption phase of the cycle. This would then be followed by the flow of cold air across the same heat exchanger tubing during the "energy recovery" phase of the cycle. The selection of a material with a high latent heat of vaporization increases the energy density of the energy storage system.

Several of the more recent and common alternative energy storage systems have been developed by companies including LightSail Energy, Inc., SustainX, Inc., RWE Power AG, Energy Storage, Power LLC, Mitsubishi, and Air Products. Many contemporary developers of energy storage systems integrate modular, sensible or latent heat energy based CAES with wind turbines or PV systems. These CAES systems store thermal energy by injecting water or foam into compressed air, or by using liquefied air directly. However, these systems tend to be expensive to manufacture, involve significant changes to the compressor or turbine designs, and are unable to achieve the high-energy storage densities available with state-of-the-art battery technology.

Other technical similarities between the common CAES systems developed include the use of a heat storage medium to store the work of compression and thus reduce the amount of energy needed during the energy recovery process to reheat the mass flow rate of air into the turbine. Storing the work of compression has the added benefit of reducing the storage temperature, which helps to increase the energy storage density. Achieving the highest energy density in a reliable manner at the lowest cost is a key objective in CAES research. However, it is also understood that the energy density, i.e. the energy per storage volume [kWh/m^3 or Btu/ft^3] of the CAES must also be lowered; an objective which directly translates into the lowering of temperatures and pressures at which a storage vessel operates. At the same time, it is critical that the inlet temperature of the turbine be kept as high as possible during the discharge phase of the energy storage system, so as to improve the turbine efficiency and increase power output.

C15.3   Modeling the proposed "Crossover" concept

A new analytical approach to modeling the performance of a CAES has been adopted, and has been successful in identifying how the energy density of the CAES can be increased at lower cost, while also increasing the turbine inlet temperature. The analysis presented in this case study enables a better understanding of the viability of integrating several forms of renewable energy to achieve the most effective thermomechanical-based energy storage system, by using state of the art battery technology that has been integrated into the CAES.

CN's proposed alternative concept (Figure C15.1) is referred to here as a "crossover" system which integrates a modular CAES with batteries storing power generated by renewable energy. After the stored energy is depleted, the proposed system continues to operate by using the heat energy from the onsite prime mover power generation system. The proposed system is thought to be suitable for use in a DPG system, as compared to the historical CAES that has been designed for utility-scale energy storage. The energy stored in the battery and/or the energy released from the combustion of biomass is used to increase the turbine inlet temperature to maintain an acceptable ERE, as defined in Equation (C15.1) earlier.

Using stored electric power in this manner allows for the use of a smaller battery during the "discharge" sequence of the energy storage cycle, and eliminates the need for power conversion and control electrical systems, such as frequency inverters. In fact, eliminating the cost of such power conversion systems is the only reason for establishing a direct connection

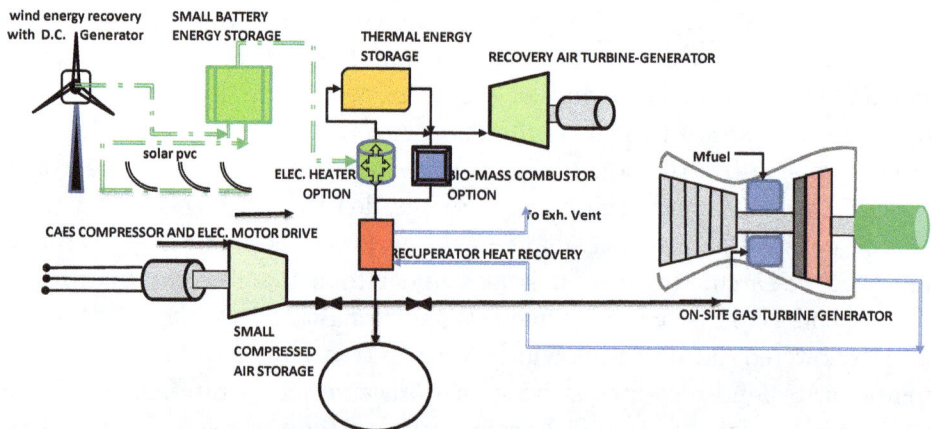

Figure C15.1.   Concepts NREC's "crossover" system, which integrates a modular CAES with battery storage of power generated by renewable energy.

between the inverted DC power and the utility grid instead of converting the stored DC electrical power. The stored energy is depleted during a typical 3-to-4-hour discharge sequence. However, with the proposed system shown in Figure C15.1, the compressor and turbine subsystems can continue to operate as a waste heat recovery system based on the operation of an open air-Brayton Cycle heat engine, as shown diagrammatically in Figure C15.1. The air-Brayton Cycle can then continue to produce useful power as necessary to meet demand for many more hours or until the power demand subsides and the next "charging" cycle for the CAES begins. Another heat input to the stored compressed air can be derived from any number of available renewable energy sources, as shown in Figure C15.1, including bio-mass or bio-gas combustion or the electrical power stored in the system's battery. In its most ambitious and novel rendition, the open air-Brayton Cycle system will use a single air compressor to serve both as the compressor for charging the CAES as well as the compressor needed as part of the GT open air-Brayton Cycle system that burns biomass or biogas.

C15.4 Analysis summary and detailed discussion

Despite the cost (~US$200–250/kWh in 2018) (Ulvestad, 2018) and limited cycle lifetime (5% per 1,000 cycles) (Crawford, *et al.*, 2018) of lithium-ion batteries, they remain the most common energy storage medium for electrical energy, due to their high energy density. The energy density of lithium-ion batteries ranges from 100 to 265 Wh/kg (=770 Wh/l). This is considerably higher than what can be stored in a simple CAES that uses a single compressor for storing the work of compression. An exergy (or availability) analysis of several common fluids, including water, at a temperature and pressure above the ambient pressure and temperature (the "dead" state) has been completed and is shown in Figure C15.2. The shaded portion of the chart represents the range of energy storage that is available from the lithium-ion battery. The temperature used for the analysis corresponds to the saturation temperature of water at the given pressure. The results indicate that water as the storage medium has a higher specific availability (Wh/kg) than air, carbon dioxide or R245fa, which is the only refrigerant studied in this textbook.

The high specific available energy for water as compared to the other fluids is due to the extremely high latent heat energy required to change water from a liquid state into a saturated vapor. Some preliminary conceptual studies have been done by the author while employed by CN, which indicate that although energy storage can be done with water, considering

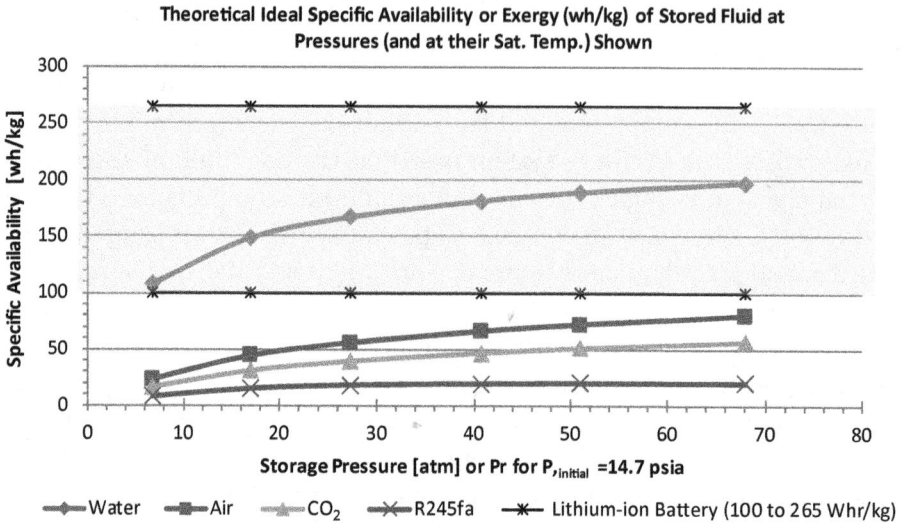

Figure C15.2.   Exergy (or availability) analysis of several common fluids.

the high latent heat energy of water, this would require a very high pressure containment vessel at temperatures above 350°C. For this case study, it can be observed from Figure C15.2 that the storage pressures for the CAES must be higher than 8 atm. For example, the compressor's compression pressure ratio (Pr) must be greater than 8 for the energy storage densities to be comparable to that available with lithium-ion batteries. However, it may also be observed that the rate of increase of the energy density decreases as the storage pressure increases. As seen in Figure C15.2, the rate of increase, or slope of the curves (Δ Availability/ΔP), appears to decay significantly at a storage pressure of 16 atm (Pr = 16). To facilitate the selection of a suitable design point for the CAES, CN developed a new way of examining the optimization of a thermomechanical energy storage system, which subsequently led to the development of the crossover energy storage system shown in Figure C15.1.

An analytical expression for the ERE (defined in Equation (C15.1)) has been derived for a CAES, assuming that air can be treated as a perfect gas. Two derivations were obtained. The first derivation expresses the ERE as a function of the storage pressure, with the assumption that no additional energy is added to the turbine inlet stream during the recovery of the stored energy. The second derivation develops an expression for ERE as a function of pressure ratio, with consideration given to the effect on the ERE, of the addition of heat energy to increase the temperature of the inlet turbine's mass flow rate of air. It is preferred that the necessary addition of heat energy to the pressurized air prior to its entry into the turbine be derived from a renewable energy source via the combustion of biomass or biogas, for example. Alternatively, it is also possible to use electric power derived from wind turbines or PV cells to heat the compressed air entering the turbine. However, although the use of electric power in this manner is cost-saving as it eliminates the need for rectifying and inverting the DC electric power, due consideration should also be given to the thermodynamics exergy destruction that results when the electric power is not returned directly to the utility grid. Both derivations use the state point temperature parameters that are defined by the air process paths, as illustrated in Figure C15.3. The ambient air (T_1) is compressed during the charging sequence and heated to T_2 by the energy of compression. The temperature of the stored energy may decrease to T_{cooled} due to heat transfer from the storage system. The magnitude of this heat transfer is dependent on the design of the storage system.

The magnitude of the temperature reduction is represented by a new parameter defined as:

$$R_c = \frac{(T_3 - T_{1.ambient})}{(T_2 - T_{1.ambient})}. \tag{C15.2}$$

The Temperature Recovery Effectiveness (Rc) is a parameter used to indicate the degree of temperature recovery (T_3) that is required before the stored fluid enters the turbine. As diagrammed in Figure C15.3, the elevated discharge temperature, T_2, from the compressor may cool to temperatures between T_2 and $T_{ambient}$ before the recovery sequence from the stored energy is initiated. This cooling is due to the loss of thermal energy to the environment if the energy is stored for a prolonged period of time. The cooling could

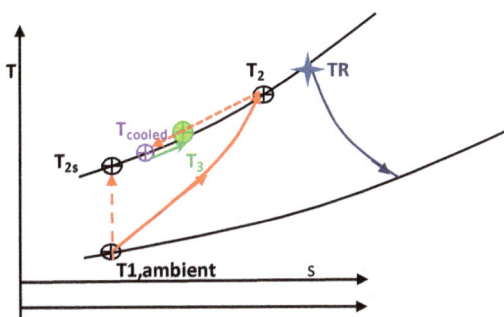

Figure C15.3.   Process path for air during the "charging" and "discharging" sequence.

also be due to the fact that the work of compression has been intentionally removed to increase the density of the stored air. Regardless of how the work of compression is lost via heat loss to the environment, the heat-energy must be restored in some manner that allows it to reheat the air before the mass flow rate of air is delivered to the turbine during the energy recovery sequence. If this work of compression is recovered completely, and there is no loss of thermal energy to the environment during the storage period, then the Rc value would be equal to 1. However, during a normal fluid extraction process, when the fluid is drawn from the storage vessel during the energy recovery sequence, the pressure of the fluid in the storage vessel drops due to the extraction of fluid. It can be demonstrated via a separate analysis that as the air is rapidly discharged from the storage vessel, the temperature of the air remaining in the tank continues to be reduced (see Case Study 16). Depending on the size of the vessel and the initial storage pressure, the value of Rc can range from 0.6 to 0.9. The lower range typically corresponds to a smaller vessel size, which may be used in Distributed Power Generation systems. The cooling of the remaining air in the storage vessel during the rapid discharge of air from it explains the need for the turbine inlet temperature to be raised during the energy recovery sequence, in order for the ERE is to be maintained at an economically acceptable level, which is usually considered to be at least 70%.

If no additional heat energy from an external source is added to the flow stream entering the turbine during the energy recovery sequence, then the instantaneous[3] ERE is calculated using the equation shown in Equation (C15.3):

$$E.R.E. = B \times \eta_{turbine} \times \left[Rc + \frac{\eta_{compressor}}{A} \right],$$

where:

$$A = (Pr, compressor)^{\left(\frac{k-1}{k}\right)} - 1; \ B = 1 - \left(\frac{1}{Pr, turbine} \right)^{\left(\frac{k-1}{k}\right)}$$

---

[3]The ERE equation is best described as providing the instantaneous value for the ERE at the end of the "charging sequence" and at the initial moments of the energy recovery sequence, when the pressure ratio (Pr) across the compressor and turbine remains constant. At all other times during the "charging" and "discharging" of the stored energy, the pressure ratios are changing. However, the ERE equation is thought to be a simple, closed-form equation that can accurately represent the transient ratio of the total work of compression and expansion that must otherwise be integrated over the "charging" and "discharging" periods as required in Equation (C15.1). An alternative is to consider using the mean effective pressure ratio, $Pr_{effect.}$, for each of the compressor and turbine, Pr, in Equation (C15.2). The same correction may also be considered for the part-load efficiencies of the compressor and turbine, which can be used to replacing the efficiencies with weighted average efficiencies during "charging" and "discharging."

with:

$$Rc < 1 \quad \text{and} \quad k = 1.41. \tag{C15.3}$$

Figure C15.4 presents the results from a parameterization of the ERE with respect to the compressor and turbine efficiencies, and Rc. Perhaps the most striking observation that can be made from Figure C15.4 is the small effect that Pr has on the ERE at pressure ratios above 8.

Thus, to maintain a competitive ERE for modular systems (e.g. thermo-mechanical energy storage systems designed for smaller, Distributed Power Generation systems), it would suffice to store energy at pressures of less than 8 atmospheres. However, to maintain or further improve the ERE during the energy recovery sequence, the advantages of the inherently high energy densities available with battery storage and/or biomass[4] combustion could be utilized.

A "crossover" system combining CAES technology with battery energy storage technology (see Figure C15.1) suggests that the typical availability of an on-site GT (or another prime mover) affords the opportunity to provide continuous recuperation of the waste exhaust gas heat by the proposed turbomachinery. Not only is the proposed turbomachinery able to provide energy storage and recovery, it also enables continuous power generation for the augmentation of power generated via the onsite prime mover.

A very interesting and important observation can be made by carefully studying Equation (C15.2) and Figure C15.4. If the compressor efficiency is equal to the Rc, then Equation (C15.2) evolves into a much simpler relationship between the ERE, turbine and compressor efficiencies, as shown here:

$$E.R.E. = \eta_t \times (R_c \equiv \eta_c). \tag{C15.4}$$

As may be observed in Figure C15.4, this equation leads to a definitive result that when compressor efficiency is made higher than Rc, the ERE is increased with a LOWER Pr. This observation serves as the motivation to achieve the highest possible turbine and compressor efficiencies using current state-of-the-art software for turbomachinery design, as well as advanced machining practices (including additive manufacturing).

To determine the effect of the amount of reheating of the stored, compressed air mass flow rate before the air enters the turbine during the energy recovery sequence on the ERE, a modification of ERE Equation (C15.2) is made by introducing another new parameter, the Recovery Factor (Rr).

---

[4]Biomass consisting of carbon–hydrogen bonds is an example of very effective energy storage.

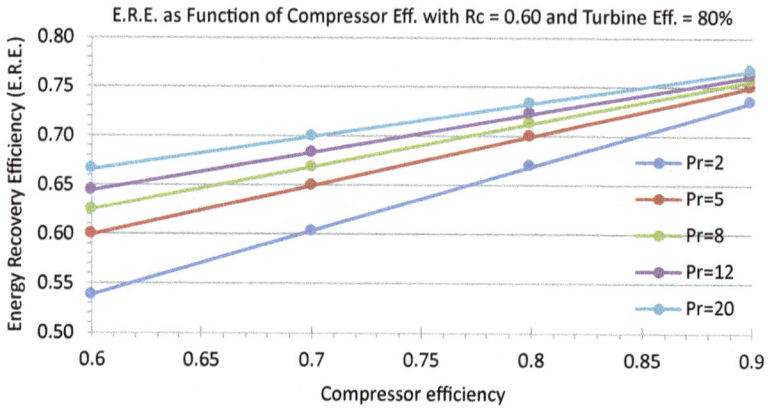

(a)

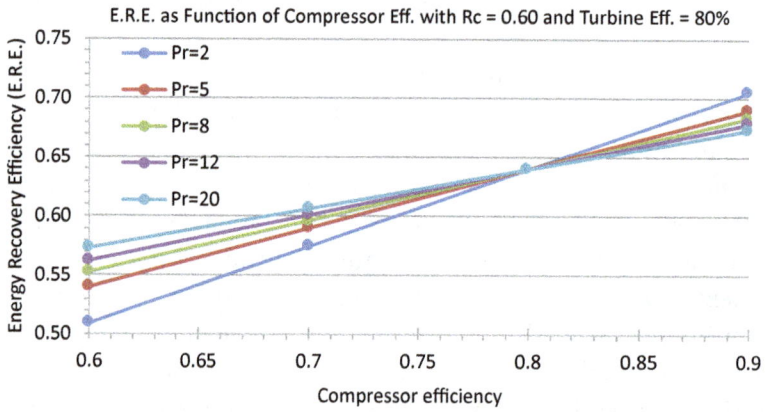

(b)

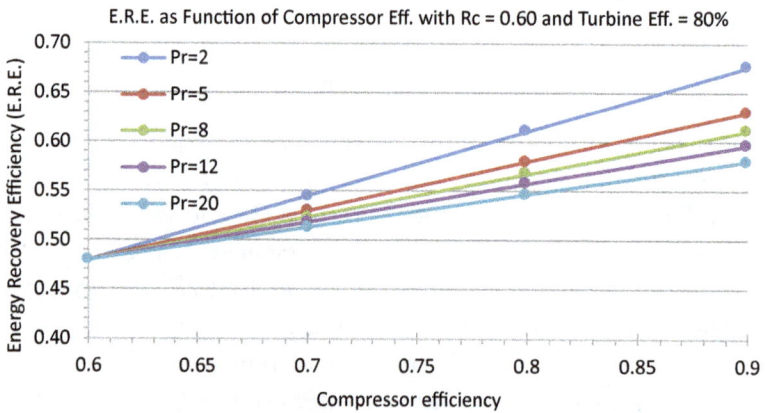

(c)

Figure C15.4. Results from a parameterization of the Energy Recovery Efficiency (ERE) as Function of Compressor Efficiency with Turbine Efficiency of 80% and (a) Rc = 1.0; (b) Rc = 0.8; (c) Rc = 0.6.

Rr is a measure of the relative magnitude of the amount of energy added to the fluid to increase the temperature, T_R, of the fluid entering the turbine during the energy recovery sequence. It is defined as shown in the equation below with respect to T_3 and T_2 in Figure C15.3:

$$Rr = \frac{(T_R - T_3)}{T_2 - T_3}. \tag{C15.5}$$

It is noted that Rr can have a range from zero (when $T_R = T_3$), to greater than or equal to 1 (when $T_R \geq T_2$). The choice of T_R is dependent on the availability of other "free" waste heat available onsite, the materials used in the turbine that may impose a constraint on the maximum inlet temperature that the turbine can sustain, or the electric resistance heater. Electric resistance heating is assumed to be used only when electric power is directly available from live wind or PV renewable energy sources at the user's site or from battery storage, and only if the power generated from the renewable energy sources is not intended to be connected to the utility grid. This results in cost-reduction by eliminating the need for complex and expensive power conditioning systems (such as voltage inverters) as well as the need for compliance with local regulatory requirements for power generation systems.

The modified ERE (net) equation that accounts for the energy added to the system during the energy recovery sequence is shown in Equation (C15.6), with respect to the same parameters previously defined:

$$E.R.E._{net} = \frac{T_R}{T_{ambient}} \times \frac{B \times \eta_{turbine} \times \eta_{compressor}}{A \times [1 + Rr - Rr \times Rc]}. \tag{C15.6}$$

Using this equation with $T_R = 900°R$, $1{,}200°R$,$1{,}400°R$ and $1{,}800°R$, the net ERE for the system shown in Figure C15.1 is obtained and displayed in Figure C15.5. As expected, this ERE_{net} is lower than the ERE obtained from using Equation (C15.3), because the energy that is added to the system to increase the inlet temperature to the turbine must be accounted for, according to Equation (C15.1). However, while Equations (C15.5) and (C15.6) are accurate in accounting for the thermodynamics concerning the additional energy input to the system during the energy recovery sequence, an argument can be made that IF the energy that is added to the system is "free" (that is, derived from a waste heat source that would otherwise not be recovered, or from a renewable energy source), the calculated ERE would be much higher, and would be independent of the inlet temperature, T_R. In such a case, the ERE would be calculated using Equation (C15.2), with T_3 equal to the desired turbine inlet temperature, T_R. The range of values of

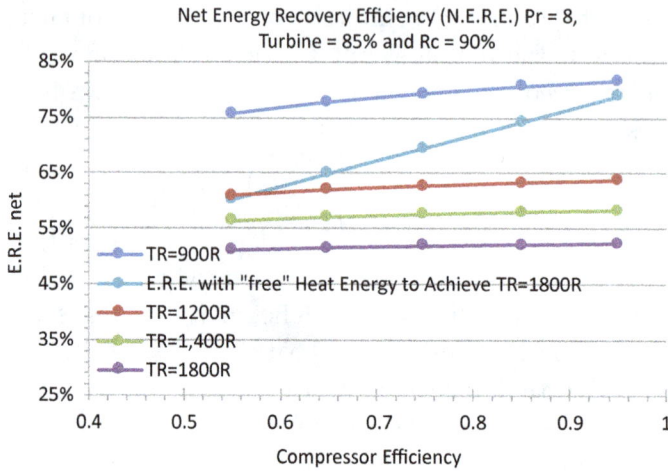

Figure C15.5.   Values of ERE for Rc = 90% and turbine efficiency = 85%.

ERE calculated using Equation (C15.2) for an Rc of 90%, with a turbine efficiency equal to 85%, and with the compressor efficiencies, is provided in Figure C15.5.

C15.4.1   The "Crossover" energy storage system: Continuous power generation after stored energy recovery

The proposed use of a modular (i.e. smaller) energy storage and recovery system for the Advanced CAES that is to be used for the DPG market using the turbomachinery components and the reheat subsystems as shown in Figure C15.1, can enable the continuous generation of power via the air-Brayton Cycle after the stored energy has been depleted. The turbo compressor subsystem can produce a net power output using the waste heat recovered from the onsite prime mover, such as a GT engine or a biomass combustor, as shown in Figure C15.1. This requires an assumption that the turbine inlet temperature can be increased above a critical minimum temperature, $T_{R,min}$, such that the efficiency of the air-Brayton Cycle calculated using Equation (C15.7) would be not only greater than zero, but large enough to make economic sense, which typically requires a Simple Payback of three years or a Return on Investment (ROI) of 10% or more.

$$\eta_{brayton\ cycle\ power\ recovery} = \frac{\left(\frac{T_R}{T_{ambient}} \times \frac{B \times \eta_t \times \eta_c}{A} - 1\right)}{R_R \times (1 - R_c)}. \tag{C15.7}$$

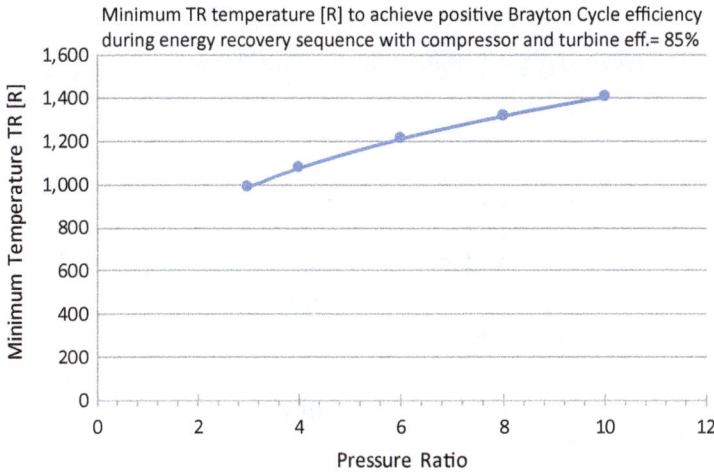

Figure C15.6.   Minimum temperature, $T_{R,min}$, to achieve Brayton Cycle efficiency during energy recovery sequence with compressor and turbine efficiency of 85%.

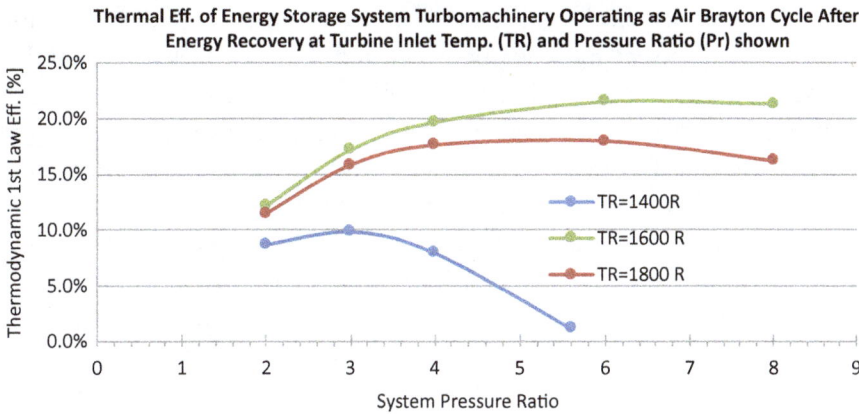

Figure C15.7.   Calculation of the cycle efficiency for several turbine inlet temperatures, T_R.

The critical minimum temperature, $T_{R,min}$, determined using Equation (C15.8), is displayed in Figure C15.6 as a function of Pr for the turbo compressor subsystem.

$$\frac{T_{R,min}}{T_{ambient}} = \frac{P_r^{\left(\frac{k-1}{k}\right)}}{\eta_t \times \eta_c}. \tag{C15.8}$$

The calculation of the air-Brayton Cycle's efficiency, using Equation (C15.7), for several turbine inlet temperatures, T_R, is shown in Figure C15.7. It is interesting to note that the optimum air-Brayton Cycle efficiency is a function of the system's pressure ratio, Pr and the temperature, T_R. This serves as an engineering design tool for selecting a design point for the system.

C15.5   Case study example

The following case study contains an application of the equations presented thus far, including estimates of the size and speed of the turbo-compressor for a DPG system that uses CAES with the continuous power generation option. That is, the air compressor used to charge the CAES is also the same compressor that is an integral part of the open air-Brayton Cycle heat engine also known as a GT power generation unit.

Consider the siting of the proposed "crossover" system consisting of the CAES-battery system and an engine following the open air-Brayton Cycle alongside a 1,500 kWe-rated wind turbine. The wind turbine is assumed to have a 65% utilization factor. That is, on average, 65% of the rated power is consumed by an electrical load that is connected to the site. The remaining 35% of the rated power may not be generated due to a lack of demand. For this case study, it is assumed that half of this capacity, which is approximately equal to 20% of the rated power, is stored using the proposed system that combines the CAES-battery system and an engine following the air-Brayton Cycle. The balance of the wind turbine's capacity is assumed to be unavailable due to the lack of wind. From Figure C15.8 it may be observed that 900 kWh of energy is available for storage. The CAES typically uses a storage pressure of only 8 atm, a very safe level by industrial standards. The compressed air is stored in five spherical vessels, each with a diameter of 10 meters. A single compressor rated at approximately 750 kW will provide the compressed air at the necessary storage pressure of 8 atm in approximately 1–2 hours. This accounts for half of the total stored energy required with the balance provided by 2,000 kg worth of lithium-ion battery. The worksheet that summarizes this data and the output from the analysis of the proposed system is shown in Figure C15.8.

Using Equations (C15.5), (C15.6) and (C15.7), the ERE for the proposed system is calculated to be 67%, assuming an Rc of 70%. This highly competitive ERE provides added confidence that the proposed system is able to provide at least half of the stored energy required, in case of failure in either the mechanical or electrical components of the system. The "crossover" CAES-battery energy storage system also enables instantaneous power to be generated via the electrical system, before the turbine part of

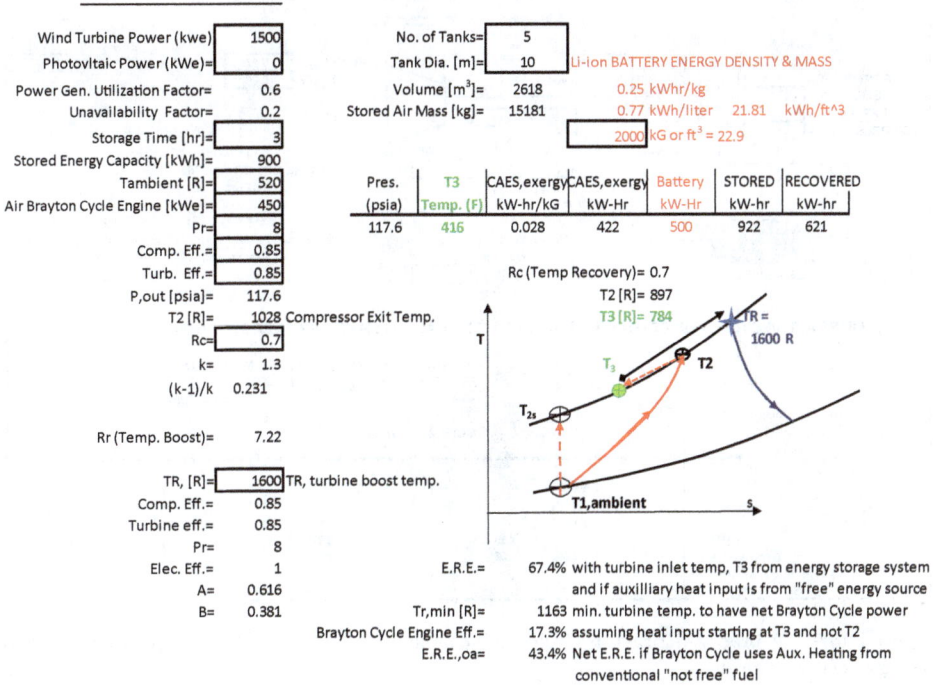

Wind Turbine Power (kwe)	1500		
Photovltaic Power (kWe)=	0		
Power Gen. Utilization Factor=	0.6		
Unavailability Factor=	0.2		
Storage Time [hr]=	3		
Stored Energy Capacity [kWh]=	900		
Tambient [R]=	520		
Air Brayton Cycle Engine [kWe]=	450		
Pr=	8		
Comp. Eff.=	0.85		
Turb. Eff.=	0.85		
P,out [psia]=	117.6		
T2 [R]=	1028	Compressor Exit Temp.	
Rc=	0.7		
k=	1.3		
(k-1)/k	0.231		
Rr (Temp. Boost)=	7.22		
TR, [R]=	1600	TR, turbine boost temp.	
Comp. Eff.=	0.85		
Turbine eff.=	0.85		
Pr=	8		
Elec. Eff.=	1		
A=	0.616		
B=	0.381		

No. of Tanks= 5
Tank Dia. [m]= 10
Volume [m³]= 2618
Stored Air Mass [kg]= 15181

Li-ion BATTERY ENERGY DENSITY & MASS
0.25 kWhr/kg
0.77 kWh/liter 21.81 kWh/ft^3
2000 kG or ft³ = 22.9

Pres. (psia)	T3 Temp. (F)	CAES,exergy kW-hr/kG	CAES,exergy kW-Hr	Battery kW-Hr	STORED kW-hr	RECOVERED kW-hr
117.6	416	0.028	422	500	922	621

Rc (Temp Recovery)= 0.7
T2 [R]= 897
T3 [R]= 784
TR = 1600 R

E.R.E.= 67.4% with turbine inlet temp, T3 from energy storage system and if auxilliary heat input is from "free" energy source

Tr,min [R]= 1163 min. turbine temp. to have net Brayton Cycle power

Brayton Cycle Engine Eff.= 17.3% assuming heat input starting at T3 and not T2

E.R.E.,oa= 43.4% Net E.R.E. if Brayton Cycle uses Aux. Heating from conventional "not free" fuel

Figure C15.8.   Summary of data and output from the system analysis.

the heat engine operating according to the air-Brayton Cycle is brought up to operating speed.

When the stored energy from the "crossover" CAES-battery energy storage system is depleted, the "compressor-turbine-motor/generator" configuration can continue to provide continuous power generation. This configuration operates according to the familiar air-Brayton Cycle, as shown in Figure C15.9, and utilizes the exhaust gas waste heat from the combined compressor-turbine system which now functions as a gas turbine. For the GT to operate in accordance to the air-Brayton Cycle during the "discharge" sequence, both compressors shown in Figure C15.9 are engaged in parallel to provide the necessary flow rate through the turbine. The waste heat energy from the proposed turbine can be augmented by an external auxiliary heat input from either the combustion of conventional fossil fuels, biomass, or the waste heat from other prime mover engines that may be installed on-site.

For this case study, a three-stage, 25,000 rpm axial turbine with a third-stage diameter of 300 mm would be used for analysis. A preliminary analysis for the compressor design indicates that a radial compressor with a

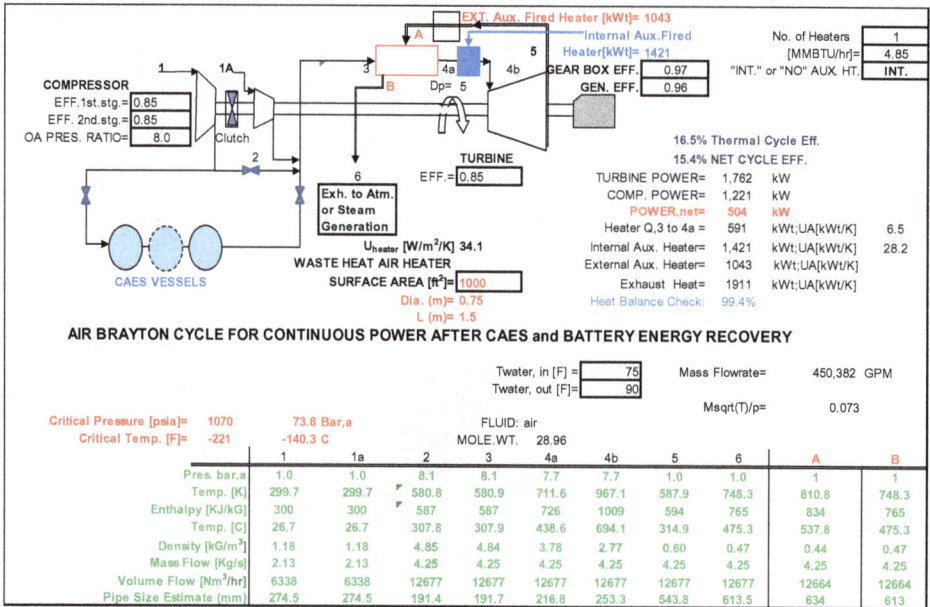

Figure C15.9.   A proposed Brayton Cycle engine as may be applied with the "crossover" CAES-battery energy storage.

diameter of 0.3 m operating at the same speed of 25,000 rpm would suit this application.

The benefit of investing in a compressor and a turbine for the purpose of energy storage and recovery is clear: these components allow the system to potentially be utilized as a continuous power generation system in place of the CAES, in the event that CAES is not available.

C15.6   Conclusion

The proposed "crossover" system combines CAES technology with battery energy storage technology and also uses renewable energy for reheating. The "crossover" system also includes a dual-purpose compressor in an air-Brayton Cycle heat engine that enables higher reliability, a fast response to transient power loads and the efficient use of renewable energy as well as heat recovery from waste heat generated by conventional onsite prime movers. The system is particularly well suited for applications in isolated DPG systems. DPG systems consist of numerous but relatively small sources or "power islands" of electric power generation. It is also proposed that the use of an induction generator, and not a synchronous generator, with the open air-Brayton Cycle, will greatly reduce the programming effort required for maintaining the cycle online. Using an induction generator, synchronizing electrical gear and the monitoring of voltage and frequency is not needed for the power system operating according to the air-Brayton Cycle, as the frequency (50 or 60 cycles/s) of the voltage and current is maintained by the local utility electrical system. Thus, the utility electrical grid controls the speed of the induction generator and eliminates the need for synchronizing the DPG to the grid. The air-Brayton Cycle operates at its maximum efficiency, depending on the availability and magnitude of the waste heat and/or bio-fuel and the design point turbine inlet temperature specification.

Similarly, the air-Brayton Cycle does not require a speed governor or fuel control valve to control the power output. The power control is simplified using a simple proportional control algorithm which will monitor the power output and turbine inlet temperature. The simplified control system will reduce the delivery of heat input, derived from stored electric energy, biofuel or exhaust gas waste heat, to the heat input heat exchanger of the air-Brayton Cycle as necessary to maintain the desired control point temperature at the inlet to the turbine.

References

Crawford, A. J., Huang, Q., Kintner-Meyer, M. C., Zhang, J.-G., Reed, M. D., Sprenkle, V. L., & Choi, D. (2018). Lifecycle comparison of selected Li-ion battery chemistries under grid and electric vehicle duty cycle combinations. *Journal of Power Sources, 380*, 185–193. doi: https://doi.org/10.1016/j.jpowsour.2018.01.080

Ulvestad, A. (2018, March 12). *A Brief Review of Current Lithium Ion Battery Technology and Potential Solid State Battery Technologies.* Retrieved from arXiv.org: arXiv:1803.04317

Numerical Solution to Compressed Air Energy Storage (CAES): An Open Cycle Example Demonstrating $\Delta U = 0$

Case Study 15 presented closed-form solutions for the equations that represent a Compressed Air Energy Storage (CAES) system. The goal of that analysis was to model the system and discern the parameters which affected its Energy Recovery Efficiency (ERE), that is, how much energy could be recovered compared to the energy input to the system. This case study has the same objective, except that it presents a solution to the problem which utilizes a numerical integration technique to solve for the first law equations for the compressor and turbine energies. This is a particularly effective technique that can be implemented using the column-and-row structure of a spreadsheet platform. To obtain precise solutions, the numerical technique uses time steps that are small enough to attain a solution within 1–2% of the exact solution.

C16.1 Case study: Numerical integration technique used to model the transient behavior during energy storage and energy recovery of a CAES

Consider the following problem. A storage vessel of $10\,\text{ft}^3$ is to be used to store 38 psia air. The air is to be drawn from the ambient air, and is compressed as it fills the fixed volume storage vessel which has an initial internal pressure of 14.7 psia and an initial storage temperature of 70°F. After the vessel reaches the desired pressure, a turbine is used to expand the stored air from its initial pressure to atmospheric pressure. Assume that both the compressor and turbine operate at 100% efficiency, and that no thermal or mechanical energy losses from the compressor-storage vessel-turbine system occur throughout the entire process. The questions to be answered are: How much energy is needed for this compressed air storage system to work? What is the maximum power required for the compressor? What is the temperature change of the air in the vessel where it is stored, and as it is released?

The choices of the start and stop pressures, as well as the fixed volume of the vessel, and the compressor and turbine efficiencies, are completely arbitrary. In the solution to the problem at hand, values for these parameters have been chosen simply because they match the specifications of a similar problems found in contemporary thermodynamics textbooks. Although the solution presented in this case study is done differently, using a more realistic time-transient approach, the textbook solved example affords an opportunity to provide a good validation of the method that is shown in this case study. The numerical solution obtained in this case study can be easily validated by comparison with other numerical solutions offered in other thermodynamics textbooks. Once this numerical solution has been validated, it can be applied to CAES systems operating with different specifications given for the size of the storage vessel, the storage pressure, and the turbine and compressor efficiencies. However, in order for the numerical solutions to approach and ultimately become identical to the "textbook solution", the numerical solution technique used in the case study requires the use of smaller and smaller time step increments. This realization is nothing more than what Sir Isaac Newton required of his Calculus mathematics wherein the derivatives of a function approach the exact slope of a function as the "... change in the independent variable, x... approaches zero." For the proposed time-transient solution in this case study, the independent variable is a differential time step or Δt. Thus, during the application of this numerical integration technique to any CAES systems with system design parameters that change with time, the time increment may need to be made smaller and smaller before the numerical solution is equal to the "closed form" solution that is the precise solution that may be obtainable by applying differential or integral Calculus. For instance, the time step of 0.2 minutes (12 seconds) used in this case study, which analyses a CAES system with a small storage volume, may be increased to 20–30 seconds (or even more) if the volume of the storage vessel were to be extremely large. This would require simply adding to or removing the number of rows that are necessary to account for the long or short time-step.

The solution employs a straightforward methodology. A similar solution was presented in Case Study 1.

Each row of the spreadsheet contains solutions to thermodynamics equations corresponding to a unique time step. The first row corresponds to time = 0 seconds. The rows that follow represent increments of constant time steps, which can be changed easily. The column headers identify the various outputs of which the numerical solution consists, which can be solved for using thermodynamics equations. The algorithm used to obtain the solutions are repeated row after row, with the solutions from row "n" used to calculate the solutions displayed in the next row "n + 1". The solutions displayed in each row may be thought of as a quasi-static solution for this problem. That is, at each time step the system is calculated "as if" the state of the pressure and temperature was in thermal and mechanical equilibrium when in fact the entire process is very much in a transient behavior until the tank is completely filled (during the energy storage sequence) or completely discharged (during the energy recovery-power generation sequence.

Figures C16.1(a) and (b) illustrate the input parameters that define the energy storage system and the outputs used in the solution of this problem. The input parameters, such as storage pressure, can be changed by the analyst to achieve, for example, the desired amount of energy for a fixed volume of the storage vessel and with a compressor that is operating at "off-design" conditions. Once these input parameters are selected, they are referred to by the equations that are used at each time step. Figure C16.1(a) describes the compression process or what is normally identified as the "charging" cycle. That is, during the "charging" cycle, energy is being stored in the storage vessel that has a fixed volume. Figure C16.1(b) describes the turbine "discharging" or expansion cycle, also known as the energy recovery cycle.

C16.1.1   *Calculations for the "charging" cycle*

The calculation proceeds as follows:

1. The initial storage pressure at time = 0 seconds is given as 14.7 psia. The state points of pressure and temperature at the next time increment at time "n + 1" are determined by applying the Perfect Gas Law to a fixed volume of the vessel, with a known mass content, and the new storage vessel temperature. The amount of mass in the storage vessel at time t_{n+1} is determined by knowing the time increment and the flow rate of the mass into the vessel at the instant of time and by multiplying the flow rate input by the time increment. For this case study, the mass flow rate is an input

(a)

		Overall Storage Eff.= 100.6%		
FILL Mass Flowrate [Lbm/hr]= 2.50		Storage Size [m³]=	0.280	
		[ft³]	10	
Tambient [F]= 70		Initial Storage Air Mass [Lbm]=	1	
		Req.d Storage Pres. [bar,a]=	2.61	
Compressor Efficiency [%]= 100%	0 Eff. Curve Factor	[psia]=	37.8	37.80 0.00
				15.18 Btu
Time Increment [min.]= 0.2	37.80	Ru/Mole.wt=	53.3	TOTAL COMP. kWh
Storage Mass Increment [Lbm]= 0.0083		Reservoir (k-1)/k=	0.291	0.004448 1.46

	Time (min)	Comp. Discharge Pres. psia	Comp. Transient Eff.	Comp. Discharge Temp.,F	Internal Energy [Btu/Lbm]	Storage Mix Instant. Temp.[F]	Comp. Instant. Power [kw]	Comp. Instant. Energy [Kwh]	Mass of Air in Storage [Lbm]	Incre. Increase Storage Pres. psi
70.00	0	14.70	1.00	70.0	90.4	70.00			0.7482	0.166
110.68	0.2	14.87	1.00	71.7	90.8	72.4	0.0	0	0.7565	0.233
148.05	0.4	15.10	1.00	74.1	91.2	74.7	0.0	0	1	0.234

(b)

				k=	1.41	0.2908	-1.291
				Turbine Power Integration [Btu]	27.44		
Max . Storage Temp.=	234		Compressor Power Integration [Btu]	##########			
Initial Steady State Storage Temp. at Start [F]=	234						
RECOVERY Mass Flowrate [Lbm/hr]=	2.4						
Initial Turbine Eff. [%]=	100%	0	Turbine Design Pressure= 1047				
Sp. Heat, Cp [Btu/Lbm/R]=	0.25						
Time Increment [min.]=	0.2	14.7	Overall Storage Eff.= 100.6%				NO
Storage Mass Increment [Lbm]=	0.0		0.00447				0.18

0.00
Btu
kWh
1.46

Mass of Air in Storage [Lbm]	Incre. Increase Storage Pres. psi	Turbine Inlet Pres. [psia]	Turbine Inlet Temp. [F]	Turbine Eff. [%]	Turbine Instant. Power [kw]	Turbine Instant. Energy [Kwh]	Mass of Air in Storage [Lbm]	Incre. Increase Storage Pres. psi	Storage Energy,after Btu/Lbm	Storage Temperature [F]
1	0.303	0	37.8	234	100.00%	0	0	1	-0.26	118.3	232
		0.2	38	232	100.00%	0	0	1	-0.29	118.0	231
		0.4	37	231	100.00%	0	0	1	-0.29	117.8	229

Figure C16.1. Illustrating (a) the editable input fields and the column headers (the parameters) for the corresponding outputs for a calculation describing the "charging" cycle when energy is stored for future recovery; (b) The input field and the column headers for the corresponding outputs for a calculation describing the turbine "discharge" cycle when power is generated from the stored energy.

to the solution by having fixed the amount of mass transferred per unit of time increment. Thus, the mass flowrate is kept constant throughout the transient charging process. The same mass flowrate is held constant during the discharge sequence. For the discharge sequence, it is possible to calculate the flowrate at each time step by applying the First Law of Thermodynamics to the flow opening and/or across a flow control device, also known as a valve. The reader will recall that a valve is one of the seven basic mechanical elements that constitute all thermodynamic processes and systems. The First Law of Thermodynamics applied in this manner is perhaps better known as the Bernoulli Equation and provides a relationship between the pressure drop, flow restriction (orifice, valve, pipe length, etc.) and velocity as shown in the following equation:

$$\Delta P = \frac{f_{friction\ coef.} \times V^2_{flow\ velocity}}{2 \times g_c} \times \rho$$

To perform this calculation, a new column for the calculation of the variable mass flow rate, $\frac{\partial m}{\partial t}$, in the Discharge Sequence Table (see Figure C16.1(b)) should be created. The mass flow rate, $\frac{\partial m}{\partial t}$ can be calculated by knowing (as another input to the solution) the diameter of the connecting pipeline and the velocity of the flow rate. Solve from the above equation with the pressure difference, ΔP, equal to the variable, tank pressure and the fixed ambient pressure at any time, t, during the discharge sequence.

2. The new vessel storage temperature, T_{n+1}, and compressor energy, $Ecompressor = \dot{W}_{compressor\ power} \times \Delta t$, is determined using the first law equation for compressor power previously stated in Chapter 3. The equation is:

$$Ecompressor = \dot{m} \times C_p \times T_{inlet} \times \frac{\left(1 - \left(\frac{P_{out}}{P_{in}}\right)^{\frac{k-1}{k}}\right)}{compressor\ eff.} \times \Delta\ time\ increment.$$

$$T_{outlet} = T_{inlet} \times \left[1 - \frac{\left(1 - \left(\frac{P_{out}}{P_{in}}\right)^{\frac{k-1}{k}}\right)}{compressor\ eff.}\right].$$

Care must be taken to account for the negative sign obtained using the *Ecompressor* equation when that value is used to calculate the total compression energy that is stored in the CAES. The absolute value of this energy is usually displayed as an output of the analysis as shown in Figure C16.2(a).

3. The new storage vessel temperature in step "n + 1" is determined by applying the First Law of Thermodynamics but now with the "rate of Change of the Control Volume Energy", $\frac{\partial E}{\partial t}$ not equal to zero, as it is often done in many "open flow, control volume" problems that have been addressed in this textbook. Starting from the full equation for the First Law of Thermodynamics, with the only assumption that the power input and output, $\dot{W}_{compressor\ power}$ for the air storage volume is zero (0), the change in temperature caused by the addition of mass to the amount of mass in step "n" is determined from the following equation:

$$\frac{\Delta m}{\Delta t} \times h_{in} = \frac{(m + \Delta m) \times u(p, T_{n+1}) + m \times u(p, T_n)}{\Delta t}.$$

It is noted that the time increment that appears on both sides of the equation cancel each other out and the incremental mass (Δm) is found as described in Step 1.

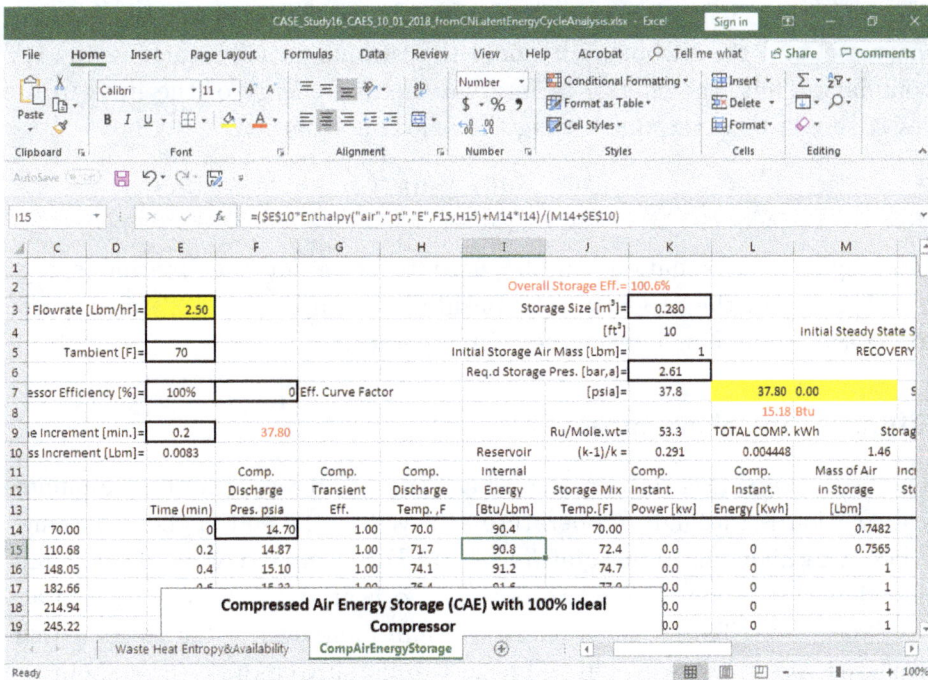

Figure C16.2. Spreadsheet containing the series of calculations as described in the text for the internal temperature of the vessel.

This equation can then be used to solve for the internal thermal energy: $u(p, T_{n+1})$.

With the internal thermal energy known, the internal temperature of the vessel can be determined from the property relationship between u, p and T. The spreadsheet that displays this series of calculations is shown in Figure C16.2.

4. Steps 1 to 3 are repeated until the desired storage pressure is achieved. The results of the calculation for storage temperature, pressure and compressor power are then plotted as shown in Figure C16.3.

C16.1.2 Calculations for the "discharging" cycle

A similar methodology is repeated for the turbine "discharging" cycle. It is assumed that the initial temperature and pressure of the storage vessel at the start of the "discharging" cycle is equal to the temperature and pressure of the storage vessel at the end of the "charging" cycle. Care must be taken in Steps 1 to 3 to apply the equations according to the direction of the flow of air. For example, during the "discharging" of the storage vessel, the inlet flow rate is zero with respect to a control volume (CV) that is drawn around the vessel and the turbine, with only the turbine outlet breaching the CV boundary. Thus, the change in internal thermal energy of the fluid in the vessel should be determined using the following equation:

$$0 = \frac{\Delta m}{\Delta t} \times h_{out} + \frac{(m - \Delta m) \times u(p, T_{n+1}) + m \times u(p, T_n)}{\Delta t}.$$

Once again, the equation can be used to solve for the internal energy $u(p, T_{n+1})$, and the property relationship between u, p and T can be used to determine the new internal vessel temperature.

C16.2   Some observations

The first observation to be made using Figure C16.3 relating to the numerical solution is the high temperature achieved at the end of the "charging" cycle. The high final temperature of 240°F is achieved with the assumption that throughout the entire process, no heat transfer from the storage vessel occurred. This is not a realistic assumption, even if the vessel is well-insulated, as some energy will still be lost to the surroundings throughout the entire transient filling of the storage vessel, before the start of the "discharging" cycle. Thus, even with compressor and turbine efficiencies of 100%, the ERE for the CAES system will very likely be less than 100%. That is, the amount of energy stored in the vessel will not be the same as the amount of energy that can be recovered from the storage vessel. A typical goal of CAES is to achieve at least an ERE of 70%.

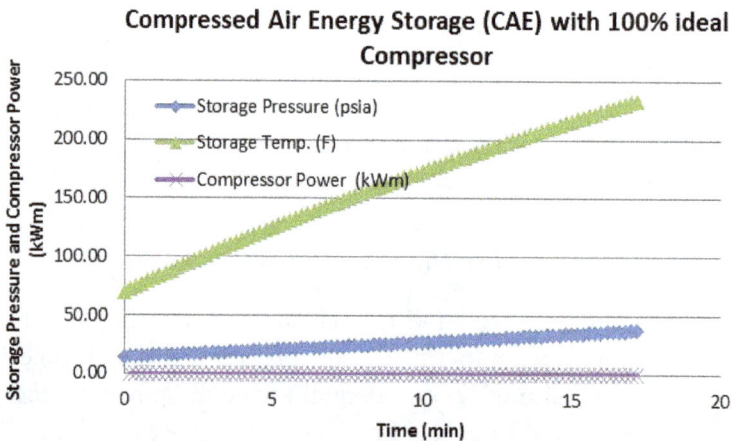

Figure C16.3.   Plots of the storage temperature (°F), storage pressure (psia) and compressor power (kW) as a function of time.

In order to compensate for the loss of energy from the storage vessel, most CAES systems either employ a combustion system, or some other means of providing heat to the air stream that is discharged from the storage vessel, before the compressed and heated air enters the turbine during the "discharge" cycle. This heat input must be accounted for and will also reduce the ERE of the CAES system, affecting the economic viability of the CAES and making the Simple Payback period longer; thus, possibly rendering the CAES as an unattractive add-on to investors or the owner of the CAES. To improve the economic viability of the CAES while also increasing the temperature of the air as it enters the turbine, some CAES sites use heat input derived from "free" energy sources such as solar renewable energy, or "renewable" energy in the form of bio-solids or bio-gas that can be generated from the decomposition of waste products.

The numerical solution to the problem also reveals a very low, instantaneous compressor power at any time step, T_n. An inspection of the spreadsheet shown as Figure C16.3 indicates a maximum power output of 29 W. This value of power was calculated at the very end of the transient "charging" period, which is when the compressor is required to perform compression work against its largest Pr of 38 psia/14.7 psia. Before the end of the "charging" cycle, the compressor was required to perform compression work against a lower Pr, and thus the power requirements were much lower. This difference in power requirement challenges the assumption that the compressor efficiency remains constant at 100% throughout the "charging" cycle. The efficiency of the compressor is lower at its off-design operating conditions. Fortunately, this can be easily accounted for by having a different compressor efficiency be used at each time step. Usually, the part-load efficiency can be obtained from the manufacturer's part-load performance maps for the compressor, as a function of Pr, flow rate and rotational speed. The same can also be done for the turbine.

The assumption of 100% efficiencies for the compressor and turbine, as well as the assumption that no loss of energy takes place during the "charging" and "discharging" cycles results in an ERE to of 100%. This means that the amount of energy used to charge the storage vessel is equal to the energy that is available for recovery. This is not a surprising conclusion, given the identity of the entire process as a Closed Mass System with a rather large CV encompassing the air entering the fixed volume storage vessel. In such a case, if there is no energy loss, it follows that the compressor power must be equal to the turbine power that would be generated from the vessel. This analysis then is equivalent to that of the Closed Mass System described in Chapter 1. It can also be stated in another way: that the change in the internal thermal energy for the "charging" and "discharging" processes must have a zero (0) change in internal thermal energy, or $\Delta U = 0$.

C16.3   Conclusion

The value of the thermodynamics model that has been described is found in its potential application in determining the numerical solution for the thermodynamics equations of almost any CAES System. Additionally, the model allows for flexibility in the input of parameters as a function of time: such as turbine and compressor efficiencies and storage vessel pressure. Furthermore, thermal energy losses due to heat transfer from the storage vessel to the surroundings can be accounted for using this model. Without the availability of this practical numerical integration model, obtaining precise solutions of the first law equations with parameters that change with time would be challenging and impractical. With this numerical solution, the various changes in parameters of the CAES system over time could be accounted for to better reflect the "real world" scenario.

Case Study 17

Concept Analysis for Increasing Storage of Hydrogen Onboard Vehicles Using a Cryogenic Matrix

The US Department of Energy (DOE) is very active in planning for a hydrogen economy. In the near future, passenger vehicles and a number of small trucks or vans will run on hydrogen fuel. Present hydrogen refueling stations have been designed to refuel a vehicle to a fuel tank pressure of 5,000 psig, in an arrangement that is illustrated in Figure C17.1. The DOE's prediction of future needs suggests that the pressure in vehicle fuel tanks will need to be increased to 10,150 psig, in order to allow an increased amount of mass to be stored in vehicle fuel tanks of a similar size to today's; thus improving the vehicles' driving range. The higher of the two discharge pressures, 10,500 psig, corresponds to the DOE's interest in storing more hydrogen mass on-board a vehicle and thus improve the driving range of a hydrogen-fueled vehicle. To achieve this, the DOE is considering the increase of storage pressures of onsite cascade storage systems from the current 6,350 psig to 12,250 psig, as shown in Figure C17.1. As a result of this pressure increase, twice as much fuel can be stored in the vehicle fuel tank.

The typical refueling sequence consists of three, on-site storage tanks, each with different storage pressures of 2,000, 4,000 and 6,350 psig, respectively. Each of these tanks are connected to the empty vehicle fuel tank and the fuel is dispensed in a controlled sequence with the lowest pressure connected first, followed by the next highest pressure, before the vehicle tank is "topped-off" by the tank with the highest pressure. The controlled sequencing of the filling of the vehicle tank maximizes the amount of hydrogen fuel that is transferred from the on-site storage tanks to the vehicles without refilling the storage tanks. This cascading of fuel from the onsite storage into the vehicle fuel tank not only enables rapid refueling (in typically less than 3 minutes), but also enables a compressor with a lower power rating

6,350 psig HYDROGEN FORECOURT REFUELING SYSTEM

Total Hydrogen Storage Vol.= 54500 scft in a 166 ft^3 Size Vessel

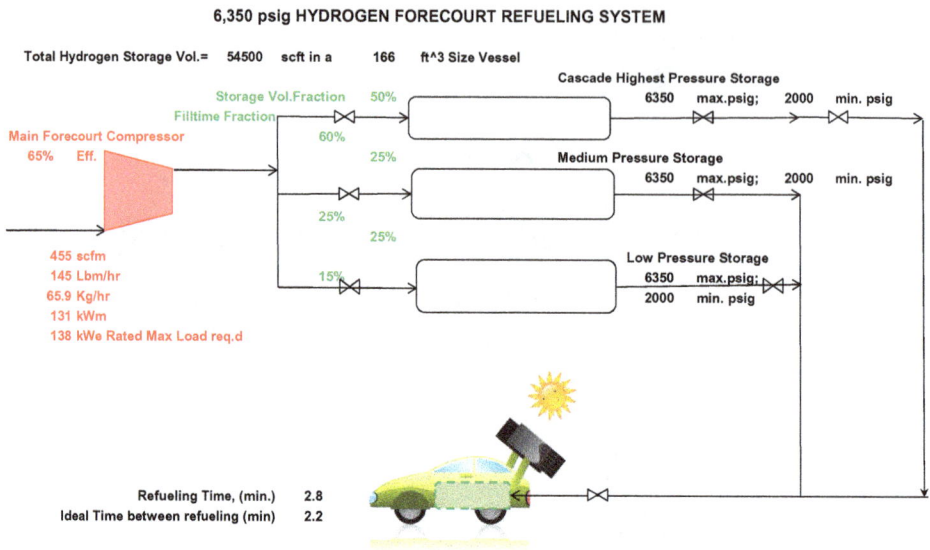

Figure C17.1. Schematic of a typical hydrogen refueling station.

to be used to achieve adequate refueling during periods of high demand. The main compressor used in conventional refueling stations is usually a positive displacement compressor with 3 or 4 stages and inter-stage cooling. Positive displacement pistons such as diaphragm-type compressors are an appropriate choice for this application where the flow rate of the dispensing fuel, hydrogen, is relatively small (less than 100 kg/h) and pressure ratio (Pr) is relatively high at 20^+:1.

Ultimately the purpose of increasing the storage pressure of a vehicle fuel tank is to increase its fuel tank capacity. The increase in fuel tank capacity increases the vehicle's driving range (i.e., drivable distance in kilometers (km)) between refueling. Unfortunately, by applying the First Law of Thermodynamics, it can be observed that the inrush of high-pressure fuel from a much larger storage tank (i.e., the on-site cascade storage system) reduces the amount of hydrogen mass that can be transferred to the closed vehicle fuel tank. The inrush of fuel causes the hydrogen mass in the vehicle tank to increase in temperature, resulting in a reduction in the density of the hydrogen in the vehicle's tank. According to the standards and codes for the material used to construct the tank, if the temperature exceeds 185°F, this increase in temperature can be significant enough to also cause damage to the tank. Figure C17.2 provides an illustration of this temperature increase caused by the inrush of fuel from an onsite cascade storage system rated at

Temperature Transients of Hydrogen Content of On-Site
Cascade and Vehicle Tank filled **without Cryogenic Matrix**

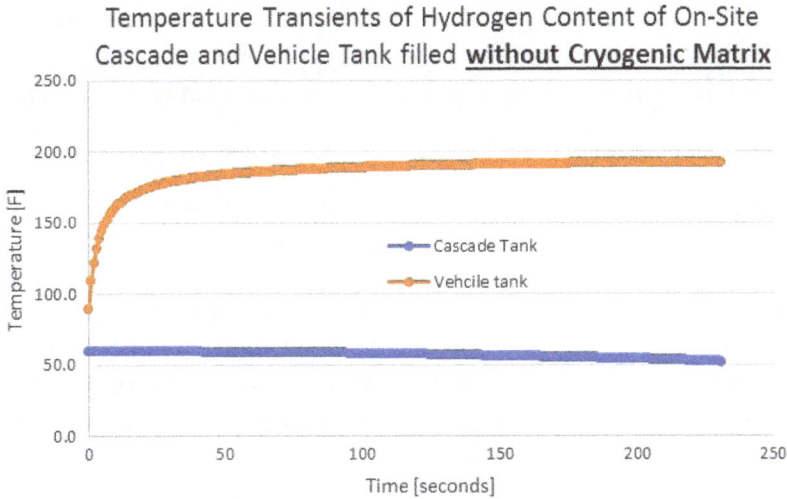

Figure C17.2. Temperature transient of the cascade and vehicle tanks without energy removal.

6,350 psig for current refueling systems, before a Cryogenic Matrix is added to the system. The use of a cryogenic matrix is proposed and analyzed in this case study as an effective method of increasing the mass transfer of hydrogen into vehicles.

The result shown in Figure C17.2 was obtained from the application of the First Law of Thermodynamics to the on-site cascade storage tank, and to the vehicle fuel tank (the receiving storage tank). This basic application of the First Law of Thermodynamics requires a transient analysis, rather than the more common steady-state analysis. A transient analysis is a numerical solution to the "charging" of the vehicle tank and the "discharging" of the cascade storage tanks with the thermodynamics properties of the hydrogen fluid determined based on the temperature at each time step. Thus, the analysis need not depend on average property values of Cp, to determine the enthalpy or Cv, to determine the oxygen for the hydrogen, resulting in a more accurate analysis of the time-history of the pressure and temperature of the cascade and vehicle tanks. A transient, numerical analysis also enables the efficiency of the compressor to be changed according to its part-load design performance if a compressor needs to be used to complete the filling of the vehicle tank.

The basic first law equation is as follows:

$$\sum \dot{Q}_{cv} + \sum \dot{M}_{in} \times h_{in} = \sum W_{shaft} + \sum \dot{M}_{out} \times h_{out} + \frac{\partial U_{cv}}{\partial t}$$

This equation can be arranged into two equations, which are simplified for use in the transient analysis for the onsite cascade storage tanks (see Equation (C17.1)) and the vehicle storage tank (see Equation (C17.2)), as follows:

$$0 = +M_{out} \times h_{out} + (M - \Delta m) \times u_{out,n+1} - (Mn) \times u_{in,n}, \quad \text{(C17.1)}$$

$$-Qloss + \Delta M_{in} \times h_{in} = (M + \Delta m) \times u_{out,n+1} - (Mn) \times u_{in,n}. \quad \text{(C17.2)}$$

Using the values obtained for internal energy, $U_{,n}$, at time, $t_{,n+1}$, the calculation proceeds by solving for the new value of internal energy, $U_{,n+1}$. The temperature at this new internal energy can then be determined by using the new specific volume, $v_{,n+1}$, as a second thermodynamics property which is necessary to define the state point where,

$$v, cascade = \frac{V_{cascade}}{(M_{cascade} - \Delta M)} \quad \text{and} \quad v, vehicle = \frac{V_{vehicle}}{(M_{vehicle} + \Delta M)}. \quad \text{(C17.3)}$$

C17.1   Feasibility analysis on the concept of using a flow-through, cryogenic matrix for hydrogen cooling

In this case study, a feasibility analysis on the concept of using a flow-through cryogenic matrix (i.e. a matrix heat exchanger in which the fluid is cooled by heat conduction from the cold material mass) that cools the hydrogen before it enters the vehicle fuel tank, has been completed using Equations (C17.1) and (C17.2). The purpose of using the cryogenic matrix is to sufficiently cool the hydrogen by thermal conduction of heat from the hot hydrogen gas to the matrix mass that has been previous cooled to cryogenic temperatures. The reduction in the final temperature of the hydrogen fuel can increase the mass transfer from the pressurized onsite cascade storage tanks.

Figure C17.3 is a schematic diagram of the major components used in a hydrogen refueling station which includes: the on-site cascade storage system of hydrogen consisting of three different tanks at three different storage pressures, the vehicle fuel tank, and the proposed flow-through, cryogenic matrix heat exchanger (also sometimes referred to as a regenerator) that can absorb some of the work of compression that is caused by the in-rush of hydrogen into a closed tank. Although the filling of the vehicle tank is only due to the very large pressure differential between the on-site storage tanks and the vehicle tank, the contents of the vehicle tank continue to be compressed and the increase in the internal thermal energy of the vehicle tank is essentially

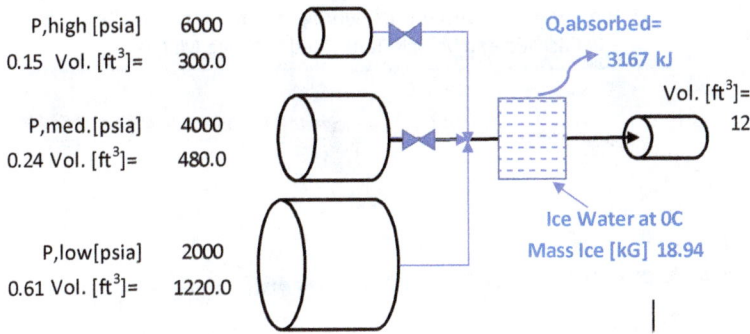

P,high [psia]	6000
0.15 Vol. [ft³]=	300.0
P,med.[psia]	4000
0.24 Vol. [ft³]=	480.0
P,low[psia]	2000
0.61 Vol. [ft³]=	1220.0

Q,absorbed= 3167 kJ

Vol. [ft³]= 12

Ice Water at 0C
Mass Ice [kG] 18.94

Figure C17.3. Schematic of an onsite cascade hydrogen storage system, the vehicle fuel tank that must be filled, and the suggested, flow-through cryogenic matrix heat exchanger.

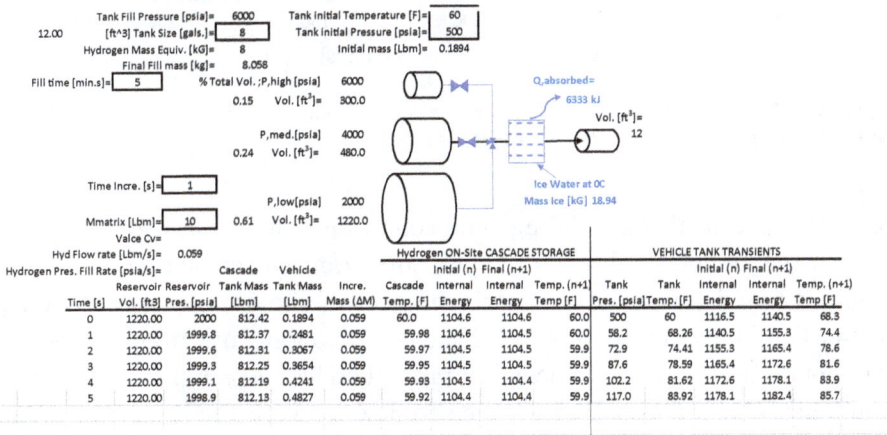

Tank Fill Pressure [psia]=	6000		Tank initial Temperature [F]=	60
12.00	[ft^3] Tank Size [gals.]=	8		Tank initial Pressure [psia]=	500
	Hydrogen Mass Equiv. [kG]=	8		Initial mass [Lbm]=	0.1894
	Final Fill mass [kg]=	8.058
Fill time [min.s]=	5	% Total Vol. ;P,high [psia]	6000

| 0.15 | Vol. [ft³]= | 300.0 |

Q,absorbed= 6333 kJ

| | P,med.[psia] | 4000 |
| | 0.24 | Vol. [ft³]= | 480.0 |

Vol. [ft³]= 12

Time Incre. [s]= 1

| | P,low[psia] | 2000 |
| Mmatrix [Lbm]= | 10 | 0.61 | Vol. [ft³]= | 1220.0 |

Ice Water at 0C
Mass Ice [kG] 18.94

Valce Cv=
Hyd Flow rate [Lbm/s]= 0.059
Hydrogen Pres. Fill Rate [psia/s]=

			Hydrogen ON-Site CASCADE STORAGE							VEHICLE TANK TRANSIENTS			
	Cascade	Vehicle					Initial (n) Final (n+1)					Initial (n) Final (n+1)		
	Reservoir	Reservoir	Tank Mass	Tank Mass	Incre.	Cascade	Internal	Internal	Temp. (n+1)	Tank	Tank	Internal	Internal	Temp. (n+1)
Time [s]	Vol. [ft3]	Pres. [psia]	[Lbm]	[Lbm]	Mass (ΔM)	Temp. [F]	Energy	Energy	Temp [F]	Pres. [psia]	Temp. [F]	Energy	Energy	Temp [F]
0	1220.00	2000	812.42	0.1894	0.059	60.0	1104.6	1104.6	60.0	500	60	1116.5	1140.5	68.3
1	1220.00	1999.8	812.37	0.2481	0.059	59.98	1104.6	1104.5	60.0	58.2	68.26	1140.5	1155.3	74.4
2	1220.00	1999.6	812.31	0.3067	0.059	59.97	1104.5	1104.5	59.9	72.9	74.41	1155.3	1165.4	78.6
3	1220.00	1999.3	812.25	0.3654	0.059	59.95	1104.5	1104.5	59.9	87.6	78.59	1165.4	1172.6	81.6
4	1220.00	1999.1	812.19	0.4241	0.059	59.93	1104.5	1104.4	59.9	102.2	81.62	1172.6	1178.1	83.9
5	1220.00	1998.9	812.13	0.4827	0.059	59.92	1104.4	1104.4	59.9	117.0	83.92	1178.1	1182.4	85.7

Figure C17.4. The first five rows of calculations in the numerical solution of the hydrogen refueling system feasibility analysis, using Equations (C17.1) to (C17.3).

the energy that would be otherwise provided by a mechanical compressor. Figure C17.4 is a screenshot of the first five rows of the calculations in the numerical solution of the hydrogen refueling system feasibility analysis, performed using an Excel calculation model.

The transient temperature history of the vehicle fuel tank during the rapid filling process is shown in Figure C17.5. The use of a cooling medium, such as a solid ice matrix, can reduce the final temperature of the hydrogen fuel in the vehicle tank from 180°F to 105°F. Reducing the temperature of the hydrogen fuel from as high as 180°F to 105°F can be achieved although the initial temperature of the matrix (cooled to just ice temperatures or extreme cryogenic temperatures), the ambient temperature, the cascade

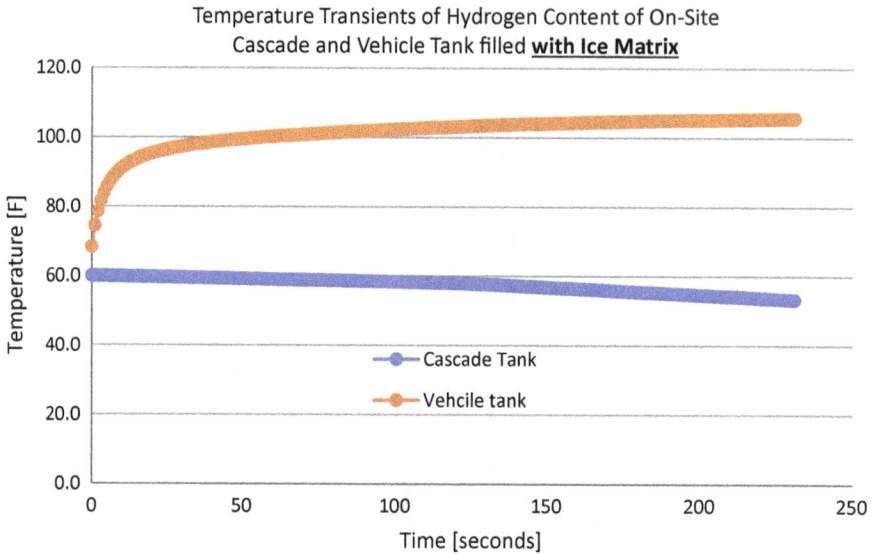

Figure C17.5. Temperature transient of the cascade and vehicle tanks with energy removal using an ice matrix with a mass of approximately 20 kg.

pressures and tank sizes, filling time constraints, and initial hydrogen pressure in the vehicle tank. The temperature transient shown in Figure C17.5 assumes that ice is used to absorb the heat of the sublimation of ice equal to 144 Btu/lbm, as well as some of the compression work from the hydrogen fuel before it enters the vehicle fuel tank. The ice mass must be arranged in a manner which allows the hydrogen to flow through the ice matrix without absorbing the water. As a result of this cooling process, approximately 13% more hydrogen can be stored in the vehicle fuel tank during refueling. This is determined by comparing the density of hydrogen at 180°F to its density at 105°F using the Perfect Gas Law for gases. This ratio is shown here:

$$\frac{\rho_{with\ cooling}}{\rho_{without\ cooling}} = \frac{(460 + 180)}{(460 + 105)}$$

$$\frac{\rho_{with\ cooling}}{\rho_{without\ cooling}} = 1.13$$

The use of an ice mass is essentially an energy storage scheme which avoids the use of a live refrigeration system during the 3-minute refueling operation, as the ice mass can be prepared "offline"; That is, the cold matrix can be initially cooled when the refueling system is not in service and when the energy needed for producing the cold matrix is at a much lower cost. This is similar to the concept of using compressed air energy storage or pumped hydro-energy storage to store energy that might be otherwise wasted. Thus, using this energy storage scheme, the cost of the power consumption of the refrigeration system that is used to freeze the water is reduced. Instead of the use of a single ice matrix, as shown in Figure C17.3, which requires 20 kg of ice per refueling sequence, an "n" number of identical ice masses could alternatively be used. The use of individual cold matrixes would require that a system of piping be designed to route the hydrogen that is being dispensed through a previously "cold charged" matrix while the depleted cold matrixes are being cold-charged again.

Compressible Gas Dynamics: Design of a Converging-Diverging Nozzle

The First and Second Laws of Thermodynamics may be applied in the analysis of the flow of fluids in all devices encountered in "real world" applications. With suitable assumptions for the seven basic mechanical systems first identified in Chapter 1 and then diagramed in Chapter 3, Table 3.8 — i.e., the turbine, compressor, pump, heat exchanger, valve, nozzle and diffuser — the energy equations derived from these thermodynamics laws can be used with confidence in the thermodynamics analysis of a single device or a system comprising of more than one of these mechanical devices. However, the thermodynamics analyses conducted can often yield some nonintuitive results. An example of this can be seen by applying the thermodynamic analysis of a converging nozzle, or a converging-diverging nozzle also known as a de Laval nozzle if the cross section is circular, as shown in Figure C18.1. At first glance, this device looks like the nozzle and diffuser system reviewed in Chapter 3 (Table 3.8), except that they appear to be placed in series with each other. But in fact, the devise in Figure C18.1 is properly and completely identified as a nozzle. That is, the flow stream through the device continues to expand in pressure from the inlet to the outlet. If the diverging portion of the nozzle shown in Figure C18.1 were actually a diffuser, the pressure would start to increase to the right of the minimum flow area. The juxtaposition of the converging and diverging parts of the conduit results in nonintuitive results when the First and Second Law of Thermodynamics are properly applied to the device. It will be found that the cross section of the conduit must start to increase, i.e., diverge, in order to account for the rapidly decreasing density of the fluid while maintaining the same mass flow rate through the conduit even as the pressure in the conduit continues to be lowered; eventually matching the exit pressure that is being maintained by some other controlling state. The most startling result is that the velocity of the flow stream continues

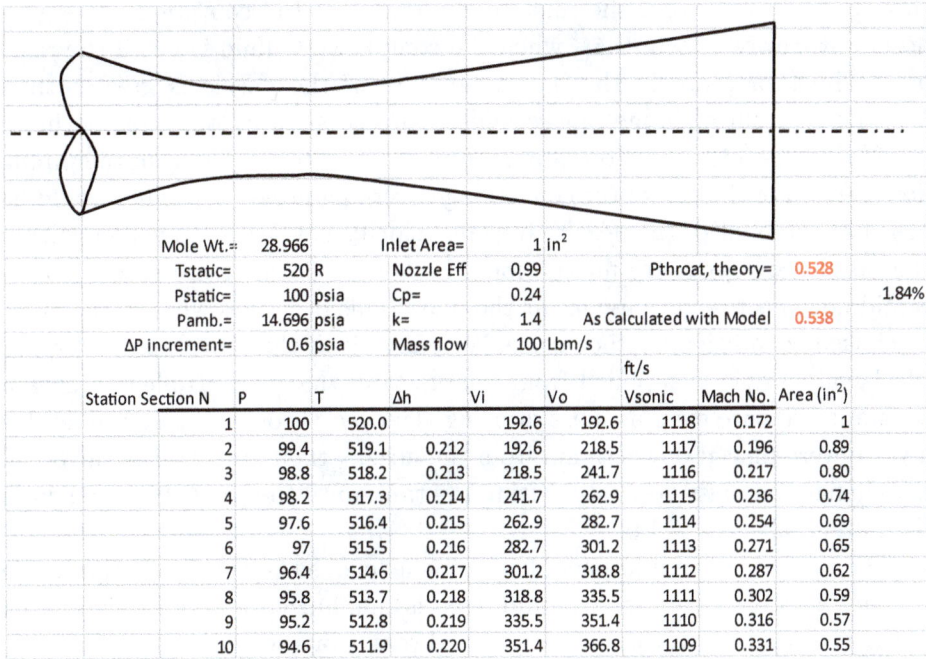

Mole Wt.=	28.966		Inlet Area=	1 in²			
Tstatic=	520 R		Nozzle Eff	0.99		Pthroat, theory=	0.528
Pstatic=	100 psia		Cp=	0.24			1.84%
Pamb.=	14.696 psia		k=	1.4	As Calculated with Model	0.538	
ΔP increment=	0.6 psia		Mass flow	100 Lbm/s			

Station Section N	P	T	Δh	Vi	Vo	Vsonic (ft/s)	Mach No.	Area (in²)
1	100	520.0		192.6	192.6	1118	0.172	1
2	99.4	519.1	0.212	192.6	218.5	1117	0.196	0.89
3	98.8	518.2	0.213	218.5	241.7	1116	0.217	0.80
4	98.2	517.3	0.214	241.7	262.9	1115	0.236	0.74
5	97.6	516.4	0.215	262.9	282.7	1114	0.254	0.69
6	97	515.5	0.216	282.7	301.2	1113	0.271	0.65
7	96.4	514.6	0.217	301.2	318.8	1112	0.287	0.62
8	95.8	513.7	0.218	318.8	335.5	1111	0.302	0.59
9	95.2	512.8	0.219	335.5	351.4	1110	0.316	0.57
10	94.6	511.9	0.220	351.4	366.8	1109	0.331	0.55

Figure C18.1. A screenshot of the output of the thermodynamics analysis of a converging-diverging nozzle.

to increase, as it must to compensate for the lower fluid density, reaching and then exceeding the velocity of sound in the fluid (more information on this in the next section). Whether the fluid is a liquid or a vapor affects the exit velocity of the converging-diverging nozzle. The density of a vapor is strongly dependent on its state point pressure and temperature, while the density of a fluid in the liquid phase is not dependent on its temperature or pressure. If the fluid is in its vapor phase, it is compressible. A compressible fluid that is expanding isentropically will behave in a manner different from an incompressible fluid expanding with the same pressure ratio.

C18.1 Sonic velocity and the "choked" condition of the nozzle

Sonic velocity is defined as the speed of sound in the fluid of choice, which is most often air or the products of combustion. The most dramatic example of the use of combustion products in a converging — diverging nozzle, where the exit velocity can achieve three times the velocity of sound in the compressible products of combustion, such as that used to propel launch vehicles. Sonic velocity, more accurately, is the velocity of the pressure wave fluctuations through the fluid, from the inlet to the outlet of the converging-diverging nozzle. It can be shown that, when the compressible fluid is air, if the ratio of the inlet pressure to the outlet pressure of the nozzle is greater than 1.9, the flow through the device will reach the sonic velocity of the fluid. During supersonic operation, if a microphone was to be placed in the piping downstream of the exit of the converging-diverging nozzle, no sound would be heard at the inlet of the nozzle. If sonic velocity is reached anywhere within the nozzle, any pressure fluctuations that occur downstream of the nozzle would not affect what happens at the inlet. This describes what is known as a "choked" condition for the nozzle. For example, if the converging-diverging (de Laval) nozzle is supplied from a large tank that is controlled to have a constant pressure even as mass is being removed through the nozzle, and if a valve were placed downstream from the nozzle, then opening the valve will not change the mass flow rate exiting the tank until the pressure ratio across the nozzle falls below 1.9.

The most significant impact of reaching a "choked" condition in the nozzle is that the mass flow rate per unit area is maximized at the smallest area of the nozzle: an area that is called the throat. For the converging-diverging nozzle shown in Figure C18.1, the throat is the point at which the nozzle starts to diverge after reaching the minimum cross-sectional area in the converging section. For a nozzle in the "choked" condition, this maximum flow rate per unit area does not change, even if the exit pressure is reduced. The flow rate can only be changed by increasing the inlet pressure.

C18.2   Thermodynamics analysis of the converging-diverging nozzle

The basic thermodynamics equations which describe the behavior of a compressible fluid as it passes through a nozzle that is not receiving or losing heat energy is shown at the end of this case study. However, at this point, the reader will find it useful to be reminded that only Equations (C18.1) and (C18.2) are needed to thermodynamically model the converging-diverging nozzle. For a more detailed review of these equations and why they apply in this analysis, the reader is referred to Chapters 3 and 5. This will give the reader more confidence that the nonintuitive behavior of the converging-diverging nozzle can, in fact, be readily analyzed by carefully applying just these equations.

The converging-diverging nozzle analysis shown in Figure C18.1 is a numerical solution obtained by applying the First Law of Thermodynamics (energy conservation) and the continuity equation (mass conservation) at each cross-section of the nozzle. The columns shown in Figure C18.1 are arranged to enable the cross-sectional area of the nozzle to be determined at each of the "n" plane sections made through the nozzle. Each row in Figure C18.1 represents a plane section through the nozzle. The calculation proceeds by using a step change (reduction) in the pressure at each section to determine the exit temperature of the plane section, "n", the change in enthalpy and, from Equation (C18.2b), the new velocity of the fluid at the exit of that cross section. With the new velocity and new density values calculated as a function of the fluid properties of pressure and temperature at the cross section, "n", the new exit cross-section area can be determined using the equation for the continuity mass flow rate shown in Equation (C18.3). This last calculation in the row, based on the mass continuity equation, is a critical step in determining the familiar "bell" shape of the converging-diverging nozzle — familiar, that is, to anyone who has carefully observed the launch of a rocket and the two or three flame spewing nozzles that appear at the bottom of the rocket engines.

The temperature of the fluid at the end of each cross-section is determined from Equation (C18.1).

$$T_{out} = T_{in} \times \left(1 - \left(1 - \left(\frac{P_{out}}{P_{in}}\right)^{\frac{(k-1)}{k}}\right) \times Nozzle\ Coef.\right). \qquad (C18.1)$$

The enthalpy change for that section is determined from Equation (C18.2a) and the result is used in Equation (C18.2b) to determine the exit velocity of the fluid.

$$\Delta h = C_p \times T_{in} \times \left(1 - \left(\frac{P_{out}}{P_{in}}\right)^{\frac{(k-1)}{k}}\right) \times Nozzle\ Coef., \qquad (C18.2a)$$

$$\Delta h = \frac{\left(V_{out}^2 - V_{in}^2\right)}{2 \times g_c} \qquad (C18.2b)$$

The exit area of the plane section can then be determined from the mass continuity equation, Equation (C18.3).

$$(\rho_{in} \times A_{in} \times V_{in}) = (\rho_{out} \times A_{out} \times V_{out}). \qquad (C18.3)$$

A plot of the area as a function of the nozzle sections is shown in Figure C18.2.

It is also interesting to note the Mach number at each of the sections. The Mach number is the ratio of the velocity of the flow stream to the sonic velocity, calculated at the pressure and temperature at that section.

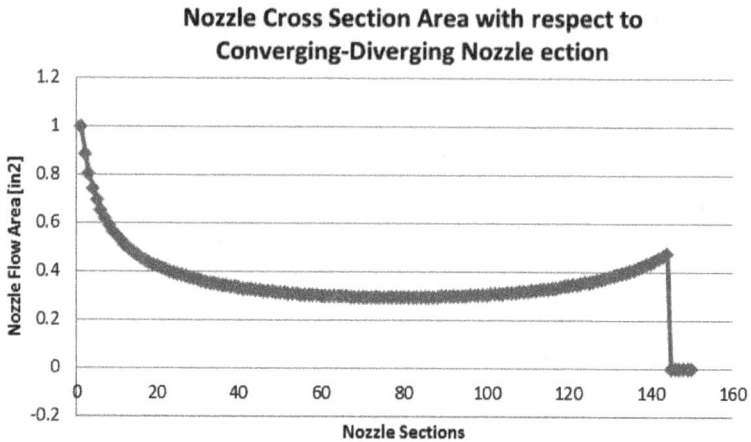

Figure C18.2. Plot of the converging-diverging area variation at each plane section. The exit area would continue to increase for higher and higher exit sonic velocities.

The sonic velocity can be shown to be a function of the partial derivative of the pressure change with respect to the density change of the fluid. This relationship is shown in Equation (C18.4a).

$$c = \sqrt{\left(\frac{\partial p}{\partial \rho}\right)_s}. \tag{C18.4a}$$

By applying this equation with a Perfect Gas Law, the sonic velocity, c, can be shown to be a single function of temperature, T, according to Equation (C18.4b).

$$c = \sqrt{g_c \times k \times \frac{R_u}{Mole\,Wt.} \times T}, \tag{C18.4b}$$

where k is the specific heat ratio for the perfect gas that has a molecular weight (MoleWt.) as shown in the equation.

Thus, the Mach number is shown in Equation (C18.4c)

$$Mach\;No.(M) = \frac{V}{c}. \tag{C18.4c}$$

The complete output of the thermodynamics model of the converging-diverging model is shown in Table C18.1. An inspection of the complete output of the nozzle design analysis shown in Table C18.1 reveals that the Mach number is 1 at the minimum area, also called the "throat", of the nozzle. With a thermodynamics model, such as shown in Figure C18.1, it is easy to demonstrate, by changing the inlet pressure and/or temperature or mass flowrate, that the minimum area is always where the velocity is equal

Table C18.1. The complete numerical output from a thermodynamics model of a converging-diverging nozzle.

Station Section N	P	T	Δh	Vi	Vo	Vsonic	Mach No.	Area (in^2)
1	100	520		192.6108771	192.6108771	1118.184851	0.172253163	1
2	99.4	519.1155881	0.212258856	192.6108771	218.4806579	1117.233546	0.195555046	0.885405379
3	98.8	518.2273394	0.213179679	218.4806579	241.6915017	1116.277299	0.216515647	0.803858228
4	98.2	517.3352139	0.214110126	241.6915017	262.9493953	1115.316052	0.235762226	0.742105828
5	97.6	516.4391708	0.215050357	262.9493953	282.6961187	1114.34975	0.253687066	0.693309222
6	97	515.5391685	0.216000535	282.6961187	301.229835	1113.378334	0.270554784	0.653535847
7	96.4	514.6351651	0.216960827	301.229835	318.7629922	1112.401745	0.28655384	0.620343155
8	95.8	513.7271176	0.217931406	318.7629922	335.4532844	1111.419924	0.30182407	0.592123638
9	95.2	512.8149824	0.218912446	335.4532844	351.4215704	1110.432809	0.316472611	0.567770478
10	94.6	511.8987152	0.219904128	351.4215704	366.7628943	1109.440338	0.330583702	0.54649347
11	94	510.9782709	0.220906635	366.7628943	381.5536003	1108.442447	0.344224999	0.527711415
12	93.4	510.0536035	0.221920154	381.5536003	395.85611	1107.439073	0.357451818	0.510986065
13	92.8	509.1246665	0.22294488	395.85611	409.7222373	1106.43015	0.37031008	0.495979918
14	92.2	508.1914123	0.223981009	409.7222373	423.1955541	1105.415611	0.382838409	0.482428293
15	91.6	507.2537926	0.225028743	423.1955541	436.3131182	1104.395387	0.395069668	0.470120321
16	91	506.311758	0.22608829	436.3131182	449.1067623	1103.369409	0.407032095	0.458885666
17	90.4	505.3652586	0.227159861	449.1067623	461.6040726	1102.337607	0.418750181	0.44858504
18	89.8	504.4142433	0.228243674	461.6040726	473.8291446	1101.299908	0.430245332	0.439103299
19	89.2	503.4586602	0.229339951	473.8291446	485.8031738	1100.25624	0.441536395	0.430344324
20	88.6	502.4984563	0.230448921	485.8031738	497.5449247	1099.206527	0.452640075	0.422227169
21	88	501.5335779	0.231570819	497.5449247	509.0711062	1098.150692	0.463571266	0.414683122
22	87.4	500.5639701	0.232705884	509.0711062	520.3966762	1097.088659	0.474343319	0.407653434
23	86.8	499.5895769	0.233854362	520.3966762	531.5350905	1096.020348	0.484968269	0.401087538
24	86.2	498.6103415	0.235016506	531.5350905	542.4985082	1094.945677	0.495457007	0.394941647
25	85.6	497.6262057	0.236192575	542.4985082	553.2979625	1093.864565	0.505819441	0.389177633
26	85	496.6371106	0.237382835	553.2979625	563.9435033	1092.776927	0.516064614	0.383762125
27	84.4	495.6429958	0.238587559	563.9435033	574.4443185	1091.682677	0.526200819	0.378665775
28	83.8	494.6437998	0.239807027	574.4443185	584.8088359	1090.581728	0.536235681	0.373862659
29	83.2	493.6394601	0.241041526	584.8088359	595.0448103	1089.473991	0.546176242	0.369329787
30	82.6	492.6299128	0.242291352	595.0448103	605.1593992	1088.359373	0.556029023	0.365046689

Table C18.1. (Continued)

Station Section N	P	T	Δh	Vi	Vo	Vsonic	Mach No.	Area (in²)
31	82	491.6150928	0.243556809	605.1593992	615.1592264	1087.237783	0.565800082	0.360995076
32	81.4	490.5949336	0.244838206	615.1592264	625.0504392	1086.109123	0.575495064	0.357158553
33	80.8	489.5693675	0.246135866	625.0504392	634.8387559	1084.973299	0.585119243	0.353522377
34	80.2	488.5383253	0.247450117	634.8387559	644.5295096	1083.83021	0.594677565	0.350073253
35	79.6	487.5017366	0.248781296	644.5295096	654.1276846	1082.679755	0.604174671	0.346799158
36	79	486.4595293	0.250129752	654.1276846	663.6379501	1081.52183	0.613614937	0.343689193
37	78.4	485.41163	0.251495843	663.6379501	673.0646888	1080.35633	0.623002494	0.340733452
38	77.8	484.3579636	0.252879935	673.0646888	682.4120231	1079.183147	0.632341253	0.337922915
39	77.2	483.2984535	0.254282407	682.4120231	691.6838379	1078.00217	0.641634922	0.335249348
40	76.6	482.2330217	0.255703649	691.6838379	700.8838011	1076.813286	0.650887029	0.332705223
41	76	481.1615881	0.25714406	700.8838011	710.0153822	1075.61638	0.660100939	0.330283642
42	75.4	480.0840712	0.258604054	710.0153822	719.0818685	1074.411333	0.669279862	0.327978275
43	74.8	479.0003876	0.260084055	719.0818685	728.0863802	1073.198025	0.678426873	0.325783305
44	74.2	477.9104522	0.2615845	728.0863802	737.0318837	1071.976332	0.687544922	0.323693377
45	73.6	476.8141779	0.26310584	737.0318837	745.9212037	1070.746128	0.696636844	0.321703554
46	73	475.7114756	0.264648539	745.9212037	754.7570343	1069.507283	0.705705371	0.319809282
47	72.4	474.6022545	0.266213076	754.7570343	763.5419489	1068.259665	0.714753139	0.318006351
48	71.8	473.4864214	0.267799943	763.5419489	772.2784096	1067.003138	0.723782697	0.31629087
49	71.2	472.3638812	0.269409649	772.2784096	780.9687749	1065.737564	0.732796517	0.314659236
50	70.6	471.2345365	0.271042719	780.9687749	789.6153082	1064.4628	0.741796997	0.31310811
51	70	470.0982878	0.272699694	789.6153082	798.2201841	1063.1787	0.75078647 1	0.311634398
52	69.4	468.9550331	0.274381132	798.2201841	806.785495	1061.885115	0.759767214	0.31023523
53	68.8	467.804668	0.27608761	806.785495	815.3132574	1060.581892	0.768741446	0.308907942
54	68.2	466.6470858	0.277819723	815.3132574	823.8054169	1059.268875	0.777711341	0.307650061
55	67.6	465.4821771	0.279578086	823.8054169	832.2638534	1057.945903	0.786679027	0.306459292
56	67	464.3098299	0.281363334	832.2638534	840.690386	1056.61281	0.795646597	0.305333503
57	66.4	463.1299294	0.283176123	840.690386	849.0867771	1055.269428	0.804616105	0.304270716
58	65.8	461.942358	0.285017132	849.0867771	857.4547366	1053.915583	0.81358958	0.303269095
59	65.2	460.7469953	0.286687061	857.4547366	865.7959258	1052.551097	0.822569021	0.302326937
60	64.6	459.5437176	0.288786638	865.7959258	874.1119607	1051.175788	0.831556406	0.301442663

(Continued)

Table C18.1. (Continued)

Station Section N	P	T	Δh	Vi	Vo	Vsonic	Mach No.	Area (in²)
61	64	458.3323984	0.290716611	874.1119607	882.4044154	1049.789467	0.840553695	0.300614813
62	63.4	457.1129077	0.292677759	882.4044154	890.6748254	1048.391943	0.849562829	0.299842034
63	62.8	455.8851124	0.294670886	890.6748254	898.9246904	1046.983018	0.858585741	0.299123079
64	62.2	454.6488756	0.296696824	898.9246904	907.1554769	1045.562487	0.867624354	0.298456798
65	61.6	453.4040571	0.298756438	907.1554769	915.3686213	1044.130142	0.876680582	0.297842134
66	61	452.1505129	0.30085062	915.3686213	923.5655318	1042.685769	0.885756341	0.297278117
67	60.4	450.88095	0.302980298	923.5655318	931.7475914	1041.229147	0.894853543	0.296763864
68	59.8	449.6166515	0.305146434	931.7475914	939.9161595	1039.760048	0.903974106	0.296298568
69	59.2	448.3360264	0.307350026	939.9161595	948.0725746	1038.278238	0.913119952	0.295881501
70	58.6	447.0460593	0.309592108	948.0725746	956.2181561	1036.783478	0.922293011	0.295512009
71	58	445.7465853	0.311873756	956.2181561	964.3542063	1035.27552	0.931495228	0.295189508
72	57.4	444.4374349	0.314196085	964.3542063	972.4820125	1033.754108	0.940728559	0.294913483
73	56.8	443.1184339	0.316560256	972.4820125	980.6028485	1032.218981	0.949994978	0.294683484
74	56.2	441.7894027	0.318967474	980.6028485	988.7179768	1030.669868	0.9592648	0.294499128
75	55.6	440.4501569	0.321418993	988.7179768	996.8286501	1029.10649	0.968635083	0.294360093
76	55	439.1005064	0.323916117	996.8286501	1004.936113	1027.528559	0.978012829	0.294266118
77	54.4	437.7402556	0.326460202	1004.936113	1013.041604	1025.935779	0.987431792	0.294217006
78	53.8	436.3692028	0.329052662	1013.041604	1021.146358	1024.327845	0.996894073	0.294212614
79	53.2	434.9871404	0.331694969	1021.146358	1029.251603	1022.70444	1.006401814	0.294252864
80	52.6	433.5938544	0.334388656	1029.251603	1037.358571	1021.065239	1.01595719	0.294337732
81	52	432.1891239	0.337135323	1037.358571	1045.46849	1019.409905	1.02556242	0.294467255
82	51.4	430.7727212	0.339936637	1045.46849	1053.582591	1017.738091	1.035219769	0.294641525
83	50.8	429.3444115	0.34279434	1053.582591	1061.70211	1016.049437	1.044931547	0.294860696
84	50.2	427.9039521	0.345710248	1061.70211	1069.828285	1014.343572	1.054700119	0.295124979
85	49.6	426.4510927	0.34868626	1069.828285	1077.962364	1012.62011	1.064527904	0.295434645
86	49	424.9855745	0.351724361	1077.962364	1086.105599	1010.878656	1.074417382	0.295790023
87	48.4	423.5071302	0.354826624	1086.105599	1094.259257	1009.118796	1.084371098	0.296191507
88	47.8	422.0154835	0.357995222	1094.259257	1102.424614	1007.340106	1.094391663	0.296639549
89	47.2	420.5103484	0.361232427	1102.424614	1110.602958	1005.542142	1.104481763	0.29713467
90	46.6	418.9914291	0.364540619	1110.602958	1118.795594	1003.724447	1.114644161	0.297677453

Table C18.1. (*Continued*)

Station Section N	P	T	Δh	Vi	Vo	Vsonic	Mach No.	Area (in²)
91	46	417.4584196	0.367922294	1118.795594	1127.003844	1001.886546	1.124881703	0.298268549
92	45.4	415.9110026	0.371380068	1127.003844	1135.22905	1000.027947	1.135197324	0.29890868
93	44.8	414.3488498	0.374916688	1135.22905	1143.472571	998.1481385	1.145594053	0.29959864
94	44.2	412.7716204	0.378535036	1143.472571	1151.735793	996.2465892	1.156075018	0.300339299
95	43.6	411.1789615	0.382238141	1151.735793	1160.020125	994.3227475	1.166643455	0.301131604
96	43	409.5705066	0.38602919	1160.020125	1168.327004	992.3760399	1.177302713	0.301976585
97	42.4	407.9458752	0.389911532	1168.327004	1176.657896	990.4058696	1.188056263	0.302875359
98	41.8	406.3046723	0.393888697	1176.657896	1185.014299	988.4116159	1.198907702	0.303829129
99	41.2	404.6464873	0.397964403	1185.014299	1193.397745	986.3926323	1.209860766	0.304839198
100	40.6	402.9708932	0.402142568	1193.397745	1201.809805	984.3482451	1.220919335	0.305906966
101	40	401.277446	0.40642733	1201.809805	1210.252089	982.2777522	1.232087448	0.307033939
102	39.4	399.5656833	0.410823056	1210.252089	1218.726248	980.180421	1.243369304	0.308221735
103	38.8	397.8351234	0.415334365	1218.726248	1227.233983	978.0554868	1.254769284	0.309472091
104	38.2	396.0852645	0.419966139	1227.233983	1235.777041	975.9021511	1.266291954	0.310786872
105	37.6	394.3155831	0.42472355	1235.777041	1244.357226	973.7195789	1.277942082	0.312168075
106	37	392.5255327	0.429612076	1244.357226	1252.976394	971.5068973	1.289724651	0.313617843
107	36.4	390.714543	0.43463753	1252.976394	1261.636467	969.2631923	1.301644876	0.315138472
108	35.8	388.8820177	0.43980608	1261.636467	1270.33943	966.9875067	1.313708214	0.316732422
109	35.2	387.0273332	0.445124284	1270.33943	1279.087338	964.678837	1.325920388	0.318402332
110	34.6	385.1498369	0.450599115	1279.087338	1287.882322	962.3361305	1.338287404	0.320151029
111	34	383.2488452	0.456238001	1287.882322	1296.726594	959.9582821	1.35081557	0.321981549
112	33.4	381.3236416	0.462048858	1296.726594	1305.622452	957.5441301	1.363511519	0.323897146
113	32.8	379.3734744	0.468040137	1305.622452	1314.572286	955.0924531	1.376382236	0.325901314
114	32.2	377.3975541	0.474220864	1314.572286	1323.578588	952.6019649	1.38943508	0.327997809
115	31.6	375.3950512	0.480600698	1323.578588	1332.643955	950.0713108	1.40267782	0.330190666
116	31	373.365093	0.487189981	1332.643955	1341.771101	947.4990613	1.4161866	0.332484229
117	30.4	371.3067604	0.493999808	1341.771101	1350.962864	944.8837077	1.42976628	0.334883174
118	29.8	369.219085	0.501042091	1350.962864	1360.222214	942.2236551	1.443629871	0.337392543
119	29.2	367.1010449	0.508329642	1360.222214	1369.552267	939.5172161	1.457719182	0.340017777

(*Continued*)

Table C18.1. (Continued)

Station Section N	P	T	Δh	Vi	Vo	Vsonic	Mach No.	Area (in^2)
120	28.6	364.9515605	0.515876257	1369.552267	1378.956295	936.7626033	1.472044561	0.342764752
121	28	362.7694904	0.523696819	1378.956295	1388.437737	933.9579212	1.486617015	0.345639828
122	27.4	360.5536262	0.531807403	1388.437737	1398.000216	931.1011569	1.501448264	0.348649889
123	26.8	358.302687	0.540225405	1398.000216	1407.64755	928.1901705	1.516550805	0.351802408
124	26.2	356.0153134	0.54896968	1407.64755	1417.383774	925.2226836	1.531937985	0.355105498
125	25.6	353.6900604	0.558060702	1417.383774	1427.213155	922.1962673	1.547624086	0.358567989
126	25	351.3253907	0.567520744	1427.213155	1437.140214	919.1083282	1.563624406	0.362199505
127	24.4	348.9196653	0.577374085	1437.140214	1447.16975	915.9560931	1.579955372	0.366010556
128	23.8	346.4711351	0.58764724	1447.16975	1457.306862	912.7365919	1.596634643	0.370012637
129	23.2	343.97793	0.598369236	1457.306862	1467.556985	909.446378	1.613681248	0.374218355
130	22.6	341.438047	0.609571916	1467.556985	1477.925914	906.082806	1.631115727	0.378641555
131	22	338.8493374	0.621290296	1477.925914	1488.419847	902.6414084	1.648960299	0.383297488
132	21.4	336.2094917	0.633562978	1488.419847	1499.045421	899.1184664	1.667239053	0.388202985
133	20.8	333.5160224	0.64643263	1499.045421	1509.809762	895.509678	1.685978163	0.393376674
134	20.2	330.7662451	0.659946546	1509.809762	1520.720536	891.8103822	1.705206137	0.398839229
135	19.6	327.9572564	0.674157296	1520.720536	1531.786012	888.0155171	1.724954106	0.404613658
136	19	325.0859085	0.689123504	1531.786012	1543.015126	884.1195717	1.745256157	0.41072565
137	18.4	322.1487803	0.704910757	1543.015126	1554.417565	880.1165306	1.766149721	0.417203976
138	17.8	319.1421441	0.72159269	1554.417565	1566.003853	875.9998093	1.787676021	0.42408098
139	17.2	316.0619263	0.73925228	1566.003853	1577.785465	871.7621796	1.809880609	0.431393155
140	16.6	312.9036621	0.757983406	1577.785465	1589.774942	867.3956811	1.832813994	0.439181842
141	16	309.6624423	0.777892735	1589.774942	1601.986043	862.8915176	1.85653238	0.447494075
142	15.4	306.3328506	0.799102014	1601.986043	1614.43391	858.2399342	1.881098567	0.456383617
143	14.8	302.9088886	0.821750883	1614.43391	1627.135274	853.4300706	1.906583011	0.465912227
144	14.2	299.3838871	0.846000351	1627.135274	1640.108693	848.4497848	1.933065129	0.476151233

to the sonic velocity. The constraint of only having a Mach number of 1 at the minimum cross section essentially limits the amount of mass flow rate that the nozzle will allow for the given pressure ratio.

Another observation that can be made from Table C18.1 is that the ratio of the pressure at the throat to the inlet pressure is 0.528. This ratio is dependent only on the specific heat ratio of the fluid, and is determined from Equation (C18.5). For air with $k = 1.4$, the ratio is 0.528. For steam with $k = 1.32$, the pressure ratio is 0.542. That ratio can be used to determine when a nozzle or an orifice that is subject to an isentropic (no entropy change, hence no heat transfer exchange) expansion from the inlet area to the outlet area.

$$\frac{P_{throat}}{P_{inlet}} = \frac{1}{\left(1 + \frac{(k-1)}{2}\right)^{\frac{k}{k-1}}}. \tag{C18.5}$$

C18.3 Parameters used in modeling the thermodynamics of compressible fluids

The numerical solution shown in Table C18.1 provides some intuition on the behavior of compressible fluids that is not readily available from an application of the energy and continuity equations. With this new-found intuition as a means of solving "real world" problems, it is now worthwhile to present some of the parameters that are relevant in modelling the thermodynamics of compressible fluids.

The stagnation enthalpy, defined in Equation (C18.6), is the first parameter that is useful in compressible fluid dynamics. Stagnation enthalpy is derived directly from the first law equation).

$$ho_1 = h_1 + \frac{V_1^2}{2 \times g_c}. \tag{C18.6}$$

This equation is useful in that it does not only include the thermal energy represented by the enthalpy parameter, h_1, but also the kinetic energy which becomes significant at high velocities.

The enthalpy difference has also been defined as $\Delta h = C_p \times \Delta T$. That is, assuming that no phase change occurs, then all of the energy transfer to or from a fluid affects only the temperature change of the fluid. This equation can then be used to define the stagnation temperature, as shown in Equation (C18.7a).

$$To_1 = T_1 + \frac{V_1^2}{2 \times C_p \times g_c}. \tag{C18.7a}$$

This expression can be reduced to a more convenient relationship with Equations (C18.4b) and (C18.4c), and with some necessary algebra and the application of the thermodynamics relationship: $C_p - C_v = \frac{R_u}{MoleWt.}$ and $\left(\frac{k-1}{k}\right) = \frac{R_u}{MoleWt.}$, resulting in:

$$T_{o_1} = T_1 \times \left(1 + \frac{(k-1) \times M^2}{2}\right). \tag{C18.7b}$$

One last parameter, the stagnation pressure, p_o, associates the thermal and kinetic energies with pressure as shown in Equation (C18.8)

$$\frac{p_o}{p_1} = \left(\frac{T_o}{T_1}\right)^{\frac{k}{(k-1)}}, \tag{C18.8}$$

It is noted that the stagnation pressure should not be confused with an expression that can also be derived from the First Law of Thermodynamics: the Bernoulli equation, which is shown below:

$$P + \frac{V^2 \times \rho}{2 \times g_c} = Constant.$$

By introducing the definition of the Mach number from Equation (C18.4c) and solving for V using the definition of sonic velocity in Equation (C18.4b), the following equation for velocity can be derived:

$$V = M \times \sqrt{g_c \times k \times \frac{R_u}{MoleWt.} \times T}.$$

This equation for velocity is substituted into Equation (C18.7a) to derive an expression with the necessary algebra manipulation, for the stagnation temperature as a function of the Mach number, which is a very useful formulation.

$$T_o = T_1 \times \left(1 + \frac{(k-1)}{2} \times M^2\right).$$

And thus, the stagnation pressure can be derived to be:

$$\frac{p_o}{p_1} = \left(1 + \frac{(k-1)}{2} \times M^2\right)^{\frac{k}{(k-1)}}.$$

From this equation for stagnation pressure, the ratio between the inlet and throat pressures can be calculated using Mach number = 1 and Equation (C18.5). In other words: even if the exit pressure from the nozzle were to be reduced, the mass flow rate through the device does not increase, unless the inlet pressure of the device is increased. This is a direct consequence of the sonic velocity achieved at the throat area, which means that there is a maximum mass per unit area that cannot be exceeded even with the decrease in exit pressure. It is also true that the presence of velocities of the fluid above Mach 1 after the throat limits any downstream disturbances from traveling upstream toward the inlet of the nozzle. Thus, if the discharge pressure is reduced below what the nozzle was designed to accommodate, that change cannot be transmitted to the inlet using the normal pressure "sound" waves that originate from such a change. That is, the pressure "sound" wave generated by the pressure change can only be swept downstream. That is not to say that the environment at the exit of the device is unaffected by the operation of a supersonic nozzle that is forced to operate off its design point pressure ratio; The exit is subject to phenomena known as Normal or Oblique Shocks, which are singularity events in which the low pressure, supersonic flow stream instantaneously undergoes an increase in pressure and entropy, while conserving momentum and energy.[1] These adjustments of the pressure at the exit of the nozzle cause both a loud and audible response as well as a very distinct visual discontinuity of the flow stream. The shocks cause the rapid increase in the pressure from the over expanded pressure ratio at the discharge of the nozzle to the higher ambient pressure outside the nozzle. The precise location of the shock in the nozzle is somewhat unpredictable. The normal or oblique shock could be observed to start anywhere downstream of the throat of a converging-diverging nozzle and proceed to the exit if the exit pressure is continually decreased. An oblique shock may be audibly and visibly observed at the exit of the device, as in the case of the exhaust of a rocket engine as it lifts the launch vehicle off the launch pad. In a rocket engine application, oblique shocks appear as very distinct diamond shapes that extend downstream from the nozzle in a standing string of diamonds.

C18.4   Relationship of mass flow rate and stagnation pressure and temperature

It is interesting to consider the combination of the velocity at the throat, the throat cross section area and the stagnation pressure and temperature that determines the maximum, mass flow rate that can be sustained by the nozzle. Equation (C18.9) shows this relationship by using Equations (C18.3)

---

[1]The thermodynamics relationships for a flow stream that is experiencing a shock phenomenon follows the continuity and energy laws in a relationship called the Fanno Line. Similarly, the same flow stream can be modeled using continuity and momentum laws in a relationship called the Rayleigh Line. The simultaneous solution to these relationships determines the inlet and exit, and enthalpy and entropy of the shock phenomenon.

and (C18.5) with Mach Number 1. Thus, the maximum mass flow rate is proportional to the stagnation pressure, p_o, and inversely proportional to the square root of the stagnation temperature. The use of Equation (C18.9) reveals that the nozzle inlet pressure,[2] P_0, must increase in order for the mass flow rate to increase above the maximum flow rate under the "choked" condition. The thermodynamics equation expressing the mass flow rate as a function of temperature, pressure and cross-sectional area of a nozzle is given in Equation (C18.9).

$$\dot{M}_{flowrate} = \left(\frac{\rho^*}{\rho_o} \times \sqrt{ \frac{g_c \times k \times T^*}{\frac{R_u}{MoleWt.} \times T_o} } \right) \times A^* \times \frac{P_o}{\sqrt{T}_o}, \qquad (C18.9)$$

The asterisk (*) denotes the throat condition, and "o" denotes the stagnation state for the parameter shown.

Recalling that the ratio, $\frac{T^*}{T_o}$, from Equation (C18.7b) is a constant for a given value of k and that the Mach number (M) = 1, the expression in the parenthesis in this equation is thus a constant. Therefore, the mass flow rate in a nozzle under the "choked" condition is observed to be directly proportional to the minimum cross-section area (A) at the throat of the nozzle, as well as stagnation pressure (see Equation (C18.8)). The mass flow rate in a nozzle under the "choked" condition is inversely proportional to the square root of the stagnation temperature (see Equation (C18.7b)).

The above expressions for the stagnation temperature, enthalpy and pressure only apply to a nozzle which undergoes isentropic expansion. Similarly, formulae can be derived for the analysis of streams flowing at sonic flow velocity which may experience a severe drop in pressure when flowing through heated (diabatic) or unheated (adiabatic) ducts that have a constant cross-sectional area. Case Study No. 21 provides an analysis of a heated duct with constant cross section area and friction heating effects. For a more complete review of compressible flow and gas dynamics, the reader is encouraged to refer to the following two excellent references: "The Dynamics and Thermodynamics of Compressible Flow" (by Ascher Shapiro) and "Gas Power Dynamics" (by Alexander Dodge Lewis).

---

[2]It is usually assumed that the inlet pressure given in this equation is equal to the stagnation pressure, Po with either zero (0) or close to zero (0) velocity (Mach No. <0.1, for example).

Case Study 19

A New Concept for a Thermal Air Power Tube Used with Waste Heat Energy Sources and Large Man-Made or Natural Landforms

This case study explores a method for power generation that enhances the recovery of concentrated solar energy in high temperature receivers while also enabling land reclamation and the effective reutilization of decommissioned open-pit mines. It combines a Rankine Cycle heat engine that is heated by concentrated solar energy and a pneumatic power tube (PPT) system, installed in a large open pit.[1] In this proposed concept, solar energy is concentrated by a plurality of heliostat mirrors placed along the embankment of the pit, which tends to be spherical in shape. Power generation is achieved using waste heat energy recovered from the Rankine Cycle heat engine, and from a variety of other sources that are either found within or in close proximity to the installation site of the system. An important feature of these open-pit mines is the availability of roadways that can be used to provide the structural support for the pneumatic tubes, thus avoiding the need to design very tall tubular columns that must support themselves. Designs for such columns tend to be extremely expensive as the height of the most effective pneumatic column must be over 1,500 ft tall. These sites, which may be deep man-made open-pits or naturally occurring chasms, also provide structural support for the PPTs installed. The air in the pneumatic tubes is heated by the recovered waste energy from the primary heat engine or from the combustion of the bio-mass available on-site until its density is sufficiently reduced, thus producing air drafts to flow upward through the vertical tubular column. The placement of wind turbines at either the bottom or the top

---

[1] Mr Jonathon Gwiazda assisted with this study as part of his 2005 graduate studies at Northeastern University, Boston, MA.

of these pneumatic tubes results in the generation of mechanical power to be converted into electrical power using suitable electric generators. Overall, the proposed system is a unique addition to the existing field of power generation technologies, as it provides a solution to the problems associated with the reclamation of depleted open pit mines as well as utilizes renewable solar energy for electric power generation.

C19.1   Production of air drafts in a pneumatic power tube

The heated air within the air column is sufficiently warmed to change its density (compared to the colder, external ambient air temperature) such that it produces a pressure differential between the bottom and top of the column. This pressure differential is augmented by the need for the vertical columns to be extremely tall, over 1,000 ft, for the air to travel into the ambient where the pressure is naturally lower by 0.25 psia and also to enable more surface area of the tube to be exposed to solar incident radiation heating.

Increasing the temperature and decreasing the density of the air in the column achieves the desired effect of establishing a pressure differential across the inlet and outlet of the pneumatic conduit. The addition of heat to increase the temperature certainly requires additional energy consumption. However, the energy can be sourced from renewable solar energy, the combustion of industrial waste, or ground source heat. The availability of such "free" energy enables the proposed system to be more economical.

Of relevance to this case study is an interesting design concept for a power generation system, suggested by Gutman and his colleagues at the Israel Institute of Technology (Gutman, Horesh, Guetta, & Borshchevsky, 2003). Their proposed power tower, called the "Aero-Electric Power Tower", is an enormous, free-standing, vertical tube that is 1,200 meters tall and 400 meters wide in diameter. Inside, dry air is cooled via water injection to induce a *downdraft* of dense, moist air. This draft is then passed through turbines at the bottom of the air column to ultimately generate electricity using generators that are coupled to the air turbines. The source of the energy for this air tower is the very dry desert air that can be saturated via the continuous injection of water. As the cooled, and hence denser, moist air flows downward through the tower, the increasing kinetic energy of the air stream is converted into electrical power via an array of turbines at the base of the tower, through which the air exits from the system. According to its inventors, as much as 380 MW of power may be generated using such an arrangement. However, it is also important to note that in the proposed concept presented in this case study, the PPTs operate by an air *updraft*, as opposed to the downdraft created in an Aero-Electric Power Tower. Additionally, the concept discussed in this case study eliminates the need for the costly engineering of a free-standing design for the air column and the need for large quantities of water. The proposed pneumatic tube for this case study is supported by either previous (man-made) constructs and/or naturally occurring landforms, and recovers the renewable energies that are available in that locale using the concept's several uniquely integrated design features, such as an air column made of self-supporting and less expensive, modular sections; the heating of the interior of the air column using the heat rejection from the concentrated solar (Rankine Cycle) power system; and surface turbines installed at the top of the open pit mine to enable easier Operation and Maintenance.

All previous attempts at producing power via an air chimney have been unsuccessful because of the need for the pneumatic tubes to be very large, self-supporting, and robust enough to withstand natural weather phenomenon such as winds, storms, earthquakes, etc.

In a very early study of ways to improve the efficiency of skyscraper-size buildings (Gregorian & Di Bella, 2001), a variety of energy generating options commonly used in skyscrapers and other freestanding structures over 1,000 ft tall were explored. One of the more radical ideas promoted in that study was the use of elevator shafts to support the controlled generation of air currents via the chimney effect caused by the difference in temperatures and densities between the hotter, less dense, internal columns of air and the cooler, denser, outside air. In that study, the use of elevator shafts as a "chimney" was a means of reducing the costs involved in constructing free-standing structures solely intended for inducing air currents by reducing the density of the interior air using solar energy or the heat rejected from another cycle. (Di Bella & Gwiazda, 2009).

C19.2   Effect of the size of pneumatic power tubes on the generation of electric power

In 2002, it was suggested that in addition to the rejected energy from the HVAC (heating, ventilation and air conditioning) and lighting energy loads required in skyscrapers, the thermal energy in the induced air chimneys could also be derived from solar energy that is focused onto the exterior surface area of the freestanding columns, half of which is exposed to solar radiation at any time. The results obtained from the second study by Gwiazda (2003) indicated that appreciable power generation (over 1 MWe) could only be obtained if the columns were very large in diameter and overall height.

The authors here recommend that a more detailed analysis be carried out, with an extension to gargantuan man-made and natural landforms, in order to investigate the scaling effect that size has on the generation of significant electric power.

C19.3 Structure of pneumatic power tubes and possible sources of thermal energy

The concept presented in this case study is the result of a series of studies presented in Di Bella & Gwiazda (2009). The concept integrates a novel design for a PPT system with a heat engine operating according to the solar Rankine Cycle. For this application, any other heat engine which produces waste heat energy may be used. In the proposed concept, a decommissioned open-pit mine (or any other very large landform, man-made or natural) is used as an economical structural support for the PPTs. The design of the proposed PPT can promote the necessary difference in the density of air in the PPT compared to the ambient air outside of the PPT. The net result of the density difference is to have the PPT produce a continuous air flow to generate a significant amount of electric power via waste heat recovery from a variety of sources, including solar energy. The novel design can eliminate the cost of constructing a freestanding column, while also providing an environmentally friendly solution to the problems associated with decommissioning such mines.

The pneumatic tube design that can be used in this application is shown in Figures C19.1 and C19.2. This novel design for the pneumatic tube includes many unique features that enhance its usefulness in producing wind energy from rejected heat. The specific features listed below are found in the PPT system components shown in Figures C19.2.

1. The integrated PPT and heat engine operating according to the Rankine Cycle is designed to generate power using the waste heat energy from the heat engine, and the ground source heat thermally stored and/or generated naturally from the earth's internal energy.
2. The pneumatic tubes can be structurally supported by the sides of the open-pit mine. This eliminates the need for the costly structural engineering required in the construction of a freestanding column.
3. The external, top surface of the pneumatic tube is constructed of reflective, flexible surfaces which enable solar energy to be focused onto the receivers.
4. The pneumatic tubes are wider at their top than at the bottom, thus functioning as a diffuser. As explained in Chapter 3, the function of a

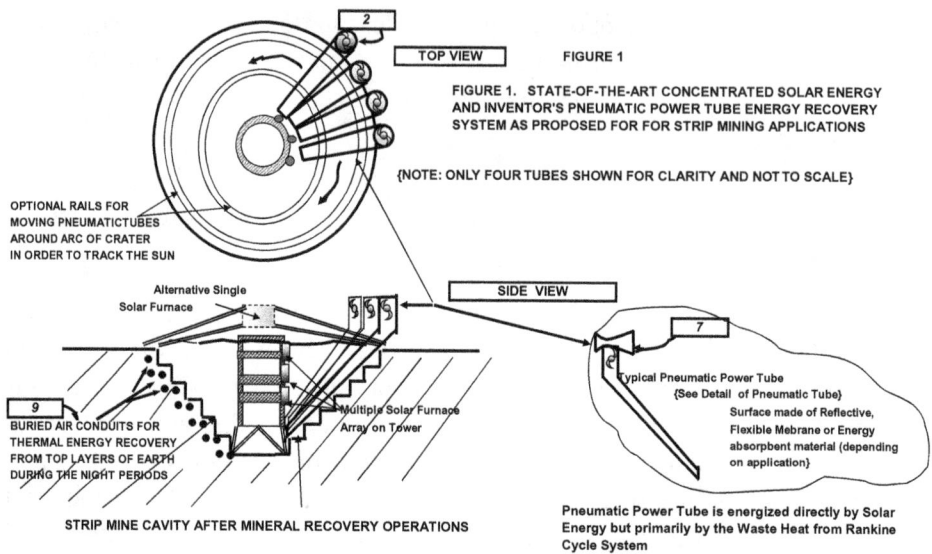

Figure C19.1. A schematic of the proposed PPT for the open-pit mine application.

diffuser is to convert the kinetic energy into a pressure differential. Thus, the diffuser designed into the PPT allows the induced air flow to recover static pressure at the inlet to the wind turbines while also maintaining the air velocity at reasonable (lower) speeds.

5. The outlet of the pneumatic tubes can serve as a cowl for the wind turbine(s) and thus increase the efficiency of the wind turbine.

6. The pneumatic tubes can be used to recover solar influx energy (i.e., transmitted energy) that is not reflected to the receivers. This energy can be used to provide additional heat to the air column, thus further inducing airflow inside the pneumatic tube.

7. The exhaust outlet of the PPT can be fitted with a converging-diverging nozzle (venturi) to produce a Bernoulli effect at the exit of the wind turbine. A Bernoulli nozzle is similar to the ejector reviewed in Case Study 14. A high wind velocity will funnel through the Bernoulli nozzle and be the motive energy to reduce the outlet pressure of the PPT wind turbine and thus increase the overall pressure ratio across the turbine. The increased pressure ratio will increase the power output of the turbine.

8. The structural support for the pneumatic tubes is provided by the sides of the open-pit mine. The interior sides of the mine are actually roadways that are wide enough to allow the passage of the large trucks that transport the mined ore from the bottom of the pit to the top of the mine. The availability of these roadways provides a means of supporting the

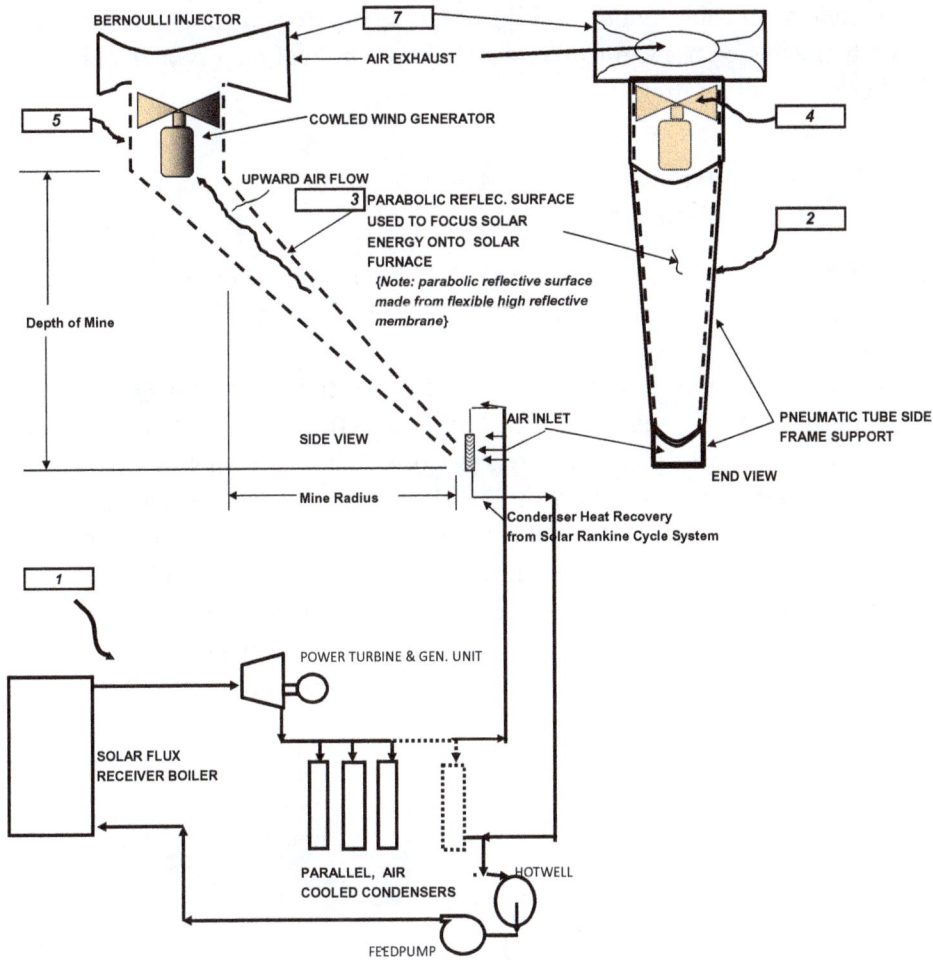

Figure C19.2.   Detail of the PPT used in the Open-Pit Mine Power recovery concept.

PPTs that are manufactured in sections. Each section can be made to be only 10 ft (3.2 m) to 20 ft (6.4 m) long, which allows the PPT sections to be transported from the manufacturing facility to the mine using public highways. The use of these assembled PPTs also enables them to be placed along the interior, inclined surface in the best orientation that maximizes the incident solar energy. The modularity of the pneumatic tubes also minimizes the manufacturing costs required.

9. The pneumatic tube design enables heat recovery from heat stored in the gravel and ground surfaces of the open-pit mine. The heat can be recovered using easily-buried air conduits by which the recovered air can be directly used in the pneumatic tubes. For example, heat pipes can be

embedded in the ground, in deep hot water springs or in volcanic locales with chasms or fissures to recover the natural heat energy available from these natural landforms.

C19.4 The concept: A pneumatic power tube concept and applications

The proposed concept combines a PPT and a Solar Power Tower (SPT) in an open-pit (ore) mine. An example of an open-pit mine is shown in Figure C19.3.

An open-pit mine is essentially a big hole in the ground as shown in Figure C19.3. The open-pit mining method extracts minerals by blasting, excavating and processing the extracted ore. The processing of the ore can include pulverizing the ore into similar sized pieces as well as drying the ore before shipment to a smelter. The shape of the footprint the mine takes on depends on what the most economical means of removing the mined mineral is, but the holes themselves generally take on a parabolic bowl shape. The sides of the mine are typically terraced with berms which prevent land-slides and provide level road surfaces that can lead miners up and out of the mine.

Figure C19.3. Photo of the Palabora mine in South Africa. This mine is a good example of a typical open-pit (ore) mine with spherical contoured topography similar to circular but planar solar collection fields.

In an SPT, solar energy is concentrated by reflectors and redirected toward receivers, which are also called heliostats. The reflectors used are similar to state-of-the-art field arrays of parabolic heliostats. Molten salt is used as the working fluid to transfer the heat from the receivers to the steam turbines. Electricity is then generated by these steam turbines. Ambient air at the bottom of the mine is used to cool the condenser that is condensing the steam in the solar Rankine Cycle system. The heated ambient air is vented into the bottom of the pneumatic tubes. The hotter air inside the PPT is therefore at a lower density than the air outside the PPT, thus creating an updraft of air flow which passes through the wind turbines in the proposed PPT to generate more electricity. During the last three decades, considerable effort has also been invested into demonstrating the feasibility and cost effectiveness of SPTs. Examples of SPTs that have been built on large, flat surfaces are those in the Solar One and Solar Two projects in Barstow, CA (Meier, 2020; Solaripedia, 2009). At this point, there has been no installation of SPTs in conically-sectioned cavities or deep vertical shafts.

The proposed concept for this case study is a unique integration of an SPT and a PPT. Figure C19.1 shows the installation of the proposed concept in an open-pit mine. An actual open pit mine is shown in Figure C19.3. The proposed concept has many attractive economic incentives as well as power generation advantages, such as the following:

1. Considerable expenditure is required to secure a decommissioned mine and reclaim it. This expenditure also includes the considerable maintenance costs required to prevent the hazards posed by the accumulation of contaminated water on the site.
2. The parabolic shape of general open-pit mines causes them to have a higher focal power than a flat plane.
3. An open-pit mine that has been in operation for decades usually has an established infrastructure that can be utilized to support site workers in charge of maintaining the new solar power generation facility.
4. The use of an existing open-pit mine eliminates the need to purchase virgin land and/or petition the local citizenry for the establishment of a solar power site. The acquisition of land is usually very expensive and time-consuming, especially if environmental permits must be obtained.
5. The PPT keeps the mine property productive after the mining life cycle and could potentially provide a means of clearing the mine of pollutants. The reutilization of an open-pit mine to generate electric power

using sustainable sources like solar power is considered a more reasonable end-use of the land as opposed to simply allowing the developed site to degrade.

The proposed concept applied to open-pit mines is seen to be an effective combination of several well-developed technologies, including: wind turbines, power generation using solar concentrated energy via the Rankine Cycle, thermal energy storage using rock and gravel as mediums, and the transportation of electric power using AC-DC-AC converters and/or microwave power generation and transmission for very remote sites. All of these technologies have been under development by a variety of research activities supported by private and public funding. The primary objective of the research funding has been to improve the efficiency of these systems while reducing costs. The deployment of a PPT in depleted and decommissioned open-pit mines is seen to be a good application of renewable energy and thermal energy storage systems that can convert a potential environmental disaster — an abandoned open-pit mine — into a functioning, clean power generation system.

C19.5   Thermodynamics analysis

The thermodynamics analysis that has been completed for the Solar powered PPT open-pit mine concept provides a first-order estimate of the power recovery potential of the solar energy-PPT-open-pit mine installation.

The analysis proceeded according to these three steps:

1. Determining the pressure differential that can be produced by a column of air that is at the bottom of an open-pit mine (i.e. 1,000 to 3,000 ft below sea level) due to the density difference between the hot air inside the PPT and the colder ambient air that is outside the PPT.
2. Determining the mass flow rate and the velocity of air that can be induced by this pressure differential, assuming that the flow is a simple Darcy-type friction flow in a large conduit. The Darcy flow equation is shown here and uses the induced pressure difference from Step 1 to determine the velocity of the flow stream.

$$\Delta P_{press.\ differential\ induced\ in\ PPT} = f \times \frac{L_{length\ of\ PPT}}{D_{diameter\ of\ PPT}} \times \frac{V^2}{2 \times g_c} \times \rho$$

3. Determining the amount of power that is therefore recoverable, assuming that the heat rejected by the condensers is used to heat the pneumatic

power tube columns. The Rankine cycle is powered by solar energy. The power of the PPT is based on the equation shown here:

$$Power_{PPT}[kW \ or \ Hp] = \frac{(\rho \times A \times V^3)}{2 \times g_c} \times (0.6 \times Turbine \ Eff.)$$
$$\times (unit \ conversion)$$

where the factor 0.6 is known as the Betz's Limit. The Betz's Limit is based on a derivation of the maximum power that can be recovered from a flow stream that must have a discharge velocity greater than zero in order to have that air flow stream move away from the turbine.

In order to estimate the power recovery potential based on the above equations, a thermodynamics model of the concentrated solar Rankine Cycle system was prepared on a spreadsheet, and a parametric study conducted to determine the effects of the mine's height and diameter on the amount of recoverable power.

The results of a parametric study are shown in Figures C19.4 to C19.7. The baseline condition is an internal air temperature of 120°F in the Pneumatic Power Pit Tube (PPPT), a solar incident flux of 300 Btu/h/ft^2 with

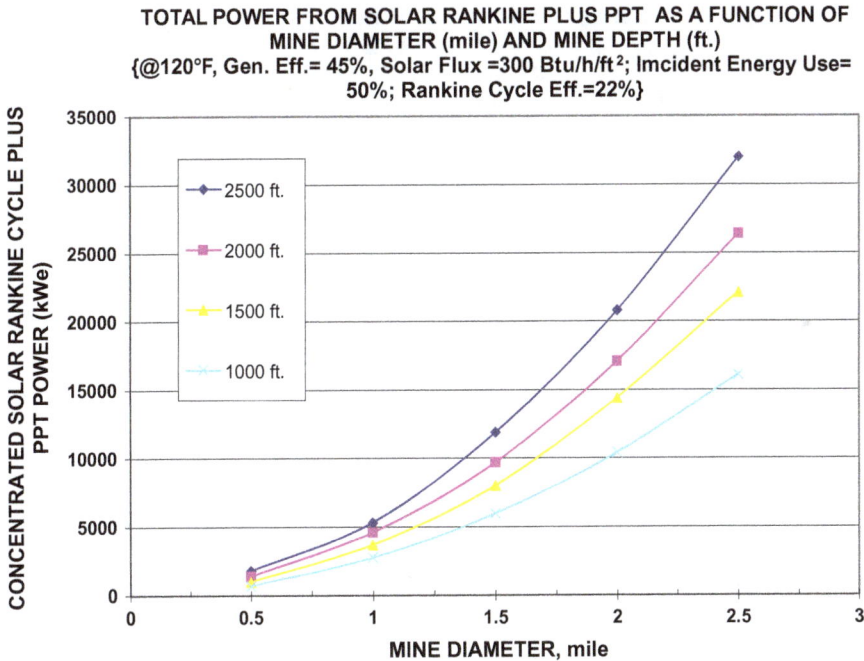

Figure C19.4.   Power generation from PPT as a function of mine depth and diameter.

RANKINE CYCLE EFF. IMPROVEMENT USING PPT SYSTEM (%)
{with Baseline Conditions given in Figure C19.4}

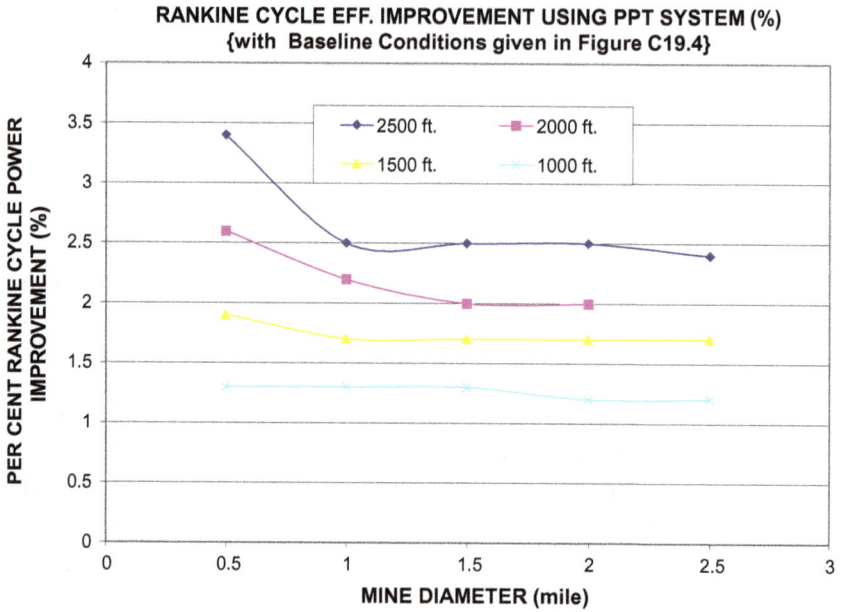

Figure C19.5. Power improvement from PPT as a function of mine depth and diameter.

POWER GENERATION FROM COMBINED SOLAR RANKINE CYCLE PLUS
PPT (kWe) AS A FUNCTION OF % INCIDENT SOLAR ENERGY USE
{Using Baseline Conditions & Mine Depth at 2,500 ft.}

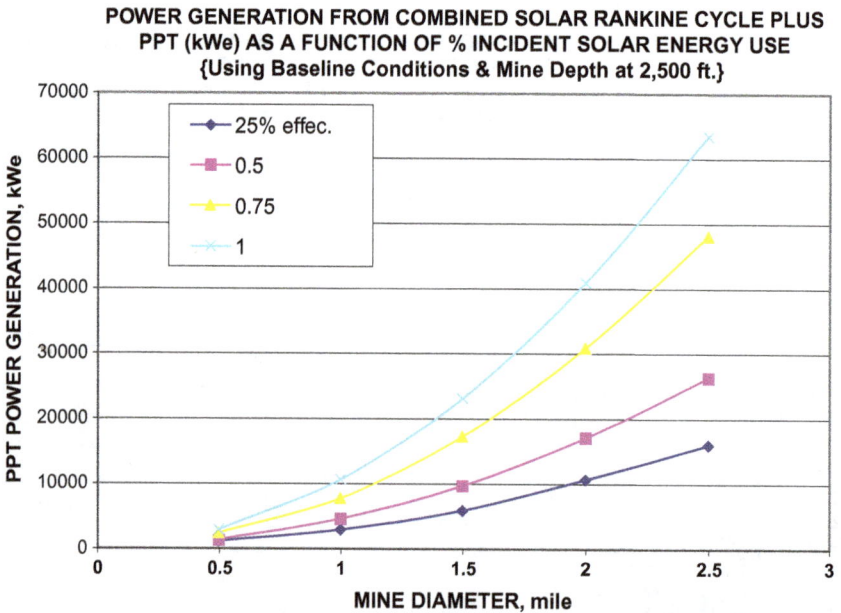

Figure C19.6. Total power generation from solar Rankine Cycle plus PPT as a function of percent incident solar energy.

POWER (kWe) GENERATION BY ONLY PPT AS A FUNCTION OF MINE DIAMETER AND INCIDENT ENERGY USEAGE (%)
{Using Baseline Conditions }

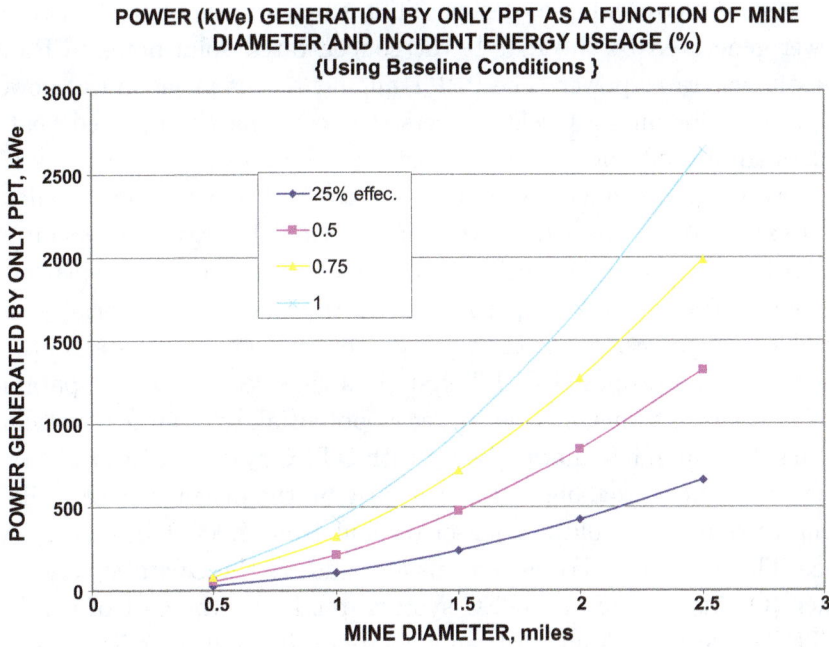

Figure C19.7.   Power generation from only the PPT as a function of incident solar energy.

an "Incident Energy Usage Factor" (IEUF) that was varied in the study from 25% to 100%, and a solar Rankine Cycle efficiency of 22% (and thus 78% of the incident solar energy into the solar Rankine Cycle is rejected to the condenser which can be used to heat the air that enters the interior of the air columns. The IEUF is a factor less than 1 that accounts for the many obstacles that could reduce the amount of incident radiation from reaching the surface of the air column. For example, the solar incident flux of 300 Btu/h/ft^2 occurs at "high noon" at sea level, at the equator and with no physical obstacles in the way such as trees, tall buildings, clouds, and poor weather. The First Law of Thermodynamics equation is then used to determine the temperature of the air that enters the PPT as shown below.

$$\dot{Q}_{condenser\ heat\ rejection} = \dot{M}_{PPT\ flowrate} \times C_p \times \Delta T_{PPT\ air\ flowrate}$$

It is also assumed that only 50% of the interior surface area of the open-pit mine is lined with PPTs. From the results, the power generation capability of the PPTs range from 1 to 30 MWe as shown in Figure C19.4, which represents an approximate 2–3% improvement in the solar Rankine Cycle's efficiency (see Figure C19.5). Additionally, approximately 50 to 800 MWe of power is generated by a solar Rankine Cycle system using concentrated solar reflectors and receivers of conventional design. That is, even without the contribution of power generated by the PPPT, the concentrated solar-powered Rankine Cycle still generates power. The PPPT augments this prime mover power as much as any Bottoming Cycle recovers the (condenser's) rejected heat and produces additional power. It is interesting to note that the efficiency of the PPT Bottoming Cycle is very comparable to the 2% overall cycle efficiency that is expected for the OTEC (Ocean Thermal Energy Conversion) System. An OTEC system is considered to be a viable alternative solar energy power generation system — perhaps primarily due to the enormity of the renewable energy "stored" in the top layers of some oceans located at or near the equator. The proposed PPPT system is thus seen to be comparable to OTEC systems in terms of power recovery potential, with far less installation problems. Common installation issues with OTEC systems concern installing the very large heat exchangers and turbines on the ocean surface while also needing to pump the cold condenser coolant from 2,000 ft below the ocean surface. The fierceness of ocean storms also adds to the Operation and Maintenance (O&M) cost of an OTEC system, rendering the cost of employing an OTEC system much higher than a comparably sized solar Rankine Cycle System with integrated PPTs that are not only entirely land based but that also utilizes an installation site that already has all of the necessary infrastructure needed to support the proposed power system. Figure C19.6 identifies the effect of the IEUF on the power generation capability of the PPPT system.

C19.6　Conclusion

A typical open-pit mine, in this case the Palabora mine in South Africa, is shown in Figure C19.3. The enormity of this open-pit mine is not unusual for such mining operations. Using the thermodynamics model that has been developed for this case study and taking into account the dimensions of the mine and its geographic location (from which the solar flux at the site can be determined), an estimate can be made of the amount of power that can be generated at the site if it were to have an integrated solar power and PPPT system installed. Using the most current specifications for the efficiency of concentrated solar energy power systems (from research supported by the US Department of Energy as noted in Romero, Buck, & Pacheco (2002)), approximately 142 MWe of electricity can be recovered from the Palabora mine once it is converted into a concentrated solar collector system using fixed (and not tracking), flat-plate (and not parabolic-shaped) reflectors.

Given that the site is already conical and perhaps even slightly parabolic, only minimal reconstruction would be required.

In place of traditional mirrored reflectors, the sloped surface of the open-pit mine could be lined with flexible membranes that have high reflectivity. Although some alterations to the shape of the mine may still be required in order for the focal point to be made more precise, other costs that could have been incurred if building an energy generator from scratch — e.g. the high costs of tracking heliostats and land acquisition, the time consuming seeking and obtaining of environmental permits to satisfy "not-in-my-back-yard" neighbors, the construction of power production systems, and the staging of building and facilities — have largely been eliminated or greatly reduced. This results in a lower cost per kW and a productive use of what may otherwise have been an unused and unattractive land area.

Similar large open-pit mines are in operation but will eventually be decommissioned in some form. For example, the Chuquicamata Mine, which is 2 by 3 km wide and 810 m deep. If converted into a PPT, it could produce as much as 734 MW of solar Rankine power plus an additional 18 MWe of power from a PPPT system. And the deepest mine in the world, the Western Deep Gold Mine in South Africa, with its depth of 4,000 m, would enable cold ventilation air from the surface to be naturally heated by the earth to over 100°F (3°C). The thermally induced air draft from this mine could serve as the largest prototype of the nature-draft wind tower concept promoted in this case study.

Whether such enormous man-made or natural landforms use the proposed wind towers as presented here, solar energy recovery methods (or even the neutralization of acidic, contaminated standing water via anode-cathode electric cells) will still be the subject of natural energy resource recovery and generation. For example, air towers made from extinct and voided volcanoes or mountains (i.e., abandoned mines) with vertical air chutes that connect $2,000^+$ ft deep chasms with the upper atmosphere may serve as a natural induced air tower.

The proposed novel design effectively recovers the kinetic and pressure potential energies of the thermally-induced air drafts using the waste heat produced by a heat engine operating according to a solar Rankine Cycle as a thermal energy input. The proposed 600^+ m by 15 m diameter PPTs can be placed at the bottom of a decommissioned, open-pit mine, which are themselves typically 1,500 m in diameter and 600 m deep. Past research by the US Department of Energy have demonstrated that thermally-induced air-drafts are able to successfully induce sufficient quantities of air to enable efficient power generation via wind turbines. This, however, requires that the chimney structures be enormously tall (typically over 2,000 ft). However, tall, freestanding air columns are expensive to construct and thus negate the economic viability of the solar power generated.

This proposed system is unique in the field of power generation in terms of its effective use of previously-developed high temperature and solar energy receiver technologies, and increased efficiency of the solar Rankine Cycle by recovering the rejected condenser heat while also providing a solution to the reclamation and utilization of depleted open-pit mines. In place of the traditional mirrored reflectors, the PPTs used to line the sloped surface of the open-pit mine will be designed with a reflective top surface to focus the solar energy received. The design of the PPTs also allows for the diffusion of the induced air flow for the recovery of both kinetic as well as potential energy, via high-efficiency wind generators designed with cowling to reduce air by-pass.

References

Di Bella, F. A., & Gwiazda, J. (2009). A Novel Thermally Induced Draft Air Power Generation System for Very Tall, Man-Made and Natural Geo-Physical Phenomena. *Proceedings of the 2002 International Joint Power Generation Conference* (págs. 987–995). The American Society of Mechanical Engineers. doi: https://doi.org/10.1115/IJPGC2002-26098

Gregorian, G., & Di Bella, F. (2001). New Paradigms in Energy Conservation and Generation. *Paper presented at 2001 Annual Conference, Albuquerque, New Mexico* (págs. 6.752.1–6.752.15). American Society of Engineering Education. Obtenido de https://peer.asee.org/9611

Gutman, P.-O., Horesh, E., Guetta, R., & Borshchevsky, M. (2003). Control of the Aero-Electric Power Station — an exciting QFT application for the 21st century. *International Journal of Robust and Nonlinear Control, 13*(7), 619–636. doi: https://doi.org/10.1002/rnc.828

Gwiazda, J. (2003). A novel application of pneumatic and solar power tower technology: Pit power tower (Master's Thesis). Boston, Massachusetts, USA: Northeastern University.

Kribus, A., Zaibel, R., Carey, D., Segal, A., & Karni, J. (1998). A solar-driven combined cycle power plant. *Solar Energy, 62*(2), 121–129. doi: https://doi.org/10.1016/S0038-092X(97)00107-2

Meier, A. C. (2020). *Solar One and Solar Two: Decommissioned Experimental Solar Facilities that Pioneered Solar Energy Technology.* Obtenido de Atlas Obscura: https://www.atlasobscura.com/places/solar-one-and-solar-two

Romero, M., Buck, R., & Pacheco, J. E. (2002). An Update on Solar Central Receiver Systems, Projects, and Technologies. *Journal of Solar Energy Engineering, 124*(2), 98–108. doi: https://doi.org/10.1115/1.1467921

Solaripedia. (2009). *Solar One and Two (Now Defunct).* Obtenido de Solarpidia: http://www.solaripedia.com/13/31/solar_one_and_two_(now_defunct).html

Case Study 20

Ground Source Heat Energy Storage for Power Generation

Imagine that you are at the beach on a very hot day. After only a few hours on the beach, you decide that you have had enough time in the sun and now need to find some shade. As you walk across the sand, you notice that the sand is uncomfortably hot. To cool your feet, you find small puddles of water or patches of grass to stand on. Soon, you find yourself hopping from one cool area to another.

The heat absorbed by the hot sand in this imagined scenario is a form of free, "natural" ground source energy storage. The Sun's radiation energy is measured to be $300\,\text{Btu/h/ft}^2$ ($\sim$1,000 W/m^2) on the surface of Earth at sea level, at high noon, without shade or clouds. The high absorptivity of the very granular sand enables it to absorb large amounts of radiated energy. The thermal energy radiated from the Sun is converted into several forms of energy including chemical energy (the most well-known being fossil fuels), potential energy (in the form of storms and picturesque waterfalls), and bio-energy (i.e. plants and other life forms). In fact, the energy from the Sun also keeps humans alive and that form of potential energy can then lead to creative new ways of transforming the radiated energy from the Sun into new and efficient ways of utilizing the Sun's thermal energy. One such "eureka" moment resulted in the realization that the thermal energy stored in sand can be used in a creative manner, as will be discussed in this case study.

The results of an analysis that utilizes only the First Law of Thermodynamics to determine the temperature of the sand as a function of depth and the time exposed to the Sun's radiated energy is shown in Figure C20.1. From Figure C20.1 it may be observed that the sand can be heated to 150°F at a depth of approximately 3 inches (an incremental change in depth that

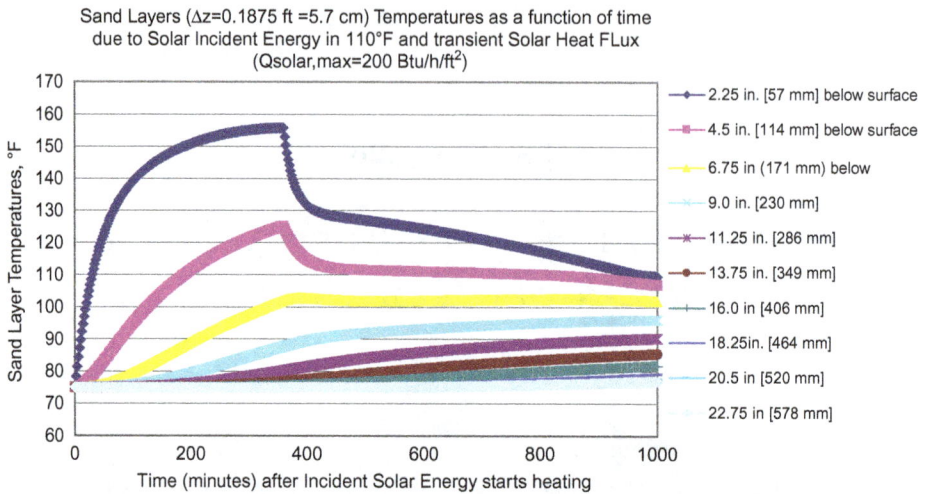

Figure C20.1.   Display of temperature gradients at different depths in static sand heated directly by solar energy. Each curve (series) represents a layer progressively deeper in the ground, up to a maximum depth of 1.8 ft.

is an input to the model) during a 12-hour exposure to the Sun. The results of Figure C20.1 were determined from a numerical solution of the thermal conduction of solar heat energy from its surface to a depth of approximately 18 inches. The numerical solution proceeds by using a spreadsheet in which each row of the spreadsheet represents a small layer of sand at a depth, h, into the ground and a unit dimension in the x and y direction. Each additional row in the spreadsheet then represents an incremental depth below the surface. The incremental depth is an input to the numerical model and may need to be very small if the numerical integration programming is to be accurate. As radiated heat energy from the Sun enters the top layer of sand, that layer of sand becomes hotter than the layer below it, thus enabling the conduction of heat from that top layer to the layer below. The top layer also has convection heat transfer from the top layer to the ambient above. The heat energy balance from the top layer is represented by the Closed Mass form of the First Law of Thermodynamics as shown here:

$$\dot{Q}_{radiative\ sun\ energy} - \dot{Q}_{convection\ to\ ambient} - \dot{Q}_{conduction\ to\ layer\ below}$$

$$= M_{sand\ layer} \times C_{specific\ heat} \times \frac{\Delta T_{sand\ layer}}{\Delta t_n}$$

The layers below the top layer only experience conduction heat transfer between the hotter top layers and the layers below. Thus, the first law equation for the Closed Mass that models each layer of sand uses the equation shown here:

$$\dot{Q}_{conductive\ heat\ from\ layer\ above} - \dot{Q}_{conduction\ to\ layer\ below}$$

$$= M_{sand\ layer,n} \times C_{specific\ heat} \times \frac{\Delta T_{sand\ layer,n}}{\Delta t_n}$$

These equations are solved for the rate of temperature per unit of time, term:

$$\frac{\Delta T_{sand\ layer,n}}{\Delta t_n}$$

The temperatures calculated using these equations result in temperatures between 100 to 150°F. This temperature range corresponds with what is recognized as "low grade" heat energy, the range at which some of the most common refrigerants such as R136a, R245fa or ammonia can be evaporated. The expansion pressure ratios of any one of these working fluids depend on the saturation pressure of the working fluid studied, at the prevalent ambient temperature. It is not uncommon for the working fluids given above to have saturation pressures above 14.7 psia. However, as noted in Chapters 3 and 4, the discharge pressure of the turbine is not relevant whereas the pressure ratio:

$$\frac{P_{out}}{P_{in}}$$

is relevant to the magnitude of power that can be generated by the turbine.

In this case study, the heating of the sand with solar energy is a means of storing the incident thermal energy. The thermal energy must first be recovered from the hot sand using a pressurized working fluid. The pressurized working fluid is stored in a storage vessel during the day, when the sun is shining and continuously radiating heat energy to the sand, albeit at different rates throughout the day. The energy stored in the sand during that period is later recovered in the evening by having this high pressure and high temperature fluid pass through the evaporator of a Rankine Cycle System. The maximum amount of power that can be generated by the Rankine Cycle will be obtained when the condenser of the Rankine Cycle can be cooled with the lowest nightly ambient temperature. The lower temperature allows the condensation of the working fluid at a much lower condenser pressure, thus keeping the turbine expansion pressure ratio shown in the equation above, as small as possible. The rotary turbine or reciprocating piston's cyclic movement enables work to be recovered from the expansion of the working fluid. This power can either be used immediately or stored using different methods, including locking springs that have been compressed by the piston or by immediately generating electric power and storing that energy in a battery. The mechanical power can also be stored, using the literal definition of work, by simply raising a weight in a gravity field. The raised weight can then be released when it is most needed and preferably when the stored energy has reached its high design point, kWh specification. The released energy can then be used to spin a shaft that is either connected to a mechanical power absorbing system or an electric generator.

For a Rankine System to be viable with only the heat absorbed by the static sand, several criteria must be taken into consideration. First and foremost, an ideal location for such a system is an open area near the equator which has good exposure to the Sun. Deserts are examples of such ideal venues due to their natural abundance of sand and the added bonus of very low ambient temperatures at nighttime. In the absence the Moon's reflection and other local human-made light sources, the absolute temperature of the ambient at nighttime can be as low as 150°R or −300°F! In a perfect arrangement, the chosen desert should also be near habitable areas, where there is a constant need for the electric power, as in the examples of Las Vegas and Dubai. Next, it is necessary to bury hundreds of feet of 1-inch diameter pipes into the sand at a depth of no more than 1 ft. Burying the pipe at some depth is necessary to maximize the utilization of the total energy stored per square foot of sand.

Renewable energy systems such as the Rankine Cycle System shown in Figure C20.2 often do not operate to their design point specification due to fluctuations in the Sun's daily radiated energy, as well as in consequence of strong wind-storms occasionally unburying the pipes. Of course, conventional wind turbines and solar photovoltaic (PV) panels are also subject to the whims of Mother Nature. As noted above, it is necessary to have as low an ambient temperature as possible in order to allow the condensation of the vapor to occur at a condensing temperature of the working fluid. This enables the highest, possible cycle efficiency per the Second Law of Thermodynamics.

As noted previously, the effective local nighttime "black sky" ambient temperature can be as low as −300°F. This value can be used to calculate the heat transfer between the condensing fluid and the ambient due to what is called the Black-Sky Radiation Effect. The Black-Sky Radiation Effect is the recognition that when the night sky is perfectly black — that is, when there is no cloud cover, no full or even partial Moon present, and no glare from local city lights — the amount of heat loss via radiation heat transfer can be more than the heat gained by convection heat transfer from a warm night air. The black night sky presents a "night sky temperature"

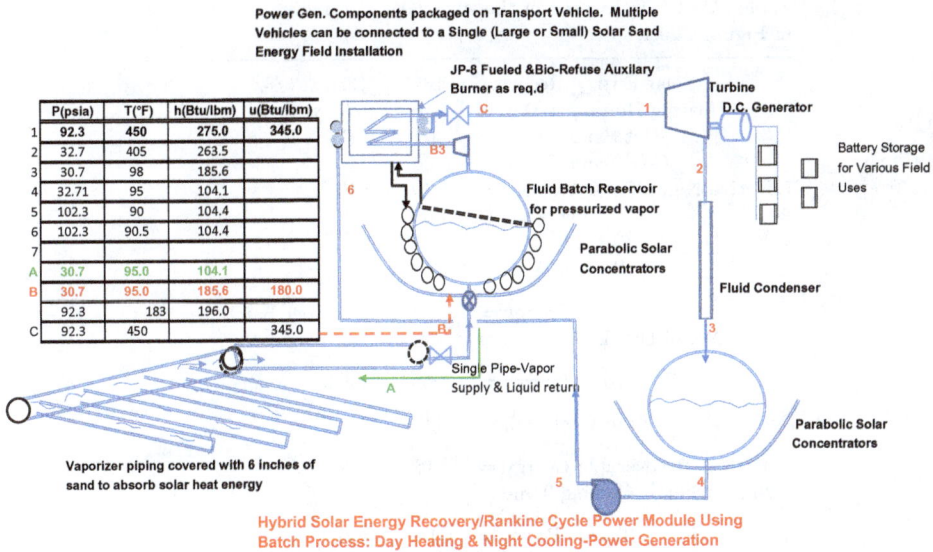

Power Gen. Components packaged on Transport Vehicle. Multiple Vehicles can be connected to a Single (Large or Small) Solar Sand Energy Field Installation

	P(psia)	T(°F)	h(Btu/lbm)	u(Btu/lbm)
1	92.3	450	275.0	345.0
2	32.7	405	263.5	
3	30.7	98	185.6	
4	32.71	95	104.1	
5	102.3	90	104.4	
6	102.3	90.5	104.4	
7				
A	30.7	95.0	104.1	
B	30.7	95.0	185.6	180.0
	92.3	183	196.0	
C	92.3	450		345.0

JP-8 Fueled &Bio-Refuse Auxilary Burner as req.d

Turbine
D.C. Generator

Battery Storage for Various Field Uses

Fluid Batch Reservoir for pressurized vapor

Parabolic Solar Concentrators

Fluid Condenser

Single Pipe-Vapor Supply & Liquid return

Parabolic Solar Concentrators

Vaporizer piping covered with 6 inches of sand to absorb solar heat energy

Hybrid Solar Energy Recovery/Rankine Cycle Power Module Using Batch Process: Day Heating & Night Cooling-Power Generation

Figure C20.2.   Diagram of a solar energy-powered Organic Rankine Cycle (ORC) system.

that can be effectively as low as 160°R when measuring the amount of heat loss by radiation from a large surface.

In this application, the evaporation and condensation of the working fluid needs to occur "out-of-phase" (i.e., at different times). In order to maximize the amount of power generated, an intermediate step of storing the evaporated, pressurized fluid in a large, well-insulated vessel must be taken in order to facilitate the expansion of the fluid through the turbine when the ambient temperature is at its lowest. A complete cycle that accomplishes the storage of vapor for use later when the condensing temperature can be at its lowest is shown in Figure C20.2. The basic cycle is recognized as a solar energy-powered Organic Rankine Cycle (ORC) system. Many ORC systems powered by waste heat have been designed using radial, axial or positive displacement-type, organic fluid turbines for various entrepreneurial applications. The electric power generated can then be stored in batteries for later use or, where live power is required, synchronized to a grid for immediate use. Figure C20.2 illustrates a diagram of this heat recovery and power generation system, which is essentially a Rankine Cycle system that operates in a batch process. That is, the heat energy absorbed by the sand during daytime when the Sun is shining is stored for later use during the evening hours. The complete cycle is thus no more than 24 hours long. Table C20.1 summarizes the data used in the analysis shown in Figure C20.2.

Table C20.1.   Data used in the solar heated Rankine Cycle shown in Figure C20.1.

Solar Energy Recovery	1675	Btu/day/ft^2
Heating Hours per Day =	10	h/day
Hot Sand Temp. =	150	°F
Cold Sand Temp. =	80	°F
Effective Night Time Ambient =	40	°F
Heating Time =	6000	s
	1.67	h
Surface Area =	100	ft^2
Total Energy =	82	kWh per batch heating
No. of Day Light Batches	4	
Total Energy =	327	kWh
Est.d Cycle Eff. =	5.0%	
Recoverable Energy =	16	kWh
Night Charge/Cooling Time =	10	h
Trickle Power =	1.6	kW

Despite the somewhat novel application, the calculation of the state points, net power and cycle efficiency shown in Figure C20.2 can be modeled using the thermodynamics of a conventional Rankine Cycle, which has been reviewed in Chapter 5.

Case Study 21

Analysis of Heat Transfer and Pressure Drop for a Heat Exchanger with the Mach Numbers of the Heated, Compressible Fluid Approaching One

Traditional texts on compressible flow gas dynamics provide considerable review of the thermodynamics and fluid dynamics of compressible flow in adiabatic nozzles and constant area ducts. There is, however, limited study provided for heated (diabatic) ducts, including the specialty category of the high speed, compressible flow of the gaseous fluid that is to be heated using a heat exchanger. Available studies on compressible flow in classic textbooks provide only nomograph solutions due to the lack of a closed-form solution for the thermodynamics and fluid dynamics equations. An analysis technique using a simple computer spreadsheet is reviewed in this case study, including a validation of the analysis. A parametric analysis is included to provide a design tool for heat exchangers that are subject to very high flow conditions during either steady state or during a short-term transient analysis. A case study example of the application of the technique is presented.

Recent interest in advancing the design of the propulsion system used in launch vehicles has brought the need for analyzing the heating of high speed "coolant" fluids in heat exchangers. However, due to the transient pressure experienced during the launch, and the inevitable reduction of the ambient pressure and discharge of the heat exchanger as the launch vehicle ascends, a heat exchanger may be prone to choking and thus experience shock waves that could damage the heat exchanger. In order to predict and avoid the choking of the heated fluid, an analysis technique that uses computer spreadsheet software was developed.

The analysis proceeded using a thermo-fluid-heat transfer model of the heat exchanger that considered the diabatic heat exchange between the hot fluid and the coolant as well as the friction heating that is also significant if the flow-through area for the coolant is very small in consequence of satisfying a constraint on the overall size of the heat exchanger. Classic textbooks on compressible flow consider diabatic heating and friction heating separately. In most "land based" applications, a heat exchanger may be classically treated as a diabatic heated conduit without consideration of the friction created by the flow rate of the coolant stream if there is not as much a concern about keeping the size of the heat exchanger small. However, for applications where the engineering specifications for the weight and overall size as well as performance is given equal priority, the assumption that the heat exchanger is only influenced by the diabatic heating of the methane is considered to be very limiting to the proper design of the heat exchanger. Therefore, a compressible fluid, a thermo-fluid-heat transfer model of the heat exchanger operation that considers the combined effect of diabatic heating, plus the heating of the methane due to the friction caused by high fluid flow rates was developed and validated.

C21.1 Compressible flow dynamics analysis applied to heat exchanger

The compressible thermo-fluids model prepared by the author was developed using Energy Conservation and Continuity equations that construct the Fanno Line in the compressible fluid flow along with the friction energy, as modeled with the Fanning friction coefficient. These equations were used with a numerical solution technique to solve for the Mach number, overall pressure drop, and stagnation temperature that will be caused by the combined heat transfer of the heat exchanger from the hot fluid plus the friction heating. The numerical solution proceeds by dividing the total length of the heat exchanger into "n" sections. Each of these sections is represented by the simple model shown in Figure C21.1.

The basic equations for each section are as follows:

$$(\dot{Q}_{hex\ input} + \dot{E}_{Fanning\ friction}) + C_{effect.avg} \times \Delta(T_1 - T_2)$$
$$+ \frac{V_1^2}{2 \times g_c} = \frac{V_2^2}{2 \times g_c} \tag{C21.1a}$$

$$\dot{M} = \rho(P_2, T_2) \times A \times V_2$$
$$\dot{M} = \rho(P_1, T_1) \times A \times V_1$$

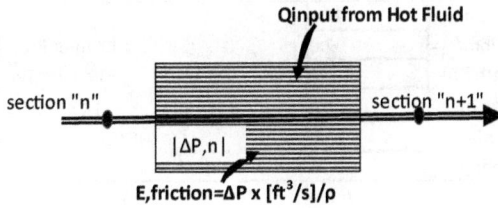

Figure C21.1.   Simple heated duct model of heat exchanger with significant pressure drop and flow friction.

Or, equivalently:

$$\dot{Q}_{heat\ input} + E_{Fanning\ friction} = C_{p,\ methane} \times (T_{stagnation,\ 2} - T_{stagnation,1})$$

$$(C21.1b)$$

Where:

$$T_{stagnation,n} = T_n + \frac{V_n^2}{2 \times g_c \times C_{p,methane}}$$

$$E_{Fanning\ friction} = 4 \times f \times \left(\frac{V_n^2}{2 \times g_c}\right) \times \rho(P_1, T_1) \times \dot{M}$$

But an expression for the velocity at State Point 2 may also be written in terms of the actual temperature, T_2 and mass flow rate, based on the familiar equation: $= \rho \times V \times \dot{Area}$ while also treating the air as a perfect gas and using the Perfect Gas Equation for the density of air as shown in Equation (C21.2).

$$V_2^2 = \left(\frac{\dot{M}_2}{A_2}\right)^2 \times \left[\frac{R_u}{MoleWt. \times P \times 144}\right]^2 \times T_2^2 \qquad (C21.2)$$

Combining Equations (C21.1a) and (C21.2) the equation takes the form of an easily solvable quadratic equation as shown in Equation (C21.3).

$$\left(\dot{Q}_{hex\ input} + \dot{E}_{Fanning\ friction}\right) + C_{p,effect.avg.} \times T_1 + \frac{V_1^2}{2 \times g_c}$$

$$= \frac{\left(\frac{\dot{M}_2}{A_2}\right)^2 \times \left[\frac{R_u}{MoleWt. \times P \times 144}\right]^2 \times T_2^2}{2 \times g_c} + C_{p,effect.avg.} \times T_2 \qquad (C21.3)$$

Thus, given a mass flow rate, an inlet temperature to the control volume, T_1 and the inlet velocity, V_1, the temperature at the discharge of the control volume section, T_2 can be determined by the solution of this quadratic

Oil Flow rate [lbm/s]=	1.50			
methane Flow rate [lbm/s]=	2.0		HEX Inlet Pres. [psia]=	60
HEX Tube Core Length [in]=	20		HEX Inlet Temp. [R]=	400
HEX Tube Free Flow Area [in²]=	3.0		HEX Tube ID. [in.]=	0.035
No. of Increments=	40	1	HEX Total Heat Transfer=	
HEX Increment [in]=	0.50			

Figure C21.2.   Input specifications for the hot and coolant flow streams and inputs for the numerical solution.

equation. The initial boundary condition for the inlet pressure and temperature of the compressible fluid is input into the model along with the geometry of the gas-tube side core as shown in Figure C21.2. The analysis proceeds with the sequential calculation of the inlet and outlet properties of each section as shown in Figure C21.1. However, the acceptance of a heat exchanger design based on the numerical analysis includes the repeating of the analysis with an adjustment of the inlet pressure to the heat exchanger until the discharge of the heat exchanger matches the boundary condition for the actual heat exchanger design specification for the discharge pressure. This last iteration effectively has the inlet pressure to the heat exchanger as a calculated "output" of the analysis via the iterative process — That is, although the inlet pressure is needed as an input to the thermo-fluid-heat transfer model, its value can be iteratively determined by the process shown in this case study.

The validation of the author's thermo-fluid-heat transfer compressible flow model was completed by replicating the results of a thermo-fluids model developed by Bandyopadhyay & Majumdar (2007). Bandyopadhyay & Majumdar used NASA's software, the Generalized Fluid System Simulation Program (GFSSP), to provide the solutions for the compressible fluid operation for a constant area duct that operates with significant friction plus diabatic heating. Figure C21.3(a) summarizes the design specifications for the heat exchanger used by Bandyopadhyay & Majumdar to achieve the results shown in Figure C21.3(b). The result of this comparison is shown in Figure C21.3. It is important to note that the referenced tech paper was only used for validation purposes. The technical paper did not include the algorithms for the GFSSP Simulation Program nor the equations for the "theoretical" solution that was also presented by the authors of the paper.

C21.1.1   *Case study application of the model*

The compressible thermo-fluids model was applied to a heat exchanger with the assumption that the new design point flow rate for the methane was

Oil Flow rate [lbm/s]=	1.50			
nitrogen Flow rate [lbm/s]=	24.9		HEX Inlet Pres. [psia]=	50
HEX Tube Core Length [in]=	3207		HEX Inlet Temp. [R]=	540
HEX Tube Free Flow Area [in²]=	28.274		HEX Tube ID. [in.]=	6
No. of Increments=	56	1	HEX Total Heat Transfer=	

(a)

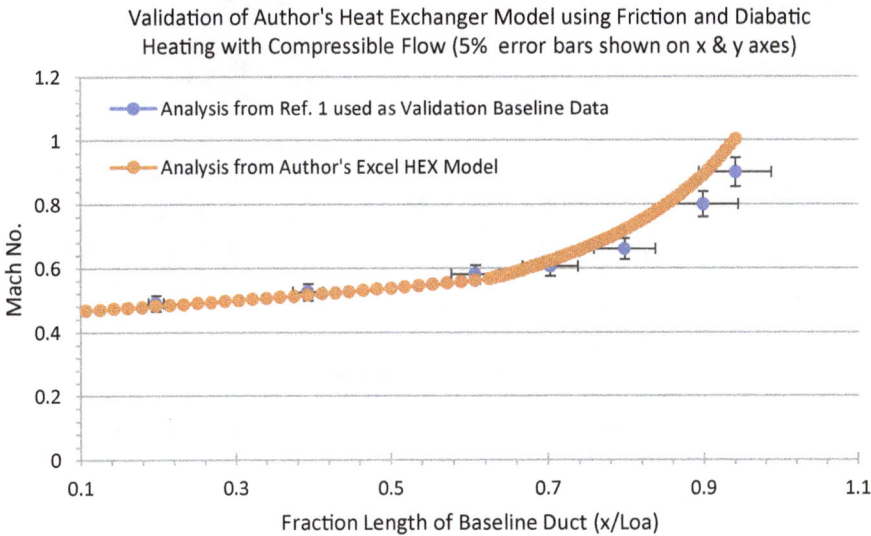

Validation of Author's Heat Exchanger Model using Friction and Diabatic Heating with Compressible Flow (5% error bars shown on x & y axes)

(b)

Figure C21.3.   Comparison of author's numerical solution with the solution presented by Bandyopadhyay & Majumdar (2007). Note: The higher density loci of data after 0.6 Mach No. is due to the reduction of the heat exchanger section to $1/60 \times L_{oa}$ (from $1/20 \times L_{oa}$) in order to improve the accuracy of the author's solution; (a) Table of Input parameters; (b) graph comparing the case study numerical solution to an independent study of Bandyopadhyay & Majumdar (2007).

2.0 lbm/s and the restricted flow area for the methane coolant is $3.0\,\text{in}^2$. The concern to be addressed is the possibility that the compressible fluid (i.e., the coolant in this case study) will achieve a Mach number greater than 1 at the exit of the heat exchanger when the ambient pressure outside the launch vehicle is reduced to a very low, sub-atmospheric pressure. A Mach number equal to or greater than 1 at the discharge of the heat exchanger upstream of the actual discharge of the coolant from the launch vehicle would cause a shock wave to develop at the exit of the heat exchanger tube core. The consequence of this is a shock-impulse that could damage the structural integrity of the heat exchanger and thus cause a failure of the heat exchanger. The results of an analysis to determine the effects of heat exchanger flow rate and coolant fluid flow rate on the Mach number are presented in Figures C21.4 and C21.5 for an inlet pressure into the heat exchanger of 60 psig and 80 psig,

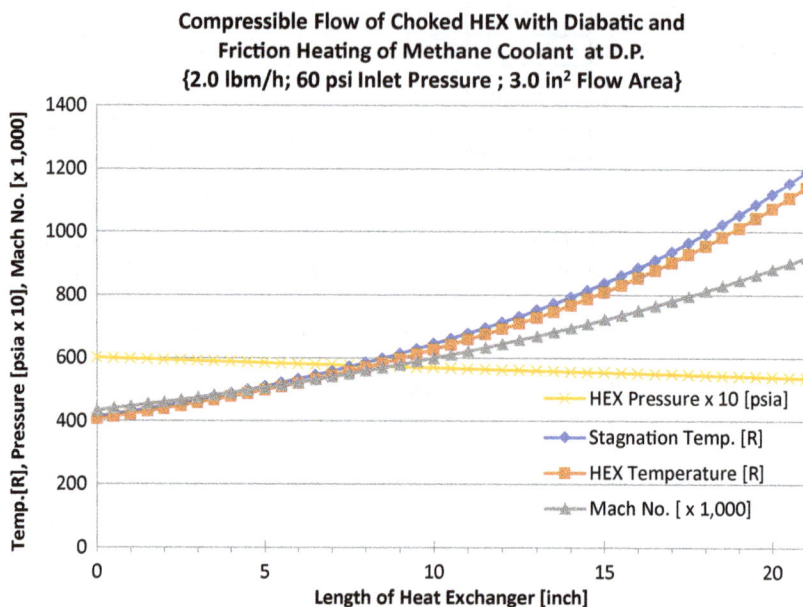

Figure C21.4.   Results of author's thermo-fluid-heat transfer model of compressible flow with diabatic and friction heating with heat exchanger inlet pressure equal to 60 psi. Inputs are shown in Figure C21.2.

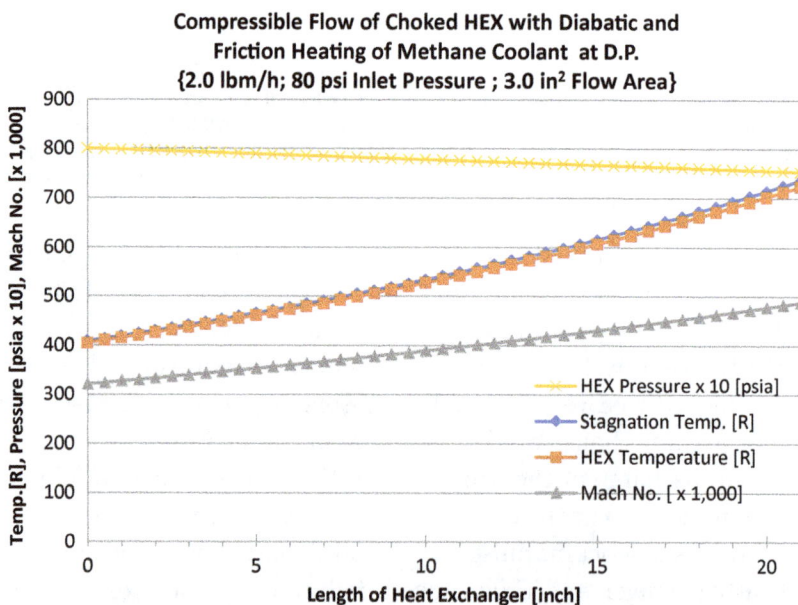

Figure C21.5.   Results of the same thermo-fluid-heat transfer model used for Figure C21.4 but with the heat exchanger inlet pressure equal to 80 psi. Note how the Mach number has been reduced from 0.85 to 0.48.

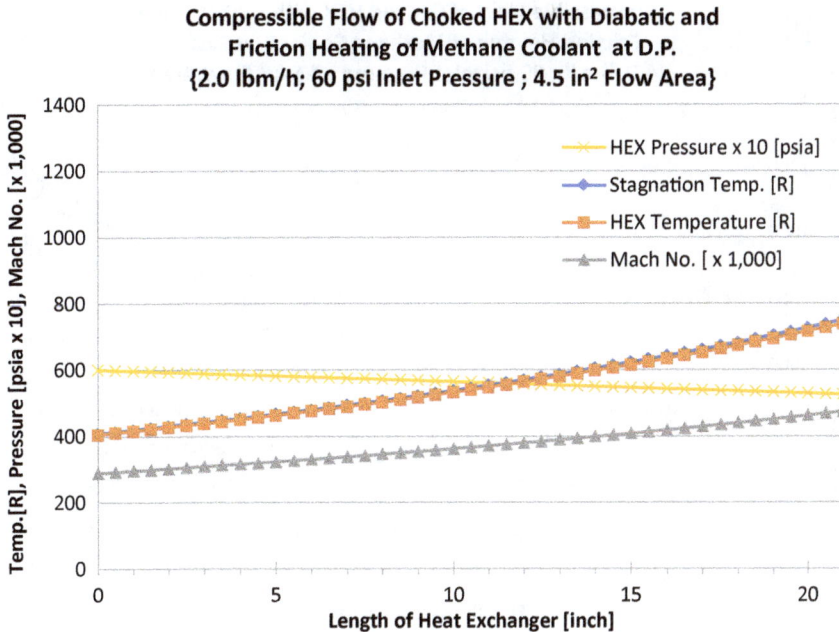

Figure C21.6.   Results of the same thermo-fluid-heat transfer model used for Figure C21.4 but with the heat exchanger free flow for the coolant increased from 3.0 to 4.5 in^2. Note how the Mach number has been reduced from 0.85 to 0.45.

respectively. As may be observed from these figures, the maximum Mach number at the exit of the heat exchanger is reduced from 0.85 to 0.48 by increasing the inlet pressure to the heat exchanger from 60 psi to 80 psi — effectively increasing the density of the coolant and thus reducing the velocity of the coolant through the heat exchanger core. The exit pressure of the heat exchanger will consequently also increase. However, maintaining this higher exit pressure can be accomplished by installing a controllable orifice, i.e., a control valve at the exit of the heat exchanger, or a fixed orifice that is choked at the discharge of the heat exchanger to maintain a pressure of 70–75 psi.

It is also necessary to determine the effect an increase in the flow rate of the compressible fluid has on the Mach number at the exit of the heat exchanger. The flow rate of the compressible fluid is entirely dependent on the flight parameters of the launch vehicle. These parameters are, at best, only predictable on a stochastic basis. For example, the flow rate of the compressible fluid may increase, even if only temporarily, by two-times the average flow rate that was used as the design point flow rate for the heat exchanger. The compressible thermo-fluid model was then used to determine the operation of the heat exchanger when the flow rate spikes to as high as

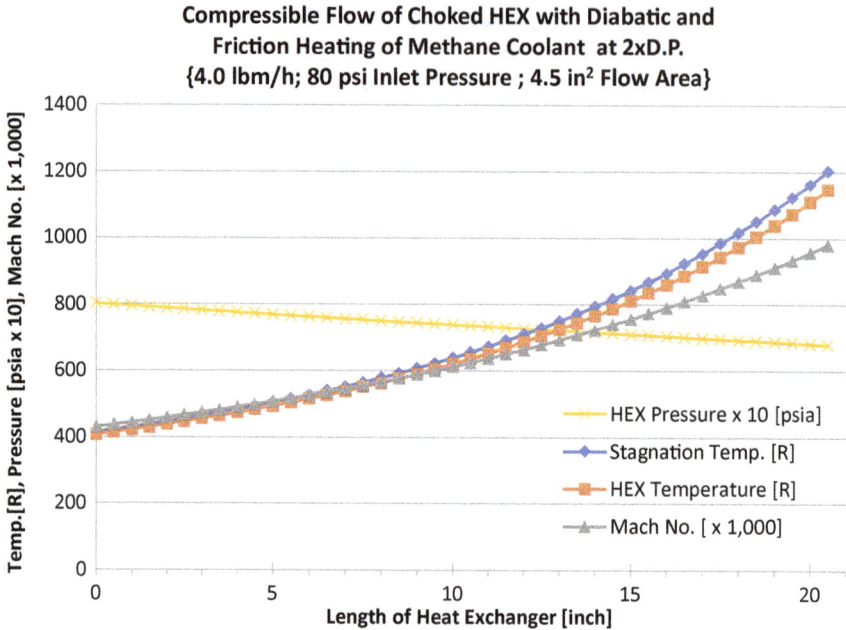

Figure C21.7.   Results of the same thermo-fluid-heat transfer model used for Figure C21.5 but with the compressible coolant flow at 2-times the design point (4 lbm/s).

two-times the design point flow rate or 4.0 lbm/s. The results of this analysis, shown in Figure C21.7, indicate that the operating pressure at the inlet to the heat exchanger will increase to 80 psi assuming the flow control orifice is designed to control the discharge pressure of the heat exchanger at the same 40–45 psi. The Mach number will be at 0.90 at the discharge of the heat exchanger. Thus, a shock wave at the discharge of the tube core is unlikely unless the pressure control orifice cannot maintain a pressure of 65–70 psi at the discharge of the heat exchanger.

It must be noted that the results shown in Figures C21.4 through C21.7 are based on ignoring the temperature profile of the hot fluid that is doing the actual heating. For example, the coolant temperatures in Figures C21.4 through C21.7 are observed to range from 700 to 1150°R. Thus, it is assumed that the heat source is available at even higher temperatures in order for the heat exchanger to actually function as required in a real application. In designing a heat exchanger for a specific application, the temperature of the heat source would be known, and it is usually a constraint to the overall performance of the heat exchanger. Typically, knowing the inlet temperatures for the heat source and the coolant, plus knowing the product, UA [kW/K or Btu/h °R], of the of the surface area (A) and heat transfer coefficient (U) for the exchanger will enable the exit temperatures for the heat source and coolant to be determined.

However, the equations used to determine these exit temperatures must now consider the additional friction heating of the coolant. The equation that is used to determine the exit temperature of the heat source flow rate is shown in Equations C21.1(a) and (b). The general effect of accounting for the friction heating is to have the same UA of a heat exchanger not perform as well to cool the heat source as when the friction heating can be ignored. A case study is offered to demonstrate this effect.

Table C21.1 summarizes the performance of the heat exchanger at the quasi-static design point with the inlet temperatures of the heat source and coolant shown. The result shown in Table C21.1 indicates that the same size heat exchanger, as represented by the UA product, but with friction would achieve only 86% of the heat transfer that could be achieved when friction energy is negligible. Table C21.2 reveals the need for a 23% larger heat exchanger in order to achieve the same amount of heat transfer from the heat source fluid to the coolant. However, given that the heat exchanger performance is virtually always transient this seemingly higher surface area requirement at this quasi-static, single design point may not be conclusive. For example, in a similar calculation at higher coolant flow rates, the smaller surface area of the heat exchanger may be adequate due to the possible reduction in the inlet temperature of the coolant during the launch application.

Table C21.1.   Steady state heat exchanger performance at design point with friction and without friction effects for the same size heat exchanger.

HEX THEORETICAL PERFORMANCE		
	HEX with Friction = 0	with Friction diabatic compressible flow
MCp)oil =	3024	3024
T, hot oil [F] =	240	240
MCp)methane =	4104	4104
T, cold, methane [F] =	−60	−60
UAFc =	1500	1500
R1 =	0.74	0.74
R2 =	0.50	0.50
Ko =	1.14	0.75
Thot, oil, out [F] =	136.1	150.1
Tcold, gas, out = Tb [F] =	16.6	82.8
Heat Exchange [Btu/hr] =	314205	271832

Table C21.2. Comparison of heat exchanger performances with and without friction heating effects. The friction heat transfer requiring 23% bigger heat exchanger (UA).

HEX THEORETICAL PERFORMANCE		
	HEX with Friction = 0	with Friction diabatic compressible flow
MCp)oil =	3024	3024
T, hot oil [F] =	240	240
MCp)methane =	4104	4104
T, cold, methane [F] =	−60	−60
UAFc =	1500	1840
R1 =	0.74	0.74
R2 =	0.50	0.61
Ko =	1.14	0.75
Thot, oil, out [F] =	136.1	136.1
Tcold, gas, out = Tb [F] =	16.6	93.1
Heat Exchange [Btu/hr] =	314205	314292

Of course, the time duration at these higher flow rates is very small, and for flow rates that are at or near the design point, average values will have greater effect on the overall cooling of the hydraulic oil.

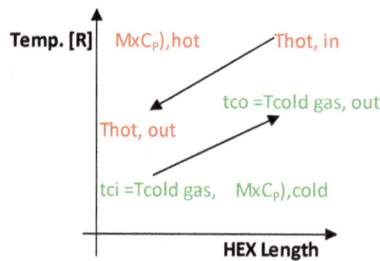

$$R_1 = \frac{\dot{M}_{hot} \times C_{p,hot}}{\dot{M}_{cold} \times C_{p,cold}}$$

$$R_1 = \frac{U_{Ht.Coef} \times A \times F_c}{\dot{M}_{hot} \times C_{p,hot}}$$

$$K_o = e^{\left(R_2 \times \left[1 - R_1 - \frac{\dot{E}_{frictiion}}{(T_{hot,\,in} - T_{hot,out})}\right]\right)}$$

$$\dot{E}_{friction} = \frac{\dot{Q}_{friction}}{\dot{M}_{cold} \times C_{p,cold}}$$

$$T_{h,out} = \frac{T_{h,in} \times (1 - R_1) - \dot{E}_{friction} + t_{c,in} \times (K_o - 1)}{K_o - R_1} \quad \text{(C21.1a)}$$

$$T_{cold,out} = T_{cold,in} + \frac{\dot{M}_h \times C_{p,hot} \times (T_{h,in} - T_{h,out})}{\dot{M}_c \times C_{p,cold}} \quad \text{(C21.1b)}$$

C21.2   Conclusion

A thermo-fluids-heat transfer model of a heat exchanger subject to the effects of fluid friction due to the compressible flow of the coolant has been completed. The model has been validated by comparing the results of the model with a similar problem solved using NASA's GFSSP software. A case study of the effects of using a compressible fluid as the coolant is provided. An equation is also derived and implemented to show how friction heating affects the performance the heat exchanger, as modeled by the product of the heat transfer coefficient and surface area, UA.

Bibliography

Bandyopadhyay, A., & Majumdar, A. (2007). *Modeling of Compressible Flow with Friction and Heat Transfer Using the Generalized Fluid System Simulation Program (GFSSP)*. Retrieved from NASA Technical Reports Server: https://ntrs.nasa.gov/search.jsp?R=20070036728

Index

a closed system, 322
A Monte, 432
ability to do work, 2
absorber, 273
absorption chiller system, 194, 195, 273, 275, 276
add-in, 235
"add-in" to the Excel spreadsheet, 235
adiabatic, 523
adiabatic chamber, 299
adiabatic flame temp., 306
adiabatic flame temperature, 306
adiabatic work, 120
advanced CAES, 488
advanced compressed, 475
advanced thermodynamics cycle, 345
aero-electric power tower, 525
aerodynamic drag, 10
air, 488, 492
air conditioners, 260
air energy system, 475
Air Fuel Ratio (AFR), 307
air intercooler, 238
air products, 479
air refrigeration cycle, 276
air-Brayton cycle, 333, 475
alchemists, 165
Allam cycle, 379, 453
Amagat's law of partial volume, 285
ambient temperature, 187
American Society of Heating, Refrigerating and Air-Conditioning Engineers (ASHRAE), 413
ammonia, 255
analysis, 135
angular kinetic energy, 47

Anthropic Principle, 177
applied thermodynamics, 321
ASHRAE (American Society of Refrigeration and Air Conditioning) Standards Handbook, 261
availability, 209
available energy, 30, 44

battery, 481, 492
battery storage systems, 476
Bernoulli equation, 497, 521
Bernoulli nozzle, 528
Big Bang, 27, 164
bio fuel heat, 303
Black-Sky Radiation Effect, 542
Boltzman's constant, 166
bottoming cycle, 331
Brake Mean Effective Pressure (BMEP), 115, 119
Brake Specific Fuel Consumption (BSFC), 115, 118
Brayton, 222
Brayton Cycle, 37, 109, 141, 230, 232, 237, 393, 410, 452, 488, 492
Brayton Cycle efficiency, 367
Btu, 2

CAES, 490, 492
calculus, 20
capture, 457
carbon dioxide (CO_2), 315, 331, 387, 452
carbon dioxide capture, 452
carbon monoxide (CO), 452
carbon–hydrogen fuels, 301
carbon–hydrogen-based compound, 301

Carlo simulation, 432

Carnot, 215

Carnot cycle analysis, 123, 124

Carnot Cycle efficiency, 185, 236, 265

Carnot Cycle equation, 207

Carnot Cycle heat engine, 175

Carnot efficiency, 114, 147, 330

Carnot efficiency equation, 23

Carnot engine, 149, 275

Carnot equation, 217, 465

Carnot heat engine, 180, 186, 196

Carnot refrigerator, 149, 176, 186, 275

Carrier, 469

cascade, 140, 268

Cascade (Economizer) cycle, 470, 473

CAT 3616 engine, 255

Caterpillar (CAT), 349

Centigrade, 51

centrifugal compressor, 278

"charging" cycle, 496

chemical oxidation, 301

chemical-thermodynamics, 305

chemically stored energy, 146

Chevrolet's Corvette, 122

"choked" condition, 511

Clausius, 176

Clausius or Kelvin Statement, 406

Clausius Statement of the Second Law of
    Thermodynamics, 260, 278

closed, 82, 253

closed mass, 89, 90, 131, 213

closed mass analysis, 103, 104, 136

closed mass system, 22, 53, 82, 89, 113,
    501

closed system, 7, 41, 169

closed systems analysis, 32

$CO_2(sCO_2)$, 452

coefficient, 227, 473

coefficient of performance (COP), 149,
    259, 265, 434, 473

Coffee Pot, 321

Cogeneration system, 254

combustion analysis, 301

combustion system, 249

combustion thermodynamics, 305

combustion' (HR-SPDC) engine, 395

combustor, 238, 363

Compressed Air Energy Storage (CAES),
    494

compressibility constant, Z, 76

compressibility factor, Z, 112

compressibility, Z, 57

compressible flow dynamics, 546

compressible flow of Choked HEX with
    diabatic and friction heating, 551

compressible gas dynamics, 510

compression Ratio (Vr), 120

compressor, 35, 144, 156, 247, 278,
    282

compressor eff., 235

compressor energy, 498

concentrated solar collector system, 381

concepts, 478

concepts NREC, 255, 394, 461

condenser, 215, 343, 462

conservation efforts, 452

constant pressure, 91

constant temperature, 91

constant temperature process, 94

constant volume (CV), 7, 67, 89, 90, 91,
    167, 220, 407, 500, 501

constant volume process, 94

control volume energy, 498

converging, 514

converging-diverging, 512

converging-diverging nozzle, 510, 512

coolant jacket, 255

cost, 341

critical pressure, 76

"crossover" system, 480, 485, 492

cryogenic matrix, 503, 505

cut-off ratio (Vc), 120

CV boundary, 170

Dalton's law of partial pressure, 285

(de Laval) nozzle, 512

dead state, 187, 202, 225

dead state pressure, 202

dehumidified, 291

dehumidifiers, 291

department of energy, 478

desalination of water, 436

desalination system, 440

dew point temperatures (T_{dp}), 295

diabatic, 523

diesel, 21, 82, 105

diesel and the Otto Cycles, 21

diesel cycle, 127, 253

diffusers, 35

"discharging" cycle, 499

distributed power generation, 351
district heating system, 435
diverging, 514
DNA and RNA, 35
DNA molecule, 89
drag and kinematic friction, 465
Dresser-Rand Synchrony Division, 429
Drinking Bird, 463
drop-in fluids, 413
dry bulb temperature, 294, 465
dry cooling tower, 263
Ds-Ns graph, 454
dual entry, 358
dual pressure, 21
dual-pressure diesel engine, 121
dual-pressure, modified diesel cycle, 107

economics of the cycle, 226
economizer, 268
economizer (cascade) system, 418
economizer System, 140
eff. corr., 118
effectiveness, 366
efficiency, 88, 458
ejector, 144, 469
energetic, 400
energetic nanoparticles, 399
Energy, 2, 44, 45, 424, 477
energy balance, 220, 221, 370
energy density of lithium-ion, 481
energy destruction, 177
energy recovery efficiency (ERE), 494
energy source, 524
energy storage, 479, 492
energy unavailability, 224
engines, 105
enthalpy, 35, 138, 168, 235, 237, 270
entropy, 19, 23, 65, 155, 164, 166, 215,
    216, 225, 236
entropy change, 171
entropy production, 224, 364
equations of state, 78
Erickson, 127
Ericson Cycles, 215
Espresso, 321
evaporative cooling, 463
evaporator, 269, 271, 343
excel spreadsheet, 38
excess air, 238
exergy, 182, 205

Exergy (or availability) analysis, 482
exergy balance, 220, 221
exergy destruction, 182, 186, 207, 224,
    329
exhaust gases, 330

Fahrenheit, 51
Fanno Line, 522, 546
feed pump, 342, 343
first and second laws, 365
first and second laws of thermodynamics,
    213
first law of thermodynamics, 6, 89, 90,
    113, 128, 131
flash tank, 268
flow methodology, 136
flow work, 135, 137
fossil-fuel electric power plants, 346
free body diagram, 67
friction, 3, 44
friction energy, 468
furnace/solar panel application, 198

Galileo, 10
gas turbine (GT), 360, 477
gas turbine cycle (GTC), 38
gas turbine engine, 245, 248
gas turbine jet, 237
gas turbines, 346
generalized fluid system simulation, 548
geothermal, 424
geothermal energy, 425
global climates, 317
global warming, 413
global warming gas, 315
global warming Potential (GWP), 281
ground source heat energy storage, 539

heat content of the fuel, 13
heat energy absorbed, 543
heat engine cycle, 109
heat engines, 213
heat exchanger, 226, 342
heat exchanger design, 227
heat exchanger sizes, 339
heat exchangers, 35
heat pipe, 424
heat pump, 149, 180, 260
heat rates, 303
heat recovery, 387

heat transfer (Q), 2, 227, 179, 545
heat transfer coefficient, 228, 431
heat transfer correction factor, 243
heat with respect to volume, 288
heated, Compressible Fluid, 545
heating, 88
heating, ventilation and air conditioning (HVAC) system, 271
heliostat mirrors, 524
high heating, 303
high heating value (HHV), 304
high-grade waste energy, 345
humidified, 291
humidifiers, 291
humidity, 289
HVAC, 271, 300, 413
Hy-Ram engine, 394
hybrid $(sCO_2)^2$ cycle, 453
hybrid ramjet, 391
hybrid ramjet propulsion system, 397
hybrid-ramjet sequential-pulse dual, 394
hydrogen, 442
hydrogen pipeline delivery system, 442
hydrogen production system, 443
hydrogen refueling stations, 503

incident energy usage factor, 535
inductors, resistors, capacitors and integrated circuits (ICs), 35
injection, 360
intercooling, 459
internal energy (E), 3, 267
internal thermal, 99
internal thermal energy, 99, 137
intrinsic thermodynamics property, 55
irreversibility, 173, 177, 181, 186, 201, 205, 208, 329
isentropic enthalpy, 236
isentropic enthalpy drop, 238
isobaric (i.e., constant pressure) compressibility, 54
isothermal (constant temperature) compressibility, 54

Japan (1997), 281
joule, 2, 46

K-promoted hydrotalcite, 458
Kelvin, K, 51
Kelvin–Planck, 176

kinetic energy, 249, 469
Kyoto, 281

L, C, R, 35
latent heat, 304
launch vehicle fuel pump, 310
law of conservation of energy, 5
Lenoir Cycle, 125
Lenoir engine, 251
let down turbine, 454, 460, 477
LightSail Energy, Inc., 479
linear kinetic energy, 47
liquid oxygen (LOX), 310
liquid-vapor phase, 217
low heating value (LHV), 304

Mach 1, 393
Mach number, 514, 546
Man-Made, 524
marine hydrokinetic (MHK), 440
mass analysis, 253
mass and mole fractions, 291
mass fraction, 290, 314
Master Brunelleschi, 324
mechanical vapor recompression system (MVRS), 143, 434, 436
mechanical work mode, 80
Meredith Effect, 251
Mitsubishi, 479
mixtures, 58
modified basic Rankine Cycle, 225
modified dual-pressure diesel engine, 106
mole fractions, 290
molten sodium salt heat transfer fluid, 382
Montreal (1987), 281
Moon–Earth system, 215
MTU Detroit Diesel, 354
multi-admission turbine, 358
multi-stage, intercooled compressor, 362

nanoparticles, 391, 400
nanoworld the entropy, 166
National Institute of Standards and Technology (NIST), 57, 78
naturally aspirated engines, 117
natural landforms, 524
Newton's Law of Gravity, 52
Newton-meters, 46
Nitric acid, 331
NO_x, 360

no heat transfer, 95
non-reacting mixture, 284, 294
normal, 522
Northeastern University, 400
nozzle, 35, 153
NREC (CN), 478
Number of Transfer Units (NTU), 242

Oblique Shocks, 522
ocean thermal energy conversion, 536
onsite power augmentation, 475
open, 136
open air-Brayton Cycle, 245, 490
open cycle, 494
open flow, 89, 135
open flow analysis, 32, 90, 134, 213
open flow system, 31, 472
open pit, 524
open-pit mine, 530, 537
open-pit mine application, 528
operation and maintenance (O&M) cost,
    226, 454
ORC system, 345, 349
ORC turbine, 358
Organic Rankine Cycle (ORC) system,
    151, 255, 330, 543
Otto, 127, 253
Otto Cycle, 21, 87, 121, 389
Otto engine, 82, 105
Otto engine cycle, 104
oxidizer to fuel ratio (OFR), 307
oxygen separation process, 379
Ozone Depletion Potential (ODP), 281,
    413

P × V, 33
P-v indicator diagram, 48
Paris COP1 Protocol, 281
payback, 246
per kW ($/kW), 341
perfect engineers, 178
perfect gas law, 16, 17, 42, 51, 56, 108,
    253, 285
perfectly insulated, 41
piston-cylinder, 59
piston-cylinder engine, 145
piston-cylinder system, 15, 104
pneumatic power tube (PPT), 524
polytrophic constant (n), 99
polytropic equation, 100

Potential (GWP), 413
potential energies (KE and PE), 147
potential energy (PE), 6, 47, 249
pound-force feet, 46
power generation, 539
power generation system, 330
Power LLC, 479
Power Tube, 524
power-generating system, 425
pressure, 395
pressure drop, 545
pressure ratio, 230
pressure swing adsorption (PSA) cycle,
    452, 457
process diagram, 106
process path, 157
process table, 59
GFSSP, 548
property diagram, 41
property table, 39, 41, 59, 67, 84, 108,
    153, 218
psychrometric, 294
psychrometric analysis, 291, 369
Psychrometric chart, 297, 467
pulsed detonation engine, 396
pumps, 35

quality, 66, 84, 263

R1234ze, 282, 413, 416
R134a, 255, 282, 413, 416
R236fa, 339
R245fa, 227, 255, 339
radiated heat energy, 540
ramjet, 387, 393
ramjet engine, 248, 389
Rankine Cycle, 214, 462, 524
Rankine Cycle system, 141, 150, 425, 426,
    469, 531
Rankine system, 542
Rankine, °R, 51
rate of temperature change, 324
rayleigh line, 522
reacting mixture, 284
reaction turbine, 358
real world, 98
reality, 11, 40
reality check, 11, 87
reciprocating engine, 387, 389
recompression sCO$_2$ system, 380

recovery efficiency (ERE), 477
recovery factor (Rr), 485
Recuperator,
recuperator, 229, 238, 343, 364
reference fluid thermodynamic and
    transport properties database
    (REFPROP), 57, 63, 78
refrigerant fluid, 279
refrigerant tons, RT, 264
refrigeration cycle, 37, 144, 257, 268, 469
refrigeration system, 406
regenerative heat exchanger, 239
regenerator, 238
reheat turbine, 458
relative, 289
relative humidity, 293, 301, 465
return on investment (ROI), 488
reverse, 410
reverse Rankine Cycle system, 406
Reversed Brayton Cycle, 276
reversible power, 186
Reversible work, 201
Rigid vessel, 41
rocket engine, 130
rotational kinetic energy, 12
rule of thumb, 185
RWE Power AG, 479

Sadi Carnot, 175
sanity, 40, 87
sanity check, 11
Santa Maria della Fiori, 324
schematic of the proposed PPT, 528
sCO$_2$ cycle's, 383, 384, 458
Second Law, 406
Second Law of Thermodynamics, 25, 26,
    45, 81, 147, 155, 164, 165, 220
sequential ignition, 391, 399
sequester CO$_2$, 457
Sequestration, 452
seven mechanical devices, 139
SI units, 46
simple, 246
Simple Payback, 488
size buildings, 526
skyscraper, 526
solar heat flux, 132
solar one and solar two projects, 531
solar power tower (SPT), 530
solar powered air conditioning system, 406

solar-powered, 406
solution process, 38
sonic velocity, 511, 514
specific, 288, 454
specific diameter (Ds), 417, 454
specific energy, 55
specific heat capacity, 4, 99
specific heat with respect to pressure, 288
specific heat, C$_p$, 234
specific humidity, 289, 292, 369
specific speed (NS), 415, 417, 454
specific volume, 17
speed of sound, 393
stagnation enthalpy, 520
stagnation pressure, p$_o$, 521
stagnation temperature, 520
standard temperature and, 395
state postulate, 51
state postulate of thermodynamics, 50
state principle, 69, 80, 300
steady state, 41
steam Rankine system, 216
sterling, 215
sterling cycle, 125, 127
sterling cycle engine, 125
stoichiometric, 309
Storage of Hydrogen, 503
STP, 395
stress (σ)–strain (ϵ), 48
supercharger, 390
supercritical, 452
supercritical CO$_2$ cycle, 141
supercritical CO$_2$ (sCO$_2$), 247, 379
superheated, 71
superheated steam table, 71
surface tension, 49
SustainX, Inc., 479
synapse, 1
System, 131

T-s diagram, 65, 156, 440, 442
T-v diagram, 62, 74, 323
temperature gradients, 540
Temperature Recovery Effectiveness (Rc),
    483
Temperature transient, 326
the "Drinking Bird" Toy, 462
the calculus, 103
the continuity equation, 512
the department of energy (DOE), 247, 478

the first and second laws of
   thermodynamics, 510
the first law, 497
the first law of thermodynamics, 87, 94,
   130, 267, 407, 504
the Second Law, 170, 406
the second law of thermodynamics, 207,
   472
The Sun-Earth system, 9
The US Department of Energy (DOE),
   503
thermal air, 524
thermal equilibrium, 41
thermal expansion work, 49
thermal internal energy, E_t, 53
thermodynamics, 365
thermodynamics cycles, 36, 213
thermodynamics model of ejector, 471
thermodynamics of moist air, 294
thermodynamics of reacting and
   non-reacting mixtures, 284
thermodynamics solution procedure, 162
thermosyphon, 424
thermosyphon heat pipe, 428
thermosyphon system, 427
throttle valve, 269
total power systems, 429
total recoverable energy cycle, 345
total recovery energy cycle (TREC)
   system, 254, 345, 349
transient analysis, 128
transient, numerical analysis, 505
TREC system, 358
turbine, 35, 226, 236
turbine-compressor, 393
turbine-generator units (TGUs), 337, 358
turbocharger, 390
two-phase, 58, 217

U_{oa}, 432
UA, 242, 336, 339
unavailable work, 177
universe, 167
useful work, 3
utopic universe, 30
utopic world, 178, 191
utopic world of perfect engineering, 191

vacuum pump, 469
values (L_{HHV}), 303
vapor compression cycle refrigeration, 268
vapor compression system, 469
vapor recompression (refrigeration) cycle,
   140

$\dot{W}_{ideal, maximum, theoretical}$, 173
$\dot{W}_{reversible\ useful}$, 205
$\dot{W}_{useful,\ dyno\ measureable}$, 205
Wartsila, 349, 351
waste heat, 524
waste treatment, 436
water, 360
water chillers, 260
water injection, 248
water injection system, 362, 363, 370
water-based, 406
watt, 46
watt-hours (Wh), 2
wet bulb (T_{wb}), 294
wet-bulb temperature, 296, 463, 465
work, 92
Wright R-1280 reciprocating engine, 387

Young's Modulus, 50

Zero Law of Thermodynamics, 286